Laboratory Manual for Physical Geology

Laboratory Manual for Physical Geology

Fifth Edition

Norris W. Jones
Emeritus Professor, University of Wisconsin–Oshkosh

Charles E. Jones
University of Pittsburgh

Featuring the Photography of Doug Sherman and Dr. Parvinder Sethi

 **Higher Education**

Boston   Burr Ridge, IL   Dubuque, IA   Madison, WI   New York   San Francisco   St. Louis
Bangkok   Bogotá   Caracas   Kuala Lumpur   Lisbon   London   Madrid   Mexico City
Milan   Montreal   New Delhi   Santiago   Seoul   Singapore   Sydney   Taipei   Toronto

LABORATORY MANUAL FOR PHYSICAL GEOLOGY, FIFTH EDITION

Published by McGraw-Hill, a business unit of The McGraw-Hill Companies, Inc., 1221 Avenue of the Americas, New York, NY 10020. Copyright © 2006, 2003, 2001, 1998, 1994 by The McGraw-Hill Companies, Inc. All rights reserved. No part of this publication may be reproduced or distributed in any form or by any means, or stored in a database or retrieval system, without the prior written consent of The McGraw-Hill Companies, Inc., including, but not limited to, in any network or other electronic storage or transmission, or broadcast for distance learning.

Some ancillaries, including electronic and print components, may not be available to customers outside the United States.

 This book is printed on recycled, acid-free paper containing 10% postconsumer waste.

3 4 5 6 7 8 9 0 QPD/QPD 0 9 8 7 6

ISBN 978-0-07-253063-6
MHID 0–07–253063–4

Publisher: *Margaret J. Kemp*
Senior Sponsoring Editor: *Daryl Bruflodt*
Senior Developmental Editor: *Lisa A. Bruflodt*
Outside Developmental Services: *Robin Reed*
Associate Marketing Manager: *Todd L. Turner*
Lead Project Manager: *Joyce M. Berendes*
Senior Production Supervisor: *Laura Fuller*
Lead Media Project Manager: *Judi David*
Designer: *Rick D. Noel*
Cover Designer: *John Rokusek/Rokusek Design*
(USE) Cover Image: ©*Dr. Parvinder Sethi/Geostock*
Senior Photo Research Coordinator: *Lori Hancock*
Supplement Producer: *Brenda A. Ernzen*
Compositor: *Prographics*
Typeface: *10/12 Times Roman*
Printer: *Quebecor World Dubuque, IA*

The credits section for this book begins on page 331 and is considered an extension of the copyright page.

www.mhhe.com

Contents

List of Tables and Maps viii

Preface x

Prologue xii

Minerals

Chapter 1
Properties of Minerals 2

Introduction 2
What is a Mineral? 2
Physical Properties 2
Hands-On Applications 13
Objectives 13
Problems 13
In Greater Depth 14

Chapter 2
Mineral Identification 17

Introduction 17
Rock-Forming Minerals 17
Accessory and Minor Minerals 21
Ore Minerals 23
How to Identify Minerals 24
Insecure About Mineral Identification? 25
Hands-On Applications 33
Objective 33
Problems 33
In Greater Depth 34

Rocks

Chapter 3
Igneous Rocks 38

Introduction 38
Igneous Rocks 38
Texture 39
Mineral Composition 40
Classification and Identification of Igneous Rocks 41
Crystallization of Magma and Bowen's Reaction Series 41

Where and How Magmas Form 47
Changes in Magma Composition 48
Hands-On Applications 49
Objectives 49
Problems 49
In Greater Depth 52

Chapter 4
Sedimentary Rocks 57

Introduction 57
Weathering 57
Erosion and Transportation 57
Deposition and Lithification 59
Sedimentary Rocks 62
Hands-On Applications 71
Objectives 71
Problems 71
In Greater Depth 74

Chapter 5
Metamorphic Rocks 79

Introduction 79
Conditions of Metamorphism 79
Types of Metamorphism 79
Changes During Metamorphism 80
Metamorphic Textures 81
Metamorphic Minerals 83
Parent Rock 83
Classification and Identification of Metamorphic Rocks 83
Zones of Metamorphism 84
Distinguishing Among Igneous, Sedimentary, and Metamorphic Rocks 85
Hands-On Applications 89
Objectives 89
Problems 89
In Greater Depth 90

Maps and Images

Chapter 6
Topographic Maps 96

Introduction 96
Map Coordinates and Land Subdivision 96
Elements of Maps 101
Topographic Maps 102

Hands-On Applications   109
Objectives   109
Problems   109
In Greater Depth   114

Chapter 7
Aerial Photographs and Satellite Images   117

Introduction   117
Visible and Infrared Light   117
Aerial Photographs   118
Satellite Images   123
Radar Images   124
Hands-On Applications   125
Objectives   125
Problems   125
In Greater Depth   129

Landscapes and Surface Processes

Chapter 8
Streams and Humid-Climate Landscapes   134

Introduction   134
Runoff and Drainage Basins   134
Landscapes in Humid Areas   139
Hands-On Applications   143
Objectives   143
Problems   143
In Greater Depth   154

Chapter 9
Groundwater and Groundwater-Influenced Landscapes   157

Introduction   157
Occurrence and Movement of Groundwater   157
Human Use   158
Groundwater in Carbonate Rocks   160
Hands-On Applications   163
Objectives   163
Problems   163
In Greater Depth   168

Chapter 10
Glaciation   173

Introduction   173
Formation, Movement, and Mass Balance   173
Erosional Landforms   174
Depositional Landforms   175
Hands-On Applications   179
Objectives   179
Problems   179
In Greater Depth   191

Chapter 11
Sea Coasts   197

Introduction   197
Erosion   197
Deposition   197
Human Interaction   198
Submergent Coasts   200
Emergent Coasts   200
Recent and Future Sea Level Rise   200
Hands-On Applications   203
Objectives   203
Problems   203
In Greater Depth   211

Chapter 12
Arid-Climate Landscapes   215

Introduction   215
Erosional Landforms   215
Depositional Landforms   217
Hands-On Applications   219
Objectives   219
Problems   219
In Greater Depth   230

Geologic Time and Sequences

Chapter 13
Geologic Age   234

Introduction   234
Relative Geologic Time   234
Numerical Geologic Time   238
Geologic Time Scale   239
Hands-On Applications   241
Objectives   241
Problems   241
In Greater Depth   246

Internal Processes

Chapter 14
Structural Geology   248

Introduction   248
Recognizing and Describing Deformed Rocks   249
Folds   252
Faults   254
Clues to the Past   256
Hands-On Applications   257
Objectives   257
Problems   257
In Greater Depth   264

Contents vii

Chapter 15
Geologic Maps 267

Introduction 267
Making a Geologic Map 267
Appearance of Geologic Features from the Air and on Maps 269
Geologic Cross Sections 274
Using Geologic Maps 274
Hands-On Applications 277
Objectives 277
Problems 277
In Greater Depth 290

Chapter 16
Earthquakes 293

Introduction 293
Seismic Waves 293
Seismograms 294
Locating Earthquakes 294
Earthquake Intensity and Magnitude 296
Using Seismic Waves to See Beneath the Surface 298
Hands-On Applications 301
Objectives 301
Problems 301
In Greater Depth 310

Chapter 17
Plate Tectonics 311

Introduction 311
General Theory of Plate Tectonics 312
Divergent Boundaries 312
Convergent Boundaries 314
Transform Faults 315
Plate Motion and Hot Spots 316
Beyond Plate Tectonics 316
Hands-On Applications 319
Objectives 319
Problems 319
In Greater Depth 327

Credits 331

Index 333

List of Tables and Maps

Tables

Scale of Hardness   4
Measurements of Interfacial Angles
  of a Crystal   15
Some Important Minerals   18
Metallic Luster or Tarnished (Dulled) Metallic
  Luster   26
Nonmetallic Luster Hardness Less than
  5.5 Streak—Colored   27
Nonmetallic Luster Hardness Less than
  5.5 Streak—White, Whitish, or Faintly
  Colored   27
Nonmetallic Luster Hardness Less than
  5.5 Streak—White, Whitish, or Faintly Colored   28
Nonmetallic Luster Hardness Greater than
  5.5 Streak—White or Whitish if Softer than Streak
  Plate   29
Nonmetallic Luster Hardness Greater than
  5.5 Streak—White or Whitish if Softer than Streak
  Plate   30
Chemical Grouping and Formulae of the Minerals in
  Tables 2.2A–F   30
Recognizing Minerals in Igneous Rocks   41
Classification of Igneous Rocks   42
Chemical Analyses of Representative Extrusive Rocks
  from the Cascade Range, Oregon   53
Common Weathering Products   58
Depositional Environments of Common Sedimentary
  Rocks   61
Recognizing Minerals in Sedimentary Rocks   64
Classification of Detrital Sedimentary Rocks   65
Classification of Chemical and Biochemical Sedimentary
  Rocks   65
Recognizing Minerals in Metamorphic Rocks   83
Classification of Metamorphic Rocks   84
G.O.E.S. Data, Mount St. Helens, 1980   129
Characteristics of Three Stream Valleys   144
Selected Data from the Gaging Station on the Minnesota River
  near Jordan, Minnesota, 1935–2003   152
Movement of Marked Rocks on Grinnell Glacier   192
Mean Ice-Surface Elevations of Segments of Profile B–B',
  Figure 10.13   195
Data for Problem 5   229
Conditions Required for Brittle versus Ductile Deformation of
  Rocks   249
Stress/Strain Conditions of Common Geologic Structures   256
Modified Mercalli Scale   297
Earthquake Intensity Survey   305
Selected Volcanoes of the Hawaiian-Emperor Chain   325

Maps

Reduced copy of the Mt. Shasta, California, 7½-minute
  quadrangle, with principal features highlighted   98
Portion of topographic map of Menan Buttes from Menan
  Buttes, Idaho, 7½-minute quadrangle   122
Outline map of area affected by ash from
  Mount St. Helens   131
Average annual precipitation in the United States and
  Canada   135
The drainage basin of Clayborn Creek and its tributaries,
  portion of the Beaver, Arkansas/Missouri quadrangle   137
The longitudinal profile along Clayborn Creek   138
Portion of topographic map of Leavenworth, Kansas, 15-minute
  quadrangle   145
Portion of topographic map of Royal Gorge, Colorado,
  7½-minute quadrangle   146
Portion of Jackson, Mississippi/Louisiana,
  1° × 2° quadrangle   147
Portion of topographic map of Kaaterskill, New York,
  quadrangle   149
Portion of Jordan West, Minnesota, 7½-minute
  quadrangle   151
Portion of Urne, Wisconsin, 7½-minute quadrangle   155
Portion of Lakeside, Nebraska, 15-minute quadrangle   165
Portion of Mammoth Cave, Kentucky, 15-minute
  quadrangle   167
Portion of Groundwater Basins in the Mammoth Cave Region,
  Kentucky   169
Portion of topographic map of Glacier National Park,
  Montana   178
Portion of Mt. Fairweather, Alaska/Canada,
  1° × 2° quadrangle, including part of Glacier Bay
  National Park   183
Portion of Glacial Map of the United States East of the Rocky
  Mountains   186–187
Portion of Jackson, Michigan, 15-minute quadrangle
  topographic map   189
Map of Grinnell Glacier, Glacier National Park, Montana   193
Portion of the Cayucos, California, 15-minute quadrangle   207
Portion of Boothbay, Maine, 15-minute quadrangle   209
Portion of the 1951 Redondo Beach, California, 7½-minute
  quadrange   212
Portion of Bennetts Well, California, 15-minute
  quadrangle   222–223
Portion of Antelope Peak, Arizona, 15-minute quadrangle   225
Portion of Upheaval Dome, Utah, 15-minute quadrangle   227
Geologic map of portion of Gateway Quadrangle,
  Colorado   280–281
Geologic map of the Black Hills, South Dakota   284

Geologic map of portion of Tazewell Quadrangle, Tennessee 288–289
Geologic map of portion of Wetterhorn Peak Quadrangle, Colorado 292
Intensity map for earthquake near Douglas, Wyoming, with isoseismal lines based on the Modified Mercalli scale 298
Map of San Francisco Bay area showing locations for which intensities have been assigned 305
Map showing the expected effects of ground shaking in the San Francisco Bay area 308
A plate boundary in the southwest Pacific Ocean near the Fuji, Samoa, and Tonga Islands 322
Hawaiian Islands and other islands and seamounts of the Hawaiian-Emperor chain 324
The San Andreas fault system 328
Map of Wallace Creak area 329

Preface

To the Student

Geology and other sciences are major parts of our lives, and many of the daily decisions we make are affected in some way by science. Because of this, all educated people (scientists or not) should understand, in a general way, how science is done, and should be able to apply the basic ideas of the scientific method. Thus, a principal objective of this laboratory manual is to show you how geologists use the scientific method by having you do it yourselves. You will find that geologists, like other scientists, operate in a very logical and straightforward manner, which differs little from the way you normally do things. You will learn how to gather and use data (for example, making measurements to use in calculations), establish relations based on these data, classify information, develop and test hypotheses or models to explain observations and relations, and use this information to draw conclusions and make predictions. This may sound a bit intimidating, but don't panic; it's not difficult. You'll be surprised at what you'll be able to do.

Laboratory sessions will be most efficient if your **read the entire chapter before coming to class.** If allowed by the instructor, students may work in groups. In a group setting, it soon becomes apparent who has prepared and can contribute to the group. As a courtesy to others, and for your own benefit, *prepare for class!* Make sure you bring those things to class that are needed for the lab, as indicated in the **Materials Needed** section. Don't depend on your classmates.

You will find that geology is relevant to our daily lives. Look around you. Most of the manufactured items you see are made of earth materials—minerals, rocks, or petroleum. We live along rivers that flood, drink water pumped from underground, and dispose of our garbage in the ground. We and our politicians have to understand the consequences of our actions. So pay attention, ask questions, and learn to think critically about these geological issues. Make sure that science-related issues are based on good science; don't let charlatans determine our public policy.

To the Instructor

This lab manual is for a college-level, introductory course in physical geology. It is based on the following assumptions: (1) A typical student is a freshman or sophomore concurrently in a physical geology class. He or she has little background in science and math. The student may have had a unit on Earth science in seventh, eighth, or ninth grade, but most likely has had little or no exposure to geology. (2) The course has no prerequisites. (3) Although a few students taking the class plan to major in geology, and more will decide to major once they learn how interesting geology is, most students are taking the class to satisfy a general requirement and will not major in geology or other sciences.

Each chapter contains more exercises than your students may have time to do, so you will have to be selective as to which problems you assign. The sequence of topics is a traditional one, but the instructor is free to choose which topics to cover as well as the order in which they are covered. The text in each chapter is intentionally brief. It is meant to serve as background for the problems rather than a "complete" treatment of the subject. An Instructor's Manual is available as PDF and Word files on-line *(www.mhhe.com/jones5e)* and is accessible with a password obtained from the publisher. We strongly recommend you take a look at it.

The first edition of this manual, by NWJ, was the outgrowth of twenty-five years of teaching physical geology and using many different lab exercises. NWJ first wrote a physical geology lab manual in 1975 for use in these classes. That manual went through many revisions, and the first edition of this manual, though much more extensive than the original, benefitted much from the comments and criticism of the earlier manual offered by colleagues and students. The second and third editions are essentially refinements of the first edition, with one major exception: new problems based on resources available on the World Wide Web were added to all but one chapter.

With each new edition, we strive to increase clarity in both the text and problems. We also try to meet changing instructional needs by keeping both text and problems as current as is necessary. We are aided in this task by the very helpful comments and suggestions from our reviewers. If you find any errors in the manual or have suggestions for future improvements, **please visit the lab manual web page** *(www.mhhe.com/jones5e).* There you will find the most up-to-date web addresses for the WWW problems, a page listing any *errata* discovered following publication, and our e-mail addresses. Our goal is to create a lab manual that meets the needs of as many instructors as possible while still maintaining its relatively slender size.

Most of the photographs in this manual were taken by geology professors Doug Sherman and Dr. Parvinder Sethi, both of whom are also professional photographers. Their expertise in both areas made our job of selecting illustrations much easier.

To Everyone: Organization and Use

This manual consists of 17 chapters that cover the usual topics in physical geology. The first five chapters deal with *Earth Materials*—minerals and rocks. Because maps, aerial photographs, and other types of remotely sensed images are used so extensively in geology, two chapters focus on *Maps and Images.* Five chapters then show how *Landscapes and Surface Processes* are studied using maps and images. One laboratory is devoted to *Geologic Time and Sequences,* and in the final chapters, *Internal Processes* and their effects on the surface are studied. Each chapter contains a list of materials needed to do the lab (**Materials Needed**); sufficient descriptive text so that all lab problems can be completed without reference to another textbook; a list of **Objectives,** which summarizes the things that a student will be expected to do or know after completing the lab; and laboratory **Problems** or exercises that are required for the lab. The **Objectives** are located immediately before the **Problems,** rather than at the beginning of the chapter, because we want students to relate them to the problems more than the text.

The problems are designed to illustrate basic principles of geology and methods of solving geological problems. A paragraph preceding the objectives in each chapter briefly indicates how the problems relate to the scientific method. As often as possible, problems were chosen to demonstrate practical aspects. Some problems require making measurements and doing simple calculations. Most questions require specific, short answers. A few questions, however, are designed so that several answers may be possible. Students must formulate several hypotheses, then determine what additional information would be needed before a single hypothesis could be chosen. Problems that require more time to complete or are more difficult are preceded by the heading *In Greater Depth.* It is up to the individual instructor to determine how many problems, or what parts of individual problems, should be assigned in his or her class.

All chapters except Chapter 5 (Metamorphic Rocks) have a **Web Problem,** which can be solved from information available on the World Wide Web. We have provided specific web addresses for all but one problem, so they don't take an unreasonable amount of time. The student can access these websites by entering the address given in the individual Web Problem or by using the links provided by McGraw-Hill at *www.mhhe.com/jones5e,* linking to *Student Center,* and selecting the appropriate chapter. We encourage you to gain access through *www.mhhe.com/jones5e,* because of the changeable nature of websites. If changes occur that affect an individual problem, they will be noted and available at *www.mhhe.com/jones5e.*

The inside of the front cover of the manual contains a plate tectonic map of the world. The inside of the back cover lists symbols of selected chemical elements, units of measurement and factors for converting between metric and English units, and a brief explanation of the use of powers of ten to abbreviate large and small numbers. Key terms are in **boldface** in the text. *Italics* are used to call attention to a word or phrase. Answer pages can be torn out and handed in for grading if necessary. Colored, tear-out, structure models at the end of the manual are needed for the problems in Chapter 14.

Acknowledgements

NWJ would like to thank his colleagues at the University of Wisconsin-Oshkosh, Bill Fetter, Gene LaBerge, Tom Laudon, Jim McKee, Brian McKnight, and Bill Mode, for their comments, suggestions, and moral support over the years. CEJ especially thanks Dana Jones, toddler Hannah, and infant clones Maya and Abby for their patientce during the long working hours of his project. We also thank the following reviewers: Tracy L. Hall—Columbus State University, Alan Lester—University of Colorado-Boulder, Christopher J. Crow—Indiana University/Purdue University, Hal Noltimier—Ohio State University, Gale Martin—Community College of Southern Nevada, James E. Barrick—Texas Tech University, Karen Savage—California State University-Northridge, Jay R. Yett—Orange Coast College, Clair R. Ossian—Tarrant County College NE, Jim Brophy—Indiana University-Bloomington, Stewart Farrar—Eastern Kentucky University, Bernard Hallet—University of Washington, and McGraw-Hill Developmental Editor Lisa Bruflodt.

Prologue: Methods of Science

Science is a systematized body of knowledge derived mostly from observation and logical inference. The ultimate goal of scientists is to discover the basic principles that determine how things behave. Scientists have the following in mind as they work: (1) natural processes are orderly, consistent, and predictable; (2) there is a cause (or set of causes) for every effect; and (3) the best explanations are usually the simplest *(Doctrine of Simplicity or Parsimony)*. Geologists and other scientists who study the past assume that physical and chemical laws, such as the law of gravity, are constant and have not varied over time *(Doctrine of Actualism)*.

Scientific advances come in different ways. Some are based on theoretical ideas about things not yet observed (for example, Einstein's theory of relativity). Others result from unplanned, fortuitous events (such as the eruption of Mount St. Helens volcano). But most advances result from a painstaking approach known as the *scientific method*.

The **scientific method** generally involves: (1) formulation of a problem or question; (2) observation and measurement to determine characteristics or properties, to measure variation, classify, and establish relations—in other words, to get a basic understanding of the situation; (3) formulation of one or more explanations, models, or **hypotheses** to explain the observations; (4) experimentation and testing to reject or modify hypotheses; and (5) acceptance of a best hypothesis that withstands all tests and the scrutiny of peers as a **theory**.

The scientific method is *not* a rigid series of steps that must always be followed to do science. For example, most scientists already have a tentative hypothesis in mind when making observations. In fact, as philosopher Karl Popper said, successful scientists often are those with the best "imaginative preconception" of the correct hypothesis.

However, to avoid being blinded by such preconception, geologist T.C. Chamberlin recommended maintaining *multiple working hypotheses* until such time as they could be reduced to only one. With this approach, the scientific process involves rejecting those hypotheses that do not withstand testing and experimentation. Commonly, the simplest hypothesis that explains the most observations is the one that survives to become a theory. However, even though a theory is supported by a great deal of evidence and seems to explain all observations, it still is subject to testing, and eventually it may be rejected or modified as a result of new observations. Theories that withstand all testing and are invariable become **scientific laws;** an example is Newton's Law of Gravitation (the law of gravity).

The scientific method may seem rather involved, but it is an approach you probably use in many situations without realizing it. Take a very simple example. Let's say you go to geology lab one day and find nobody in the classroom. That *observation* should raise your curiosity. Why is no one there? Several possible explanations, or *hypotheses,* come to mind: (1) the professor is ill and the class was canceled; (2) no class was scheduled for that day; (3) you are there at the wrong time or the wrong day; (4) the class has gone on a field trip that you forgot; (5) the class is hiding in the room, ready to leap out and yell "Happy birthday"; or (6) everyone has been lifted magically from the room. As a good scientist and rational thinker, you rank them in order of which is the most likely, and quickly eliminate those that are impossible or too unlikely. The last one (6) is not testable and is out of the realm of science, because it calls on a supernatural phenomenon for an explanation. A quick check of the time and date allows you to eliminate number 3. It's not your birthday, so you forget about number 5. Three reasonable hypotheses remain to be tested.

One way to test a hypothesis is to assume it is true, make a prediction based on that assumption, then see if the prediction comes true. If the professor is ill (hypothesis number 1), you might predict that a note to that effect would be posted on the door or the blackboard, but there is none. You might also expect to see other curious students about, but there are none. These negative results strongly suggest that hypothesis 1 is incorrect, although they do not necessarily prove it wrong. After all, it is possible that the professor was not able to have a notice put in the classroom, and it is possible that all the other students decided not to come to class that day. Those are pretty complex and improbable explanations, however. Hypotheses 2 and 4 are simpler and more reasonable. An easy way to test these is to look at the syllabus for the class.

Let's assume that no class is scheduled, according to the syllabus. Then you can establish a *theory*. The theory is that the classroom is empty because there is no class that day. Note that this is a **scientific theory** (although theories in science usually require much more time to develop), one that can be tested in many ways, with the same result. For example, you could check with the Geology Department office or look at another student's syllabus. The public tends to use the word *theory* in quite a different sense, almost disdainful, as something that has little basis in fact. As you can see from the example, a *scientific theory* is firmly established and as close to truth as scientifically possible. However, even a scientific theory is always subject to more testing and possible revision in light of new evidence.

That is how much of science works. There is nothing mysterious about it. It is simply a logical, open-minded approach to a question. Conclusions are not made without considerable thought and testing. It is a rational approach to all kinds of problems or questions, not just scientific ones, and undoubtedly is the approach you have used many times yourself. As you do the exercises in your geology laboratory, you will be using the scientific method as applied by geologists.

PART I
Minerals

Cubic crystals of pyrite at three times actual size

CHAPTER 1

Properties of Minerals

Materials Needed
- Pencil
- Hand lens
- Calculator
- Set of mineral samples
- Equipment for determination of mineral properties: streak plate, copper penny, glass plate or steel knife, magnet, dilute hydrochloric acid, contact goniometer (optional)

Introduction

To understand most geologic processes, we must understand rocks. And to understand rocks we must first understand minerals, which are the components of rocks. Minerals differ from each other in chemical composition and atomic arrangement, and these factors produce distinctive physical properties that enable minerals to be identified. The most useful physical properties for identifying minerals are examined in this chapter. In Chapter 2, these properties are used to identify minerals.

What is a Mineral?

A **mineral** is a naturally occurring compound or chemical element made of atoms arranged in an orderly, repetitive pattern. Its chemical composition is expressed with a chemical formula **(symbols for important elements are given on the back endsheet).** Both chemical composition *and* atomic arrangement characterize a mineral and determine its physical properties. Most minerals form by inorganic processes, but some, identical in all respects to inorganically formed minerals, are produced by organic processes (for example, the calcium carbonate in clam shells). A few naturally occurring substances called **mineraloids** have characteristic chemical compositions, but are amorphous; that is, atoms are *not* arranged in regular patterns. Opal is an example.

The precise chemical composition and internal atomic architecture that define each mineral also directly determine its outward appearance and physical properties. Thus, in most cases, general appearance and a few easily determined physical properties are sufficient to identify the mineral.

Physical Properties

Color, luster, streak, hardness, cleavage, fracture, and crystal form are the most useful physical properties for identifying most minerals. Other properties—such as reaction with acid, magnetism, specific gravity or density, tenacity, taste, odor, feel, and presence of striations—are helpful in identifying certain minerals.

Color

Color is the most readily apparent property of a mineral, but BE CAREFUL. Slight impurities or defects within the crystal structure determine the color of many minerals. For example, quartz can be colorless, white, pink, purple, green, gray, or black (Fig. 1.1). Color generally is diagnostic for minerals with a metallic luster (defined in next section) but may vary quite a bit in minerals with a nonmetallic luster. **Check the other properties before making an identification.**

Luster

Luster describes the appearance of a mineral when light is reflected from its surface. Is it shiny or dull; does it look like a metal or like glass? Most minerals have either a **metallic** or **nonmetallic** luster. As you will see, the first thing you must determine before a mineral can be identified using the tables in Chapter 2 is whether its luster is metallic or nonmetallic.

Minerals with a metallic luster look like a metal, such as steel or copper (Fig. 1.2A, B). They are opaque, even when looking at a thin edge. Many metallic minerals become dull looking when they are

× 0.2

FIGURE 1.1
Color is NOT a reliable criterion for the identification of many minerals, as illustrated here by three varieties of the mineral quartz: amethyst (purple), rock crystal (clear), and smoky (black). The × 0.2 below the picture is a shorthand notation to indicate the actual size of the mineral samples. It means that the samples are 0.2 times the size in the picture; that is, the picture shows them one-fifth (0.2) their actual size. This notation will be used throughout this lab manual.

FIGURE 1.2

Types of luster: Metallic (A. galena [dark mineral] and B. pyrite); Nonmetallic, vitreous (C. muscovite and D. biotite); E. Nonmetallic, dull (kaolinite); F. Nonmetallic, pearly (talc); G. Nonmetallic, waxy (jasper); H. Nonmetallic, resinous (sphalerite).

A. Metallic (galena) × 0.3

B. Metallic (pyrite) × 0.5

C. Nonmetallic, vitreous (muscovite) × 0.5

D. Nonmetallic, vitreous (biotite) × 0.5

E. Nonmetallic, dull (kaolinite) × 0.5

F. Nonmetallic, pearly (talc) × 0.5

G. Nonmetallic, waxy (jasper) × 0.5

H. Nonmetallic, resinous (sphalerite) × 0.5

exposed to air for a long time (like silver, they tarnish). To determine whether or not a mineral has a metallic luster, therefore, you must look at a recently broken part of the mineral.

Minerals with nonmetallic luster can be any color. At first, you may have difficulty determining whether some black minerals have metallic or nonmetallic luster. However, thin pieces or edges of minerals with nonmetallic luster generally are translucent or transparent to light, and even thick pieces give you the sense that the reflected light has entered the mineral a bit before being reflected back. There are several types of nonmetallic lusters. *Vitreous luster* is like that of glass (Fig. 1.2C, D). Remember that glass can be almost any color, including black, so don't be fooled by the color. A *dull luster* has an earthy appearance caused by weak or diffuse reflection of light (Fig. 1.2E). Other nonmetallic lusters include: *pearly luster,* like a pearl or the inside of a fresh clam shell (Fig. 1.2F); *greasy luster,* as though covered by a coat of oil; *waxy luster,* like paraffin (Fig. 1.2G); and *resinous luster,* like resin or tree sap (Fig. 1.2H).

Hardness

Hardness is the resistance of a smooth surface to abrasion or scratching. A harder mineral scratches a softer mineral, but a softer one does not scratch a harder one. To determine the hardness of a mineral, something with a known hardness is used to scratch, or be scratched by, the unknown. The minerals in the Mohs Hardness Scale (Table 1.1) are used as standards for comparison. The Mohs hardnesses are also given for some common items, which you can use to determine the hardness of an unknown mineral if you don't happen to have a pocketful of Mohs minerals.

To determine hardness, run a sharp edge or a point of a mineral with known hardness across a smooth face of the mineral to be tested (Fig. 1.3). Do not scratch back and forth like an eraser, but press hard and slowly scratch a line, like you are trying to etch a groove in glass. Make sure that the contact points of both minerals are the minerals you intend to test, and not impurities. Also, make sure that the mineral has actually been scratched. Sometimes powder of the softer mineral is left on the harder mineral and gives the appearance of a scratch on the harder one. Brush the tested surface with your finger to see if a groove

TABLE 1.1
Scale of Hardness

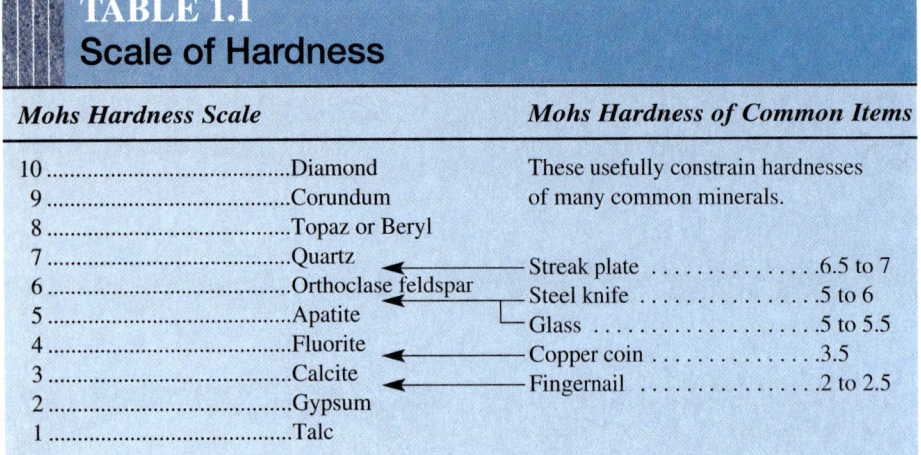

Mohs Hardness Scale	Mohs Hardness of Common Items
10 .... Diamond	These usefully constrain hardnesses of many common minerals.
9 .... Corundum	
8 .... Topaz or Beryl	
7 .... Quartz	Streak plate .... 6.5 to 7
6 .... Orthoclase feldspar	Steel knife .... .5 to 6
5 .... Apatite	Glass .... .5 to 5.5
4 .... Fluorite	Copper coin 3.5
3 .... Calcite	Fingernail .... 2 to 2.5
2 .... Gypsum	
1 .... Talc	

or scratch remains. You may need to use a hand lens or magnifying glass to see whether a scratch was made. To double-check your results, try scratching the substance of known hardness with the mineral of unknown hardness. If two minerals have the same hardness, they may be able to scratch each other.

A piece of window glass is commonly used in geology labs as a standard for determining hardness. There are several reasons for this: 1) it's easy to see a scratch on glass; 2) the hardness of glass (5 to 5½) is midway on the Mohs scale; and 3) glass is inexpensive and easily replaced. In fact, you will discover that the hardness of a mineral when compared to glass is one of the principal bases in identifying a mineral using the determinative tables in Chapter 2.

Put the piece of glass on a stable, flat surface such as a tabletop. Then rub the mineral on the glass, and check to see if the glass was scratched. DO NOT TRY TO SCRATCH THE MINERAL WITH THE GLASS, because glass chips easily.

More sophisticated methods than the scratch test have been developed to determine hardness; Figure 1.4 illustrates how some of the minerals of the Mohs scale compare with their hardnesses, as determined by another method. As you can see, Friedrich Mohs, who devised the scale in 1822, did a pretty good job of selecting minerals so that the intervals between adjacent minerals on his scale were consistent.

Streak

Streak is the color of the mineral when finely powdered; it may or may not be the same color as the mineral. Streak is more helpful for identifying minerals with metallic lusters, because those with nonmetallic lusters generally have a colorless or light-colored streak that is not very diagnostic. Streak is obtained by scratching the mineral on an unpolished piece of white porcelain called a streak plate (Fig. 1.5). Because the streak plate is harder than most minerals, rubbing the mineral across the plate produces a powder of that mineral. When the excess powder is blown away, what remains is the color of the streak. Because the streak of a mineral is usually the same, no matter what the color of the mineral, streak is commonly more reliable than color for identification.

FIGURE 1.3
The harder white mineral (calcite) scratches the softer one (gypsum).

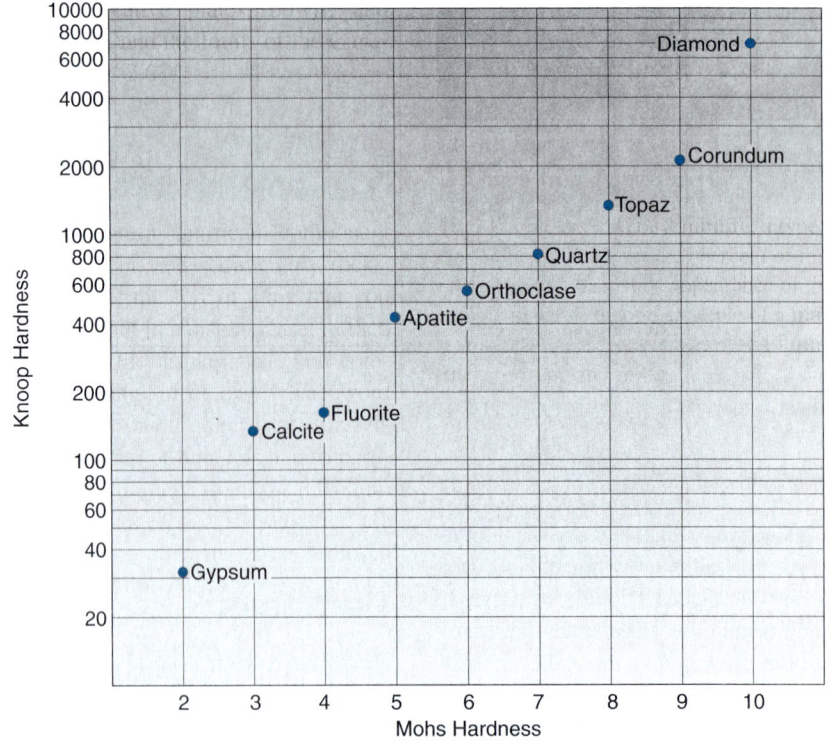

FIGURE 1.4
Comparison between Mohs-scale hardness and "actual" hardness (Knoop hardness). Note that "actual" hardness is shown with a logarithmic scale. Talc (Mohs hardness of 1) was not included in the study that produced these data.

FIGURE 1.5
The streak of this dark gray mineral (hematite), obtained by rubbing it on the white streak plate, is reddish brown.

Cleavage and Fracture

The way in which a mineral breaks is determined by the arrangement of its atoms and the strength of the chemical bonds holding them together. Breakage occurs where chemical bonds are relatively weak and where there are no atoms. Because these properties are unique to the mineral, careful observation of broken surfaces may aid in mineral identification. A mineral that exhibits **cleavage** consistently breaks, or *cleaves*, along parallel flat surfaces called **cleavage planes**. A mineral **fractures** if it breaks along random, irregular surfaces. Some minerals break only by fracturing, while others both cleave and fracture.

The mineral halite (sodium chloride [NaCl], or salt) illustrates how atomic arrangement determines the way a mineral breaks. Figure 1.6 shows the arrangement of sodium and chlorine atoms in halite. Notice that there are planes with atoms and planes without atoms. When halite breaks, it breaks parallel to the planes with atoms but along the planes without atoms. Because there are three *directions* in which atom density is equal, halite has three **directions of cleavage,** each at 90° to each other. The *number of cleavage directions* and the angles between them are important in mineral identification because they reflect the underlying atomic architecture that helps define each mineral.

Cleavage planes, as flat surfaces, are easily spotted by turning a sample in your hand until you see a single flash of reflected light from across the mineral surface. Individual cleavage surfaces may extend across the whole mineral specimen (Fig. 1.7) or, more commonly, they may be offset from each other by small amounts, as illustrated in Figure 1.8. Even though they are offset, they work as tiny mirrors that create the single flash seen in Figure 1.9. Figures 1.10 and 1.11A also show offset cleavage surfaces.

Cleavage quality is described as *perfect, good,* and *poor.* Minerals with a *perfect* or excellent cleavage break easily along flat surfaces and are easy to spot. Minerals with *good* cleavages do not have such well-defined cleavage planes and reflect less light. *Poor* cleavages are the toughest to recognize, but can be spotted by small flashes of light in certain positions. All cleavages illustrated here are perfect or good.

Minerals have characteristic numbers of cleavages. This number is determined by counting the number of cleavage surfaces that are *not* parallel to each other. For example, the mineral in Figure 1.7 has two cleavage surfaces that are visible plus the one lying on the table. However, each of these cleavage surfaces is parallel to the other, so this mineral is said to have only **one cleavage direction.** Minerals with one cleavage are often said to have a *basal cleavage*.

Two cleavage directions are illustrated in Figure 1.10. In Figure 1.10A, the cleavage surfaces or planes are horizontal and vertical: they intersect each other at 90°. In Figure 1.10B, the cleavage surfaces are horizontal and steeply inclined: they do not intersect at 90°, but at 56° and 124°. The fronts and backs of each specimen are irregular fracture surfaces. Counting only *nonparallel cleavage surfaces,* each sample shows two cleavage directions, one horizontal and one vertical (Fig. 1.10A) or inclined (Fig. 1.10B). Intersecting cleavages may define an elongate geometric object

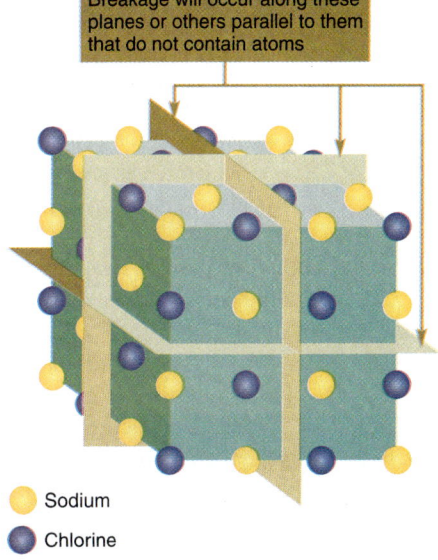

×1

FIGURE 1.6
The illustration shows that atoms of sodium and chlorine in the mineral halite are parallel to three planes that intersect at 90°. Halite breaks, or *cleaves,* most easily between the three planes of atoms, so it has three *directions* of cleavage that intersect at 90°. The photograph of halite illustrates the three directions of cleavage.

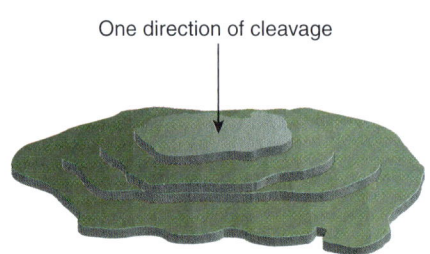

FIGURE 1.7
Biotite mica has one direction of cleavage (basal cleavage).

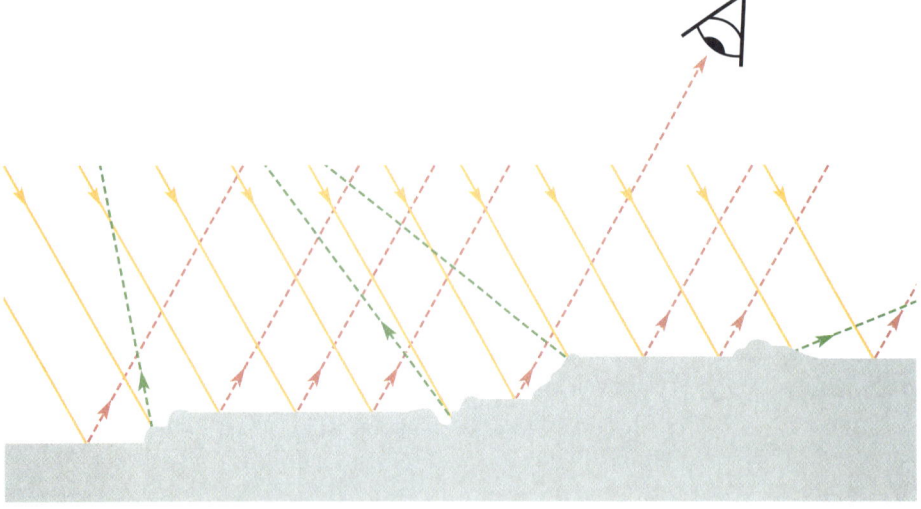

FIGURE 1.8
When incoming light rays (yellow lines) strike parallel cleavage surfaces (seen here in cross section), the rays are all reflected in the same direction (pink lines), even though the surfaces are at different elevations. When viewed from that direction, the mineral shines. Rays striking irregular fracture surfaces reflect in many different directions (green lines), causing fracture surfaces to appear duller.

FIGURE 1.9
The parallel cleavage planes in hornblende shine because the light is reflected in the same direction.

called a *prism;* such minerals are said to have a *prismatic cleavage.* When there are two cleavages, you should note the angle between them. Most commonly, cleavage angles are at or close to 90° (Fig. 1.10A) or 60° and 120° (Fig. 1.10B).

With **three cleavage directions,** a mineral can be broken in the shape of a cube if the three cleavages intersect at 90° (called a *cubic cleavage;* Fig. 1.11A) or a rhombohedron if the angles are not 90° (called a *rhombohedral cleavage;* Fig. 1.11B).

Minerals with **four** or **six cleavage directions** are not common. Four cleavage planes can intersect to form an eight-sided figure known as an octahedron *(Fig. 1.12).* Fluorite is the most common mineral with an *octahedral cleavage.* Six cleavage directions intersect to form a *dodecahedron,* a twelve-sided form with diamond-shaped faces. A common mineral with *dodecahedral cleavage* is sphalerite (Fig. 1.13).

When counting cleavage directions it is essential that you count surfaces on just *one* mineral crystal. The photographs shown here used single, large, broken crystals to illustrate cleavage. In nature you often find that a single hand-sized sample contains a large number of crystals grown together (see following discussion under "Crystal Form"). If you count up cleavage surfaces from more than one crystal, a wrong number is likely.

Finally, **fracture surfaces** can cut a mineral grain in any direction. Fractures are generally rough or irregular, rather than flat, and thus appear duller than cleavage surfaces (Fig. 1.14A). Some minerals fracture in a way that helps to identify them. For example, quartz has no cleavage but, like glass, it breaks along numerous small, smooth, curved surfaces called *conchoidal*

Chapter 1 Properties of Minerals 7

FIGURE 1.10
Two directions of cleavage: A. Cleavages in potassium feldspar intersect at 90°; B. Prismatic cleavages in hornblende intersect at 56° and 124°.

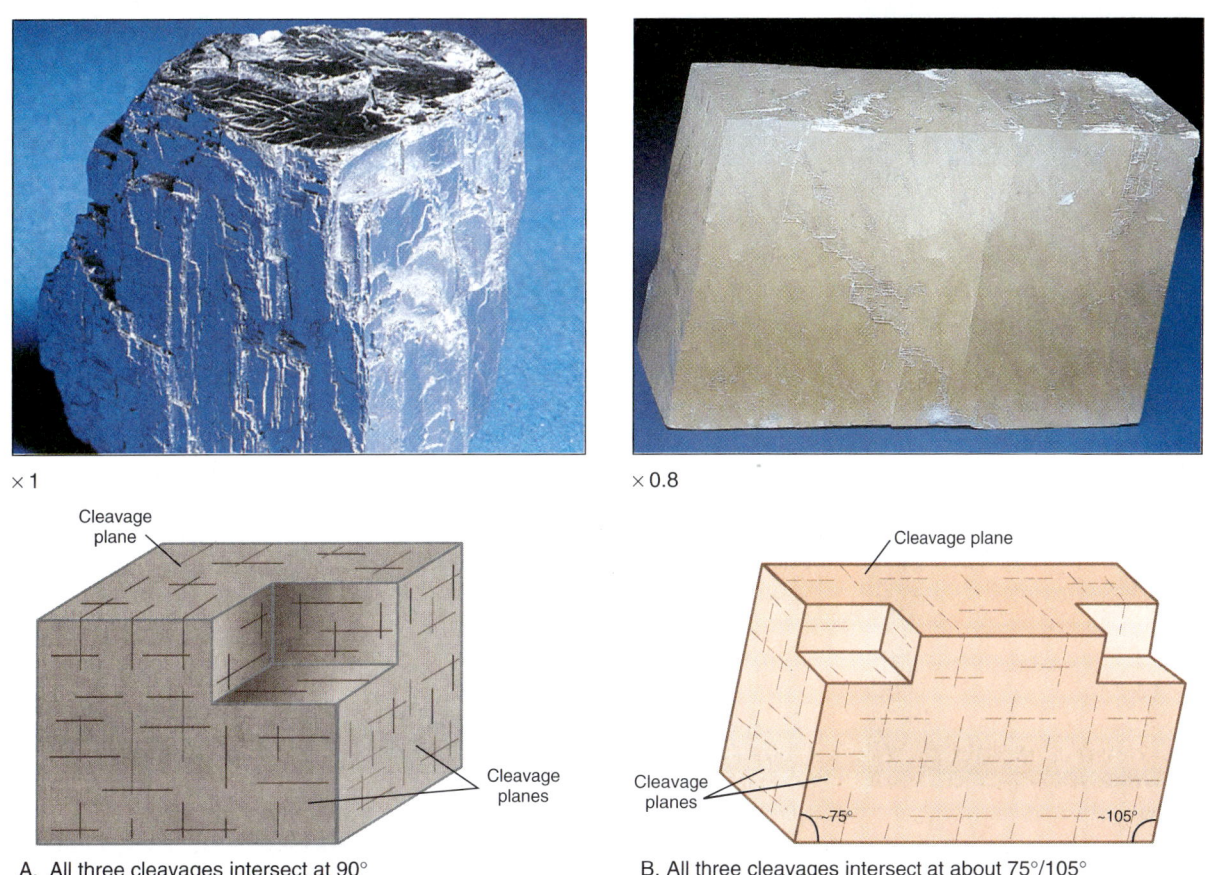

FIGURE 1.11
Three directions of cleavage: A. Cleavages in galena intersect at 90° (cubic cleavage). Note the metallic luster! B. All three cleavages in calcite intersect at either 75° or 105° (rhombohedral cleavage).

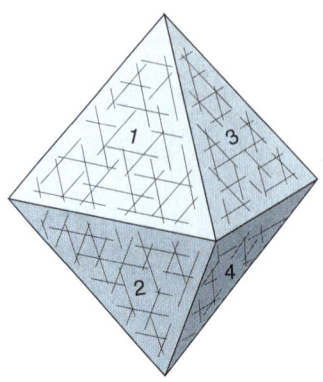
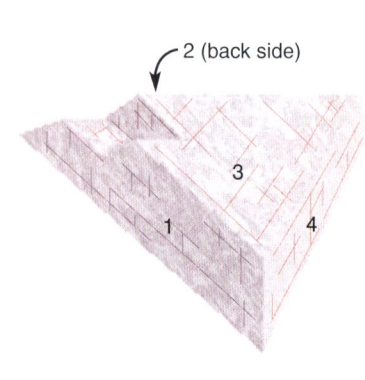

FIGURE 1.12
Four directions of cleavage (octahedral cleavage) in these samples of fluorite. Numbers indicate different cleavage directions.

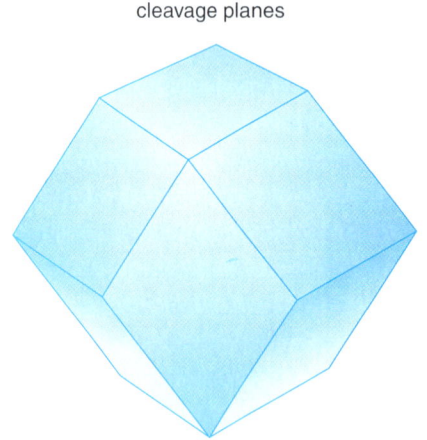

All faces are cleavage planes

FIGURE 1.13
Six directions of cleavage (dodecahedral cleavage) in a sample of sphalerite.

A. Irregular fracture × 1

B. Conchoidal fractures × 1
(See also Figure 3.3.)

FIGURE 1.14
Two types of fracture: A. An irregular or uneven fracture surface on the end of a potassium feldspar cleavage fragment (note the two cleavages at 90° to each other). B. Numerous conchoidal fractures mark the broken surface of this large quartz crystal. Figure 3.3 shows a much larger conchoidal fracture in obsidian.

fractures (Fig. 1.14B). Other kinds of fracture have descriptive names such as fibrous, splintery, or irregular.

In the field you will often have to break samples into pieces to observe cleavages and fractures on fresh surfaces. While it is instructional to hammer some mineral samples yourself, do not break the lab samples without your instructor's approval! Samples cost money and in most cases have already been broken to show characteristic features.

Crystal Form

A **crystal** is a solid, homogeneous, orderly array of atoms and may be nearly any size. Some crystals have smooth, plane faces and regular, geometric shapes (Fig. 1.15); these are what most people think of as crystals. However, a small piece broken from one of these nicely shaped crystals is also a crystal, because the atoms within that small fragment have the same orderly arrangement throughout. When examining minerals, and especially when determining cleavage, you must determine whether you are looking at a single crystal with well-developed crystal faces, a fragment of such a crystal, or a group of small, irregularly shaped intergrown crystals.

Cleavage surfaces may be confused with natural crystal faces; in fact, cleavage planes are parallel to possible (but not always developed) crystal faces. They can be distinguished as follows: (1) Crystal faces are normally smooth, whereas cleavage

×1

FIGURE 1.15
These crystals of potassium feldspar have smooth, flat faces and regular, geometric shapes.

×1

FIGURE 1.16
Crystal of quartz. Sides form a hexagonal prism that is capped with pyramid-like faces. Note the fine grooves on some crystal faces.

A. × 0.8

B. × 0.8

FIGURE 1.17
Crystals of pyrite in the form of (A) a cube and (B) pyritohedron. Note the fine grooves on the faces of the cubes.

planes, though also smooth, commonly are broken in a step-like fashion (see drawing in Fig. 1.10B or photograph in 1.11A). (2) Some crystal faces have fine grooves or ridges on their surfaces (Figs. 1.16, 1.17), whereas cleavage planes do not. Similar-looking, very thin, parallel grooves, or *striations,* are seen on plagioclase cleavage surfaces, but these features persist throughout the mineral and are not surficial, as described below. (3) Finally, unless crystal faces happen to coincide with cleavage planes, the mineral will not break parallel to them.

The arrangement of atoms in a mineral determines the shape of its crystals. Some minerals commonly occur as well-developed crystals, and their crystal forms are diagnostic. A detailed nomenclature has evolved to describe crystal forms, and some of the common names may be familiar. For example, quartz commonly occurs as *hexagonal* (six-sided) *prisms* with pyramid-like shapes at the top (Fig. 1.16); pyrite occurs as *cubes* (Fig. 1.17A) or *pyritohedrons* (forms with twelve pentagonal faces, Fig. 1.17B); calcite occurs as *rhombohedrons* (six-sided forms that look like cubes squashed by pushing down on one of the corners; Fig. 1.18A) or more complex, twelve-faced forms called *scalenohedrons* (Fig. 1.18B).

A. ×0.5

B. ×0.3

FIGURE 1.18
Crystals of calcite (yellow) in the form of (A) a rhombohedron and (B) a scalenohedron (yellow).

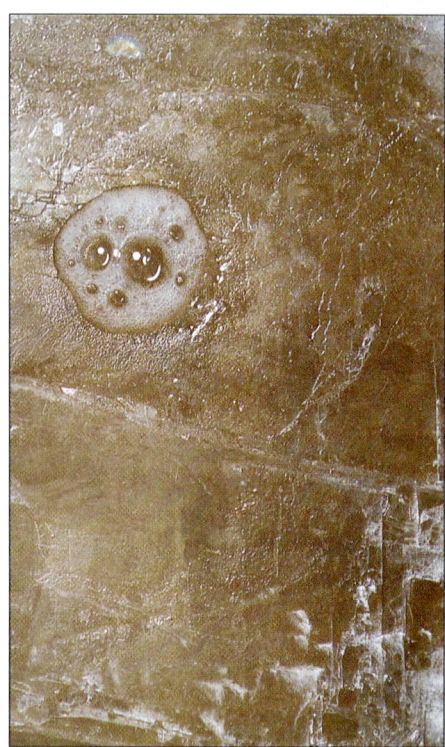
×1.5

FIGURE 1.19
Calcite reacting to a drop of acid.

Other Properties

Special properties help identify some minerals. These properties may not be distinctive enough in most minerals to help with their identification, or they may be present only in certain minerals.

Reaction with Acid

Some minerals, especially carbonate minerals, react visibly with acid. (Usually, a dilute hydrochloric acid [HCl] is used.) The acid test is especially useful for distinguishing the two carbonate minerals calcite and dolomite. When a drop of dilute hydrochloric acid is placed on calcite, it readily bubbles or effervesces, releasing carbon dioxide (Fig. 1.19). When a drop of acid is put on dolomite, the reaction is much slower unless the dolomite is powdered first; you may even have to look with a hand lens to see the bubbles, or, if the acid is weak, there may not be any. BE CAREFUL when using the acid—even dilute acid can burn your skin or put a hole in your clothing. Only a small drop of acid is needed to see whether or not the mineral bubbles. When you finish making the test, wash the acid off the mineral immediately. Should you get acid on yourself, wash it off right away, or if you get it on your clothing, rinse it out immediately.

Magnetism

Some minerals are attracted to a hand magnet. To test a mineral for magnetism, just put the magnet and mineral together and see if they are attracted. Magnetite is the only common mineral that is strongly magnetic.

Striations

Plagioclase feldspar can be positively identified and distinguished from potassium feldspar by the presence of *very thin, parallel grooves* called **striations** (Fig. 1.20). The grooves are present on only one of the two sets of cleavages and are best seen with a hand lens. They may not be visible on all parts of a cleavage surface. Before you decide there are no striations, look at all parts of all visible cleavage surfaces, moving the sample around as you look so that light is reflected from these surfaces at different angles.

Until you have seen striations for the first time, you may confuse them with the small, somewhat irregular, differently colored intergrowths or veinlets seen on cleavage faces of some specimens of potassium feldspar (Fig. 1.21). However, these have variable widths, are not strictly parallel, and are not grooves, so they are easily distinguished from striations.

Specific Gravity and Density

The **specific gravity** of a mineral equals its weight divided by the weight of an equal volume of water. The specific gravity of water (at 4°C) equals 1.0, by definition. **Density** is the mass of an object divided by its volume. Specific gravity and density give almost identical numerical values even though specific gravity, being a ratio of weights, is a unitless number, whereas density is commonly given in units of grams per cubic centimeter. Most of the rockforming minerals (see Chapter 2) have specific gravities of 2.6 to 3.4; the ore minerals (Chapter 2) are usually heavier, with specific gravities of 5 to 8. If you compare similar-sized samples of different minerals with sufficiently different specific gravities, the one with the higher specific gravity will feel heavier. For example, the mineral barite has an unusually high specific gravity (S.G=4.5) for a mineral with a nonmetallic luster. When compared with a similar-sized piece of, say, calcite (S.G=2.7), the barite will feel much heavier. For most minerals specific gravity is not a particularly noteworthy feature, but for some, such as barite or galena, high specific gravity is distinctive.

Taste, Odor, Feel

Some minerals have a distinctive taste (halite is salt, and tastes like it), some a distinctive odor (the powder of some sulfide minerals, such as sphalerite, a zinc sulfide, smells like rotten eggs), and some a distinctive feel (talc feels slippery).

Tenacity

The **tenacity,** or toughness, of a mineral describes its resistance to being broken. *Brittle* minerals, such as quartz, shatter when broken; *flexible* minerals, like chlorite, can be bent without breaking, but will not resume their original shape when the pressure is released; *elastic* minerals, such as the micas, can be bent without breaking and will spring back to their original position when the pressure is released; *malleable* minerals can be hammered into thin sheets (examples are gold and copper); *sectile* minerals can be cut with a knife (for example, gypsum).

× 0.5

FIGURE 1.20
Striations are visible on the upper surface of this sample of plagioclase.

× 1.5

FIGURE 1.21
The thin veinlets seen in some potassium feldspars should not be confused with striations in plagioclase.

Hands-On Applications

Geologists, like other scientists, must learn to be good observers. Observations provide the basic data on which most science is based. Good observational skills are acquired through training, and experience teaches you which kinds of observations are most important. Observation may mean looking without doing anything else, it may require making some measurements, or it may mean observing the results of a test or experiment. The problems in this first exercise illustrate the kinds of observations needed to identify minerals.

Objectives

If you complete all the problems, you should be able to:
1. Determine the color, luster, streak, and hardness of a mineral.
2. Determine whether a mineral has a cleavage or a fracture.
3. Determine the number of cleavages and the angular relations between the cleavages, if a mineral has more than one cleavage.
4. Distinguish between calcite and dolomite using the "acid test."
5. Determine whether a mineral is magnetic.
6. Distinguish between minerals on the basis of specific gravity.
7. Determine whether a sample of feldspar has striations.

Problems

1. Determine whether the luster of each sample provided for you is metallic or nonmetallic. If nonmetallic, indicate whether it is vitreous, pearly, greasy, waxy, resinous, or dull.

2. Determine the streak of each of the samples designated by your instructor.

3. Arrange the designated samples in order of increasing hardness. Determine the Mohs hardness of each.

4. Determine whether each of the samples provided has a cleavage. If it does, determine whether it has 1, 2, 3, 4, or 6 cleavages. If it has 2 or 3 cleavages, estimate whether or not adjoining cleavages intersect at right angles (90°). If only a fracture is present, indicate whether it is a conchoidal fracture.

5. Compare the samples provided by your instructor and list them in order of increasing specific gravity. This is done most easily by holding equal-size samples in each hand and estimating, on the basis of feel, which is the heavier; if the samples are the same size, the heavier one has the higher specific gravity. Alternative methods for determining specific gravity are described in Problem 12.

6. Determine which of the samples provided is attracted to a magnet.

7. Describe the feel of the samples provided.

8. Describe the odor of the mineral supplied by the instructor after rubbing it on a streak plate.

9. Put a small drop of dilute hydrochloric acid (HCl) on the designated samples. Describe what happens and explain why.

10. You have so far seen minerals only under visible light. What about invisible light? It turns out that ultraviolet (UV) light not only darkens skin and causes cataracts, but it also induces some spectacular effects in certain minerals. UV light is relatively safe at longer wavelenghts and is found at some dance parties in the form of black lights. Visit the Amethyst Galleries website *http://mineral.galleries.com/* or get to it through *www.mhhe.com/jones5e*—see Preface, click on *Interesting Groupings*, and check out *Fluorescent Minerals* to find out which minerals will look good at a dance club and why.

   a. What is fluorescence?

   b. What is phosphorescence and how does it differ from fluorescence?

   c. Triboluminescence is something you may have observed ripping open Band-Aids™ or crushing Certs™ candies in the dark, but it also occurs with minerals. What is triboluminescence?

   d. Explain whether minerals can always be identified with certainty based on their fluorescence.

   e. You have probably never heard of willemite, but it is a pretty cool mineral. Look it up (click on *By Name*, then *W*) and describe its fluorescence and why specimens from New Jersey excite collectors.

   f. There are two types of UV lights. What are they? Go back to the *Flourescent Minerals* page and find out if the Franklin, New Jersey, samples would look good at a party equipped with a typical black light.

In Greater Depth

11. People have long wondered what causes crystals to have perfect shapes. If you ever pulled a nice quartz crystal out of some weathered rock, you may have wondered at how something so perfectly ordered came from something so filthy and disordered. The alchemists tried to unravel such mysteries of nature, but their work was unscientific and they could not replicate each other's results. Their data and conclusions were therefore worthless. Nicolaus Steno in 1669 was the first to make objective, scientific measurements related to minerals. He wanted to see if evidence supported the hypothesis that there are underlying 'building blocks' that make up each mineral and determine its basic shape. To do this, he obtained quartz crystals of various sizes and shapes from around the known world. He then measured the angles between adjacent crystal faces and compared them to comparable angles measured on the other samples.

   NOTE: The following exercise can be done with a set of quartz crystals, plus a few amethyst and smoky quartz crystals, or with calcite rhombohedra. The calcite rhombohedra allow a test of whether the angles between cleavage planes vary by size or specimen.

   a. Examine a specimen and select four interfacial angles to compare among all the specimens. Pick some angles that are distinctly different from others on the same sample. Sketch below a cross section and/or profile of a representative sample and number your selected angles. Be sure to label whether each sketch is a top or side view!

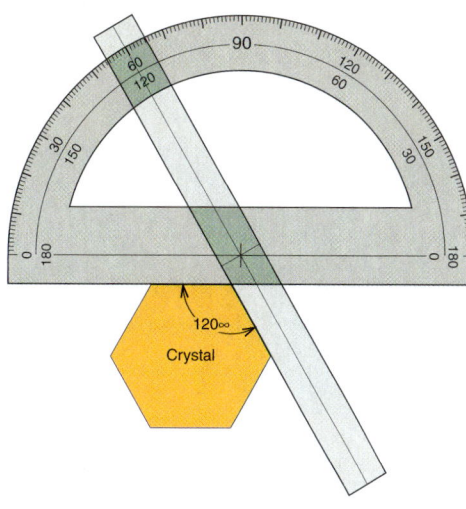

FIGURE 1.22
A simple contact goniometer consists of a protractor and a rotatable arm. The angle between adjacent crystal faces is measured as shown.

b. Using a simple contact goniometer as shown in Figure 1.22, carefully measure the interfacial angles at the four places you indicated. You may want to make several measurements of a single angle to be sure you have a good number. Record these data in the table below.

Angle Number	Sample	Sample	Sample	Sample	Sample	Sample	Sample	Sample	Average
1									
2									
3									
4									

To calculate the average, add the numbers for the same angle on each sample and divide by the total number of samples. For example, if you had four samples on which angle number 1 measured 44, 48, 43, and 45 degrees, the average is (44 + 48 + 43 + 45)/4 = 45°.

c. Steno's Law states that comparable interfacial angles are constant from specimen to specimen. Given the scatter in *your* data, do your data support or contradict Steno's Law? You have to decide whether the scatter in your data reflects measurement errors stemming from your simple measurement tool or from real differences that are clearly beyond your measurement error.

Part I Minerals

d. Compare the clear quartz crystals with amethyst and smoky quartz crystals. Their color is different from clear quartz, but the crystal shapes look similar. Measure your set of reference interfacial angles on the amethyst and smoky quartz to see if there are any significant differences between the colored and clear crystals.

Angle Number	Sample	Sample	Sample	Sample	Average
1					
2					
3					
4					

e. What do your data suggest about the underlying chemical composition and atomic architecture of the colored crystals as compared to clear quartz? If we had sophisticated lab equipment (an X-ray diffractometer), we could test your hypothesis.

12. Your instructor may want you to determine the specific gravity of a mineral from the results of a few simple measurements.

    Method 1 1. Weigh a dry sample of the mineral to be tested. 2. Weigh the dry dish to be used to catch the overflow in step 3. 3. Put the mineral into a beaker that is brim-full of water, causing the water to overflow into the previously weighed dish. 4. Weigh the dish with the overflow water in it. 5. Subtract the weight of the dry dish from the weight of the overflow water; the result is the weight of water that has a volume equal to that of the mineral. 6. Calculate the specific gravity of the mineral by dividing the weight of the dry mineral by the weight of the equal volume of water.

    Method 2 A more accurate method utilizes Archimedes' principle that "a body floating or submerged in a fluid is buoyed by a force equal to the weight of the fluid displaced." It was this discovery that caused Archimedes to run naked through the streets shouting "Eureka," because it enabled him to prove that the King's crown did not have a high enough specific gravity to be pure gold. A simple instrument that utilizes this principle is called a Jolly balance. Using a Jolly balance, first weigh a sample in air, then submerge it in water, and weigh it again. The specific gravity is calculated as:

    $$\text{specific gravity} = \frac{\text{weight in air}}{\text{weight in air} - \text{weight in water}}$$

CHAPTER 2

Mineral Identification

Materials Needed
- Pencil
- Hand lens
- Set of mineral samples
- Equipment for determination of mineral properties: streak plate, copper penny, glass plate or steel knife, magnet, dilute hydrochloric acid

Introduction

Most of the Earth is made of rocks, and rocks are made of minerals. Just as you had to learn the alphabet before you could read, you need to learn about minerals before you can understand rocks. The first step is to learn how to identify minerals. Chapter 1 shows that minerals have a number of rather easily determined physical properties, and that different minerals have different properties. This chapter explains how to use these properties to identify minerals. First, the text describes the most important (common) minerals found on Earth, and then it prepares you to go from observing the physical properties of an unknown mineral to an accurate identification of that mineral. You will see these same minerals over and over again through the next three labs, so it is worth taking extra care to learn them.

Some of the mineral samples you will see in your laboratory, or that are shown in these pages, may also be considered rocks. They are rocks if they are *aggregates* of crystals of the mineral rather than single crystals. Because large single crystals are very rare or nonexistent for some minerals, the only available samples may be aggregates. In common usage the term *mineral* is used whether referring to a single crystal or aggregates of crystals of the same mineral.

More than 3000 minerals are known to exist, yet only a comparative few are common. The **rock-forming minerals** are the most abundant minerals found in rocks at the Earth's surface. **Minor minerals** may be important constituents of certain rocks, but they are not common in most rocks. **Accessory minerals** are commonly present in rocks, but generally in only small amounts. **Ore minerals** have economic value when concentrated and are mined for the elements they contain. Table 2.1 lists some of the more important minerals within each of these groups; descriptions follow.

Rock-Forming Minerals

Feldspars (K, Na, and Ca Al-Silicates)

As a group, the feldspars are by far the most abundant minerals in the Earth's crust. They are the principal constituents of many kinds of igneous and metamorphic rocks and are abundant in some kinds of sedimentary rocks, as discussed in Chapters 3, 4, and 5. Feldspar is fairly easy to identify due to its two good cleavages that intersect at essentially 90°, a Mohs hardness of 6, and a vitreous luster. There are two principal varieties of feldspar—plagioclase and potassium feldspar.

Plagioclase has a variable chemical composition, ranging from sodium aluminum silicate ($NaAlSi_3O_8$) to calcium aluminum silicate ($CaAl_2Si_2O_8$). **Note: Abbreviations of common chemical elements are inside the back cover!** The color varies as well: sodium-rich varieties tend to be light colored—commonly white, cream to buff, or light gray—whereas calcium-rich varieties tend to be gray to dark gray (Fig. 2.1). However, **striations,** not color, are the distinctive feature of plagioclase (see Fig. 1.20, Chapter 1). Unfortunately, not every plagioclase cleavage shows striations.

Potassium feldspar is a general name that includes several varieties of $KAlSi_3O_8$ (e.g., *orthoclase, microcline, sanidine*). It comes in white, cream, light gray, and even green, but salmon pink is a common and often distinctive color (Fig. 2.2). Some varieties contain semi-parallel veinlets (Fig. 1.21), which, as discussed in Chapter 1, are not to be confused with the striations in plagioclase.

The principal use of feldspar, especially potassium feldspar, is in the production of porcelain and ceramics. Some varieties of Ca-rich plagioclase (called labradorite) show a play of dark blue iridescent colors as the angle of reflected light changes. These are used for ornamental purposes, as is an unusual green variety of microcline known as *amazonite*.

TABLE 2.1
Some Important Minerals

Rock-Forming Minerals*	Accessory and Minor Minerals	Ore Minerals
Potassium feldspar	Pyrite	Native Copper
Plagioclase feldspar	Magnetite	Graphite
Quartz	Corundum	Sulfur
Hornblende (an amphibole)	Halite	Galena
Augite (a pyroxene)	Fluorite	Sphalerite
Biotite	Apatite	Chalcopyrite
Muscovite	Garnet	Pyrite
Chlorite	Sillimanite	Hematite
Talc	Kyanite	Magnetite
Clays (e.g., kaolinite)	Topaz	Chromite
Olivine	Staurolite	Goethite
Calcite	Epidote	Malachite
Dolomite	Beryl	Azurite
Gypsum	Tourmaline	Barite
Halite	Serpentine	
	Opal	

*In order of decreasing abundance in the Earth's crust.

Quartz (SiO$_2$)

Quartz is one of the most abundant minerals in the crust of the Earth and is common in soils, rocks, and even atmospheric dust. It is hard and tough, so it survives well on the Earth's surface, and it occurs in a wide variety of colors and forms.

Distinctive properties are hardness (H = 7), conchoidal fracture, and vitreous to somewhat greasy luster, but color varies. Quartz comes in several varieties. Well-formed crystals of colorless *rock crystal quartz,* purple *amethyst,* and dark grey *smoky quartz* are common in rock shops, but not in your backyard. Masses of coarsely intergrown quartz crystals with no well-defined crystal faces occur as white (or iron-stained) *milky quartz* (Fig. 2.3), which is common in many areas, or the rarer pinkish *rose quartz* or grayish *smoky quartz* (Fig. 2.3). Extremely fine-grained varieties of quartz, collectively

×0.5 ×0.5 ×0.5

FIGURE 2.1
Various colors of **plagioclase.** Light-color varieties are usually richer in sodium. Darker varieties are richer in calcium.

×0.5

FIGURE 2.2
Microcline, a variety of **potassium feldspar.**

×0.5

FIGURE 2.3
Rose **quartz,** common milky quartz, and smoky quartz.

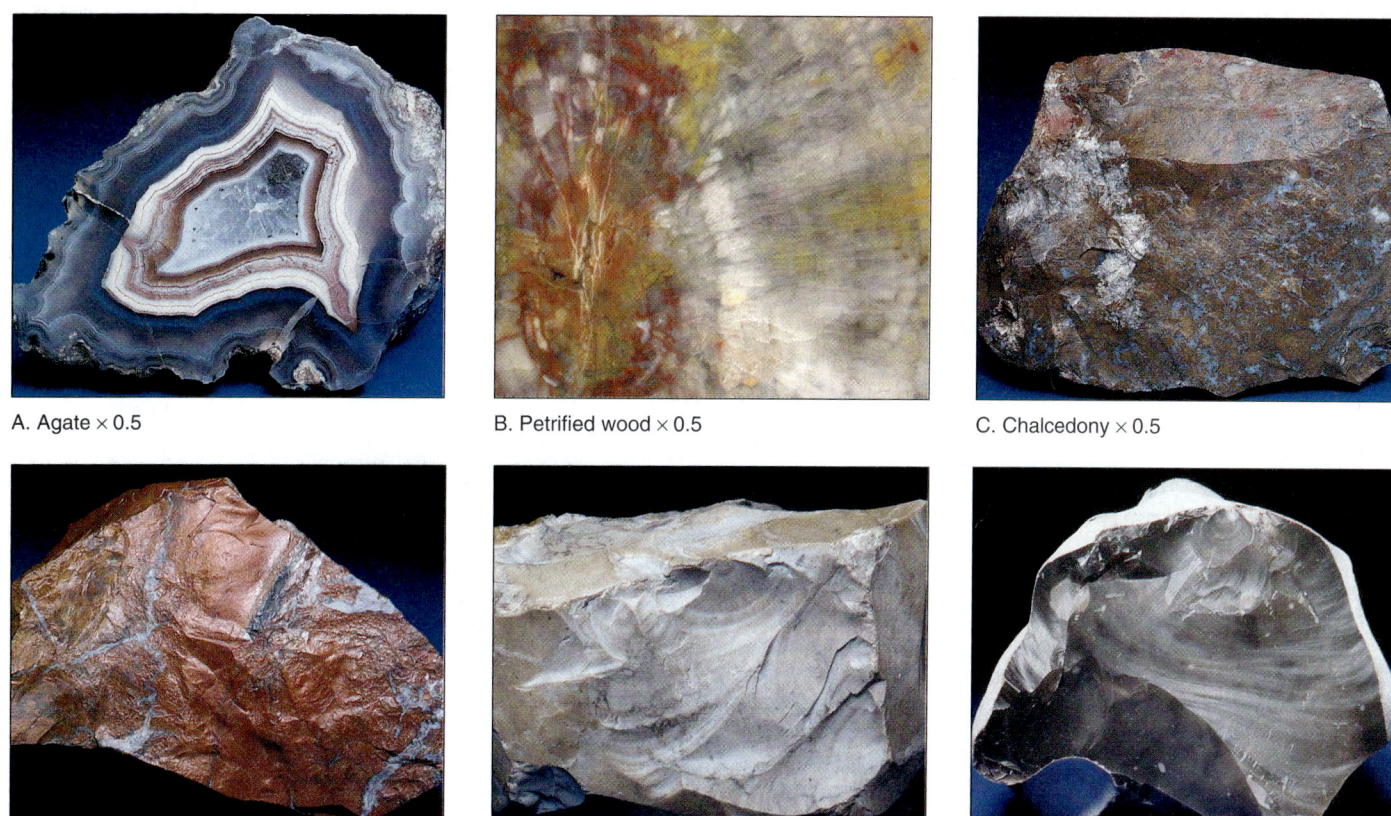

FIGURE 2.4
Microcrystalline varieties of SiO$_2$: A. agate; B. petrified wood; C. chalcedony; D. jasper; E. chert; F. flint. Note the color variations in A–F, and the conchoidal fractures in E and F.

termed *microcrystalline quartz,* include varicolored banded varieties such as *agate* and *onyx* and more homogeneous varieties including *chalcedony* (brown or gray, and translucent), *jasper* (red), *chert* (gray, white, or pink), and *flint* (dark gray to black) (Fig. 2.4). *Petrified wood* typically is composed of microcrystalline quartz (Fig. 2.4).

Quartz has many uses. It is a major constituent in glass, including the world's best microscope and telescope lenses. Some varieties are used in jewelry (rock crystal, amethyst, tiger's eye (yellowish brown with a wavy or silky sheen), citrine (transparent, yellow to orange-brown), and both chert and flint were used to make arrowheads, spear points, and other sharp tools. Quartz watches keep good time because a quartz crystal, when hooked to a battery, vibrates at a constant rate. As sand grains or parts of rocks, quartz is also used in concrete, mortar, and building stone.

Amphiboles (Complex Hydrous Na, Ca, Mg, and Fe Al-Silicates)

Hornblende is the most common amphibole. As you can see in Figures 1.9 and 1.10B (Chapter 1), hornblende is a black mineral with two directions of good cleavage that meet at roughly 60° or 120°; the cleavage gives it a splintery look, which helps to identify it. Hornblende occurs most commonly as small crystals in combination with other minerals in igneous and metamorphic rocks, but even there, it is distinctive when viewed with a hand lens.

The principal use of amphibole is for a small subset of the group with a fibrous habit that, as with some varieties of serpentine, makes it useful as asbestos. While airborne asbestos fibers can become imbedded in various parts of the lung or abdomen and cause asbestosis, only one variety (the amphibole crocidolite) is known to be especially carcinogenic. Asbestos is a poor conductor of heat and electricity and can be made into a variety of fabrics. These properties make it useful in many industrial applications, including cements, flooring, friction material, roofing, coatings, packings, gaskets, and protective clothing.

Pyroxenes (Ca, Mg, and Fe Silicates)

Like the amphiboles, most pyroxenes occur as small, crystalline components of rocks, especially certain igneous rocks. The most common pyroxene is *augite,* a greenish-black to black mineral that is easily confused with hornblende. Figure 2.5 shows that augite has two cleavages that intersect at about 90°, and it is not splintery looking like some hornblende. The cleavages may be difficult to see and are not nearly as good as those on hornblende. Another pyroxene, *diopside,* is commonly light green. A few varieties of pyroxene, such as *jadeite,* are prized for their ornamental value.

×1

FIGURE 2.5
Augite, a variety of pyroxene. Pyroxenes have two good-to-poor cleavages at 90°, but they can be tough to spot.

×0.5

FIGURE 2.6
Inuit carving of walrus mother and baby in soapstone, a rock composed mostly of **talc,** resting on a platform made of talc. See Figure 1.2F for a more typical lab sample of talc.

Micas (Hydrous K, Mg, and Fe Al-Silicates)

The two most frequently encountered micas, *biotite* and *muscovite,* are both common constituents of igneous and metamorphic rocks. Biotite is black, and muscovite commonly is colorless to pale gray, as you can see in Figures 1.2 and 1.7 (Chapter 1). Both have one perfect cleavage and a vitreous luster, so they are usually easy to identify. Muscovite is an excellent insulator for both heat and electricity and is used widely in the electrical industry. Big sheets were once used as windows on old wood stoves and in early automobiles (isinglass).

Chlorite (Hydrous Mg and Fe Al-Silicate)

This dark green mineral exhibits one perfect cleavage, but unlike the micas, thin pieces don't spring back to their original position after being bent; pieces are flexible, but not elastic. Most chlorite occurs as small grains in metamorphic rocks, and you may have to use a hand lens to see that it looks like a dark green mica.

Talc (Hydrous Mg Al-Silicate)

As number one on the Mohs hardness scale, talc is soft enough for use even on a baby's soft bottom. Talcum powder is talc. Talc crystals, if large enough, show one perfect cleavage, but in most cases they are too small to see. Instead, the pearly luster (Fig. 1.2F), greasy feel, and softness are distinctive. Talc forms by the alteration (e.g., metamorphism) of preexisting minerals such as olivine, and is the major ingredient in the rock soapstone. You may be familiar with the soapstone carvings of the American Indians (especially peace pipes), Kenyans (decorative plates, bowls, and candle holders), and the Inuits of the Canadian Arctic (Fig. 2.6).

Clays (Hydrous Al-Silicates)

Clay minerals are a significant component of soils, sediment, and sedimentary rocks. But they are not the sort of thing that you would want to put in a display case; in fact, when you dust the display case, you commonly are getting rid of clays! The word *clay* is used in two different ways: it refers to a specific group of minerals and to particles that are extremely small (<0.004 mm [<0.0002 in] diameter). Clay minerals are usually clay-sized particles. Clays are used for all kinds of things, from the manufacture of china and bricks, to stopping leaks in dams, to fillers in paper, paint, medicine (the clay mineral kaolinite is used in Kaopectate™), and chocolate (doesn't that make your mouth water?).

Clays are recognized by their extremely small crystal size, dull luster, softness, and earthy odor. Their colors vary, but the variety most commonly encountered in

×0.5

FIGURE 2.7
Microscopic crystals of **kaolinite,** a type of clay mineral, comprise 100 percent of this rock. The homogeneous appearance of this sample makes it a good example of a **massive** texture: There are no visible crystal forms or intergrown crystals.

physical geology labs is the white clay *kaolinite* (Fig. 2.7).

Olivine (Mg, Fe Silicate)

Figure 2.8 illustrates a typical sample of olivine. It is yellowish green to green and is granular, being made of a multitude of small grains (crystals). You have to look at individual crystals to see that it has a conchoidal fracture but no obvious cleavage. And if you check the hardness, make sure that little pieces aren't breaking off, because it is hard (H = 6.5 to 7). Most olivine occurs in dark-colored igneous rocks. There is a gem variety of olivine (*peridot*), and olivine is used in some heat resistant bricks.

Calcite ($CaCO_3$) and Dolomite ($CaMg(CO_3)_2$)

These two carbonates are the major minerals in the common sedimentary rocks limestone and dolostone, and their metamorphosed equivalent, marble. Calcite is also one of the most common minerals to form in veins and open cavities in rocks. When formed in such places, large cleavable pieces and beautiful crystals may develop, such as those shown in Figure 1.18 (Chapter 1). Calcite and dolomite are used for many things, among them waste treatment, animal feed, fertilizer, cement, and the manufacture of glass, paper, glazes, enamels, and industrial chemicals.

If you have a piece of calcite that shows cleavage (chances are you will), the three directions of cleavage not at 90° to each other are a giveaway. Add to that the Mohs hardness of 3, and calcite is easy to identify. The easiest way to distinguish calcite from dolomite is to put a drop of dilute hydrochloric acid (HCl) on them to see if they bubble; calcite bubbles strongly, dolomite not so much, or not at all unless powdered.

Gypsum ($CaSO_4 \cdot 2H_2O$)

Gypsum ranges from a massive, white to gray, carvable variety known as *alabaster* to a white variety with a silky or satiny luster and distinctive fibrous fracture known as *satin spar* to clear crystals or cleavage fragments known as *selenite* (Fig. 2.9). Its hardness—softer than your fingernail—gives it away. The clusters of crystals can be quite beautiful, but they are so fragile that you have to display them in a safe place. Gypsum is used extensively as the basis for plaster and drywall. If you have a clay-rich soil in your garden, you may have used it as a soil conditioner. Gypsum is a common evaporite mineral found as thick layers in sedimentary rocks.

Halite (NaCl)

This is an easy mineral to identify. Just taste it—it's salt. If you don't like the idea of licking where others have licked before, check the other properties. It is generally colorless or white, although impurities may add color. It commonly displays perfect cubic cleavage (see Fig. 1.6, Chapter 1) and is soft (H = 2.5). Halite occurs in sedimentary rocks that formed from the evaporation of seawater and commonly is associated with other *evaporite* minerals, such as gypsum. Its principal use is not for a flavoring or for melting ice, but as a source of sodium and chlorine for the chemical industry.

Accessory and Minor Minerals

Fluorite (CaF_2)

Fluorite is a pretty mineral that ranges from colorless to pale yellow, light green, or purple, and frequently occurs as either octahedral cleavage fragments (Fig. 1.12, Chapter 1) or as large cubic crystals. It would make a great gemstone except for its softness (Mohs hardness = 4) and ready ability to cleave. It is used primarily in making steel and for the manufacture of hydrofluoric acid. Hydrofluoric acid is one of the few substances that dissolves silicate minerals.

×1

FIGURE 2.8
Olivine. This sample is a mass of irregular, intergrown crystals.

A. ×0.5   B. ×0.5

FIGURE 2.9
Gypsum. A. Satin spar variety. B. Selenite variety showing one perfect and two poor cleavages.

×1

FIGURE 2.10
Serpentine asbestos. Note the fibers.

× 0.8

FIGURE 2.11
Twelve-sided crystals (dodecahedrons) of different varieties and colors of **garnet.**

Serpentine (Hydrous Mg Silicate)

The most familiar type of serpentine is serpentine asbestos (Fig. 2.10), but massive (no crystal forms), green serpentine is most common. The greasy to waxy luster (and feel) of massive serpentine is distinctive; its Mohs hardness (H = 3–5) serves to distinguish it from massive talc (H = 1). Massive-veined serpentine is commonly used as a decorative building stone. More than 95 percent of the asbestos produced in the world is serpentine. Studies have shown that health risks from serpentine asbestos are considerably less than from certain types of amphibole asbestos, although the distinction is rarely recognized by the public.

Garnet (Ca, Mg, Fe, and Al Silicate)

Garnet is a metamorphic mineral that almost always occurs as a perfect crystal (Fig. 2.11). However, some crystals are so large that the sample you see in the lab may be just a piece of a crystal. Mohs hardness (H = 6.5–7.5), vitreous to resinous luster, and crystal form help to identify garnet, but be careful of color. A common color is brownish red, but garnet can be black, brown, yellow, green, or other colors. Garnet is used for abrasives (such as in garnet sandpaper) and gemstones.

×2

FIGURE 2.12
Crystals of **staurolite.** Intergrown crystals on right are *twinned.*

Staurolite (Hydrous Fe Al-Silicate)

Like garnet, staurolite almost always occurs as "well-formed" crystals in metamorphic rocks. Dark brownish crystals (Fig. 2.12) are characteristic of this mineral. Intergrown crystals called *twins* are common and are sometimes sold as "fairy crosses."

Kyanite (Al_2SiO_5)

This metamorphic mineral is light blue and forms blade-shaped crystals (Fig. 2.13). It is used in the manufacture of spark plugs, among other things.

×1

FIGURE 2.13
Kyanite varies from sky blue to blue grey, but frequently shows its *bladed* crystal habit. The mineral surrounding the kyanite is quartz.

Ore Minerals

Hematite (Fe_2O_3)

The red to red-brown streak is the key to identifying hematite, because, as shown in Figure 2.14, it occurs as several strikingly different forms. One form is a reddish brown, soft, earthy material. Another is dull black and hard. A third type, known as *specular hematite*, has a metallic luster and occurs as a mass of small, mica-like scales. However, scratch any hematite, or rub it across a streak plate, and you will see the same red powder. Most hematite occurs in sedimentary rocks (or their metamorphic equivalents) called *iron formations*. It is the most important ore of iron, which is the principal ingredient in steel.

Magnetite (Fe_3O_4)

The only magnetic mineral you are likely to see in this lab is magnetite (Fig. 2.15). But just in case you don't have a magnet, other properties (such as black color and streak, metallic luster on fresh breaks, and moderately high specific gravity) should identify it. Magnetite is a common accessory mineral in igneous and metamorphic rocks. It is an important ore of iron in iron formation and where concentrated by igneous processes.

Limonite (Various Fe oxides)

Limonite is a general name for a mixture of poorly crystalline, hydrated and non-hydrated iron oxides. Although really a rock, it was once considered to be a distinct mineral. The name applies to yellow, yellow-brown, red or black, dull to metallic, soft to hard material. An example is shown in Figure 2.16. It usually forms by oxidation of iron-bearing minerals, so in essence, it is a rust. It has a yellow-brown streak, which serves to distinguish it from hematite, and is a minor ore of iron.

Galena (PbS)

Figures 1.2A and 1.11A (Chapter 1) show that galena has an obvious metallic luster and three perfect cleavages that intersect at 90°. It also has a very high specific gravity (SG = 7.5). Galena is almost the only ore of lead, which is used in batteries and metal products, and occurs most commonly in veinlike deposits.

Sphalerite (ZnS)

This may be the only mineral you will ever see with six directions of cleavage, although you probably will have a difficult time seeing them all (see Fig. 1.13, Chapter 1). If the cleavage is hard to see, look for the resinous luster (like tree sap); the yellow, brown, or black color; and the yellow-brown streak with its faint, rotten-egg-like odor of hydrogen sulfide. Sphalerite is the most important ore of zinc and commonly

× 0.5

FIGURE 2.14
Three varieties of **hematite**: specular hematite (left), oolitic hematite, and botryoidal hematite (right).

× 0.5

FIGURE 2.15
The lodestone variety of **magnetite** is a natural magnet strong enough to hold the paper clip. The brownish areas are oxidized ("rusted").

× 0.5

FIGURE 2.16
Limonite. This mixture of iron oxides was once considered a mineral, but is really a rock.

Pyrite (FeS$_2$)

Pyrite is pale brass-yellow, has a metallic luster, and is called *Fool's gold* by those in the know (who, presumably, are not themselves fools). Crystals in the form of cubes and pyritohedrons (see Fig. 1.17, Chapter 1) are common, but so are irregular masses that show the characteristic color and uneven fracture. Given these properties, a Mohs hardness of 6 to 6.5 will enable you to distinguish it from other yellow sulfide minerals such as chalcopyrite or, should the chance arise, gold. It is a common accessory mineral in igneous, sedimentary, and metamorphic rocks and is concentrated in veins or irregular masses in many places. Pyrite is mined along with the more valuable sulfide minerals with which it occurs, and some of it is used in the production of sulfuric acid.

Chalcopyrite (CuFeS$_2$)

This brass-yellow mineral is distinguished from massive pyrite by its slightly more yellow color and its Mohs hardness of 3.5 to 4. It is one of the most important ores of copper. Copper is used for electrical conductors (such as wire), brass (an alloy with zinc), bronze (an alloy with tin), fittings, sheathings, and other manufactured items. Chalcopyrite occurs in veins and various types of more massive (unlayered, no obvious structure) deposits.

Graphite (C)

Graphite is dark gray, silver gray, or black, and is greasy, soft, and soils your fingers or whatever it touches. Mixed with clay, it forms the "lead" in pencils, and mixed with oil, it is a good lubricant. It has other uses as well, including generator brushes, batteries, and electrodes. It is usually found in metamorphic rocks.

Bauxite (Various Al oxides)

Bauxite is really a rock made of several hydrous Al oxide minerals, but it once was regarded as a mineral. It is characterized by pea-shaped structures (Fig. 2.17); it may be white, gray, or red, and usually is soft, dull, and earthy. It forms by extensive weathering in tropical to subtropical climates. It is important because it is the only ore of aluminum, which is used wherever a light, strong metal is needed.

How to Identify Minerals

Warning: The samples you see in lab may, at first glance, look quite different from the photos in the manual. Nature is expert at providing infinite variation in color and shape, and these are the features most easily captured by camera. In addition, certain photos show museum-quality specimens (for example, pyrite in Fig. 1.17), which are too rare and expensive to use in labs. Thus, you will have to identify your mineral specimens using the physical properties that you learned in Chapter 1 and the tables that follow (Tables 2.2A through F). These tables are designed to allow you to use basic observations to quickly narrow the possibilities to just a few choices. In terms of the scientific method, you will first make some observations and form one or several hypotheses regarding the identity of the mineral. Then you will closely examine the descriptions in the tables to guide you toward making additional observations that eliminate certain possibilities and help you settle on the best mineral identification. Finally, check to see that the mineral description correctly matches your sample in all important regards.

To streamline the process, we've included a flow chart (Fig. 2.18) to guide you from your unknown mineral to the table that contains its correct identification. For example, suppose you are given a mineral that has the following properties:

Luster	Hardness	Streak
nonmetallic (vitreous)	3.5 to 5	white

Cleavage	Color	Other
four perfect	clear	purplish tinge

According to Figure 2.18, the first determination is luster. Table 2.2A contains minerals with metallic luster; Tables 2.2B through 2.2F contain minerals with nonmetallic luster. Because the unknown mineral has a nonmetallic luster, Table 2.2A is eliminated.

Following the flow chart (Fig. 2.18), the next thing to determine is whether a mineral is harder than glass (H>5.5) or softer than glass (H<5.5). Since our mineral is softer than glass (H is between 3.5 and 5), it must be in Tables 2.2B through 2.2D.

The third observation needed is streak (Fig. 2.18). The white streak eliminates Table 2.2B, leaving only 2.2C and 2.2D.

The last major distinction is based on cleavage. Table 2.2C is for minerals with no cleavage, and 2.2D is for minerals with one or more cleavages. Because our mystery mineral has four good cleavages, it must be in Table 2.2D. Examination of the mineral descriptions in this table shows only one mineral with four cleavages: fluorite.

NOTE: The most useful properties for identifying a given mineral are underlined in the tables. "H" in the descriptions refers to *hardness,* and "S.G." to *specific gravity.* Some minerals appear in more than one table because they may or may not show a particular property in the sample you examine. For example, hematite appears in Table 2.2A, because one variety has a metallic luster, and in Table 2.2B, because another variety does not. Hornblende, augite, and bauxite are listed in two places, because their hardness may be greater or less than 5.5.

×1

FIGURE 2.17
Bauxite with pea-shaped structures. This rock was once considered a mineral.

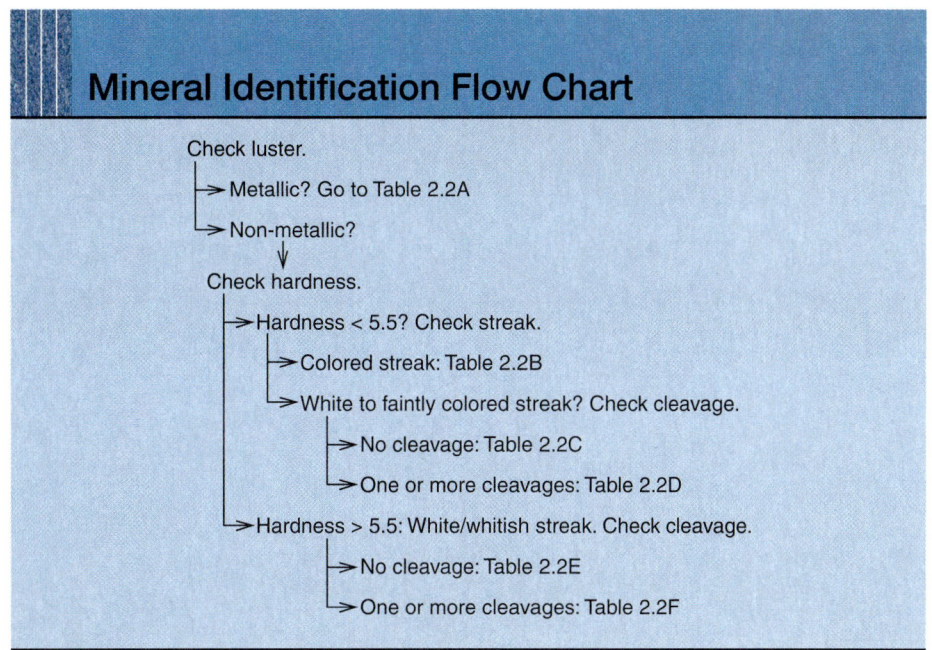

FIGURE 2.18
A flow-chart guide to the mineral identification tables.

Since it may be useful in future chapters, Table 2.3 gives the chemical groupings and formulae of the minerals listed in Tables 2.2 A–F.

Insecure About Mineral Identification?

As you work through this lab and the next few that require rock identification, you may feel some insecurity in your mineral identifications. How can you be *sure* that you are correct? Well, all geologists feel this unease at first. With experience, however, confidence and skill grow. This is why every geologist tries to see as many different samples as possible. If you continue in geology, you will also encounter more sophisticated tools used to make more definitive identifications. Geology majors use expensive petrographic microscopes to identify unknown minerals. Graduate students and other researchers use advanced technology to help identify pesky but potentially important unknown minerals. For example, electron microprobes zap samples with electron beams to work out their chemical composition. If more information is needed, an X-ray diffractometer uses X-rays to map out the atomic structure of unknown minerals. Such techniques either provide a definitive mineral name or, with luck, uncover a mineral that is new to science!

TABLE 2.2A
Metallic Luster Or Tarnished (Dulled) Metallic Luster*

Hardness	Streak	Cleavage	Other	Mineral
Less than 5.5	Black, greenish black brownish black, or gray	One (but not always visible)	Gray to black; H = 1; S.G. = 2.5; may have dull luster; greasy feel, soils paper and fingers. **See p. 24.** Also in Table 2.2B.	**Graphite** C
		Three (excellent cubic cleavage)	Silvery gray; H = 2.5; S.G. = 7.5; cubic crystals common. **See p. 23.**	**Galena** PbS
		None	Black to brownish black; H = 5.5; S.G. = 4.6; luster may be almost nonmetallic; brownish-black streak; weakly magnetic.	**Chromite** $FeCr_2O_4$
		None	Golden yellow or greenish yellow; H = 3.5 to 4; S.G. = 4.2; massive. **See p. 24.**	**Chalcopyrite** $CuFeS_2$
	Yellow-brown	None	Dark brown to black; H = 5 to 5.5; S.G. = 4; massive or rounded forms with radiating, fiberlike layers.	**Goethite** FeO(OH)
	Copper-red	None	Copper to brown, may have green coating; H = 2.5 to 3; S.G. = 8.9; malleable, sectile.	**Native Copper** Cu
Greater than 5.5	Greenish black to black	None	Brass-yellow; H = 6 to 6.5; S.G. = 5.0; massive or as crystals (cube, pyritohedron). **See p. 24.**	**Pyrite** FeS_2
	Black	None	Dark gray to black; H = 6; S.G. = 5.2; may have dull luster if surface not fresh; attracted to magnet. **See p. 23.**	**Magnetite** Fe_3O_4
	Red-brown	None	Steel gray to dull red; H = 6; S.G. = 5.0; may be micaceous (tiny flakes) or massive. **See p. 23.** Also listed in Table 2.2B.	**Hematite** Fe_2O_3

Underlined properties are most useful in mineral identification. H = hardness (Mohs). S.G. = specific gravity.

*Like metals, many minerals with a metallic luster oxidize or tarnish as they react with moisture and various compounds in air. This transforms metallic lusters into dull lusters. Geologists in the field break open rocks to expose fresh surfaces (that's why we carry rock hammers!). In this lab you are probably not allowed to smash the samples, so if you have a sample with a dull luster that looks like it could almost be metallic, take a streak and check out the distinctive properties listed in this table.

TABLE 2.2B
Nonmetallic Luster • Hardness Less Than 5.5 • Streak—Colored

Streak	Cleavage	Other	Mineral
Gray-black	One (not always visible)	Gray to black; $\underline{H = 1; S.G. = 2.5;}$ may have metallic luster; greasy feel; soils paper and fingers. **See p. 24.** Also listed in Table 2.2A.	**Graphite** C
Red-brown	None	$\underline{\text{Red to reddish brown}}$; H = 1.5 to 5.5; S.G. = 5.0; dull luster; earthy or oolitic (containing spherical structures having diameters of 0.25 to 2 mm) masses. **See p. 23.** Also listed in Table 2.2A.	**Hematite** Fe_2O_3
Yellow-brown	None	$\underline{\text{Yellow-brown, orange-brown}}$ to dark brown; H = 1.5 to 5.5; S.G. = 3.6 to 4.0; dull luster, earthy, may be a rustlike coating. Not a true mineral. Limonite is a general name for several rustlike, hydrous iron oxides. **See p. 23.**	**Limonite** Various hydrous iron oxides
Yellow	None	$\underline{\text{Yellow}}$; H = 1.5 to 2.5; S.G. = 2.1; resinous or vitreous luster.	**Sulfur** S
Yellow	$\underline{\text{Six}}$-may be difficult to count	Light yellow, yellowish brown, black; H = 3.5 to 4; S.G. = 4.0; $\underline{\text{resinous or vitreous luster}}$; faint odor of hydrogen sulfide (rotten eggs). **See p. 23.**	**Sphalerite** ZnS
Blue	None	$\underline{\text{Azure blue}}$; H = 3.5 to 4; S.G. = 3.8; occurs as coatings, masses, or tiny crystals, $\underline{\text{commonly with malachite,}}$ and commonly as an accessory mineral.	**Azurite** $Cu_3(CO_3)_2(OH)_2$
Green	None	$\underline{\text{Bright green}}$; H = 3.5 to 4; S.G. = 4.0; occurs as coatings, masses, or tiny crystals, $\underline{\text{commonly with azurite,}}$ and commonly as an accessory mineral.	**Malachite** $Cu_2CO_3(OH)_2$

$\underline{\text{Underlined}}$ properties are most useful in mineral identification. H = hardness (Mohs). S.G. = specific gravity.

TABLE 2.2C
Nonmetallic Luster • Hardness Less Than 5.5
• Streak—White, Whitish, Or Faintly Colored • No Apparent Cleavage

Cleavage	Other	Mineral
None or Indistinct	White, may have brownish tint; $\underline{H = 1 \text{ to } 2.5;}$ S.G. = 2.6; dull luster; greasy feel, earthy odor, powdery. **See p. 20.**	**Kaolinite** $Al_2Si_2O_5(OH)_4$
	White, gray, or apple green; pearly luster; H = 1 (or 2 in some impure varieties); S.G. = 2.7 to 2.8; $\underline{\text{greasy feel;}}$ may have light gray streak. **See p. 20.** Also listed in Table 2.2D.	**Talc** $Mg_3Si_4O_{10}(OH)_2$
	White; $\underline{H = 2;}$ S.G. = 2.3; Satiny or silky luster and fibrous fracture indicates $\underline{\text{satin spar}}$ variety of gypsum. An opaque, white to gray, massive variety is $\underline{\text{alabaster.}}$ **See p. 21.** Also in Table 2.2D.	**Gypsum** $CaSO_4 \cdot 2H_2O$
	Reddish brown to brown color; H = 1 to 6; S.G. = 2.0 to 2.5; dull to earthy luster; massive or with $\underline{\text{pea-shaped structures;}}$ may scratch glass with difficulty. Not a true mineral, but a rock consisting of several similar minerals. May have faint, red-brown streak. **See p. 24.** Also listed in Table 2.2E.	**Bauxite** Mixture of hydrous aluminum oxides
	Multicolored green, gray, black; H = 3 to 5; S.G. = 2.5 to 2.6; dull to $\underline{\text{greasy luster;}}$ $\underline{\text{slight greasy feel;}}$ massive to fibrous ($\underline{\text{asbestos}}$). **See p. 22.**	**Serpentine** $Mg_3Si_2O_5(OH)_4$
	$\underline{\text{Light green to medium green}}$, brown, yellow; vitreous luster; $\underline{H = 5;}$ S.G. = 3.2; six-sided crystals common; may show one poor cleavage.	**Apatite** $Ca_5(PO_4)_3(F,Cl,OH)$
	Buff, gray, white, pinkish; H = 3.5 to 4; S.G. = 2.8 to 2.9; small, rhombohedral crystals or massive; three cleavages, not at 90°, may be indistinct in massive varieties; $\underline{\text{reacts slowly or not at all with dilute hydrochloric acid unless powdered.}}$ **See p. 21.** Also listed in Table 2.2D.	**Dolomite** $CaMg(CO_3)_2$

$\underline{\text{Underlined}}$ properties are most useful in mineral identification. H = hardness (Mohs). S.G. = specific gravity.

TABLE 2.2D
Nonmetallic Luster • Hardness Less Than 5.5
• Streak—White, Whitish, Or Faintly Colored • One Or More Cleavages

Cleavage	Other	Mineral
One	Light apple green, gray, white; pearly luster; H = 1; S.G. = 2.7 to 2.8; cleavage not evident in finely crystalline, massive varieties; may have light gray streak; greasy feel. See p. 20. Also listed in Table 2.2C.	**Talc** $Mg_3Si_4O_{10}(OH)_2$
	Green to blackish green; dull to vitreous or pearly luster; H = 2 to 2.5; S.G. = 2.6 to 3.3; may have faint green-yellow streak; cleavage flakes are flexible but not elastic; finely crystalline aggregates common. See p. 20.	**Chlorite** Hydrous Mg-Fe-Al silicate
	Black to brownish black; vitreous luster; H = 2.5 to 3.0; S.G. = 2.8 to 3.2; may have faint brown-gray streak; individual crystals commonly are small and cleavage surfaces are wavy; perfect cleavage; transparent, flexible and elastic in thin sheets. See p. 20.	**Biotite** Hydrous K-Mg-Fe-Al silicate
	Colorless, silvery white, brownish silvery white; vitreous luster; H = 2.0 to 2.5; S.G. = 2.8 to 2.9; perfect cleavage; transparent, flexible, and elastic in thin sheets. See p. 20.	**Muscovite** Hydrous K-Al silicate
	Clear, white, light gray; H = 2; S.G. = 2.3; vitreous to pearly luster; brittle sheets; one perfect cleavage, two poor cleavages indicate the *selenite* variety of gypsum; *alabaster* is massive, *satin spar* is fibrous. See p. 21. Also listed in Table 2.2C.	**Gypsum** $CaSO_4 \cdot 2H_2O$
Two	Black; H = 5 to 6; S.G. = 3.0 to 3.4; vitreous luster; two perfect cleavages meet at 124° and 56°; cleavage faces stepped rather than smooth; splintery appearance; may have greenish black to black streak. An amphibole. See p. 19. Also listed in Table 2.2F.	**Hornblende** Hydrous Na-Ca-Mg-Fe-Al silicate
	Black to dark green; H = 5 to 6; S.G. = 3.2 to 3.4; vitreous to dull luster; two imperfect cleavages meet at nearly 90°. Tends to look 'blocky,' not splintery. A pyroxene. Another pyroxene, *diopside,* is similar but is light grayish green. See p. 19. Also listed in Table 2.2F.	**Augite** Ca-Mg-Fe silicate
Three	Clear to gray to red; H = 2.5; S.G. = 2.2; three perfect cleavages meet at 90° (cleavage surfaces may dull and partially dissolve with prolonged exposure); salty taste. See p. 21.	**Halite** NaCl
	Clear, white, light gray; H = 2; S.G. = 2.3; vitreous to pearly luster; brittle sheets; one perfect cleavage, two poor cleavages may be evident in clear pieces. See p. 21. Also in Table 2.2C.	**Gypsum** $CaSO_4 \cdot 2H_2O$
	Colorless or white; H = 3 to 3.5; S.G. = 4.5—heavy for a nonmetallic; tabular crystals, roselike array of crystals, or massive; three cleavages meet at 90°.	**Barite** $BaSO_4$
	Clear, white, less commonly other colors; vitreous luster; H = 3; S.G. = 2.7; three perfect cleavages form rhombic cleavage fragments; double image seen through clear pieces; reacts strongly with dilute hydrochloric acid. See p. 21.	**Calcite** $CaCO_3$
	Buff, gray, white, pinkish; H = 3.5 to 4; S.G. = 2.8 to 2.9; small, rhombohedral crystals or massive; three cleavages, not at 90°, may be indistinct; reacts slowly or not at all with dilute hydrochloric acid unless powdered. See p. 21. Also listed in Table 2.2C.	**Dolomite** $CaMg(CO_3)_2$
Four	Purple, green, blue, yellow, clear; H = 4; S.G. = 3.2; vitreous luster; perfect cleavage up to four directions may yield octahedral cleavage fragments; may occur as cubic crystals. See p. 21.	**Fluorite** CaF_2

Underlined properties are most useful in mineral identification. H = hardness (Mohs). S.G. = specific gravity.

TABLE 2.2E
Nonmetallic Luster • Hardness Greater Than 5.5
• Streak—White Or Whitish If Softer Than Streak Plate • No Cleavage

Cleavage	Other	Mineral
None	Brown, pink, blue, gray; H = 9; S.G. = 4.0; six-sided prismatic crystals; *ruby* (red) and *sapphire* (commonly blue) are gem varieties.	**Corundum** Al_2O_3
	Black, pink, blue, green, brown; vitreous luster; H = 7 to 7.5; S.G. = 3.0 to 3.3; slender crystals with triangular cross sections and striated sides.	**Tourmaline** Complex hydrous silicate
	Reddish brown, yellowish tan; vitreous to resinous luster; H = 6.5 to 7.5; S.G. = 3.6 to 4.3; twelve-sided crystals common. Broken surfaces may resemble cleavage in some samples. **See p. 22.**	**Garnet** Ca-Mg-Fe-Al silicate
	Red-brown to brownish black; vitreous, resinous, or dull luster; H = 7 to 7.5; S.G. = 3.7; prismatic and X- or cross-shaped crystals. **See p. 22.**	**Staurolite** Hydrous Fe-Al silicate
	Coarsely crystalline varieties: clear, milky, white, purple, smokey, pink; transparent to translucent; vitreous luster; H = 7; S.G. = 2.7; conchoidal fracture; usually massive; sometimes occurs as six-sided crystals; milky = *milky quartz,* purple = *amethyst,* smoky = *smoky quartz,* pink = *rose quartz*. Microcrystalline varieties: *chert* (gray, dull luster), *flint* (black, dull luster), *chalcedony* (brown to gray, translucent, waxy luster), *agate* and *onyx* (varicolored bands, vitreous luster). **See p. 18 and 19.**	**Quartz** SiO_2
	Colorless, white, or pale shades of yellow, green, red, or blue; may show play of colors; vitreous to resinous luster; H = 5 to 6; S.G. = 2.0 to 2.3; rounded forms common, but also massive; conchoidal fracture. A mineraloid.	**Opal** $SiO_2 \cdot nH_2O$
	Olive green to yellow green; vitreous to dull luster; H = 6½ to 7 but difficult to test because granular; S.G. = 3.3 to 4.4. **See p. 20.**	**Olivine** $(Mg,Fe)_2SiO_4$
	Reddish brown to brown color; H = 1 to 6; S.G. = 2.0 to 2.5; dull to earthy luster; massive or with pea-shaped structures; may scratch glass with difficulty. Not a true mineral, but a rock consisting of several similar minerals. **See p. 24.** Also listed in Table 2.2C.	**Bauxite** Mixture of hydrous aluminum oxides

Underlined properties are most useful in mineral identification. H = hardness (Mohs). S.G. = specific gravity.

TABLE 2.2F
Nonmetallic Luster • Hardness Greater Than 5.5 • Streak—White Or Whitish If Softer Than Streak Plate • One Or Two Cleavages

Cleavage	Other	Mineral
One	Colorless, yellow, brown, pink, bluish; H = 8; S.G. = 3.4 to 3.6; vitreous luster; elongate crystal prisms with pointed ends and striated side faces.	**Topaz** $Al_2SiO_4(OH,F)_2$
	Bluish green, yellow, white, pink; H = 7.5 to 8; S.G. = 2.7 to 2.8; elongate, six-sided crystal prisms with flat ends common.	**Beryl** $Be_3Al_2(Si_6O_{18})$
	Light blue to greenish blue; vitreous luster; H = 5 parallel to long direction of crystal, seven across crystal; blade-shaped crystals; one cleavage. See p. 22.	**Kyanite** Al_2SiO_5
	White, pale green, brown; H = 6 to 7; S.G. = 3.2; long, slender crystals, commonly as groups of parallel crystals.	**Sillimanite** Al_2SiO_5
Two	Salmon-pink, white, gray, green; vitreous luster; H = 6; S.G. = 2.5 to 2.6; two cleavage directions meet at nearly right angles; no striations. See p. 17.	**Potassium feldspar** $KAlSi_3O_8$
	White to dark gray, sometimes buff; vitreous luster; H = 6; S.G. = 2.6 to 2.8; two cleavage directions meet at nearly right angles; some cleavage faces have perfectly straight, parallel striations, which show up in reflected light. See p. 17.	**Plagioclase feldspar** $NaAlSi_3O_8$ to $CaAl_2Si_2O_8$
	Pistachio green, yellowish green; H = 6 to 7; S.G. = 3.3 to 3.5; elongate crystals or finely crystalline masses.	**Epidote** Hydrous Ca-Fe-Al silicate
	Black; H = 5 to 6; S.G. = 3.0 to 3.4; vitreous luster; two perfect cleavages meet at 124° and 56°; cleavage faces stepped rather than smooth, giving splintery appearance; may have faint greenish-gray streak. An amphibole. See p. 19. Also listed in Table 2.2D.	**Hornblende** Hydrous Na-Ca-Mg-Fe-Al silicate
	Black to dark green; H = 5 to 6; S.G. = 3.2 to 3.4; vitreous to dull luster; two imperfect cleavages meet at nearly 90°. Tends to look 'blocky,' not splintery. A pyroxene. Another pyroxene, *diopside*, is similar but is light gray to light grayish green. See p. 19. Also listed in Table 2.2D.	**Augite** Ca-Mg-Fe silicate

Underlined properties are most useful in mineral identification. H = hardness (Mohs). S.G. = specific gravity.

TABLE 2.3
Chemical Grouping And Formulae Of The Minerals In Tables 2.2A–F

Chemical Group	Mineral Name	Chemical Formula
Elements	Copper	Cu
	Sulfur	S
	Graphite	C
Sulfides	Galena	PbS
	Sphalerite	ZnS
	Chalcopyrite	$CuFeS_2$
	Pyrite	FeS_2

Category	Mineral	Formula
Oxides and hydroxides	Corundum	Al_2O_3
	Hematite	Fe_2O_3
	Magnetite	Fe_3O_4
	Chromite	$FeCrO_4$
	Goethite	$FeO(OH)$
Halides	Halite	$NaCl$
	Fluorite	CaF_2
Carbonates	Calcite	$CaCO_3$
	Dolomite	$CaMg(CO_3)_2$
	Malachite	$Cu_2CO_3(OH)_2$
	Azurite	$Cu_3(CO_3)_2(OH)_2$
Sulfates	Barite	$BaSO_4$
	Gypsum	$CaSO_4 \cdot 2H_2O$
Phosphates	Apatite	$Ca_5(PO_4)_3(F,C,OH)$
Silicates[1]		
Isolated SiO_2 tetrahedra	Olivine	$(Mg,Fe)_2SiO_4$
	Garnet	$(Mg,Fe,Ca)_3(Al,Fe)_2(SiO_4)_3$
	Kyanite and sillimanite	Al_2SiO_5
	Topaz	$Al_2SiO_4(OH,F)_2$
	Staurolite	$Fe_2Al_9O_6(SiO_4)_4(O,OH)_2$
Double SiO_2 tetrahedra	Epidote	$Ca_2(Fe,Al)Al_2O(SiO_4)(Si_2O_7)(OH)$
Rings of SiO_2 tetrahedra	Tourmaline	$(Na,Ca)(Li,Mg,Al)(Al,Fe)(BO_3)_2(Si_6O_{18})(OH)_4$
	Beryl	$Be_3Al_2(Si_6O_{18})$
Single chains of SiO_2 tetrahedra (Pyroxene group)	Augite	$Ca(Mg,Fe)(SiO_3)_2$
Double chains of SiO_2 tetrahedra (Amphibole group)	Hornblende	$(Ca,Na)_{2-3}(Mg,Fe,Al)_5Si_6(Si,Al)_2O_{22}(OH)_2$
Sheet of SiO_2 tetrahedra	Serpentine	$Mg_3Si_2O_5(OH)_4$
	Kaolinite	$Al_2Si_2O_5(OH)_4$
	Talc	$Mg_3Si_4O_{10}(OH)_2$
	Muscovite	$KAl(AlSi_3O_{10})(OH)_2$
	Biotite	$K_2(Mg,Fe)_3(AlSi_3O_{10})(OH)_2$
	Chlorite	$(Mg,Fe)_3(Si,Al)_4O_{10}(OH)_2 \cdot (Mg,Fe)_3(OH)_6$
Framework of SiO_2 tetrahedra	Quartz	SiO_2
	Opal	$SiO_2 \cdot nH_2O$
	K-feldspar	$KAlSi_3O_8$
	Plagioclase feldspar	$NaAlSi_3O_8$-$CaAl_2Si_2O_8$

[1]The silicates are grouped according to the way in which the silica (SiO_2) tetrahedra are arranged in their crystal structures. Many of the silicates have variable chemical compositions, and the formulae given in the table may be representative only.

Hands-On Applications

Just as bacteriologists need to be able to identify different kinds of bacteria, geologists need to know what the most common minerals look like and how to identify ones they don't already know. In Chapter 1, you learned how to determine some of the physical properties of minerals. The present chapter illustrates how the basic physical properties, arranged in the form of an identification key or determinative tables, provide a rational, straightforward method for mineral identification. Identification keys or tables, like the ones used for minerals, are a means of organizing large amounts of information. They can be used to identify trees, flowers, birds, animals, or any group of things that can be organized on the basis of similarities and dissimilarities. Organization of knowledge is one of the ways in which science progresses; it would be terribly inefficient if no general procedures were available to identify things like minerals.

Objectives

When you have completed this laboratory, you should be able to identify those minerals specified by your instructor.

Problems

1. Determine the physical properties of the mineral samples provided by your instructor and identify each sample using the mineral identification tables. You may find it helpful to use the tear-out Mineral Identification Worksheet, which is the last page of this chapter.

2. Distinguish among similar minerals:
   a. Both pyrite and chalcopyrite have a metallic luster, a greenish-black streak, and no cleavage; they are heavy and can be very difficult to distinguish on the basis of color. If no crystals are visible, how would you tell them apart?

   b. Magnetite and hematite can both be dark gray to black and look similar. How could you tell them apart if you did not have a magnet?

   c. Hornblende and augite commonly are black, have the same hardness, and exhibit two directions of cleavage. They are difficult to distinguish when they occur as small crystals in a rock. What would you look for to tell them apart?

   d. Calcite, halite, and fluorite all have perfect cleavages, and they can all be the same color. How would you distinguish among them?

   e. Talc and serpentine can be the same color, have similar greasy to pearly lusters, and both can have a greasy feel. How do they differ?

   f. If chlorite and biotite occur as small crystals in rocks, it may be difficult to tell them apart. What properties might be helpful?

   g. What single property is most useful for distinguishing between potassium feldspar and plagioclase?

3. Go to *http://mineral.galleries.com* (or link to it through *www.mhhe.com/jones5e*—see Preface) and link to *Birthstones* through *Interesting Groupings*. Look up your birthstone.

   a. What is your birthstone?

   b. What is its chemical formula?

   c. What is its color? Luster? Hardness? Cleavage? Streak?

   d. With what other minerals does it commonly occur?

   e. Where are its notable occurrences?

   f. Is it a variety of another type of mineral or a member of a mineral group?

   Other websites of interest include *www.theimage.com*, which contains pictures of, and information on, many minerals; and *www.gemstone.org/*, a site that has pictures and lots of information about gems.

In Greater Depth

4. Acid mine drainage is a problem that affects many rivers and streams in the coal mining regions of the Appalachians (especially Pennsylvania and West Virginia) and in watersheds surrounding coal and metal mines across the United States, Canada, and the rest of the world. Do a World Wide Web search using the phrase "acid mine drainage" to answer the questions below. *www.google.com* is a good search engine. If you are interested in acid mine drainage in a particular state or country (and you are pretty sure that it has coal or metal ore mines in it), add the name to your search.

   You will get a lot of hits from an open search. Read the questions below and pick a site or two that seems informative and authoritative.

   a. What is the name and web address of the site or sites you selected?

   b. Indicate whether it was set up by a government organization, an independent environmental organization, an environmental company, a university research group, or another type of individual or organization.

   c. What mineral or minerals are causing the acid mine drainage?

   d. How does acid mine drainage affect the chemistry of the streams it flows into?

   e. What is one visually obvious symptom of acid mine drainage?

   f. The question of how to remediate (fix) the problems associated with acid mine drainage is interesting, but takes us too far beyond the scope of this lab. Instead, given the owner of the website and the way the information was presented, do you think the information was wholly objective and reliable, or do you think it was slanted by a political or economic agenda?

Mineral Identification Worksheet

Specimen Number	Luster	Hardness	Streak	Cleavage	Other Properties	Mineral Name	Chemical Grouping (Table 2.3)*
A	Metallic	<5.5	Silver/Gray	3	Very shiny + silvery	Galena	
B	Nonmetallic	>5.5	Gray/Black	0	magnetic	Magnetite	
C	Metallic	<5.5	Brassy-Yellow	0	Fools Gold / cubic crystals	Pyrite	
D	Nonmetallic	>	White	3	Smushed (Rhombus)	Calcite	
E		>5.5		0	Hexagonal Crystals	Quartz	
F		<2.5		1	Flexible / Pale Thin Black / Dark Black	Biotite	
G					Can crack it with nail	Gypsum	
H				3	Tasted like salt	Halite	
I				2 90°	Usually Dark Green Black like	Augite	
J				1	White/Flexible	Muscovite	
K		>5.5		2 90°	Striations	Plagioclase Feldspar	
L		>5.5		2 90°	No striations	Potassium Feldspar	
M		>5.5	Oil	no/no	Olive green color	Olivine	
N					Black or dark green	Hornblende	
O			Red		Red Streak	Hematite	

*This is useful information to remember if you want to continue in a geology or environmental program.

Mineral Identification Worksheet

Specimen Number	Luster	Hardness	Streak	Cleavage	Other Properties	Mineral Name	Chemical Grouping (Table 2.3)*

*This is useful information to remember if you want to continue in a geology or environmental program.

PART II
Rocks

**Beach pebbles, Lake McDonald,
Glacier National Park, MT**
×2

CHAPTER 3

Igneous Rocks

Materials Needed
- Pencil
- Hand lens
- Calculator
- Samples of igneous rocks

Introduction

Most **rocks** are aggregates of crystals or grains of one or more minerals. The individual mineral particles in rocks are generally small, with average dimensions less than 1 centimeter (cm). Even so, the minerals are identifiable. They have the same physical properties as the larger mineral specimens you have already studied.

Deciphering the Earth's history begins with rocks because they bear testimony to their origin and subsequent history. Once you have completed the next three chapters, you will be able to use easily observable details to infer at least some of the history reflected in the rocks commonly found all around you.

Rocks are classified as igneous, sedimentary, and metamorphic based on how they formed: **igneous rocks** solidified from melted rock, **sedimentary rocks** form from material (sediment) deposited on the Earth's surface, and **metamorphic rocks** form when preexisting rocks recrystallized, as solids, under high temperature and pressure beneath the Earth's surface. These basic rock types are linked by a set of processes that routinely converts igneous rocks to sedimentary and metamorphic rocks, sedimentary rocks to igneous and metamorphic rocks, and metamorphic rocks to sedimentary and igneous rocks. This global rock recycling program is called the **rock cycle.**

Igneous Rocks

The most spectacular way in which igneous rocks form is by volcanic eruption. This happens when molten rock, called **magma,** rises to the surface. The magma then either simply flows onto the surface as **lava** or violently explodes at the surface due to rapid bubble formation within. The forming bubbles increase the volume of the magma so dramatically that it shoots up and out of the volcano. Igneous rocks that cool on the surface are called **extrusive rocks.** Extrusive rocks cool quickly because air, rain, and snow rapidly remove a lot of heat.

Most rising magmas never make it to the surface. They stall deep in the crust and slowly cool. Igneous rocks that cool beneath the surface are termed **intrusive.** The trapped magma cools slowly because the surrounding rocks form an insulating jacket that helps the magma to retain its heat. Because temperature increases with depth in the Earth, the surrounding rocks are themselves warm and this, too, promotes slow cooling.

Igneous rocks are widely studied for many reasons. First, many people live close enough to volcanoes to be killed by explosive eruptions or huge landslides triggered by eruptions. Geologists study the ancient deposits of individual volcanoes to understand their likely future eruptive style. Second, much of the Earth's crust, both continental and oceanic, is made of igneous rocks; thus, we look to igneous rocks to understand how our crust formed and why our planet is so different from the other rocky planets. Third, igneous rocks commonly form at tectonic plate boundaries; hence, the study of ancient igneous rocks tells us a great deal about the history of the Earth. For example, igneous rocks tell us that about one billion years ago North America partially split along a zone stretching from Lake Superior to northeastern Kansas. Finally, hot magmas drive the circulation of a lot of hot water. These hot fluids pick up a number of important metals and can deposit them to create metal ore deposits. These deposits of gold, copper, silver, and a host of other rare metals form the basis of our modern technological economy.

Although all igneous rocks by definition cool from a liquid magma, they form under such diverse conditions that they differ widely in texture, mineral composition, and appearance. Let's cover these basic differences first.

Texture

Texture refers to the size, shape, and arrangement of the crystals or grains composing the rock. The texture of a rock reflects its composition and the way in which it formed.

With the exception of some volcanic rocks, all igneous rocks have a **crystalline** texture (Fig. 3.1), in which the various mineral crystals are interlocked with one another. This texture develops when crystals grow together as a magma solidifies to form solid rock.

We illustrate rock textures using not only hand specimen photographs but also microscopic views using photomicrographs (Fig. 3.1). Photomicrographs—in this case, photographs of very thin (0.03mm), transparent slices of rock called **thin sections**—are taken through a microscope equipped with polarizing filters. These filters produce distinctive colors and other optical properties that enable individual minerals in a rock to be confidently identified. The filtered colors are *not* natural, but textures are clearly revealed.

The rate at which a magma cools has the greatest effect on the sizes of the crystals in an igneous rock. In general, the more slowly a magma cools, the larger the mineral crystals will be, because slow cooling provides more time for the chemical constituents to migrate to the growing mineral. As you will see shortly, water and gas content can also profoundly affect rock textures.

Textures Based on Crystal Size

Coarse-grained (phaneritic). An igneous texture in which nearly all crystals are large enough to be seen without a hand lens (Fig. 3.1). For convenience, this is taken to be crystals greater than 1mm (0.04 in). This texture suggests slow cooling of the magma.

Fine-grained (aphanitic). An igneous texture in which most crystals are too small to recognize with the unaided eye (< 1 mm) (Fig. 3.2). This texture suggests fast cooling of the magma or lava.

Glassy. Describes igneous rocks that are made of glass and lack crystals (Fig. 3.3). Glassy rocks cool so fast that crystals do not have time to form. Thus, glasses do not have the regular atomic structures of crystalline solids; they are amorphous. Some igneous rocks are part crystals and part glass.

Pegmatitic. Describes igneous rocks with exceptionally large crystals (> 3 cm [1.2 in]). Such rocks are called *pegmatites* (see Fig. 3.7). Large crystals result not only from slow cooling but also from extra water in the magma. Water promotes rapid crystal growth by speeding migration of chemical elements to the growing crystals.

Porphyritic. An igneous texture in which crystals of two different sizes are present in the same rock (Fig. 3.4). The larger crystals, called **phenocrysts,** are surrounded by many smaller crystals, collectively called the **groundmass.** The groundmass can be either coarse grained

FIGURE 3.1
A coarse-grained crystalline texture clearly shows intergrown mineral crystals. Potassium feldspar, quartz, biotite, and plagioclase are visible in this sample of **granite.** The inset photomicrograph emphasizes how tightly intergrown the crystals are. **The colors in this and most subsequent photomicrographs are not natural; they have been changed by polarizing filters placed on the microscope.**

FIGURE 3.2
A fine-grained, or aphanitic, texture is typical of **basalt.** Its crystalline texture is visible under a microscope (inset photomicrograph).

Obsidian × 1

FIGURE 3.3
Obsidian is easy to identify because of its glassy texture and vitreous (glassy) luster. The excellent conchoidal fracture produces edges that are sharper than any made from metals. This characteristic made obsidian a favorite among Native Americans for arrows and cutting tools and, more recently, it has even been used for very delicate eye surgery. It is most commonly black (due to inclusions of tiny crystals of magnetite) but may be reddish (inclusions of hematite). *Snowflake obsidian* has white spots that formed where the glass devitrified (crystallized slightly), usually to a variety of silica (SiO_2).

× 1

FIGURE 3.4
Porphyritic texture in hand specimen of **basalt.** Large light-colored crystals (plagioclase phenocrysts) are set in a fine-grained, dark-colored groundmass.

or fine grained, but the phenocrysts are typically coarse grained. Although both intrusive and extrusive rocks can be porphyritic, it is more typical of extrusive rocks. A porphyritic texture suggests an initial period of slow cooling followed by faster cooling. For example, phenocrysts may grow in a slowly cooling magma beneath a volcano. If the volcano then erupts, rapid cooling of the remaining liquid would produce a fine-grained groundmass.

Additional Extrusive Rock Textures

Extrusive rocks can develop a range of additional textures if their magmas have high water and gas concentrations. A rising magma experiences a dramatic drop in pressure as it reaches the surface. This release of pressure allows bubbles to form in the magma (or lava), much as bubbles form in soda pop just after the bottle is opened.

Vesicular. Bubbles that form in lava and neither pop nor escape can be frozen into the lava. The cavities left behind are called **vesicles;** the texture is called **vesicular.** Vesicles range from a few millimeters to several centimeters in diameter. Some rocks contain just a few vesicles; others are more vesicle than rock. Some water and gas-rich magmas form so many bubbles that the lavas froth out onto the surface, solidify very rapidly, and form a rock with lots of millimeter-sized vesicles. *Pumice* and *scoria,* described under Fig. 3.16, are examples.

Pyroclastic. Explosive volcanic eruptions expel a lot of gas, glassy volcanic ash, lava, and fragments of rocks and minerals. All of this material comes to rest on the ground following either a free fall through the air or a powerful gas-charged surface flow down the side of the volcano. Either way, the resulting deposit contains a lot of shattered volcanic glass, rock, and mineral fragments, collectively termed **pyroclasts,** that define a **pyroclastic texture.** An extrusive igneous rock with a pyroclastic texture, like that shown in Figure 3.5, is called a **tuff.** Air-fall tuffs frequently show layering, which is otherwise unusual in igneous rocks. Thin pyroclastic deposits formed from flows cool rapidly and, because of their mode of formation, produce unusually soft and

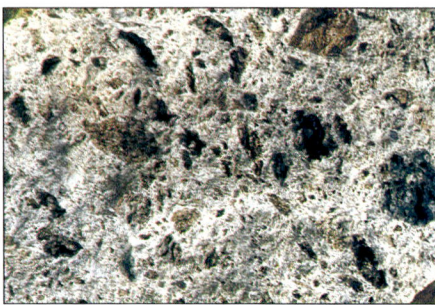

Rhyolite tuff × 1

FIGURE 3.5
Pyroclastic texture in hand specimen of **rhyolite tuff.**

lightweight rocks with rather open, porous textures. Thick, hot pyroclastic deposits can retain enough heat for the glassy fragments to fuse (weld) together to form a dense, hard, compacted tuff known as a *welded tuff.*

Mineral Composition

Relatively few minerals make up most igneous rocks. To correctly identify the rock, you must identify the major minerals. This is fairly easy for coarse-grained rocks, more difficult (but not necessarily impossible) for fine-grained rocks, and impossible for glassy rocks (unless phenocrysts are present). Table 3.1 and the following may be helpful:

1. Most dark-colored igneous rocks are rich in calcium plagioclase and ferromagnesian (iron-magnesium) minerals such as pyroxene or olivine. The word **mafic** refers to such rocks. **Ultramafic** igneous rocks are composed entirely of ferromagnesian minerals.

2. Light-colored or **felsic** igneous rocks commonly contain potassium feldspar, sodium plagioclase, and quartz, and only minor amounts of other minerals.

3. **Intermediate** igneous rocks are neither dark nor light and generally contain light-colored minerals (feldspars, minor quartz) and dark minerals such as hornblende or biotite.

4. Quartz has a vitreous luster and conchoidal fracture.

TABLE 3.1
Recognizing Minerals In Igneous Rocks

Mineral	Properties
Potassium feldspar p. 17	Usually white or pink Two cleavages at 90° Equidimensional crystals
Plagioclase feldspar p. 17	Usually white (Na-plagioclase) or gray (Ca-plagioclase) Two cleavages at 90° Elongate crystals Striations
Quartz p. 18	Colorless to gray Glassy with conchoidal fracture Irregular crystals in intrusive rocks Equidimensional phenocrysts in extrusive rocks
Biotite p. 20	Shiny and black One perfect cleavage Thin crystals
Muscovite p. 20	Shiny and silvery white One perfect cleavage Thin crystals
Hornblende (amphibole) p. 19	Black with shiny, splintery appearance Two cleavages at 56° and 124° Elongate crystals
Augite (pyroxene) p. 19	Black, greenish black, or brownish black Vitreous, but rather dull, luster Two cleavages at 90° Blocky crystals
Olivine p. 20	Light green to yellow-green Glassy luster Small (few millimeters), equidimensional crystals

5. A pink feldspar is usually potassium feldspar; white or gray feldspars may be either potassium feldspar or plagioclase—if striations are present, it is plagioclase.

6. Cleavage and general appearance help to identify amphiboles and pyroxenes. Amphibole cleavages intersect at about 60° and 120°, whereas pyroxene cleavages intersect at 90°; amphiboles typically are elongate and have a splintery appearance, whereas pyroxenes look blocky.

Classification and Identification of Igneous Rocks

Igneous rocks are classified by their texture and mineral content. The classification scheme in Table 3.2 lists common textures in the left-hand column and typical mineral composition in the top row. Although rock names occupy distinct spaces in the table, understand that the dividing lines between rock types are artificially imposed; both mineral abundance and composition change gradationally from one rock type to the next.

When identifying a rock, it makes no difference whether texture or mineral content is determined first. Since texture is recognized more easily, first determine the average grain size and follow the top flow chart (texture) in Figure 3.6. This will lead you to one of the texture rows in Table 3.2. Next identify the major minerals and follow the lower flow chart (composition) in Figure 3.6. This will take you to the proper 'pigeonhole' in Table 3.2 that incorporates both texture and composition.

Some common igneous rocks are shown in Figures 3.7 through 3.16. **Remember, the photographs are representative examples only, and your samples may differ *considerably* in color and general appearance. The texture and mineral content** should be similar, so concentrate on those, not on color.

Crystallization of Magma and Bowen's Reaction Series

At high temperatures, a magma is completely liquid, but as the temperature drops, crystals begin to form. They don't all form at once, however. At first, crystals of only one or two minerals begin to grow. As the temperature continues to drop, these early-formed crystals may grow to become nicely shaped, they may react with the magma and be partially or wholly dissolved, or new minerals may begin to grow around them, preventing further growth. Continued crystallization increases the proportion of crystals and decreases the amount of liquid. Crystals grow up against one another and form an interlocking, crystalline network. Other minerals may crystallize with further temperature decreases; these will grow in the liquid remaining between earlier-formed crystals.

Because minerals form in a sequence, it is possible to figure out that sequence by carefully studying the textural relations in the rock. This kind of process was done by N.L. Bowen during the early part of the 20th century, but he carried it a giant step further by duplicating the crystallization process in the laboratory. He paid particular attention to reactions between minerals as indicated by textural relations. For example, he commonly found rounded crystals of olivine surrounded by pyroxene (Fig. 3.17) and concluded that olivine formed first, then at a lower temperature, reacted with the liquid (magma) to form pyroxene. He called this a *discontinuous reaction,* because it results in the formation of a completely different mineral (pyroxene). As the pyroxene grows around the olivine, it sometimes prevents further reaction from taking place, and a partially reacted-upon, rounded olivine is left surrounded by pyroxene.

Bowen also observed that plagioclase feldspars gradually changed their chemical composition as magma temperature decreased and plagioclase crystallized. He attributed this to a *continuous reaction* between the magma and the growing crystals of plagioclase: as the

TABLE 3.2
Classification of Igneous Rocks*

Texture \ Mineral Composition	Felsic	Intermediate	Mafic	Ultramafic
	• > 10% quartz, • > 50% feldspar with K-feldspar > Na-plagioclase, • ± < 15% biotite, muscovite, hornblende	• 0–10% quartz, • > 50% Na/Ca-plagioclase, • 0–10% K-feldspar • < 50% hornblende (especially), biotite, augite	• 20–85% Ca-plagioclase, • 15–50% augite, • 0–35% olivine	• Olivine, and/or pyroxene
Pegmatitic	Granite pegmatite	Diorite pegmatite	Gabbro pegmatite	
Coarse grained (phaneritic)	Granite	Diorite	Gabbro	Peridotite (dunite, if mostly olivine)
Coarse grained and porphyritic	Porphyritic granite	Porphyritic diorite	Porphyritic gabbro	
Fine grained (aphanitic)	Rhyolite	Andesite	Basalt	
Fine grained and porphyritic	Porphyritic rhyolite	Porphyritic andesite	Porphyritic basalt	
Fine grained and vesicular		Vesicular andesite	Vesicular basalt	
Pyroclastic	Rhyolite tuff or breccia	Andesite tuff or breccia	Basalt tuff or breccia	

Texture	Glass Equivalents of Felsic and Intermediate Rocks		Glass Equivalents of Mafic Rocks	
Glassy	Obsidian			
Glassy with many small vesicles	Pumice		Scoria (may have some crystals)	

*For simplicity, names of less common rocks are omitted and their spaces are left blank.

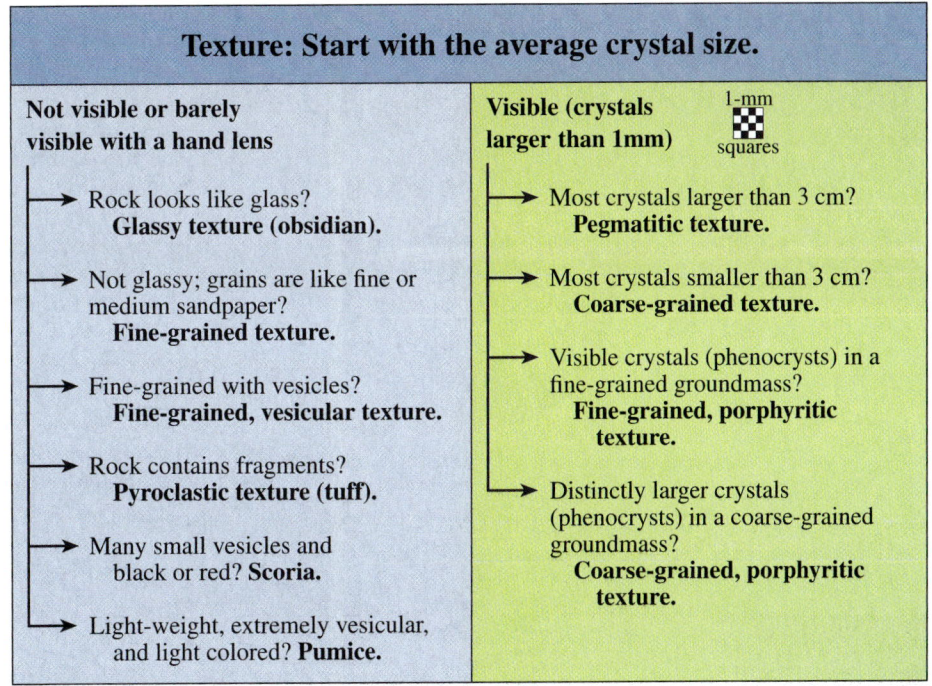

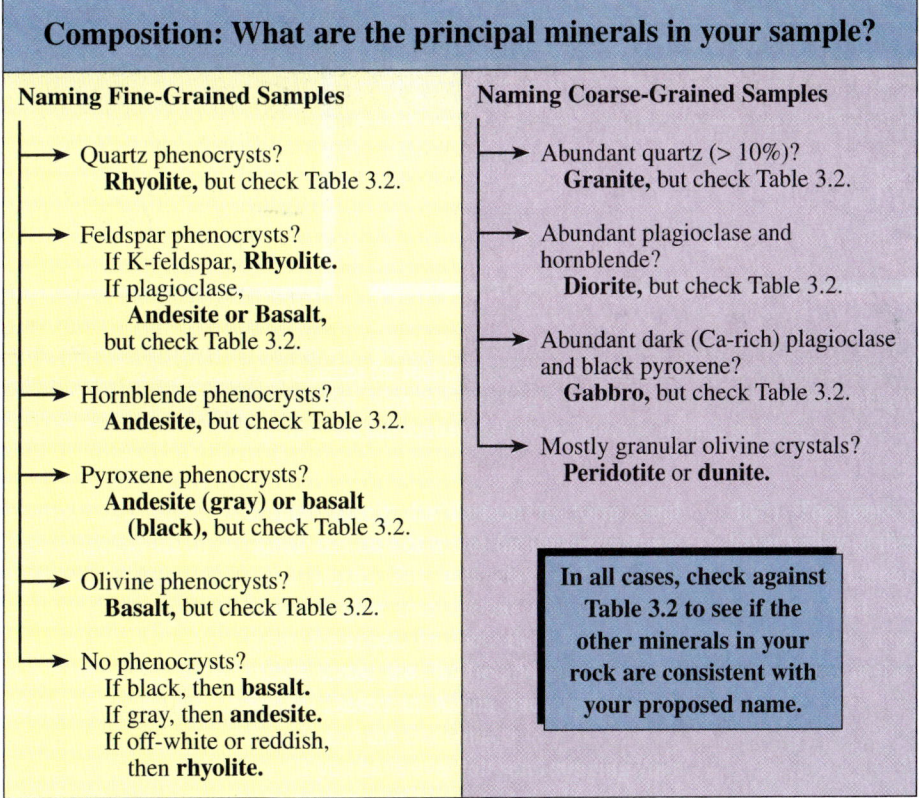

FIGURE 3.6
Flow charts for the identification of igneous rocks. The top chart refers to igneous textures whereas the bottom one refers to composition. Both charts should be used together with the more comprehensive information in the text and Table 3.2.

Granite × 1
Photomicrograph × 10

FIGURE 3.7
Granites are coarse-grained igneous rocks that make up the bulk of the large bodies of intrusive rock called batholiths. Most are light gray or pinkish, but gray and red granite are common too. Gray, glassy-looking quartz is an essential mineral, as is potassium feldspar. Potassium feldspar is commonly pink but may be white. If feldspars of two colors are present, the pink one is probably potassium feldspar, and the white or light gray one is probably sodium plagioclase. Most granites contain some black biotite or hornblende, and some contain muscovite. The fine-grained equivalent of granite is rhyolite. Some granites are very pretty when cut and polished, so they are quarried for use in monuments and buildings.

Granite pegmatite

FIGURE 3.8
Pegmatites are very coarse crystalline intrusive rocks that typically occur as small intrusions. The **granite pegmatite** shown here is light colored and has large, irregular-shaped crystals of gray quartz and crystals of salmon-colored potassium feldspar up to 30 cm long. Some varieties contain white, sodium-rich plagioclase. Big "books" of both biotite and muscovite mica are common. Some pegmatites contain rare minerals such as the silicate minerals tourmaline and beryl, both of which have several gem varieties (for example, emerald is a variety of beryl).

Diorite × 1
Photomicrograph × 10

FIGURE 3.9
Diorite is a coarse-grained, intrusive rock usually made of white to light gray sodium-calcium plagioclase and black, splintery-looking hornblende, as shown here. Finely crystalline diorites look like a mixture of salt and pepper. Some diorites contain biotite as well as hornblende, and possibly a little quartz. The photomicrograph confirms the fully crystalline texture of this rock. Diorite is common in both large (batholith) and small intrusive bodies. Andesite is the fine-grained equivalent of diorite.

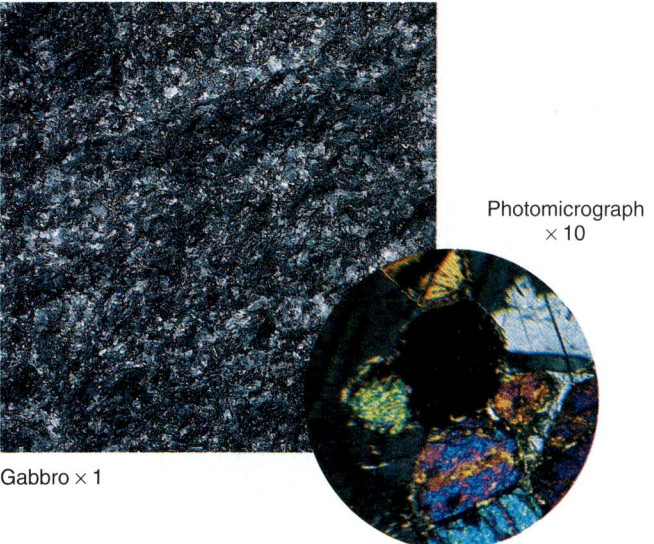

FIGURE 3.10
Gabbro is a coarse-grained intrusive rock composed of light to dark gray calcium plagioclase and black pyroxene. Olivine may be present as well. With so many dark minerals it can be tough to tell one from the other. However, in many gabbros, the plagioclase is easy to identify, because the striations are readily visible on the large elongate crystals. Gabbro is the coarse-grained equivalent of basalt, and it commonly forms thin (dikes and sills) or irregular-shaped, small bodies of intrusive rock. The plagioclase in some gabbros causes a peculiar iridescent blue play of colors that makes the rock valuable for monument or building purposes.

FIGURE 3.11
Peridotite is generally a dark green to black rock, because its principal constituents are pyroxene and olivine. Peridotite consisting almost entirely of olivine is called *dunite,* which usually has the green color of olivine. In fact, the mineral sample of olivine you saw in the lab was probably a piece of dunite (see Figure 2.8). Peridotite rich in pyroxene may have fairly large crystals; in many peridotites, however, the crystals are small, though large enough to identify. Although peridotite is not abundant in the crust, it is the most abundant rock beneath the crust, in the upper part of the Earth's mantle.

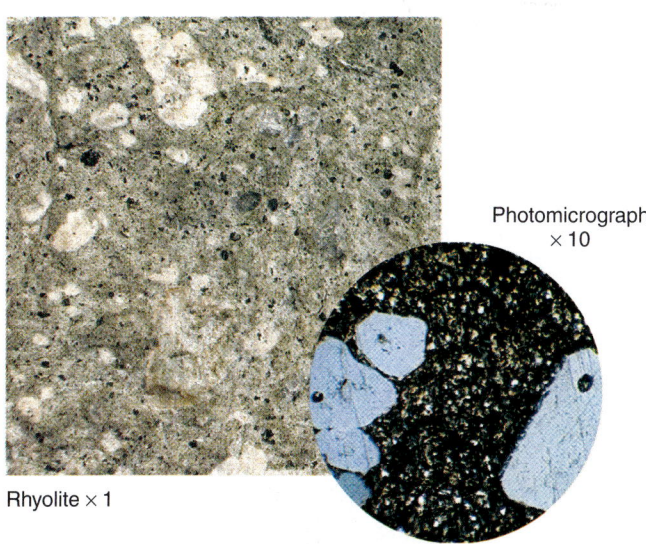

FIGURE 3.12
Rhyolite is a fine-grained extrusive rock that comes in a variety of colors. It is commonly light gray or pink, but red or even black rhyolite is not rare. Many rhyolites, like the one shown here, are porphyritic, so they are identifiable by the glassy quartz phenocrysts and white to salmon potassium feldspar phenocrysts in an aphanitic or even glassy groundmass. In the absence of phenocrysts, a light color suggests, but does not prove, rhyolite. The photomicrograph shows a few phenocrysts (gray) surrounded by a groundmass (black) with tiny crystals (brown to white). Granite is its coarse-grained equivalent.

FIGURE 3.13
Andesite is the common extrusive rock of volcanoes formed at convergent plate boundaries. It is typically gray and porphyritic, with phenocrysts of white to light gray plagioclase and black hornblende or biotite. The photomicrograph shows a few phenocrysts (white, yellow, brown) surrounded by a glassy groundmass containing very tiny crystals. It is the fine-grained equivalent of diorite.

Basalt × 0.5

FIGURE 3.14
Basalt is the most abundant extrusive igneous rock, forming the base of the seafloor and large plateaus on land. It is dark gray and fine grained, as in Figure 3.2, and can be vesicular, as shown here. Porphyritic basalt may have phenocrysts of calcium plagioclase, olivine, or both. Olivine phenocrysts are usually small (1 to 5 mm [0.04 to 0.2 in]) and are recognizable by their green color and glassy luster. Basalt is the fine-grained equivalent of gabbro.

Rhyolite tuff × 1

FIGURE 3.15
Tuff is a textural name indicating a pyroclastic texture and particle size less than 6.4 cm (2.5 in). A similar rock with particles bigger than 6.4 cm (2.5 in) is a volcanic breccia. Compositional varieties of tuff or breccia are indicated by using rhyolite, andesite, or basalt as prefixes when possible; for example, rhyolite tuff, basalt breccia. Tuffs vary widely in appearance. Some are made entirely of compacted volcanic glass shards, others have abundant mineral crystals, and others consist of many angular fragments of extrusive rock, especially pumice; most commonly, all three components are present. Just as components and composition are variable, so is color, although many tuffs are light colored, like the rhyolite tuff shown here.

A. Pumice × 1

B. Scoria × 1

FIGURE 3.16
A. Pumice is the rock that floats because it has so many tiny holes in it. It is a froth of volcanic glass. Look at it with a hand lens to appreciate its character. Pumice is commonly, but not always, white or light gray. It is a good abrasive and is used in Lava® soap and in blocks used to scrape calluses off your feet or to clean griddles. **B. Scoria** is a basaltic rock with many small holes. You can think of it as being extremely vesicular with small vesicles. It is dark in color, ranging from black to reddish brown. It rarely looks vitreous, although microscopic examination of fresh samples shows that they formed as a bubbly glass. Although it looks as if it might float in water, it is usually too dense. Scoria is sold under the general name "lava rock" for outdoor decorative purposes.

temperature of the magma drops and plagioclase crystals grow, they continuously change their chemical composition and become richer in sodium.

Bowen summarized his studies with a diagram, now called *Bowen's Reaction Series,* which includes all the major rock-forming minerals of common igneous rocks (Fig. 3.18). The *discontinuous reaction series* contains the common iron-magnesium (ferromagnesian) silicate minerals. Plagioclase makes up the *continuous reaction series.* The three minerals at the bottom do not form by reaction. They simply crystallize last, sometimes in the sequence in which they are listed and sometimes simultaneously.

The reaction series shows that when a natural magma cools, certain minerals crystallize earlier and at higher temperatures than others. The common rock-forming minerals form in a regular sequence, with olivine and calcium-rich plagioclase at the top of the list, crystallizing at temperatures some 500–600°C higher than quartz, which is at the bottom of the list. The crys-

FIGURE 3.17
This photomicrograph shows rounded olivine crystals (green and red) surrounded by pyroxene (yellow). The early-formed olivine reacted with the magma to form some of the pyroxene that now surrounds it (× 10).

Bowen's reaction series		Corresponding rocks		
		General	Intrusive	Extrusive
High temperature, first to crystallize — Discontinuous series: Olivine, Pyroxene, Amphibole, Biotite / Continuous series: Plagioclase feldspar Ca-rich → Na-rich		Mafic	Gabbro	Basalt
		Intermediate	Diorite	Andesite
Low temperature, last to crystallize — Potassium feldspar ± Muscovite + Quartz		Felsic	Granite	Rhyolite

FIGURE 3.18
Bowen's Reaction Series. Minerals at top crystallize first from a magma at high temperatures (1200–1300°C); those at bottom crystallize last at lower temperatures (600–900°C). Ferromagnesian minerals are related by discontinuous reactions with the magma as temperature decreases, whereas plagioclase reacts continuously to form crystals progressively richer in sodium. For names of *Corresponding Rocks*, see Problem 5a.

tallization temperatures of the plagioclase feldspars fall in the same range as the ferromagnesian minerals. Note, however, that the more calcium-rich plagioclases crystallize at higher temperatures than the less calcium- and more sodium-rich plagioclases. The most sodium-rich one crystallizes at the lowest temperature.

Cautionary note: Although the reaction series is instructive, composition and many other factors influence the precise crystallization history of a magma. The *Reaction Series* experiments focused on what happens when one type of basaltic magma cools. While the series summarizes the common order of crystallization and common reactions, the overlapping arrows in the discontinuous series suggest and allow for significant variations in the crystallization sequence.

Where and How Magmas Form

Most volcanoes are associated with tectonic plate boundaries (Fig. 3.19). What happens at plate boundaries that allows magmas to form? The first part of the answer involves the concept of **partial melting.** For example, what would happen if a rock made of quartz, potassium feldspar, sodium-plagioclase, biotite, and hornblende began to melt? We can use Bowen's Reaction Series to predict the results. The minerals lowest in the series, such as quartz and potassium feldspar, would melt at lower temperatures than minerals higher in the series. Imagine that the temperature got just high enough to melt the quartz and feldspar, but not high enough to melt the rest of the minerals. Because quartz and potassium feldspar are the main minerals in granite, the *partial melt* would have a felsic composition similar to that of granite. The liquid, because it is less dense than the surrounding rock, would migrate upward. Drops of liquid would eventually coalesce to form a body of magma, which would continue to move upward until it either erupted on the surface or crystallized below the surface.

The second part of the answer is that partial melting takes place at plate boundaries because it is here that rock either moves up toward the surface or down below the surface. For example, at *divergent plate boundaries* (where plates move away from one another), hot, buoyant peridotite rises in the mantle as the plates move apart (Fig. 3.19). As it rises to shallower depths, pressure decreases until at some point it is low enough to cause the hot periodite to partially melt. The partial melting of ultramafic periodite produces a mafic magma. This magma rises and intrudes into the surrounding lithosphere to form gabbro or is extruded onto the seafloor as basalt (Fig. 3.19).

At *convergent plate boundaries* (where plates move toward one another), the cold oceanic seafloor is dragged down into the mantle under the overriding plate (Fig. 3.19). At some depth, the down-going oceanic plate releases its water-rich fluids into the ultramafic rocks of the overlying mantle. The addition of water to hot rocks lowers the melting points of some minerals and partial melting takes place. Mafic magmas are produced that rise through the mantle and enter the crust above. As these magmas

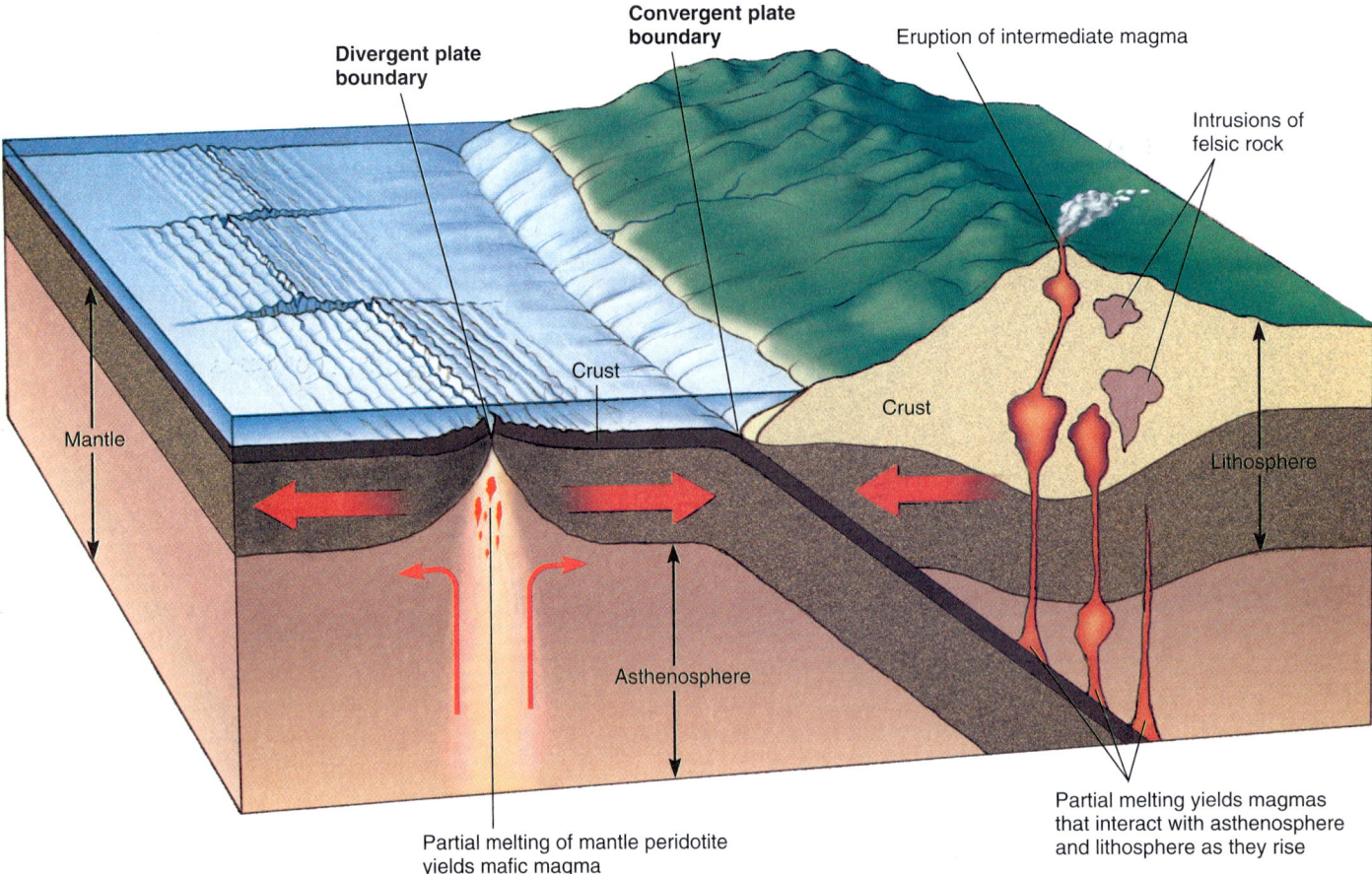

FIGURE 3.19
Most igneous activity takes place at divergent and convergent plate boundaries. Partial melting of peridotite mantle at divergent boundaries yields mafic magmas. Convergent boundaries are more complex; magmas formed by partial melting may interact with a variety of rocks to produce abundant intermediate and felsic magmas.

rise, they may give off enough heat to partially melt the overlying rocks to produce a secondary magma, which also may rise in the crust. Because of the nature of partial melting, the new magma likely will be intermediate or felsic in composition.

Changes in Magma Composition

Most magmas undergo some kind of change in composition between the time they form and the time they solidify. Three types of processes seem to be particularly effective: fractional crystallization, assimilation, and magma mixing.

Fractional crystallization is a process whereby growing crystals are physically separated from the liquid magma. One way this can happen is by the sinking or floating of crystallizing minerals in the magma chamber. For example, as a mafic magma rises, temperature and pressure decrease and eventually olivine or other high-temperature minerals begin to crystallize. Because olivine is more dense than the magma, it tends to sink to the bottom of the magma chamber and make the magma above it more felsic.

A rising magma can react with and partially melt pieces of the surrounding rock. This process, called *assimilation,* changes the composition of the magma. When a mafic magma passes through continental crust, which generally has an intermediate composition (like diorite), assimilation may change the composition of the magma to a more intermediate magma, one capable of erupting andesite.

The third process, *magma mixing,* takes place when two or more magmas come together and mix to form a single magma. The magmas may have been derived from different parents, for example, a mafic magma derived from the mantle and a granitic magma formed by partial melting of the continental crust. The composition of the mixed magma depends on the compositions and proportions of each of the parents.

At divergent plate boundaries, the composition of the mafic magma may not be changed very much because it does not rise very far nor through rocks with strikingly different compositions. On the other hand, opportunities for change abound for those magmas rising some 100 km through continental crusts. Because of these differences, volcanoes that erupt at divergent plate boundaries are mainly basaltic, whereas those on continental crust are highly diverse in composition, ranging from basalt to andesite to rhyolite.

As you work through this lab, keep in mind that intrusive rocks that you examine are the frozen remains of long-dead magma chambers, many of which probably fed ancient volcanic eruptions. The characteristics of the extrusive rocks you see depend not only on processes in the magma chamber but also on the type of eruption, whether explosive or quiet.

Hands-On Applications

When a geologist picks up a rock, she or he wants to know what kind of rock it is and how it formed. This information is the basis for most geological endeavors, whether it be a search for mineral resources or an environmental impact study. To "read" the rock, it is necessary to identify its constituents and interpret the basic textural relations. This lab will show you how to identify igneous rocks and how to use their textures to interpret their mode of formation. You also will learn how it is possible to make logical inferences or predictions from existing information. This is one of the most important things that geologists, or any scientists, do.

Objectives

If you complete all the problems, you should be able to:

1. Recognize the following textures in igneous rocks: glassy, aphanitic, phaneritic, pegmatitic, porphyritic, vesicular, and pyroclastic.
2. Explain how each texture formed.
3. Distinguish between intrusive and extrusive igneous rocks on the basis of texture.
4. Recognize quartz, potassium feldspar, plagioclase, biotite, hornblende, augite, and olivine in an igneous rock.
5. Explain how igneous rocks are classified.
6. Identify igneous rocks.
7. Determine the relative temperature (high, intermediate, low) at which a common igneous rock formed, using Bowen's Reaction Series.
8. Relate the common igneous rocks to their plate-tectonic setting.

Problems

1. Which of the rock samples provided for you have the following textures? (Some of these may not be available to you.)

   Glassy: I-G, I-H, I-F

   Fine grained (aphanitic): I-C, I-L, I-K

   Coarse grained (phaneritic): I-I, I-B, I-m, I-A

   Pegmatitic: I-J

   Porphyritic: I-E, I-D, I-F

   Vesicular: I-K, I-H

   Pyroclastic: I-F, I-H

Part II Rocks

2. Of the rocks with glassy, fine-grained, coarse-grained, pegmatitic, and porphyritic textures, which:

    a. Formed from the most rapidly cooled magma? Glassy (I-G)

    b. Probably resulted from two stages of cooling—slow, then rapid? Porphyritic (E, D, F)

    c. Crystallized from a water-rich magma? Pegmatitic (J)

3. Which of the rocks in the previous question are likely to be extrusive, and which are likely to be intrusive igneous rocks? Why?

    Extrusive - Glassy - cool quickly (G)
    Intrusive - Pegmatitic, Porphyritic - cool slower (J)
    Both: D, E, F

4. Identify the samples of igneous rocks provided by your instructor. You may find it helpful to use the tear-out Igneous Rock Identification Worksheet, which is the last page of this chapter.

5. Use Bowen's Reaction Series (see Fig. 3.18) to answer the following.

    a. Based on the mineral content and texture of the following rocks, place them in the appropriate blank spaces under *Corresponding Rocks* in Figure 3.18: granite, gabbro, andesite, diorite, rhyolite, basalt. Which forms at the highest temperature, granite or gabbro? Andesite or basalt?

    b. Based on your knowledge of the colors of rock-forming minerals, are rocks at the top of Bowen's Reaction Series likely to be darker- or lighter-colored than those at the bottom? Which of the rocks in 5a are felsic? Intermediate? Mafic?

    c. If you identified quartz in a sample of igneous rock, what other minerals might you expect to find in the rock?
    Muscovite, Feldspar (both), Biotite, Quartz

    d. If you identified olivine in a sample of igneous rock, what other minerals might you expect to find in the rock?
    Pyroxene, Plagioclase feldspar

    e. Would you expect to find olivine and quartz in the same igneous rock? Explain your answer.
    No, quartz is Felsic + Olivine is mafic

6. Most igneous rocks form at divergent and convergent plate boundaries. Of granite, gabbro, andesite, diorite, rhyolite, and basalt, which are most likely to form at a divergent boundary?

    At a convergent boundary?

7. Go to **Volcano World** (*http://volcanoworld.org/*, or link to it through *www.mhhe.com/jones5e*—see Preface). Use the links under *Volcanoes* to find information on Mount Rainier, a volcano near Seattle, Washington, and answer the following questions. Some links have **highlighted** terms in their text that lead to a glossary link. The glossary may also be useful for text that does not provide glossary links. Its address (at present) is *http://vulcan.wr.usgs.gov/Glossary/framework.html*.

   a. In what mountain range is Mount Rainier?

   b. Is Mount Rainier a shield volcano, composite volcano (also called stratovolcano), or cinder cone?

   c. Does it have a caldera at its summit?

   d. Volcanoes grow when the rate of addition of new volcanic materials exceeds the rate of erosion. Is Mount Rainier currently getting larger or smaller?

   e. When did Mount Rainier first become active?

   Click on *Radar Image of Mount Rainier* at the bottom of the page and answer the next two questions.

   f. When was the most recent eruption?

   g. Mount Rainier is considered a dangerous volcano. Why?

   Hit the *Back* button and click on the link to the *Cascade Volcano Observatory*. At the bottom of the page, under *Other Menus of Interest,* click on *Cascade Range Volcanoes and Volcanics Menu (http://vulcan.wr.usgs.gov/Volcanoes/Cascades/framework.html).* Click on DESCRIPTION: *Cascade Range Volcanoes and Volcanics* and answer the following questions.

   h. Which states host the Cascade Range volcanoes?

   i. List several other volcanoes in the Cascade Range.

   j. Explain, in terms of plate-tectonic theory, why Mount Rainier and the other Cascade volcanoes are there.

   k. What rock compositions (such as basalt or granite) are the most common?

   Another volcano website of interest is *http://satftp.soest.hawaii.edu/space/hawaii/virtual.field.trips.html,* which features virtual field trips of the Hawaiian Islands. If you would like to see very nice photomicrographs of igneous and metamorphic rocks, try *www.geolab.unc.edu/Petunia/IgMetAtlas/mainmenu.html.*

Part II Rocks

In Greater Depth

8. Indicate which chemical elements increase in abundance toward the bottom of Bowen's Reaction Series (refer to the chemical compositions given in Table 2.3).

9. Chemical analyses of representative extrusive rocks from the Cascade Range in Oregon are given in Table 3.3. Notice that SiO_2 is by far the most abundant oxide and that it makes up from about 50 to 75% of these rocks. These values are typical of most igneous rocks. The amount of SiO_2 can be used to identify and name extrusive rocks, many of which are too fine grained to identify on the basis of mineral content. For one category of extrusive rocks, the following names are assigned on the basis of percent SiO_2:

Percentage SiO_2	Name of Extrusive Rock
<52	Basalt
52–57	Basaltic andesite
57–63	Andesite
63–70	Dacite
>70	Rhyolite

    a. A common way to express a table of numbers such as Table 3.3 is to make a graph. A graph not only summarizes the data in a visually effective manner, it shows trends that otherwise may not be evident. In this case, you will make a graph to show how the oxides FeO, MgO, CaO, Na_2O, and K_2O vary with SiO_2. Figure 3.20 will be used for this purpose. Figure 3.20 already has the X-axis (horizontal) labeled as "Percent SiO_2" and the Y-axis (vertical) as "Percent Oxides". Examine the table with the weight percent data and note that the SiO_2 numbers range from about 48 to 74%. To make an easily-read graph, label the X-axis with values ranging from 45 to 80% along the whole bottom. The other oxide data show numbers ranging from 0.2 to 10.6%. Label the Y-axis with values ranging from 0 to 12%.

    b. Now plot the data from the table. First plot SiO_2 (X-axis) versus FeO (Y-axis) for each sample. Connect the points and label the line at one end as "FeO". Then do the same with MgO, CaO, Na_2O, and K_2O.

    c. Finally, to see how these data compare to the rocks you have seen in lab, draw vertical lines at each of the percent SiO_2 boundaries that divide each of the rock types: 52, 57, 63, and 70% SiO_2. In the spaces between these lines, label with the rock types listed above (basalt, basaltic andesite, etc.) along the top of the graph.

    d. Which elements decrease with increasing abundance of SiO_2 and which ones increase?

    Looking at Bowen's Reaction Series, what could explain the changes in the FeO, CaO, and MgO concentrations going from low to high SiO_2 lavas?

    Why would the Na_2O and K_2O concentrations change?

    e. Now that you can put a rock name to each sample, and using what you know from Bowen's Reaction Series, list the sample numbers in the order in which these rocks should have formed, from earliest to last.

TABLE 3.3
Chemical Analyses Of Representative Extrusive Rocks From The Cascade Range, Oregon

Source: Data from D. J. Geist et. al., 1985, GPP (Geochemical Program Package), Center for Volcanology, University of Oregon.

	1	2	3	4	5	6	7	8
SiO_2	48.60	50.68	63.81	59.85	53.50	69.36	74.12	55.92
FeO	10.31	9.27	4.05	5.39	8.31	2.32	1.26	7.32
MgO	6.73	6.12	2.28	3.58	5.42	1.14	0.16	4.04
CaO	10.65	9.67	4.87	5.95	7.63	3.07	1.23	6.77
Na_2O	2.47	2.70	3.72	3.65	3.18	3.92	4.08	3.49
K_2O	0.25	0.48	1.96	1.28	0.73	3.02	4.47	0.97
Other	20.99	21.08	19.31	20.30	21.23	17.17	14.68	21.46

Numbers are weight percentages of constituents in the rock.

FIGURE 3.20
Graph for Problem 9.

Part II Rocks

10. The *density* of a substance is its mass divided by its volume ($\rho = m/V$, or density = mass/volume) and is commonly expressed in grams per cubic centimeter (g/cm^3). *Specific gravity* is the weight of a substance divided by the weight of an equal volume of water, and it has no units. The numerical values of density and specific gravity are effectively the same, because the density of water is 1 g/cm^3. For most scientific purposes, it is more convenient to use density than specific gravity.

   a. Calculate the density of an igneous rock with the following volume percentages of minerals: 18% quartz, 57% plagioclase, 11% hornblende, 14% biotite. Assume the following densities: quartz = 2.65 g/cm^3, plagioclase = 2.69 g/cm^3, hornblende = 3.20 g/cm^3, and biotite = 3.00 g/cm^3. Show your calculations in outline form.

   b. What would be the total mass of this rock, in kilograms, if it were used to face the lower 7 m of a 75-m by 45-m building? Assume the facing stone is 2 cm thick. Show calculations.

   c. You may be more familiar with pounds than kilograms. If 1 kg = 2.2046 pounds, what is the weight of all the facing stone in 10b in pounds? Show calculations.

Igneous Rock Identification Worksheet

Sample Number	Texture	Mineral Composition	Other Properties (e.g., color)	Rock Name	Inferred Rock Origin/History*
A	Coarse Grained	Felsic, Quartz, Biotite, Muscovite		Granite Pegmatite	
B	Coarse Grained	Mafic		Gabbro	
C	Fine Grained	Mafic		Basalt	
D	Porphyritic	Intermediate		Porphyritic Andesite	
E	Porphyritic	Felsic		Porphyritic Rhyolite	
F	Porphyritic			Porphyritic Rhyolite Tuff	
G	Glassy	Mafic		Obsidian	
H	Vesicular & Glassy	Felsic		Pumice	
I	Coarse grained	Intermediate		Diorite	
J	Pegmatitic	Felsic, Quartz, Biotite, Muscovite		Granite Pegmatite	
K	Vesicular	Mafic		Vesicular Basalt	

*For example, intrusive/extrusive; single- or two-step cooling history; lava flow or pyroclastic deposit; probable gentle versus violent eruptive style.

Igneous Rock Identification Worksheet

Sample Number	Texture	Mineral Composition	Other Properties (e.g., color)	Rock Name	Inferred Rock Origin/History*
C	Fine Grained	Felsic		Rhyolite	
M	Coarse Grained	Intermediate Pyroxene, Amphibole, Biotite		Mafic Granite	

*For example, intrusive/extrusive; single- or two-step cooling history; lava flow or pyroclastic deposit; probable gentle versus violent eruptive style.

CHAPTER 4

Sedimentary Rocks

Materials Needed
- Pencil
- Hand lens
- Metric ruler
- Glass or knife to test hardness
- Dilute hydrochloric acid
- Binocular microscope (optional)
- Samples of sedimentary rocks to be identified
- Samples of sediment
- Rock samples illustrating cross-bedding, ripple marks, graded beds (or jars of sediment in water), and mud cracks (optional)
- Samples of weathered igneous rocks

Introduction

Sedimentary rocks form from sediment that accumulates on the Earth's surface. The components of the sediment, derived from weathering and erosion of preexisting rocks, are transported as solid particles or chemical ions to their site of deposition. Sedimentary rocks contain many kinds of clues about their history and the history of the Earth's surface. In this lab, you will learn how to recognize and interpret some of these clues and to identify common sedimentary rocks.

Sedimentary rocks cover most of the Earth's surface, so they are the kind of rock you are most likely to encounter. They have much to offer us. They contain most of the oil and gas in the world, and coal is a sedimentary rock. They are the source of most or all of our iron, fertilizer, salt, and other important commodities, and they provide us with many kinds of construction materials. They can tell us what surface conditions were like in the past and when seas were or were not present in an area, and their fossils tell us about the changing pageant of life through time. Sedimentary rocks can provide more details about the geologic past than other kinds of rock if we know how to read their pages, so let's learn how to do that.

Rocks and soils at the Earth's surface disintegrate and decompose as they interact with the surface environment. The solid particles and *ions* in solution (atoms or molecules with an electrical charge) freed by this action are carried by water, wind, or glaciers from their source to the place where they are deposited, the **basin of deposition.** There, under the influence of gravity, they accumulate as layers, or **strata,** on the Earth's surface (though commonly underwater). This collection of loosely packed, unconsolidated minerals or rock fragments is **sediment.** In time, sediment is buried, consolidated, and hardened to form **sedimentary rock.**

The mineral composition and texture of many sedimentary rocks provide clues to the:

1. original source of the sediment;
2. type and extent of the weathering processes by which the source rock was broken down;
3. type of agent (water, wind, ice) that transported the sediment and, sometimes, the duration of transport;
4. physical, chemical, and biological environment in which the sediment was deposited; and
5. changes that may have occurred after deposition.

By carefully examining the clues in the rock and knowing how to interpret them, it is possible to learn a good deal about the history of the Earth's surface environments and processes.

Weathering

Weathering is the process by which rocks at the surface disintegrate or decompose; it produces most of the components of sedimentary rocks and the inorganic components of soils. **Mechanical weathering**—for example, by frost action or wind abrasion—causes rocks to disintegrate into pieces or loose mineral grains. During **chemical weathering,** chemical reactions cause the rock to decompose. As a result, the ions of such elements as calcium, sodium, potassium, and magnesium are freed from the rock and dissolve into water. These reactions also produce new minerals, principally clay minerals and iron oxides, and release nonreactive ones, especially quartz, from the rock. Table 4.1 lists the common mineral products of weathering.

Erosion and Transportation

As a rock is broken down by weathering, the freed rock fragments and minerals move downhill due to gravity and are eroded and transported away by water, glaciers, or wind. As transport distance increases, these transport agents change the **size,**

TABLE 4.1
Common Weathering Products

Mechanical Weathering	Chemical Weathering
Quartz	Secondary Minerals:
K-feldspar	Clay minerals
Na-plagioclase	Iron oxides
Muscovite	Fine quartz
Fragments of preexisting rock	Aluminum oxides
	Dissolved in water:
Other silicate minerals are less common because they undergo rapid chemical weathering.	Ca^{2+}, Mg^{2+}, Na^+, K^+ HCO_3^-, CO_3^{2-}, SO_4^{2-}, Silicic acid
	Nonreactive minerals (e.g., quartz)

shape, and **sorting** of the sedimentary particles. Geologists study these characteristics in modern sediments, in which the transporting process is known, in order to understand ancient sedimentary rocks, for which transport processes must be inferred.

Particle or Grain Size

To make scientific communication precise, geologists have assigned specific **grain sizes** to familiar words for sediments (Fig. 4.1A). For example, *sand* refers to particles between 0.06 and 2 mm in diameter, and *clay* refers to particles smaller than 0.004 mm. Note that "clay" can refer to either a grain size or a group of minerals (of which kaolinite, for example, is a member). Many clay mineral grains are in fact clay sized. Particles larger than those shown in Figure 4.1 include cobble (diameter of 6.4 to 25.6 cm [2.5 to 10.25 in]) and boulder (diameter greater than 25.6 cm [10.25 in]).

Grain Shape

Mechanical weathering, especially by freeze-thaw fracturing, produces highly angular particles. When moved by wind or water, the corners and sharp edges of angular grains are gradually smoothed and rounded away as the particles bump against each other and other rocks. Relative terms such as *angular, sub-rounded,* and *well-rounded* describe the overall degree of **roundness** of sedimentary grains (Fig. 4.1B). Note that well-rounded does not imply that the particle is a sphere, only that there are no sharp or irregular edges. Particles larger than about 1 mm tend to become rounded fairly quickly as they bang into each other in turbulent streams. Smaller, lighter grains require more energetic collisions (greater turbulence) or longer transport distances in order to become rounded. Sand-sized particles are most effectively rounded in the crashing waves of a beach or in sand dunes formed by the wind, which often blows sand grains fast enough to sting your skin.

Sorting

Whereas grain size and shape refer to individual grains, **sorting** refers to the similarity in size of all grains in a rock or handful of sediment. If all grains are essentially the same size, they are *well-sorted* (Fig. 4.1C). If they show a wide range of sizes, they are *poorly sorted*. Poorly sorted sediments may display particles of a variety of different sizes, as in Fig. 4.1C, or they may show a set of larger **grains** (particles) imbedded in a **matrix** (Fig. 4.1D). The **matrix** is a mass of smaller particles filling the spaces between larger grains.

Transporting Agents

Sedimentary particles are produced by weathering and erosion. More and larger sedimentary particles are produced in mountainous areas. Rocks break loose from cliffs, roll or slide down steep slopes, and either land on ice (mountain glaciers) or roll into fast, turbulent streams. Both carry the sediment down the valley. Gravity is less effective at removing loose rock from low relief areas. Because loosened rock tends to stay in place in flatter areas, there is more time for chemical weathering to convert the rock into a thick layer of clay minerals and other fine-grained particles collectively known as soil. The erosion of soils in flat regions supplies fine-grained particles for water, ice, and wind transport.

Water

Water moves more sediment than any other transporting agent. Fast, turbulent streams more effectively erode the landscape and transport larger particles than do slow-moving streams. It's like when you clean off pavement: you do a faster, better job with a hose with a nozzle than one without. Thus, steep, fast mountain streams carry and deposit much more coarse gravel than do the slower lowland streams. If you find an ancient deposit of nonglacial pebbles, cobbles, and boulders, you are likely to be near an ancient area of high elevation and the source of the sediment. The slower, less turbulent streams and rivers found in flatter areas do not move much gravel, but they transport their finer particles great distances by pushing sand along the bottom and carrying silt and clay in suspension in the water column. This suspended sediment is what gives some rivers their muddy look. River sediments are deposited along the river banks, in deltas, or ultimately in the oceans. Dissolved elements from chemical weathering (Table 4.1) are transported regardless of stream slope as long as there is water to carry them.

River transport rapidly rounds particles larger than about 1 mm. Thus, angular gravel, cobbles, and boulders tend to be found only very near the ridges and mountains that produced them. The greater the transport distance from these sediment sources, the rounder these coarse particles become. Particles finer than 1 mm (sand, silt, and clay) are not effectively rounded by river transport. Instead, sand is most effectively rounded by the pounding waves of the beach and by wind. If you find rounded sand grains in river sediment, it was likely eroded from ancient sandstones deposited as beach sediments or sand dunes.

Fast mountain streams contain very poorly sorted sediment because they move everything from clay up to gravel or even boulders. As they move downstream to flatter areas, the streams slow and thus deposit their larger particles. This causes sorting to improve in lowland river sediments. However, river sediments are generally more poorly sorted than the clean sand found in beaches and wind-blown sand dunes. Beach waves sort sediment by moving a certain maximum grain size toward the beach (larger waves move larger grains) and by washing away finer particles.

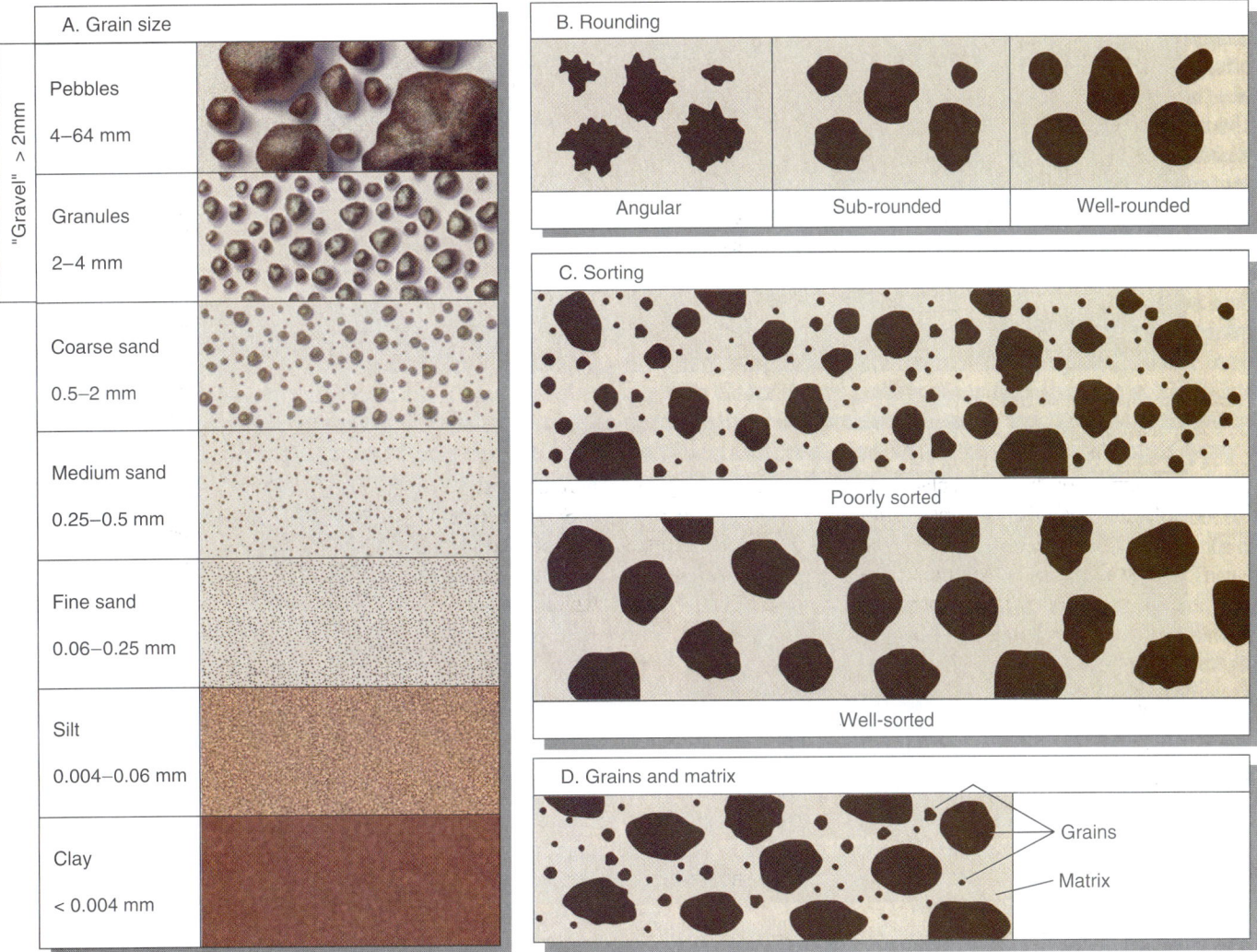

FIGURE 4.1
Characteristics of sedimentary grains. A. Scale of grain-size names for sediment and sedimentary rocks; not shown are cobbles (64–256 mm) and boulders (> 256 mm). B. Angular, sub-rounded, and well-rounded particles. Note that rounding is not synonymous with, nor does it imply, sphericity. C. Poorly sorted and well-sorted sediment. D. Larger particles, or *grains,* are surrounded by a *matrix* of much smaller particles.

Glaciers

Moving glaciers act as huge conveyor belts that carry any size particle that gets trapped in the ice. The moving ice can polish rocks at the base of a glacier but otherwise does not further round its particles. When the ice melts, the sediment is deposited on the spot or transported further by melt-water streams. Thus, the particles in the glacial sediments that cover most of Canada and the northern United States are characteristically poorly sorted, ranging from clay to huge boulders. Glacial melt-water streams leave behind deposits of better sorted, sub-rounded sands and gravel.

Wind

Wind transports only sand, silt, and clay-sized particles. Sand grains generally bounce along the surface to form sand dunes and do not travel far during a single wind storm. Silt and clay particles get blown higher into the air and can travel great distances. In fact, dust from China blows across the Pacific and dust from the Sahara Desert in Africa ends up in south Florida and the Bahamas.

If you have ever felt sand stinging your legs during a windy day at the beach or on sand dunes, you understand the high energy impacts that sand grains make with each other. These high-energy impacts rapidly create well-rounded sand grains.

Wind-blown sediments are remarkably well sorted. Wind bounces sand along to form sand dunes but carries away the silt and clay. The wind-blown silt can also form thick deposits known as loess (pronounced 'lus'). The loess of the upper Mississippi River valley of the United States was blown in from the barren expanses of fresh, loose sediment that were left behind roughly 11,000 years ago as the great ice sheets melted back at the end of the last Ice Age.

Deposition and Lithification

Types of Sediment

So far, we have spoken mainly of **detrital** (or **clastic**) **sediments,** which consist of rock fragments, mineral grains, or clay minerals that have been transported to a site of deposition as clay, silt, sand, and gravel. These almost synonymous names come from *detritus,* which refers to loose solid

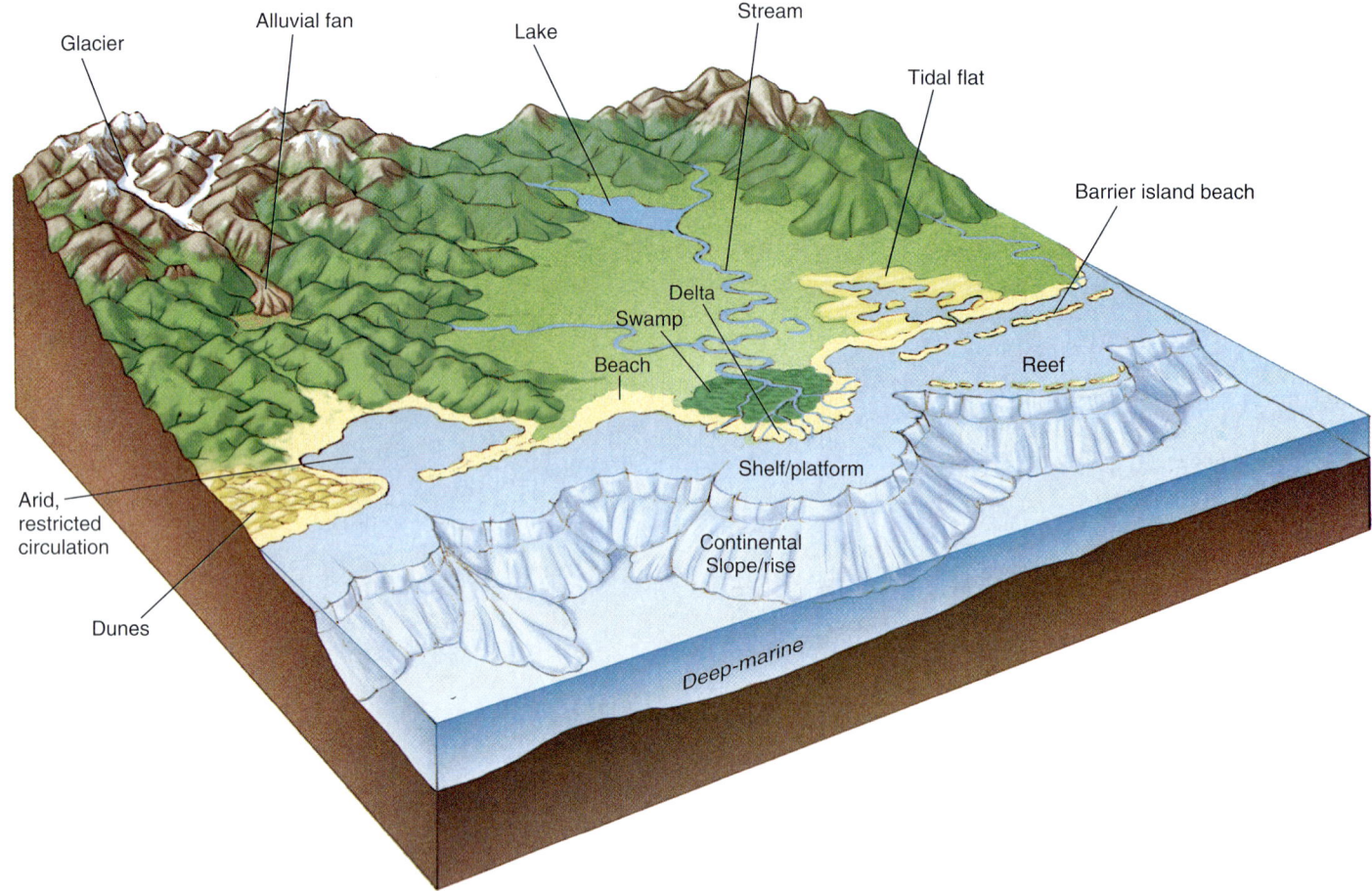

FIGURE 4.2
Typical sedimentary depositional environments.

material resulting from rock disintegration and *clasts,* which are broken pieces of pre-existing rocks and minerals. Detrital sediment deposition takes place when water or wind currents slow enough to allow the particles to settle or when the glacier ice melts. The rate of settling in water or air depends mainly on particle size and density: large, heavy particles settle faster than small, light ones. Silt and larger particles settle rapidly as currents slow. Clay-sized particles can remain suspended in fresh water for long periods before settling, which is why slow-moving rivers are commonly muddy. When fresh water mixes with salty ocean water, the clay-sized particles clump together to form larger particles that settle rapidly near the river mouth.

Chemical sediments form when dissolved ions (such as Ca^{2+}, HCO_3^-, SO_4^{2-}, Na^+, and Cl^-) join together to form a solid. This solid, usually a mineral crystal, is called a *chemical precipitate. Inorganic chemical precipitation,* which occurs without the help of organisms, takes place when the chemistry of the water becomes appropriate. For example, the evaporation of seawater causes the concentration of ions in the remaining liquid to become so high that minerals precipitate. This is how rock salt (halite) and rock gypsum are formed. Such precipitation is most likely to take place in shallow basins that are almost closed off from the main oceans, especially if they occur in hot subtropical regions where rainfall is low and evaporation is high. This includes a number of famous areas around the Persian Gulf.

Biochemical sediments form through biological chemical precipitation: plants and animals extract calcium carbonate or silica from seawater to build such skeletal structures as shells, exoskeletons, or, in the case of calcareous algae, tiny calcium carbonate crystals that help support soft tissues. When these organisms die, most of the soft, carbon-rich tissue rots away and leaves the hard skeletal parts behind to become part of the bottom sediment. In some cases, organic carbon accumulates so rapidly that it does not rot completely away. The rapid accumulation of plant materials in swamps gives rise to peat deposits, which, following burial, ultimately become coals. Rapid accumulation of dead algae on the sea floor produces black muds that can, with sufficient burial and heating, form oil and natural gas.

Environments of Deposition

One of the most important goals of geology is to learn exactly how, and under what conditions, a particular rock formed. If we know these things, we can learn or predict many others. For example, we can reconstruct the history of the Earth, we can predict where we might find ore deposits or fossil fuels, we can see how climatic conditions have varied in the past, and we can even formulate hypotheses about the future.

We learn how and under what conditions ancient sedimentary rocks formed by studying the environments in which mod-

TABLE 4.2
Depositional Environments of Common Sedimentary Rocks

Depositional Environment			Characteristic Sedimentary Rocks
Continental	River/Stream	Channel	Pebble conglomerate and sandstone with current ripple marks, crossbeds, and discontinuous beds
		Flood plain	Shale, possible mud cracks, root traces, soil horizons
	Base of cliff/steep slope		Pebble to boulder breccia
	Alluvial fan		Pebble to boulder conglomerates, arkosic sandstone, poor sorting, cross-bedded
	Dune		Sandstone, well-sorted, large crossbeds
	Glacier	Till	Tillite—poorly sorted, unlayered conglomerate; particles angular to rounded; some may have scratches
		Stratified drift	Sandstone and conglomerate, similar to stream channel deposits
	Swamp		Coal
	Freshwater Lake		Shale, freshwater limestone away from shore; sandstone, conglomerate near shore
	Playa lake (desert lake; frequently dries up)		Evaporites, mudstone; mudcracks
Transitional	Delta		Complex association of marine and nonmarine sandstone, siltstone, and mudstone, possibly with crossbeds and ripple marks; coal and plant debris common
	Beach/barrier island		Fine- to medium-grained, well-sorted sandstone; crossbeds common
	Tidal flat		Mudstone, siltstone, and fine-grained sandstone; ripple marks, bidirectional crossbeds, mud cracks; evaporites possible
Marine	Shelf	Inner shelf (near shore, shallow water)	Fine-grained sandstones; crossbeds and ripple marks common
		Middle shelf (deeper water)	Siltstones, claystones; marine fossils common
		Outer shelf/platform (may build up to sea level)	Limestones, possibly fossiliferous, or reefs (see below)
	Reef		Fossiliferous limestone, massive; coral and algae fossils common
	Continental slope and rise		Mudstone, graywacke (graded bedding common)
	Deep marine		Chert, chalk, micritic limestone, mudstone
	Evaporite basin (hot, arid, restricted access to sea)		Evaporites (rock gypsum, anhydrite, rock salt)

ern sediments accumulate. The most important depositional environments are illustrated in Figure 4.2; Table 4.2 is a summary of the sedimentary rocks and their characteristics for each of these environments.

Sedimentary Structures and Their Significance

Sedimentary structures are features in sedimentary rocks that formed during or after deposition of the sediment, but before lithification (see next page), and are large enough to be visible in the field. They are important because they provide evidence of the transporting agent and the environment of deposition. Figures 4.3 through 4.7 illustrate some common sedimentary structures.

FIGURE 4.3
Stratification (**bedding** or **layering**) is the most distinctive feature of sedimentary rocks. Strata accumulate by deposition on horizontal or gently inclined surfaces. The strata shown here, exposed near Arkadelphia, Arkansas, are tilted from their original horizontal position. If the characteristics of the sediment being deposited remain constant for a long enough period, a thick stratum, or **bed** (1 cm or thicker), is formed; if characteristics fluctuate, thin strata, or **laminations** (less than 1 cm thick), accumulate. The top or bottom surface of a bed or lamination is a **bedding plane**.

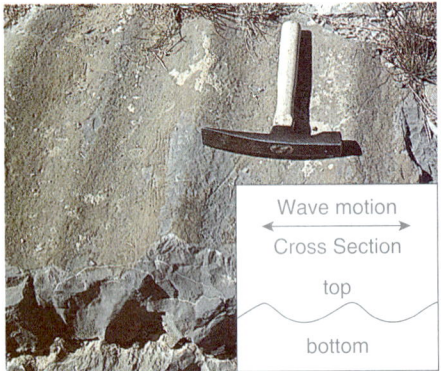

FIGURE 4.4
Ripple marks are wave-like features found on bedding planes. A. **Current ripple marks,** like these in Utah, form when wind or water currents moving in one direction shape the loose sediment into asymmetrical wave forms whose gentle slope is on the side from which the current came. In this example, the current moved from right to left as shown in the schematic cross section. B. **Oscillation ripple marks** form where the back-and-forth motion of water waves shapes the bottom sediment into symmetrical wave forms (see cross section). This occurs in relatively shallow water—less than 10 m for normal waves and up to 200 m for large storm waves. (Sierra Catorce, San Luis Potosí, Mexico.)

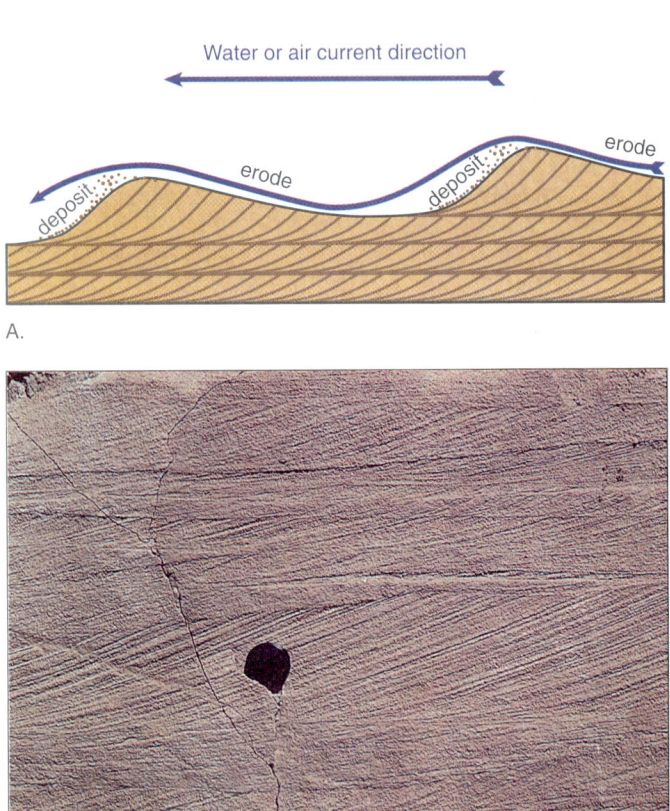

FIGURE 4.5
A. Wind and water currents create ripples and dunes in sand. Both move down-current because sand is eroded off one side and, where the current 'jumps' off the ripple crest, is dumped down the steep face of the ripple or dune. The successive burial of the steep faces is recorded as a succession of **cross-beds** that always tilt down-current. Cross-beds are preserved in layers only when erosion removes just the top parts of the ripple or dune. Geologists use cross-beds in sedimentary rocks to infer the flow direction of ancient rivers, wind systems, deltas, and beaches. In addition, many cross-beds form concave-up curves; geologists use these to tell if sedimentary layers in complexly deformed regions have been turned upside down. B. The **cross-bedding** (or *cross-stratification*) in the Moenkopi Formation in Canyonlands National Park, Utah, indicates that it was deposited by a current moving from right to left. In this case, the current was due to running water and the cross-beds were formed in a stream some 225 million years ago. Six-cm lens cap (black circle) provides scale.

Lithification

Lithification is the conversion of sediment to sedimentary rock. It is accomplished by *compaction* caused by the weight of the overlying sediment, *cementation* by new minerals that fill the open spaces between grains (Fig. 4.8), and various kinds of *recrystallization* processes that cause grains to interlock.

Sedimentary Rocks

Classification

The two types of sediment provide the basis for two categories of sedimentary rocks: (1) **detrital,** which are made of solid particles derived from outside the basin of deposition, and (2) **chemical and biochemical,** which form by precipitation of ions or accumulation of organic materials within the basin of deposition. Classification within these groups is based on texture and mineral composition.

Texture
Most **sedimentary rocks** have a **clastic texture** (Fig. 4.9), which is characterized by discrete clasts or grains of rocks, minerals, or fossils. The originally loose grains are not intergrown with each other, but are generally bound together or cemented by a chemical precipitate of silica, calcite, dolomite, or iron oxide. This **cement** generally has a crystalline texture (Fig. 4.8). If the clastic texture is due to abundant fossils or fossil fragments, the rock is called **bioclastic.**

Although many chemical and biochemical sedimentary rocks have clastic textures, some have **crystalline** textures: careful examination reveals a rock made of interlocking crystals. A primary crystalline texture forms during or just after deposition, as in the cases of rock gypsum and rock salt. Secondary crystalline textures result from the recrystallization or replacement of primary grains or minerals by new minerals. This occurs during or after lithification, as in the case of many crystalline limestones and dolostone.

Chapter 4 Sedimentary Rocks

× 0.3

FIGURE 4.6
In **graded beds,** larger grains on the bottom usually grade to finer ones on the top, as shown here. Graded beds may form when a sediment-laden (turbidity) current slows after moving down an underwater slope; larger grains settle first, smaller ones last. The irregular contact between the coarse sand and the underlying mud reflects sinking of the coarse sand into the still-soft mud. As the sand sank, the mud injected itself into the sand, creating several *flame structures*. (Fincastle Conglomerate, near Roanoke, Virginia.)

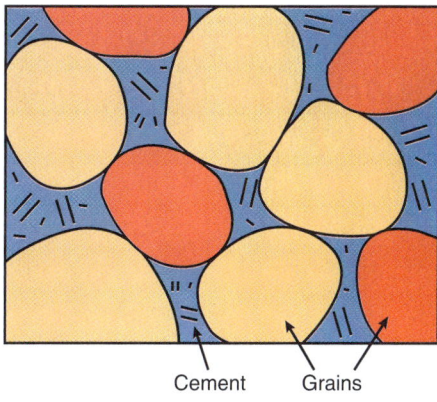

FIGURE 4.8
Lithification by cementation occurs when larger grains are "glued" together with crystalline **cement,** as shown in this "microscopic" view.

The textures seen in extremely fine-grained rocks depend on what they are made of. Chert (Fig. 2.4E), flint (Fig. 2.4F), and some limestones represent recrystallized biochemical rocks with crystals so tiny that they can be seen only under a microscope. These rocks are termed **microcrystalline.**

The texture seen in fine-grained detrital rocks (clay- and silt-sized particles) depends on the amount of compaction during lithification. When first deposited, clay-sized particles of platy or flat minerals, such as clays and micas, mimic a pile of loose leaves: most are roughly horizontal, many are not. Shallow burial and moderate compaction of clay and silt creates a **mudstone,** which breaks into characteristically blocky pieces. Mudstones may show layering or have a **massive** texture (no obvious layers or color variations). With deeper burial, compaction forces the platy minerals to flatten until they become parallel with the bottom. This fully lithified rock, called a **shale,** easily splits into sheets that are parallel to the flat faces of the tiny compacted clays. A sedimentary rock that splits into such sheets is said to be **fissile.**

Mineral Composition

The major mineral components of sedimentary rocks, and their salient properties, are listed in Table 4.3.

Identification of Sedimentary Rocks

Common sedimentary rocks are listed in Table 4.4 (Detrital Sedimentary Rocks) and Table 4.5 (Chemical and Biochemical Sedimentary Rocks). Like igneous rocks, sedimentary rocks are classified according to texture and composition. To identify an unknown rock, first identify its texture (Fig. 4.10) and then use this texture to select the appropriate flow chart within Figure 4.10. For reference, Figure 4.1 summarizes clast size and shape and Table 4.3 summarizes the diagnostic properties of common minerals in sedimentary rocks.

Some common sedimentary rocks are shown in Figure 4.9 and Figures 4.11 through 4.21. The photographs are representative examples only, and your samples may differ *considerably* in color and general appearance from those illustrated here.

× 0.3

FIGURE 4.7
Mud cracks form when mud (fine-grained sediment) dries and shrinks. In side view, the cracks are widest at the top and taper to a point. From above, most ancient mud cracks appear as rough polygons formed when new sediment washed into the cracks and defined a color or textural contrast with the original mud. Mud cracks indicate such depositional environments as desert lakes, tidal flats, or other places where periodic wetting and drying occurs. (Rome Shale Formation, Southwestern Virginia.)

Conglomerate × 1

FIGURE 4.9
Clastic texture in **conglomerate.** Conglomerates contain rounded granules, pebbles, cobbles, even boulders, in a matrix of finer-grained particles. The largest fragments can be any kind of rock that is durable enough to be transported by water or ice. Some conglomerates contain a variety of large fragments that are made up principally of one rock type. Granule- and pebble-size pieces in some conglomerates are made entirely of quartz because it is so resistant to abrasion. Common matrix minerals are sand-size quartz and feldspar. Conglomerates form from gravels, and common depositional environments are alluvial fans, stream channels, and beaches along rocky coasts.

TABLE 4.3
Recognizing Minerals In Sedimentary Rocks

Mineral	Properties
Quartz p. 18	White to light gray Exterior may be frosted; broken grains glassy Hardness = 7
Chert (white, gray, pink) Flint (black) Jasper (usually red, also brown, yellow) p. 19	Color varieties of microcrystalline (extremely fine-grained) SiO_2 Smooth, conchoidal fracture Hardness = 7
Feldspar p. 17	K-feldspar most common White or pink Two cleavages Commonly angular
Biotite p. 20	Shiny and black (gold if altered) One perfect cleavage
Muscovite p. 20	Shiny and silvery One perfect cleavage
Clay (e.g., kaolinite) p. 20	White, gray, green, red, black Clay-sized particles (microscopic) Soft (H = 1–2.5)
Iron oxide p. 23	Yellow, orange, red, brown—strong coloring agent Usually very small particles Mainly a cement or surface stain
Calcite p. 21	White, gray, black Hardness = 3 Bubbles readily in acid
Dolomite p. 21	Gray, buff Hardness = 3.5 Bubbles slowly in acid if powdered
Gypsum p. 21	White to gray Soft (H = 1)
Halite p. 21	White, light gray, pinkish white Hardness = 2.5 Tastes like salt

TABLE 4.4
Classification of Detrital Sedimentary Rocks

Grain Size	Comments		Rock Name	
"Gravel" (>2 mm)	Rounded grains. Rock name may be modified by type and size of dominant grains (e.g., quartz-pebble conglomerate) or composition of grains and matrix (e.g., arkosic conglomerate). Clastic texture.		Conglomerate	
	Angular grains. Rock name may be modified by type of dominant grains (e.g., chert breccia). Clastic texture.		Breccia	
Sand (0.06–2 mm)	Mostly (>95%) quartz grains. Grains commonly well rounded and well sorted. Clastic texture.		Quartz sandstone	Sandstone
	Mostly quartz and >25% feldspar. Grains commonly angular and poorly sorted. Commonly reddish. Clastic texture.		Arkose	
	Composed of quartz, feldspar, rock fragments, and clay. Grains angular and poorly sorted. Dark color common. "Graywacke" is an imprecise general term for such sandstones. May require hand lens to see clastic texture.		Graywacke	
Silt (0.06–0.004 mm)	Commonly massive. Feels gritty on teeth. Fine grained.	Grains too small to identify composition. Mudstone and shale are made of silt and clay. Mudstone breaks with conchoidal fracture; shale is fissile.	Siltstone	Shale (fissile) or mudstone
Clay (<0.004 mm)	Commonly laminated. Feels smooth on teeth. Fine grained.		Claystone	

TABLE 4.5
Classification of Chemical and Biochemical Sedimentary Rocks

Composition		Comments	Rock Name
Calcite (CaCO$_3$)	All effervesce in dilute HCl	Crystalline texture.	Crystalline limestone*
		Microcrystalline. Breaks with conchoidal fracture.	Micritic limestone
		Sand-size spheres with concentric layers (ooids). Clastic texture.	Oolitic limestone
		Abundant fossils or fossil debris cemented together with a fine-grained or crystalline matrix. Bioclastic texture.	Fossiliferous limestone*
		Angular to rounded grains that are neither ooids nor clearly bioclastic. Fine-grained or crystalline matrix. Clastic texture.	Clastic limestone
		Poorly cemented shell fragments; little or no matrix. Bioclastic texture.	Coquina
		Fine-grained, massive, earthy, poorly cemented, generally white.	Chalk
		Banded, finely crystalline to microcrystalline.	Travertine
Dolomite (CaMg(CO$_3$)$_2$)		Effervesces weakly with dilute HCl; powder effervesces more strongly. Commonly crystalline. May contain small, crystal-lined cavities.	Dolostone†
Quartz (SiO$_2$)		Dense, microcrystalline texture, hardness = 7. See Table 4.3.	Chert/Flint/Jasper
Gypsum (CaSO$_4$ • 2H$_2$O)		Massive, crystalline, soft, generally white to gray.	Rock gypsum
Halite (NaCl)		Massive, crystalline, tastes like salt, generally white, clear, gray, reddish.	Rock salt
Plant debris (mostly C)		Black, shiny to earthy luster, low density. Plant fossils may be visible. Combustible.	Coal

*Limestones can display more than one texture, sometimes mixed together and sometimes in separate layers. For example, fossiliferous limestones commonly have a crystalline texture. The important thing is to recognize the components and name the limestone so that communication is clear.

†Since dolostones are recrystallized and chemically altered limestones, there are crystalline, microcrystalline, oolitic, and fossiliferous dolostones. If you recognize a component, name it!

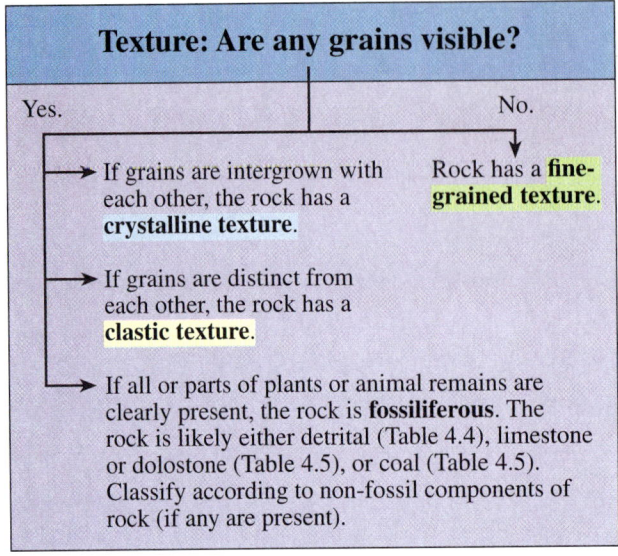
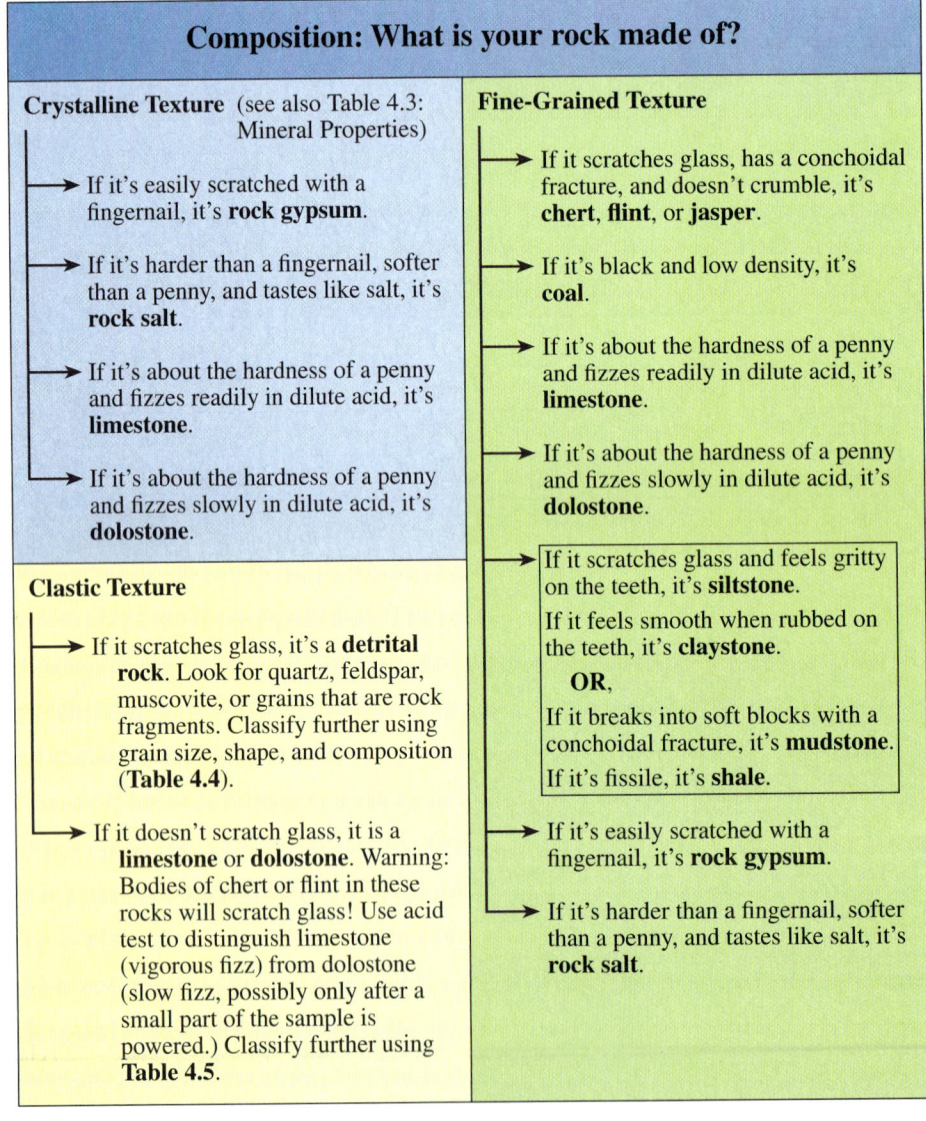

FIGURE 4.10

Flow charts for the identification of sedimentary rocks. The top chart refers to rock textures and the bottom one refers to composition. Both charts should be used together with the more comprehensive information in the text and Tables 4.4 and 4.5.

Chapter 4 Sedimentary Rocks 67

Breccia × 1

FIGURE 4.11
Breccia is similar to conglomerate (Fig. 4.9), but has angular, rather than rounded, particles. Because the angular pieces have not moved far from their source, they commonly are all of the same rock type, rather than a mixture of rock types. The matrix of a breccia may be finer particles or cement with a crystalline texture. The angular particles imply a nearby source, so common depositional environments include the bases of cliffs or steep slopes, landslide areas, or places where caves have collapsed.

Quartz sandstone × 1
Photomicrograph × 10

FIGURE 4.12
Quartz sandstone is made almost entirely of sand grains of quartz. The quartz grains are commonly rounded (see photomicrograph) and cemented together by calcite, iron oxide, or secondary quartz. Some quartz sandstone is white, but more commonly it is buff or rusty brown because of a small amount of iron oxide cement. Relatively pure quartz sand accumulates only when other minerals have weathered away, or where quartz is the only material available in the source area. Depositional environments are beach and near-shore areas, extensive sand-dune fields, and some stream channels.

Arkose × 1
Photomicrograph × 10

FIGURE 4.13
Arkose. While most sandstones contain a small amount of feldspar, arkose is unusual because feldspar is a prominent component along with quartz. The grains are commonly less rounded and more poorly sorted than in quartz sandstone (see inset). The combination of pink potassium feldspar and rusty iron oxide gives arkoses their typical reddish color. Many arkoses were derived from erosion of granite in steep terrains and were deposited in nearby alluvial fans or beach-near-shore settings.

Graywacke × 1
Photomicrograph × 10

FIGURE 4.14
Graywacke is a general field term for gray to dark gray sandstones with fine-grained matrices. Common components include quartz, feldspar (especially plagioclase), rock fragments (especially volcanic rocks), and micas. The angular grains tend to be fine sand, rather than coarse, and the combination of small grain size and even finer-grained matrix give graywacke a rather nondescript appearance. A hand lens may be needed to recognize the clastic texture. Graywackes are common sedimentary rocks in the vicinity of volcanic island arcs or off the margins of continental shelves, where they were deposited from dense, sediment-laden (turbidity) currents that periodically flow off the shelf edge into the deep ocean.

A. Shale × 1

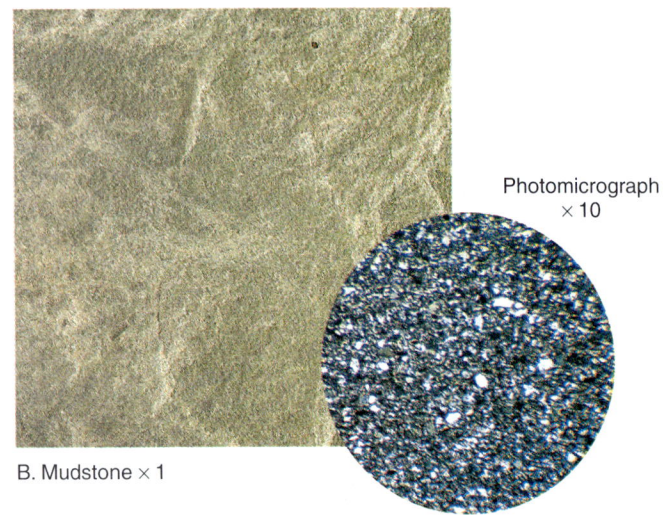
B. Mudstone × 1

FIGURE 4.15
(A) **Shale** and (B) **mudstone.** The finest particles end up as components of these mud rocks. Shale is fissile; mudstone is not. Both tend to break up easily in outcrops, but mudstone is more likely to break into larger pieces with a conchoidal fracture. If rich in organic matter, shales and mudstones are black; if not, they are gray, red, or green. Some geologists distinguish two types of mudstone—claystone and siltstone—based on particle size. Siltstone feels gritty when chewed; claystone does not. (So, if you like, take a chew.) Mudstone and shale are by far the most abundant kinds of sedimentary rocks. They accumulate in the quiet water of floodplains, lagoons, and lakes, and offshore in shallow and deep marine environments.

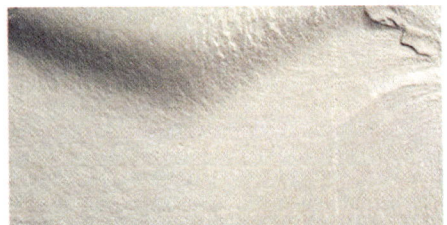
A. Micritic limestone × 1

B. Crystalline limestone × 1

C. Oolitic limestone × 1

D. Fossiliferous limestone × 1

E. Coquina × 1

FIGURE 4.16
Limestone is made of calcite, so its hardness of 3 and ability to bubble readily in acid are distinctive. Only a few of the many types of limestone are mentioned here. *Micritic* or *microcrystalline limestone* (A) has such tiny crystals that it often forms smooth fracture surfaces. Most samples represent the recrystallized remains of billions of microscopic particles released by the decay of calcareous algae. *Crystalline limestone* (B) is made of small but recognizable interlocking calcite crystals. Many samples formed when coarse bioclastic fossil debris was extensively recrystallized during or after lithification. Few obvious fossils remain. *Oolitic limestone* (C) (the *oo* is pronounced like the two o's in moo) consists of sand-sized spheres of calcite (called *ooids*) cemented together. The interior of an ooid shows concentric layers of calcite formed around a central tiny bioclastic particle. Ooids form like tiny snowballs in warm shallow waters in response to wave or tidal currents that wash them back and forth. *Fossiliferous limestones* (D) contain abundant calcareous (made of calcium carbonate) fossils or fossil fragments that are cemented together with microcrystalline or crystalline calcite. Such limestones frequently formed in shallow continental shelves; many contain impurities in the form of clay and silt-sized clastic particles. *Coquina* (E) consists almost entirely of poorly cemented, coarse, bioclastic debris. There is little or no fine-grained matrix so the rock is very open and porous. Coquinas generally form near beaches where waves wash away the finer grains. *Chalk* is a white, soft limestone composed of microscopic calcareous fossils that accumulate in deeper ocean waters. In general, most limestones form in warm, shallow seas far from shore, where the water is clear and free from land-derived detritus. Although today confined to small parts of the outer continental shelves, the geologic record of limestones tells us that in the past such seas covered vast areas of the continents.

Dolostone × 0.5

FIGURE 4.17
Dolostone is made of the mineral dolomite. It is distinguishable from limestone by its lesser reaction with acid. It is sometimes necessary to use a hand lens to see that it actually bubbles, or to scratch the rock (to make a powder) before testing with acid. Dolostone has a crystalline texture and commonly is light gray or buff. A common, but not invariable, feature is the presence of crystal-lined cavities, or *vugs*. The sample shown here contains a crystal of clear quartz (a "Herkimer diamond"). Most dolostone is formed by replacement of limestone.

Chert × 1

FIGURE 4.18
Chert (shown here), **flint,** and **jasper** are made of microcrystalline quartz. They break with a conchoidal fracture and have smooth surfaces. They most resemble micritic limestone, but their hardness of 7 gives them away. See Table 4.3 and Figure 2.4D, E, and F for characteristic colors. Some chert forms by replacement of limestone and may retain some of the textural elements, such as fossils or ooids. Most chert forms by biochemical extraction of silica from ocean water. Such chert consists of microscopic plant (diatom) or animal (radiolaria and sponge spicules) fossils.

Rock gypsum × 1

FIGURE 4.19
Rock gypsum is usually white, light gray, or pale shades of red or orange. It has a crystalline texture and can be scratched with your fingernail. Dark impurities may be present as layers or outlines around irregular areas. Gypsum is an evaporite mineral; when seawater evaporates, gypsum is one of the minerals that precipitates. Thick deposits accumulate in bays in arid regions that have restricted access to the open ocean; deposition also occurs in salty inland seas or lakes.

Rock salt × 1

FIGURE 4.20
Rock salt is made of halite. It has a crystalline texture and is white, light gray, or tinged with red or orange. It forms in the same manner and depositional environment as does gypsum. The reddish mineral here is sylvite (KCl).

Coal × 1

FIGURE 4.21
Coal is brown to black, depending on the amount of carbon it contains. Brown to brownish black *lignite* contains less carbon than black *bituminous coal,* which, in turn, contains less carbon than metamorphic coal, *anthracite.* Coal is noncrystalline and has a luster that ranges from dull to vitreous as carbon content increases. Coal forms by burial and compaction of vegetation that accumulated in swamps.

Hands-On Applications

This lab continues the process of learning to identify and interpret rocks by focusing on sedimentary rocks. You also will learn how observation, measurement, and comparison help to establish general relations between characteristics of rocks and their history of transport and depositional environment.

Objectives

If you complete all the problems, you should be able to:

1. Determine whether a sedimentary rock has a clastic, bioclastic, or crystalline texture.
2. Distinguish grains from matrix in a clastic sedimentary rock.
3. Identify a sedimentary rock and the major minerals contained in it.
4. Estimate grain size, degree of rounding of grains, and degree of grain sorting of a clastic sediment or sedimentary rock.
5. Recognize the following sedimentary structures in a rock sample or a picture: stratification, cross-bedding, ripple marks, graded bedding, and mud cracks.
6. Study selected sedimentary rock samples and suggest possible depositional environments in which they may have formed.

Problems

1. To help you learn to recognize the principal textures of sedimentary rocks, sort the samples provided for you into the following textural categories:

    Clastic

    Bioclastic

    Crystalline

    Microcrystalline or very fine grained

2. Identify the samples of sedimentary rocks provided by your instructor. You may find it helpful to use the tear-out Sedimentary Rock Identification Worksheet at the end of this chapter.

3. Examine the sample of loose sediment provided for you, using a hand lens or a binocular microscope.
    a. Identify the major minerals and estimate their proportions.

    b. By comparison with the roundness scale (Fig. 4.1B), determine the roundness of the grains. Are some minerals rounder or more angular than others? If so, which are most angular and which are most rounded? Suggest an explanation for the similarities or differences.

    c. Use a metric ruler or graph paper with 1 mm squares to measure the diameters of several grains in millimeters. What is the range in size, and what sizes are the most prevalent? What grain-size names (for example, sand or granules) best describe the sediment (see Fig. 4.1A)? (For an alternative and more accurate way to estimate grain-size distribution, see Problem 8.)

Part II Rocks

d. Describe the degree of sorting. For the purposes of this lab, consider the sediment to be well sorted if most grains have the same grain-size name, and poorly sorted if they do not.

e. Make the unlikely assumption that all of the grains were freed from the source area by mechanical weathering (that is, no chemical weathering took place). Based on the mineral content of the sediment, what igneous rock(s) could have been the source of this sediment?

f. What was the probable transporting agent (wind, water, ice) for the materials in the sediment? Explain.

g. Is the sediment detrital or chemical/biochemical? Why?

h. What kind of sedimentary rock would be formed by lithification of this sediment?

4. Examine samples of the following sedimentary structures, and answer the questions.
   a. Graded bedding.
      (1) Before looking at the rock sample, shake a jar of water containing a mixture of clay, silt, sand, and granules; let it sit on the lab bench while you do something else for 5 or 10 minutes.
      (2) Look at the sediment that has settled on the bottom of the jar. How are different-size grains distributed? Are similar-size grains segregated in layers, or are they distributed randomly throughout the sediment?

   If in layers, are the layers parallel or perpendicular to the bottom of the jar?

   Does the grain size change in any regular manner from top to bottom or from one side of the jar to the other?

   Is there still sediment in suspension that has not settled? If so, what grain-size term would describe it most aptly?

      (3) Next, look at the rock sample, paying particular attention to the distribution of its grains. Make a rough sketch of the sample showing stratification and emphasizing grain-size differences within or across strata. By analogy with the sediment in the jar, determine the original top of the sample and label it in your sketch.

   b. Ripple marks. Sketch the sample to show the shape of the ripple marks in cross section, and identify the type of ripple mark. For current ripple marks, indicate with an arrow the current direction at the time the ripples formed. For oscillation ripple marks, label the top and bottom if you can. Explain your answer.

c. Cross-stratification. Sketch the sample to show the strata in cross section and indicate with an arrow the current direction at the time of deposition. Explain your answer.

d. Mud cracks. Sketch the original top of the sample. How did you know which side was the original top?

5. Go to *www.kaibab.org/geology/gc_layer.htm* (or link to it through *www.mhhe.com/jones5e*—see Preface), which contains descriptions of the rocks exposed in the Grand Canyon.

   a. What kinds of Paleozoic sedimentary rocks occur in the Grand Canyon (types of rocks, not names of formations)?

   b. At the bottom of the web page is a drawing summarizing the rocks seen in the Grand Canyon. Use the maximum thickness given for each of the Paleozoic formations to calculate the relative proportions of the following rock types: carbonates (limestone plus dolostone), shale, and sandstone.

   c. A global average of the relative proportions of the different rock types reveals the following proportions: carbonates = 15%; shales = 65%; sandstones = 20%. When you gaze out across the Grand Canyon (see links below), are you seeing a set of rocks comparable in composition to the world average?

   d. Based on their descriptions and Table 4.2 in this lab manual, what are the depositional environments of the following:
      (1) Kaibab Limestone

      (2) Coconino Sandstone

      (3) Hermit Shale

74   Part II   Rocks

   (4) Supai Formation

   (5) Temple Butte Limestone

 You may be interested in seeing more of the Grand Canyon. Try *www.kaibab.org* and click on *armchair travelers*. You can take a virtual *13 Day River Trip,* make a *Virtual Visit,* and explore several other links showing photos of the many things you can see in and around the Grand Canyon.

In Greater Depth

6. Examine the samples of partially weathered igneous rocks provided in the lab.
   a. Arrange and list the rocks in order of susceptibility to weathering, assuming the samples were all subjected to the same degree of chemical weathering.

   b. Next, indicate which *minerals* seem to be weathered the least, and which the most, in each sample; list the minerals in order of increasing susceptibility to weathering.

   c. How does the order of your list compare with the sequence of crystallization according to Bowen's Reaction Series (Chapter 3)? Suggest a relation between crystallization temperature of igneous minerals and susceptibility to chemical weathering.

7. Following are descriptions of some common sedimentary rocks, possibly similar to ones provided in lab. Using the descriptions, suggest possible depositional environments for each. (See Fig. 4.2 and Table 4.2)
   a. Reddish rock made principally of angular to sub-rounded, coarse sand- to granule-size grains of quartz and feldspar. In the field, structures like those in Fig. 4.5 are evident, and the sandstone is seen to be interbedded with pebble and cobble conglomerate.

   b. Gray rock with numerous fossil shells cemented firmly together (similar to Fig. 4.16D). Rock readily fizzes with dilute HCl.

   c. White rock with crystalline texture. Very soft, does not fizz with acid or taste distinctive.

   d. A conglomerate with grain sizes ranging from clay to boulder. No apparent bedding, unsorted; particles range from angular to rounded, and some have flat, scratched faces.

   e. Light tan rock with well-rounded, 0.5 mm (0.02 in), frosted quartz grains. Well sorted, with large crossbeds.

f. Black, shiny rock with some poorly preserved plant fossils.

g. Reddish-gray, very fine-grained, detrital rock, with features like those in Fig. 4.7.

h. Dark gray rock that looks somewhat like basalt, but with a hand lens, the texture appears to be clastic, not crystalline. Some white feldspar grains are present, as are dark grains of something unidentifiable; the matrix is too fine to recognize anything. One sample has thin beds that show a progressive decrease in grain size going from bottom to top.

8. Sedimentary geology is most interesting when you examine a whole series of rocks deposited in layers one on top of the other because, like pages in a book, they tell you a story. On a hike up a long hill you find the following succession of rock layers: (1) At the bottom are gray shales containing plant fossils and a thin layer of coal. (2) Then comes a transition to a quartz sandstone displaying abundant crossbeds and even some ripple marks. The grains are medium-sized at the base but become fine near the top. (3) This is followed by a transition to more shales, but no plant fossils are visible and there is no coal. Instead, there are ancient seashells of various types. (4) Finally, at the top of the hill come various layers of gray and light tan limestones, some of which display various marine fossils, including abundant corals. How did the environments of deposition change at this locality? What one thing could have occurred to cause these changes?

9. Visual estimates of grain size and sorting of sediments are *subjective* and not appropriate for most scientific purposes. For this reason, *objective* methods of measuring these properties must be used, and the results must be reproducible to be valid. A common way to measure size distribution is to pass the sediment through a series of screens or sieves with progressively smaller, premeasured openings. Grains pass through holes larger than themselves but are caught in sieves whose openings are smaller. The proportion of grains falling within a certain size range is determined by weighing the amount of sediment trapped by each sieve and comparing that to the weight of the entire sediment sample. The activities required for this problem can be done as a group.

   a. Weigh the sample of dry sediment provided by your instructor.

   b. Pass the sediment through a series of sieves and weigh the sediment trapped in each sieve. Along with the weight of sediment, record the opening sizes of the sieve that trapped the sediment and the sieve with the next largest opening. The grain size of the sediment is between these two values.

   c. Determine the percent of grains in each size range by dividing the weight of each by the weight of the entire sample and multiplying by 100. Make a histogram or bar graph of these data by plotting grain-size increments on the horizontal (x) axis and weight percent on the vertical (y) axis.

   d. Sorting can be evaluated numerically from these data, but that aspect will not be pursued here. Instead, use your histogram to determine whether a few bars are markedly higher than others. If so, the sample is well sorted; if not, it is poorly sorted.

   e. What factors could affect the accuracy of your results?

Sedimentary Rock Identification Worksheet

Sample Number	Texture Clastic=C Crystalline=X Fine-grained=F	Mineral Composition	Sorting and Roundedness (if applicable)	Fossils (leaves, clams, corals, etc.)	Condition of Fossils (largely whole or mostly broken up?)	Rock Name	Inferred Depositional Conditions*
S-A	C		Well sorted			Shale	low
S-B	C	Potassium Feldspar, Quartz	Well sorted			Arkose	Moderate
S-C	C	Quartz	Well sorted			Quartz Sandstone	Moderate
S-E	C		Not sorted, Round			Conglomerate	High
S-K	C		Not sorted			Breccia	High
S-M	C	Quartz, Potassium Feldspar	Well sorted			Arkose	Moderate

*Grain size indicates current strength. Gravels=very strong currents or near sediment source; sands=strong currents; shales/mudstones=quiet conditions. Fossils may indicate a land or marine setting. Limestones: most are marine and form in middle to outer shelf (including reefs); grain size indicates whether they experienced strong, moderate, or gentle currents from tides or waves. Broken fossils indicate higher currents than whole fossils. See Table 4.2 and text for more details on these and other depositional settings.

Sedimentary Rock Identification Worksheet

Sample Number	Texture Clastic=C Crystalline=X Fine-grained=F	Mineral Composition	Sorting and Roundedness (if applicable)	Fossils (leaves, clams, corals, etc.)	Condition of Fossils (largely whole or mostly broken up?)	Rock Name	Inferred Depositional Conditions*
S-G		Calcite	Well sorted Rounded			Oolitic Limestone	
S-H		Gypsum				Rock Gypsum	
S-I		Dolomite				Dolostone	
S-J		Quartz				Chert	
S-T		Calcite				Fossiferous Limestone	
S-L		Carbon				Coal	

*Grain size indicates current strength. Gravels=very strong currents or near sediment source; sands=strong currents; shales/mudstones=quiet conditions. Fossils may indicate a land or marine setting. Limestones: most are marine and form in middle to outer shelf (including reefs); grain size indicates whether they experienced strong, moderate, or gentle currents from tides or waves. Broken fossils indicate higher currents than whole fossils. See Table 4.2 and text for more details on these and other depositional settings.

CHAPTER 5

Metamorphic Rocks

Materials Needed
- Pencil
- Hand lens
- Dilute hydrochloric acid
- Glass or knife to test hardness
- Samples of metamorphic rocks to be identified

Introduction

As a result of heat and/or pressure, rocks can undergo a process called metamorphism. Metamorphism can yield new crystals of the same mineral or, by reorganization of chemical constituents, crystals of new minerals. In the process new textures are developed. The characteristics of the original rock, and the type, intensity, and duration of metamorphism, determine how drastic the change is. This chapter will introduce some of the common varieties of metamorphic rocks, illustrate their typical textural and mineralogic features, show how they relate to the original or parent rock, and introduce the idea of metamorphic zones.

In some ways, metamorphic rocks are the most mysterious of the three major types of rocks. We can observe volcanoes and see how some igneous rocks form, and watch as sediment accumulates on the Earth's surface. But metamorphic rocks typically form well below the surface where we can't see what happens. To understand them, we must infer what goes on, or run experiments in the lab at high temperatures and pressures. We now think we understand how and under what conditions they form and can, in turn, use actual metamorphic rocks to understand what happens deep beneath the Earth's surface, especially in the cores of mountain belts and underneath volcanic regions, and to unravel the geologic past represented by metamorphic rocks found around our planet.

Metamorphic rocks are crystalline rocks that form from other rocks. **Metamorphism** takes place at varying depths within the Earth's crust, where both temperature and pressure are higher than at the surface. The results of metamorphism include the creation of new minerals, the development of bands or layers of like minerals (for example, layers of quartz or mica), and the parallel alignment of new and old mineral crystals.

Conditions of Metamorphism

Metamorphism takes place at temperatures and pressures that fall between those in which sediments are lithified and those in which rocks begin to melt to form magmas. Figure 5.1 shows approximate lower and upper boundaries of metamorphism. A rock will be metamorphosed if temperature and pressure conditions place it within the areas labeled "metamorphism" in Figure 5.1, and if there is sufficient time for the appropriate reactions to occur. Some rocks, such as granite or quartz sandstone, are not affected much by metamorphism until temperatures and pressures are well within the field of metamorphism on Figure 5.1; others, such as porous tuffs consisting of shards of volcanic glass, are highly reactive and undergo changes at temperatures and pressures near those of the field labeled *lithification of sediment*. Similarly, melting begins at different temperature-pressure conditions for different rocks.

Water and such fluids as carbon dioxide are generally present during metamorphism, because they are either contained in the rocks undergoing metamorphism or are released by metamorphic reactions. These fluids are important in the metamorphic process. Among other things, they allow ions to move about more readily, thereby speeding up the chemical reactions that enable the recrystallization and growth of mineral crystals.

Types of Metamorphism

There are two common types of metamorphism: regional and contact.

Regional metamorphism occurs over areas of hundreds or thousands of square kilometers. Regionally metamorphosed rocks were buried beneath thick accumulations of other rocks at some time during their history (Fig. 5.2). During this period of burial, they were subjected to the higher temperatures and pressures found beneath the surface (Fig. 5.1). The weight of the

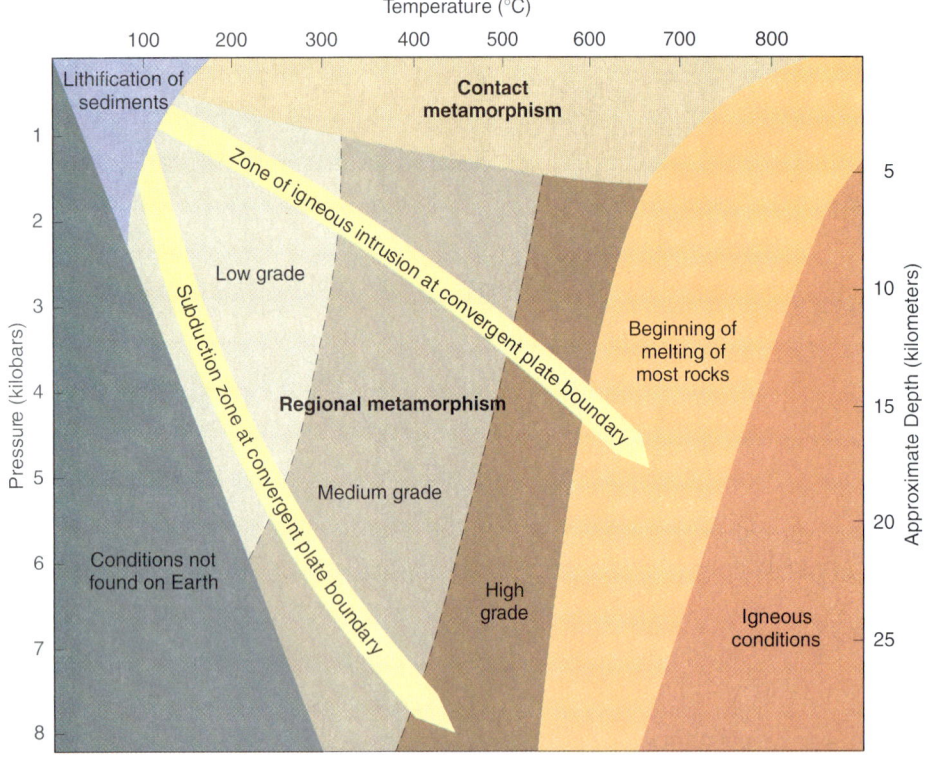

FIGURE 5.1
A. Temperature and pressure (or depth) conditions for lithification, contact metamorphism, grades of regional metamorphism, and melting. Arrows show how temperature changes with depth at convergent plate boundaries. B. The locations of the "Zone of igneous intrusion at a convergent plate boundary" and the "Zone of subduction at a convergent plate boundary" in A are shown in the cross section in B.

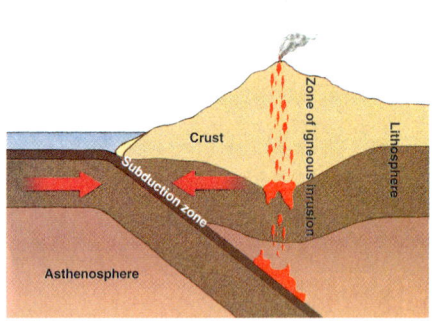

FIGURE 5.2
Cross section illustrating a possible setting for regional and contact metamorphism. The regional metamorphic rocks shown are those that would form from a shale parent in the "Zone of igneous intrusion at convergent plate boundary" of Figure 5.1. They have a near vertical foliation reflecting a horizontal directed pressure (squeezing) during metamorphism.

overlying rocks provided most of the pressure, but collisions between tectonic plates, such as occurs at convergent boundaries, also creates pressure directed from the side.

Heat from an intruding body of magma causes **contact metamorphism** of the surrounding rocks (Fig. 5.2). Fluids released from the magma or the surrounding rocks may accentuate the changes and, in some cases, enable the formation of ore deposits. As suggested in Figures 5.1 and 5.2, contact metamorphism is most evident around igneous intrusions that formed within a few kilometers of the surface. Because temperature increases with depth, it becomes increasingly difficult below a few kilometers to distinguish contact and regional metamorphism.

Changes During Metamorphism

The important changes that may take place during metamorphism are:

1. Recrystallization of existing minerals, especially into larger crystals (Fig. 5.3);

A. Quartz sandstone × 10

B. Quartzite × 10

FIGURE 5.3
These photomicrographs (× 10; polarized light) compare (A) quartz sandstone and (B) quartzite. The original sand grains are evident in the sandstone, but recrystallization has obliterated them in the quartzite.

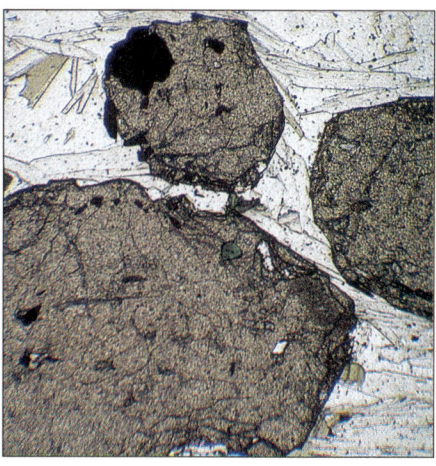

FIGURE 5.4
A new mineral, garnet (brown in photograph), formed during metamorphism, as seen in this photomicrograph of garnet in schist (× 10).

2. Development of new minerals and disappearance of some old ones (Fig. 5.4); and
3. Deformation and reorientation of existing mineral crystals and growth of new ones with a distinctive orientation (Fig. 5.5).

The net result is a rock with a different texture and, commonly, a different mineral content.

Recrystallization and development of new minerals (and disappearance of old ones) take place during both contact and regional metamorphism. In general, the crystal size and kinds of minerals in metamorphic rocks indicate the intensity, or **grade,** of metamorphism (see Fig. 5.1). **High-grade metamorphic rocks** are the most intensely metamorphosed. They are characteristically coarsely crystalline and contain minerals that are stable under higher temperatures and pressures. **Low-grade metamorphic rocks** are the least intensely metamorphosed. They are generally finely crystalline (although crystal size also depends on the size of the grains or crystals in the rock prior to metamorphism) and contain minerals that are stable under lower temperatures and pressures.

Minerals are deformed and reoriented primarily as the result of **directed pressure,** pressure that is greater in one direction than in others (Fig. 5.5). For example, directed pressure is created at convergent boundaries as the tectonic plates collide and compress each other. A subducting oceanic slab pushes hard against the overriding plate, and the collision of two continental plates can create huge mountain belts such as the ancient Appalachians and modern Himalayas.

Simple deep burial of rocks causes only **lithostatic pressure,** which is a pressure felt equally in all directions. If you were in a deeply buried barrel, the weight of overlying rocks would cause the rocks along the side of the barrel to push in with the same force as the rocks above and below the barrel. Since the pressure during contact metamorphism is usually lithostatic, it normally results in little deformation or reorientation of metamorphic minerals.

Metamorphic Textures

Metamorphic rocks have crystalline textures and are further subdivided into two main textural groups: **nonfoliated** and **foliated. Foliation** is the arrangement of mineral crystals in parallel or nearly parallel planes. An example is shown in Figure 5.5, where the flat crystals of mica are aligned in the same direction.

Nonfoliated metamorphic rocks lack flat or elongated mineral crystals with parallel alignment. Instead, crystals typically are all about the same size and are interlocked in a crystalline texture (see Fig. 5.3B). Quartz, feldspar, and calcite com-

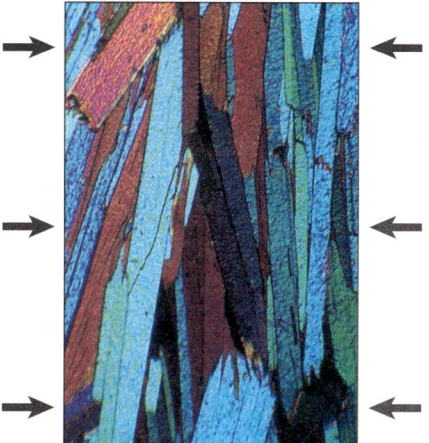

FIGURE 5.5
This photomicrograph (× 10) of schist shows the distinctive orientation of the mica crystals (viewed edge-on) caused by directed pressure during metamorphism. The black arrows show the sense of squeezing. Examination of similar foliations in ancient rocks helps geologists work out the history of tectonic plate collisions. The photograph was taken in polarized light, which changes the natural mineral colors.

monly produce nonfoliated textures. Examples of nonfoliated rocks include quartzite (Fig. 5.6) and marble (Fig. 5.7).

Foliated metamorphic rocks contain minerals that are aligned in parallel planes (Figs. 5.8 through 5.11). The **foliated texture** commonly results from the parallelism of micas, but crystals of other flat or elongate metamorphic minerals, such as chlorite or hornblende, also can form parallel planes. Elongate minerals such as hornblende also may be aligned parallel to each other in the plane of foliation (this alignment is called *lineation*). Foliated rocks commonly appear to be layered or banded. In addition to the parallel arrangement of minerals, the crystals are grown together in an interlocking fashion, as in other rocks with crystalline textures. Foliation results from directed pressures usually present during regional metamorphism; the pressures were perpendicular to the planes of foliation (Fig. 5.5).

Warning: Some metamorphic rocks preserve banding that is not considered foliation. Metamorphosed sandstones can preserve cross-bedding and metamorphosed limestones can display banding that reflects layered impurities in the original rock. In both cases, the rocks are dominated by interlocking mineral grains that are all about the same size (they are nonfoliated), and the banding is defined only by color or traces of tiny mineral grains. Foliation requires a substantial fraction of the rock to consist of aligned mineral grains.

Four distinct types of foliation are recognized primarily on the basis of crystal size. The following list is arranged from low (slaty) to high (gneissic) grade metamorphism.

Slaty (Fig. 5.8): a foliation in metamorphic rocks made of platy (flat) minerals too small to be seen without a microscope. Such rocks readily split or cleave along almost perfectly parallel planes and are said to have a **slaty cleavage.**

Phyllitic (Fig. 5.9): a foliation in metamorphic rocks made of platy minerals ranging in size from visible with a 10-power hand lens to just barely visible to the unaided eye. Phyllitic rocks have a shiny or glossy luster. The rock cleavage is along nearly parallel surfaces, but these surfaces may be wrinkled to varying degrees. This type of foliation represents a state of textural development intermediate between slaty and schistose.

Schistose (Fig. 5.10): a foliation in rocks composed of mineral crystals large enough to be seen with the unaided eye. Rock cleavage is parallel to the foliation, but cleavage surfaces are usually irregular. Platy minerals commonly predominate, but other minerals, such as quartz, are also present.

Quartzite × 1

FIGURE 5.6

Quartzite is a nonfoliated rock made mostly of quartz. It may be white to various shades of purple, green, gray, or pink. Its crystalline texture is evident in hand specimen, but a microscopic view (Fig. 5.3) better illustrates how the original grains of quartz sand have become intergrown. Impure quartzites contain minerals other than quartz, such as muscovite or kyanite. In many quartzites, original sedimentary features, such as bedding, cross-bedding or ripple marks, are still evident, having survived metamorphism.

Marble × 1

FIGURE 5.7

Marble, a nonfoliated rock made mostly of calcite or dolomite, has a variety of colors and textural patterns. It has long been used for decorative purposes, but because of its susceptibility to acidic rain, it does much better in an interior setting. Its hardness (3 to 3.5), reaction with acid, and commonly coarse crystalline texture aid in identification. Some impure marbles, especially those formed by contact metamorphism, contain spectacular crystals of unusual minerals.

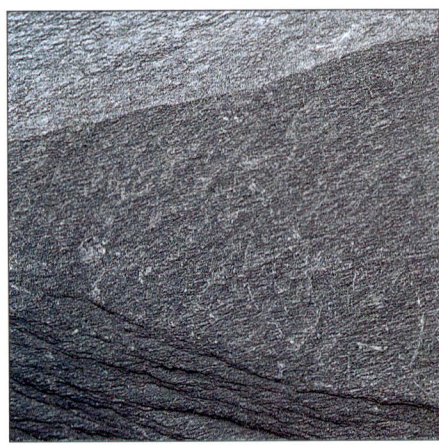

Slate × 1

FIGURE 5.8

Slaty cleavage is the most distinctive feature of **slate.** Slate is so finely crystalline—mineral crystals are indistinguishable—and breaks along such flat surfaces that the really good slate was used for blackboards before about 1945. It makes a durable roof too, and is still used for that purpose. Slate comes in a variety of colors, but is commonly black, bluish black ("slate blue"), gray-green, purple, or dull red. Slate forms from metamorphism of shale and is usually distinguishable from it by its slightly shinier luster.

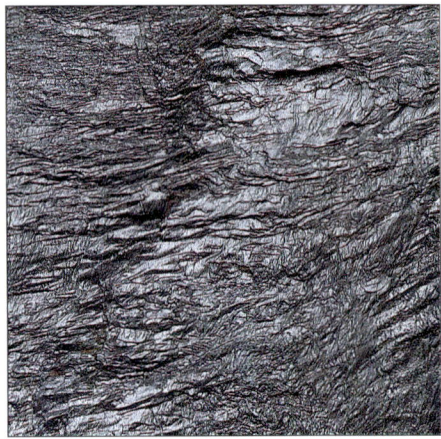

Phyllite × 1

FIGURE 5.9

Metamorphic intensity slightly higher than that required to form slate turns shale into **phyllite,** a rock characterized by its phyllitic foliation. On the low-metamorphic-grade side, phyllite is distinguished from slate by a shinier, more satin-like luster and by crystals that are large enough to give just a hint of their presence. On the high-grade side, phyllite is distinguished from schist by crystals that are almost, but not quite, large enough to identify. Phyllite has a good rock cleavage, but the cleavage surfaces are commonly wrinkled and the cleavage is not as good as slate's.

Gneissic (Fig. 5.11): a coarse foliation in which like minerals are more or less segregated into roughly parallel bands or layers. A common example is a rock in which nonfoliated layers made of quartz and feldspar alternate with foliated layers made of biotite. Mineral crystals are coarse and can be identified with the unaided eye.

During very high-grade metamorphism, local partial melting of quartz and feldspar may result in the formation of a **migmatite,** a mixed igneous-metamorphic rock with enlarged pods of quartz and feldspar in the midst of swirling light and dark bands. With further melting, the foliation would be destroyed and an igneous rock formed.

Metamorphic Minerals

Metamorphic rocks contain many of the same minerals found in igneous and sedimentary rocks. Certain minerals, however, form almost exclusively in metamorphic rocks. In addition, some minerals of metamorphic rocks are present only in rocks of a specific metamorphic grade. Table 5.1 summarizes the properties of common metamorphic minerals, and Figure 5.12 shows the metamorphic grades at which they occur.

Parent Rock

Metamorphic rocks are formed by the effects of temperature and pressure on a preexisting rock, the *parent*. It is always helpful in working out the geologic history of a metamorphic area to determine what the parent rock was, although this is not always possible. Common parent rocks and their metamorphic derivatives are listed in Table 5.2.

Classification and Identification of Metamorphic Rocks

Metamorphic rocks are classified on the basis of texture and the kinds and proportions of minerals present. To identify a metamorphic rock, first determine whether it is foliated or nonfoliated (Fig. 5.13). Foliated rocks are named according to the type of foliation they display and, when possible, by their principal mineral components (Fig. 5.13, Table 5.2). Nonfoliated and weakly foliated rocks are named based on composition or the recognition of parent rock textures (Fig. 5.13, Table 5.2). Examples of the latter include *metaconglomerate, metagraywacke,* and *metachert,* whose parent rocks are conglomerate, graywacke, and chert, and *greenstone*. Greenstones are identified by their greenish gray or greenish black color, which comes from tiny (frequently microscopic) chlorite or epidote crystals, and the preservation of the parent

TABLE 5.1
Recognizing Minerals In Metamorphic Rocks

Mineral	*Properties*
K-feldspar p. 17	Usually white or pink Two cleavages at 90° Equidimensional to ovoid crystals
Plagioclase p. 17	Usually white (Na-plagioclase) or gray (Ca-plagioclase) Two cleavages at 90° Elongate to equidimensional crystals Striations may or may not be present
Quartz p. 18	Colorless to gray Glassy with conchoidal fracture Irregular crystals or lens-shaped masses
Biotite p. 20	Shiny and black One perfect cleavage Thin crystals parallel to foliation
Muscovite p. 20	Shiny and silvery white One perfect cleavage Thin crystals parallel to foliation
Hornblende (amphibole) p. 19	Black with shiny, splintery appearance Two cleavages at 56° and 124° Elongate, commonly parallel, crystals
Garnet p. 22	Pink, red, reddish brown Vitreous to resinous luster Equidimensional, twelve-sided crystals with diamond-shaped faces are common
Staurolite p. 22	Brown, red-brown to brownish black Vitreous to dull luster Prismatic and X- or cross-shaped crystals
Kyanite p. 22	Light blue to greenish blue Vitreous luster Blade-shaped crystals
Talc p. 20	White, gray, apple green Pearly luster Soft (H = 1), with greasy feel
Chlorite p. 20	Green to blackish green One good cleavage Crystals commonly small, flaky, and parallel to foliation
Serpentine p. 22	Multicolored black, green, grey Hardness = 3 to 5 Dull to greasy luster, with a slightly greasy feel Commonly dominates composition of rock (serpentinite)
Calcite p. 21	White, gray, pink Three cleavages H = 3 Bubbles readily in acid
Dolomite p. 21	White, gray, pink Three cleavages H = 3.5 Bubbles slowly in acid if powdered

A. Quartz-biotite schist × 1

B. Garnet-muscovite-staurolite schist × 1

FIGURE 5.10
Schists have crystals that are large enough for individual minerals to be identified. Their distinctive schistose foliation is caused by the presence of platy minerals such as talc, chlorite, biotite, and muscovite. The larger crystal size and the presence of other minerals, such as quartz and feldspar, cause schists to split along rough, irregular surfaces that more or less parallel the foliation. Schists derived from the metamorphism of shale may contain typically metamorphic minerals such as garnet, staurolite, kyanite, or even sillimanite, depending on the metamorphic grade. Schists derived from other parent rocks contain other minerals, such as blue, green, or black amphibole. Because the mineral content of schists is variable, the characterizing minerals are used in the rock name: (A) quartz-biotite schist, and (B) garnet-muscovite-staurolite schist are examples. More intensely metamorphosed schists may be somewhat banded due to segregation of dark and light minerals. Commercially valuable minerals such as talc and kyanite are obtained from schists in which these minerals are sufficiently concentrated.

rock textures of basalt, gabbro, or andesite. **Hornfels** is a general name used for contact metamorphic rocks that are nonfoliated and too fine grained for individual minerals to be readily recognized. Hornfels rocks are often identified in the field by their close association with intrusive igneous rocks (e.g., Fig. 5.2).

Zones of Metamorphism

During regional metamorphism, rocks subjected to higher temperatures and pressures are squeezed and cooked more than those exposed to less intense conditions. The result is that high-grade metamorphic rocks differ from intermediate- and low-grade rocks in texture and mineral content (Fig. 5.12). For example, slate, phyllite, schist, and gneiss are the metamorphic products of progressively more intense temperature and pressure conditions.

TABLE 5.2
Classification of Metamorphic Rocks

Foliated Metamorphic Rocks

Parent rock	Comments	Rock name
Shale or mudstone	Low metamorphic grade. Slaty cleavage, cleavage surfaces dull to slightly shiny. Very fine grained. Generally dark (e.g., black, green, red).	Slate
	Low metamorphic grade. Phyllitic foliation. Fine grained. Silky, shiny luster.	Phyllite
	Intermediate metamorphic grade. Schistose foliation. Medium to coarse grained. Name modified by mineral name (e.g., quartz-biotite schist).	Schist
Shale, mudstone, granite	High metamorphic grade. Gneissic foliation. Medium to coarse grained. Light and dark layers common. Name modified by mineral names.	Gneiss
	Border between igneous and metamorphic rock. Pods of feldspar or quartz in midst of swirling light and dark bands. Medium to coarse grained.	Migmatite

Nonfoliated or Weakly Foliated Metamorphic Rocks

Parent rock	Comments	Rock name
Limestone or dolostone	Variable metamorphic grade. Chiefly calcite or dolomite; H = 3–4; effervesces in dilute HCl. Commonly coarsely crystalline.	Marble
Quartz sandstone	Variable metamorphic grade. Chiefly quartz; H = 7. Crystalline.	Quartzite
Coal	Low to medium metamorphic grade. Shiny, dark gray to black. Conchoidal fracture.	Anthracite coal
Basalt or gabbro	Low to medium metamorphic grade. Chlorite, green amphibole, epidote, and Na-plagioclase. Greenish gray or black. Fine to coarse grained.	Greenstone
Basalt or gabbro	Medium to high metamorphic grade. Hornblende, Na-Ca plagioclase, ± garnet. Dark gray to black. May be foliated. Medium to coarse grained.	Amphibolite
Peridotite	Low to medium metamorphic grade. Chiefly talc. Gray to dark greenish gray. Soft; can be carved (see Fig. 2.6). Fine grained.	Soapstone
	Variable metamorphic grade. Mostly fine-grained serpentine; some may be fibrous. Greenish; commonly mottled or streaked.	Serpentinite
Conglomerate, graywacke, chert	Low to medium metamorphic grade. Parent rock easily recognizable.	Metaconglomerate, metagraywacke, metachert
Various	A low-grade, contact-metamorphic rock. Commonly fine grained and dark color.	Hornfels

One way to learn how varying intensities of metamorphism affect rocks is to find a place where erosion has stripped away the overlying rocks to expose a sequence of metamorphic rocks ranging from low to high grade. It is then possible to study the entire sequence in the field and the laboratory to see exactly what textural and mineralogic changes have occurred, as illustrated by the following hypothetical example.

Figure 5.14A is a map of an imaginary area showing rock *outcrops,* places where the bedrock is not covered by soil (circled and lettered areas). Most of the rocks in this area are metamorphosed to varying degrees, but the *parent rock* for all of them was shale. Furthermore, the whole area was tilted to the northeast before it was eroded to a nearly flat surface. Thus, the rocks that were most deeply buried during metamorphism are now exposed in the southwest part of the area. Unmetamorphosed shale, which was closest to the surface during metamorphism, is present only in the northeast corner. In between are rocks that were buried to intermediate depths. Note that the texture changes from northeast to southwest, with more coarsely crystalline rocks reflecting more intense metamorphism.

The minerals present in rocks from localities A through J are given in the box in Figure 5.14A. Those that are underlined have special significance, because they formed by chemical reactions that take place only under certain temperature and pressure conditions. Figure 5.12 shows that some minerals are restricted to a narrow range of metamorphic conditions whereas others form over wider ranges. Thus, the presence or absence of these *index minerals* in a rock formed by the metamorphism of a shale indicates whether or not certain temperatures and pressures were attained during metamorphism (see "Metamorphic Reactions" box, on p. 87).

In Figure 5.14B, lines are drawn to outline zones based on the appearance of the index minerals. Because these lines approximate the same metamorphic grade, they are called **isograds.** This metamorphic map gives a clearer picture of the metamorphism of the shale. For example, it is easy to see at a glance that the metamorphic grade increases from northeast to southwest, and the index minerals indicate just how intense the metamorphism was.

A. Gneiss × 1

B. Gneiss × 0.3

FIGURE 5.11

Gneissic foliation, or banding, characterizes **gneiss.** For example, *granitic gneiss* has light-colored layers of quartz and feldspar that alternate with dark layers of biotite and/or hornblende. Other minerals such as garnet or sillimanite may be present as well. Layers are a few millimeters to a few centimeters thick. Some gneisses and migmatites are beautiful when cut and polished and are used extensively for monuments and decorative building stones.

FIGURE 5.12

Most of the common minerals of metamorphic rocks form at different metamorphic grades. Thick lines indicate the range of metamorphic grade through which each mineral is possible.

Distinguishing Among Igneous, Sedimentary, and Metamorphic Rocks

After finishing this lab, you will have looked at a number of specific examples of igneous, sedimentary, and metamorphic rocks. What if you were handed a rock and simply asked to tell whether it was igneous, sedimentary, or metamorphic? Could you do it? This seems like an easy thing, yet there are times when it is very difficult. Let's first review some of the major characteristics of each rock type, then consider some specific cases.

Most igneous rocks have nonfoliated, crystalline textures and are composed of several of the common minerals—feldspar, quartz, olivine, augite, hornblende, biotite, or muscovite—in varying proportions.

Texture: Is the rock foliated or nonfoliated?

Foliated

- Slaty foliation (with crystals too small to see): **Slate**
- Phyllitic foliation (with barely visible crystals): **Phyllite**
- Schistose foliation (with crystals large enough to identify): **Schist**
- Gneissic foliation (layers of visible crystals mm to cm thick): **Gneiss**
- Enlarged pods of quartz/feldspar in the midst of swirling light and dark bands: **Migmatite**

Composition: What minerals are present in your foliated rock?

Names of foliated rocks are modified when possible by mineral names, as in quartz-biotite schist. Some use igneous rock names to describe gneisses with similar mineral compositions (e.g., granititic gneiss).

Nonfoliated or weakly foliated.

Composition: What is the non-foliated/weakly foliated rock made of?

- Chiefly calcite or dolomite (H = 3-4, fizzes in acid): **Marble**
- Chiefly quartz (scratches glass): **Quartzite**
- Dark gray to black, shiny, relatively light weight, conchoidal fractures: **Anthracite coal**
- Gray, dark gray, greenish, or black: **Greenstone, amphibolite, soapstone, serpentiinite, or hornfels.** See Table 5.2.
- Somewhat recrystallized versions of conglomerate, graywacke, or chert: **Metaconglomerate, metagraywacke, or metachert.** See Table 5.2.

FIGURE 5.13 (Above)

Flow chart for the identification of metamorphic rocks. Foliated rocks are identified following the charts on the left while nonfoliated rocks follow the charts on the right. These charts should be used together with the more comprehensive information in the text and Table 5.2.

Most common sedimentary rocks have clastic textures. They may contain some of the same minerals as igneous rocks (quartz and feldspar are the most common), but they also may contain non-igneous minerals such as clay or calcite. Sedimentary rocks with crystalline textures are composed of minerals, such as calcite, dolomite, gypsum, or halite, not usually found in igneous rocks. Foliated metamorphic rocks are distinguished from igneous and sedimentary rocks by foliation. Some contain typically metamorphic minerals like garnet, staurolite, or kyanite. Nonfoliated metamorphic rocks usually consist mostly of one mineral, such as quartz or calcite.

Some specific "problem rocks" are the following:

Shale vs. slate. Because a slate forms by low-grade metamorphism of a shale, these two rocks are very similar. In general, a slate is shinier and a bit harder and tougher than a fissile shale. Thin pieces of slate tend to "ring"

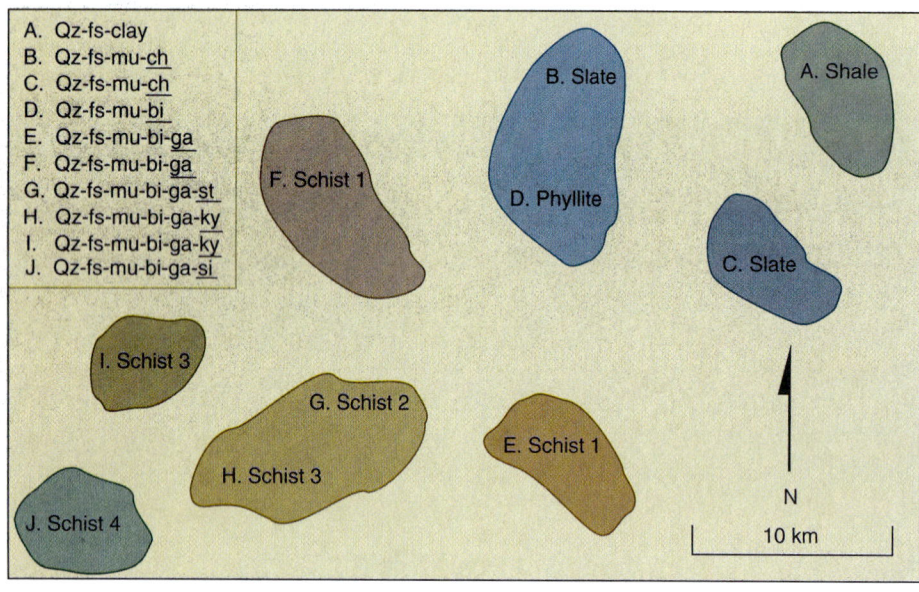

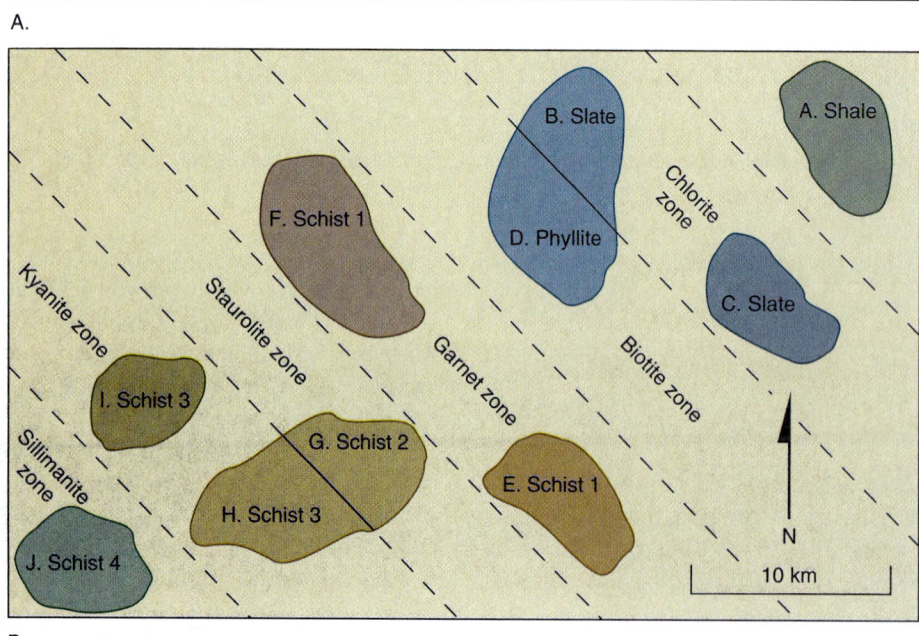

FIGURE 5.14

A. Map of outcrops indicating rock types and mineral assemblages. Qz = quartz; fs = feldspar; mu = muscovite; ch = chlorite; bi = biotite; ga = garnet; st = staurolite; ky = kyanite; si = sillimanite. Index minerals are underlined. B. Same map showing isograds and mineral zones.

when dropped on a table, whereas shale tends to "thunk."

Quartz sandstone vs. quartzite. Like shale and slate, a complete range, or gradation, can be seen from sandstone to quartzite, depending on the degree of metamorphism. One way to distinguish the two is based on the way the rock breaks. If the break cuts through individual quartz grains, the rock generally is quartzite; if it breaks around grains, it is quartz sandstone.

Limestone or dolostone vs. marble. Again, a complete range exists from limestone/dolostone to marble, depending on the degree of metamorphism. Marbles have readily distinguishable crystalline textures, with individual crystals large enough to see rather easily. Although some limestones have crystalline textures, the crystal size generally is quite small. Sometimes, however, it is nearly impossible to distinguish limestones from marbles in hand specimens. They usually can be distinguished in the field by looking at the surrounding rocks; if they are sedimentary, it is likely a limestone; if they are metamorphic, it is likely a marble. Marbles are *not* harder than limestones; they are equally hard. Why?

Graywacke vs. basalt. A dark gray, fine-grained graywacke may be difficult to distinguish from basalt. However, with a hand lens, you generally can see the clastic texture in graywacke and the crystalline texture in basalt. Graywackes may contain quartz, whereas basalts do not.

Mudstone vs. tuff. Some tuffs are so fine grained that they look like mudstones. Tuffs are commonly lighter in color and weight than mudstones, but examination with a microscope may be necessary to distinguish the two.

Conglomerate or sedimentary breccias vs. volcanic breccia. Volcanic breccias or tuffs with pyroclastic fragments may look like conglomerate or sedimentary breccia. The volcanic rocks usually contain only fragments of other volcanic rocks, whereas the particles in conglomerate and sedimentary breccia may be any rock type or a mixture of rock types. Tuffs commonly contain small crystals of common igneous minerals and almost always contain tiny bits of broken volcanic glass or shards (these may be difficult to see). Many tuffs and some volcanic breccias are lighter weight than conglomerates.

Metamorphic Reactions

Metamorphic reactions that have been studied in the laboratory can be used to estimate the actual conditions under which metamorphic rocks formed. This, in turn, provides information about the geologic history of the area. An example of a metamorphic reaction is illustrated in Figure 5.15. That figure shows the temperature and pressure (or depth) conditions at which the reaction *muscovite + quartz = potassium feldspar + sillimanite + H_2O* can occur. Muscovite and quartz can occur together without reacting, as long as the combination of temperature and pressure is to the left of the reaction curve; for example, at 500° and 3 kilobars. If the temperature at that pressure were higher, so that it was to the right of the reaction curve (for example, 700°), then muscovite and quartz could not occur together without reacting to form potassium feldspar, sillimanite, and water. When the rock cools again, the reverse reaction (forming muscovite + quartz) generally does not take place in nature, because the water necessary for the reaction has escaped. Because water and other fluids are driven off during metamorphism, the minerals present in metamorphic rocks generally are those formed at the highest temperature and pressure conditions that were attained.

It is sometimes possible to find rocks that contain both products and reactants. For the reaction shown in Figure 5.15, that rock would contain quartz, muscovite, potassium feldspar, and sillimanite. Such a rock would have been subjected to conditions right on the reaction curve (for example, about 630° at a pressure of 3 kilobars). If a series of outcrops with the same products and reactants could be located, they would all represent the same metamorphic grade. A line on a map connecting them would be an isograd, a line of equal metamorphic grade, generally defined by a specific reaction or the appearance of a specific mineral.

By using numerous reactions like the one shown in Figure 5.15, it is possible to determine the temperature and pressure conditions that must have existed at the time of metamorphism. Because pressure, especially, can be related approximately to depth (see Fig. 5.1), it is also possible to figure out how deep the rocks must have been when they were metamorphosed.

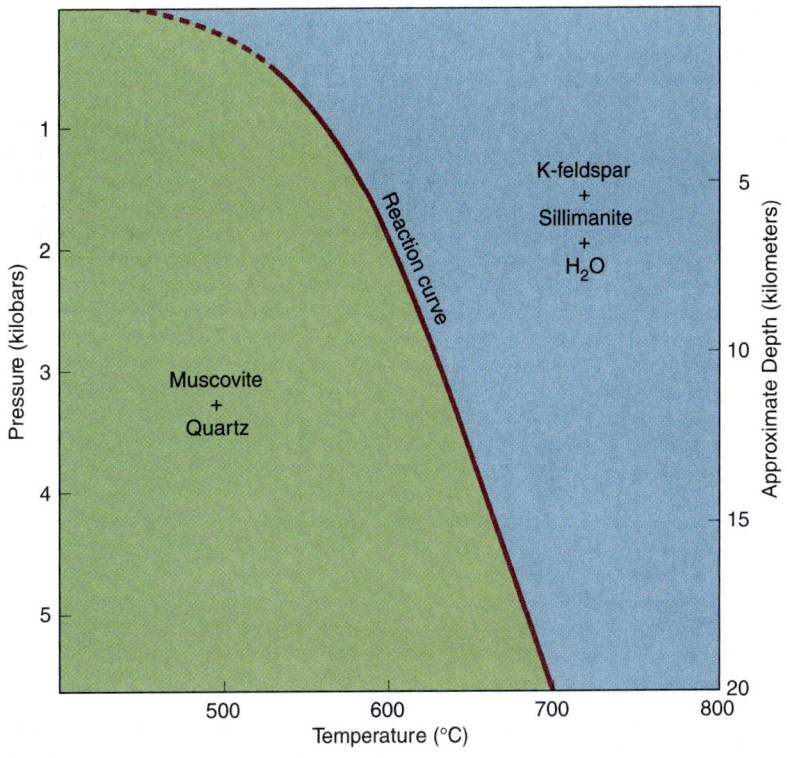

FIGURE 5.15

The reaction *muscovite + quartz = K-feldspar + sillimanite + H_2O* occurs when the combination of temperature and pressure (depth) fall on the reaction curve. If temperature and pressure are left of the curve, muscovite and quartz are present; if they are right of the curve, K-feldspar, sillimanite and H_2O are present.

Hands-On Applications

This lab continues the application of the scientific method to the study of rocks. Problems focus on identification and interpretation based on observation of metamorphic rock samples. You will see how correct identification will enable you to infer the parent of the rock you have identified and the approximate temperature-depth conditions during metamorphism.

Objectives

If you complete all the problems, you should be able to:

1. Identify the major minerals in metamorphic rocks.
2. Identify a metamorphic rock.
3. Distinguish between high- and low-grade metamorphic rocks.
4. Distinguish between foliated and nonfoliated metamorphic rocks.
5. Distinguish between slaty, schistose, and gneissic foliations.
6. Suggest a probable parent for a given metamorphic rock.
7. Determine whether a rock is igneous, sedimentary, or metamorphic.
8. Construct isograds on a map, given the appropriate data.

Problems

1. Metamorphic rocks are subdivided into two major textural groups: FOLIATED and NONFOLIATED. Divide the samples provided for you into these two categories.

    Foliated

    Nonfoliated

2. Which of the samples with foliated textures have the following specific types of foliation?

    Slaty

    Phyllitic

    Schistose

    Gneissic

3. Many foliated metamorphic rocks form by the metamorphism of shale; the more intense (or *higher grade*) the metamorphism, the coarser the crystals contained in the rocks.
    a. Select from among your samples those that may have been derived from shale, and arrange them according to metamorphic grade (see Table 5.2).
        Lowest grade:

Intermediate grade: B

Highest grade: B

b. Assuming that the rocks you selected formed in the vicinity of a magmatic zone at a convergent plate boundary, what were the approximate temperatures and depths of each grade at the time of metamorphism (see Fig. 5.1)?

4. What are the *parent rocks* of the samples specified by your instructor (see Table 5.2)?

5. Identify the samples of metamorphic rocks provided for you. You may find it helpful to use the tear-out Metamorphic Rock Identification Worksheet, which is at the end of this chapter.

6. How can you distinguish between the following similar-looking rocks?
    a. Marble and quartzite

    b. Slate and phyllite

    c. Schist and gneiss

In Greater Depth

7. When confronted with scattered outcrops of a variety of metamorphic rocks, a geologist must try to make some sense of their distribution in order to understand their origin. One of the first things to do is to make a map of the outcrops and indicate what rocks and minerals are present in each. Figure 5.16 shows outcrops of several kinds of rocks metamorphosed to varying degrees; colors indicate the type of parent rock. Between outcrops, the metamorphic rocks are covered with soil and other surficial sediment.
    a. Construct isograds and label the distinct metamorphic mineral zones using the geologic map and brief rock descriptions of Figure 5.16.

    b. Where on the map was metamorphism most intense? Least intense?

    c. Which parent rocks best record differences in metamorphic grade?

FIGURE 5.16
This map and rock descriptions are to be used with Problem 7. The map shows the distribution of outcrops with different colors representing different parent rocks. The numbers correspond to the samples described in the list below the map.

Metamorphic Rock Identification Worksheet

Sample Number	Texture	Mineral Composition	Other Properties (e.g., color)	Rock Name	Metamorphic Grade/ Probable Parent Rock(s)
A	Foliated Slaty	Too fine to ID	Breaks into sheets	Slate	Shale/mudstone low grade
B	Foliated Slaty	Quartz Biotite Muscovite Biotite	Sparkles	Quartz Biotite schist Muscovite Biotite schist	Shale/mudstone intermediate
C	Both	Hornblende	Needle like hornblend crystals	Phyllite	Shale/mudstone intermediate
D	Gneissic	Dark layers hornblend Light layers: quartz k-spar	Light + dark bands	Gneiss	Shale/mudstone granite
E	Non-Foliated	Quartz	Scratches glass	Quartzite	Quartz Sandstone
F	Non-Foliated	Calcite	Reacts to acid, softer than glass	Marble	Limestone/Dolostone
G	Non-Foliated	Quartz	Scratches glass	Quartzite	Quartz Sandstone

Metamorphic Rock Identification Worksheet

Sample Number	Texture	Mineral Composition	Other Properties (e.g., color)	Rock Name	Metamorphic Grade/ Probable Parent Rock(s)

PART III
Maps and Images

Portion of the Cumberland, MD, 7.5' topographic quadrangle, shaded to emphasize topography

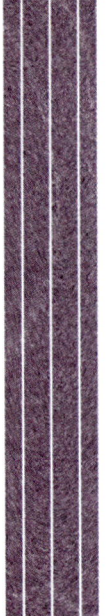

CHAPTER 6

Topographic Maps

Materials Needed
- Pencil and eraser
- Calculator
- Metric ruler
- Topographic quadrangle map (provided by your instructor)

Introduction

Maps have many uses in everyday life as a means by which data of diverse kinds can be succinctly displayed. In geology, we are interested in the natural features on or below the Earth's surface. This chapter describes the basic features of maps, including the bases for describing locations and land areas; the standard map elements, such as scales and symbols; and the use and interpretation of contours to show the third dimension.

A map is a convenient representation of an area or group of features. A **road map** typically shows roads, political boundaries, cities, and such natural features as rivers and lakes. A **topographic map** shows **topography** (the three-dimensional shape of a surface, including mountains, hills, valleys, and so forth) and frequently includes such human-designed **cultural features** as roads, buildings, dams, and political boundaries. **Geologic maps** show the distribution of different rock bodies exposed at the Earth's surface and frequently include features found on topographic maps. **Land-use maps** combine elements of topography, geology, and soil type to indicate areas of potential flooding, landslides, or sensitivity to earthquakes; areas of good or poor soil; zones impacted by hazardous waste contamination; or areas that are ecologically sensitive or unique. As you will see in the rest of this manual, maps are extremely useful to people interested in how the Earth works and how human projects impact or are impacted by the natural world. In addition, they are used by hikers, bikers, hunters, climbers, and others who take their outdoors recreation seriously. If you continue in a geology or environmental program, you may encounter **GIS (Geographic Information Systems).** GIS software enables academic, government, and private sector users to combine the information from a variety of maps, aerial photographs, and satellite images into one easily understood graphic. This and the next chapter thus lay the basic groundwork necessary for the rest of this course and for further studies in the field of GIS.

Map Coordinates and Land Subdivision

Latitude-Longitude System

Because most maps show only a small portion of the Earth, they have to be located with respect to the rest of the Earth. This is done by establishing a coordinate system or network of lines. A system used the world over is based on east-west lines called lines of latitude and north-south lines called lines of longitude.

Latitude measures distance north or south of the equator. The lines of latitude, also called **parallels,** form a series of parallel circles running eastwest (horizontally) around the globe. The equator represents the 0° latitude line. Other parallels are set at angular intervals measured north or south of the equator, as shown in Figure 6.1A. A latitude line 40° north of the equator is

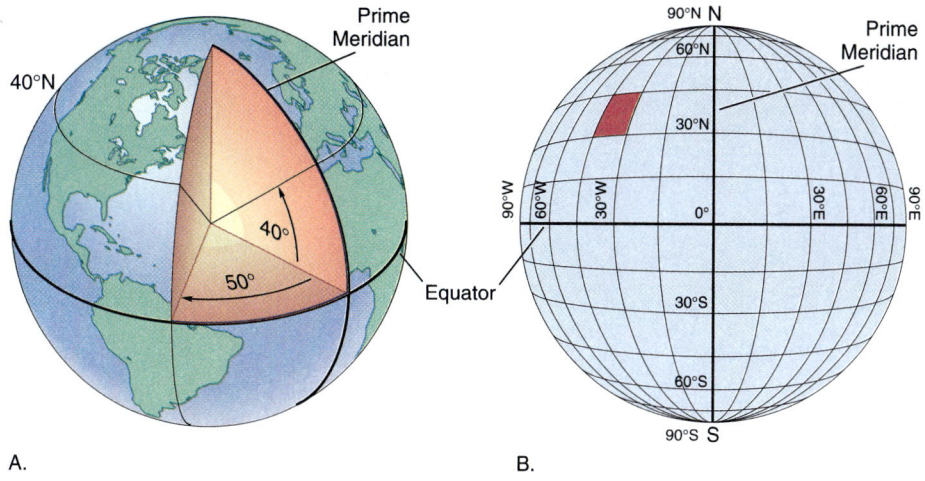

FIGURE 6.1
A. Cutaway view of the Earth. The east-west line 40° north of the equator is latitude 40° N; the north-south line 50° west of the prime meridian is longitude 50° W. B. Lines of latitude parallel the Earth's equator, while lines of longitude intersect at north and south geographic poles. The shaded area, bounded by lines of latitude and longitude, is a quadrangle.

termed latitude 40° N. The **geographic poles** are at latitudes 90° N and 90° S.

Longitude measures distance east or west of the **Prime Meridian.** Lines of longitude, also termed **meridians,** form a series of circles running north-south (vertically) and intersecting at the geographic poles. The **Prime Meridian** is the north-south line passing through the Royal Observatory in Greenwich, England; it is defined as 0° longitude. The other meridians are set at angular intervals east or west of the Prime Meridian, as shown in Figure 6.1A. A longitude line 50° west of the Prime Meridian is termed longitude 50° W; one 90° east of the Prime Meridian, or one-quarter the way around the globe, is longitude 90° E. The east and west meridians meet at 180° on the opposite side of the planet. The 180° meridian corresponds to the International Date Line, except for where the date line shifts to avoid causing time-zone problems in islands or island groups cut by the 180° line. The latitude and longitude lines make a single grid network that covers the globe (Fig. 6.1B).

The measurement of angles from the equator (latitude) and from the Prime Meridian (longitude) is most commonly done in *degrees* (symbol °). A circle is divided into 360°. One degree is divided into 60 *minutes* (60'), and each minute into 60 *seconds* (60"). Maps also may indicate angular measurements in *mils*. One mil = 1/6400 of 360°, or 0.05625°. When measured on the Earth's surface, one degree of latitude (as measured along a meridian) is approximately 111 km (69 miles), and one degree of longitude (as measured along a parallel) varies from about 111 km at the equator to 0 km at the poles, where the meridians intersect.

The maps used most frequently show small segments of the network of latitudes and longitudes. Because the Earth is nearly spherical, these segments (like the shaded one in Fig. 6.1B) are actually curved surfaces that are represented on maps as flat surfaces. To do this, the curved, three-dimensional surfaces must be projected onto a two-dimensional sheet of paper. Many ways have been developed to accomplish this—names such as "Mercator projection" or "polyconic projection" may be familiar to you—but all unavoidably result in some sort of distortion.

The U.S. Geological Survey (USGS) has made most of the accurate maps of this country. These maps are bounded by latitudes and longitudes, both of which are usually separated by 1°, ½° (= 30'), ¼° (15'), or ⅛° (7½') intervals. Such maps cover rectangular-shaped areas called **quadrangles,** which are generally named after the largest town or most prominent geographic feature in the area (for example, Mt. Shasta, California 7½' Quadrangle). The numerical values of the latitudinal and longitudinal boundaries are given on each corner of the map (Fig. 6.2). In addition, intermediate values are indicated on the margins of a map (Fig. 6.2).

A point on a map can be located by referring to the latitude and longitude of the point (for example, latitude 43° 5½'N, longitude 132° 15½'W). By convention, latitude (north-south) is given first, longitude (east-west) second.

Note that minutes and seconds can also be expressed as decimal equivalents of a degree. This is commonly done on global positioning system (GPS) receivers and in software applications. A latitude of 43° 5' 8", for example, can be converted to a decimal as follows: $5' = 5/60^{th}$ of a degree or $0.083°$, and $8" = 8/3600^{th}$ of a degree = $0.002°$ (there are $60 \times 60 = 3600$ seconds in a degree). Thus $43° 5' 8" = 43° + 0.083° + 0.002° = 43.085°$.

Two other coordinate systems are commonly used in the United States. The Universal Transverse Mercator (UTM) system is extremely useful for precisely locating points on quadrangle maps both in the United States and around the world. The U.S. Public Land Survey System was designed to efficiently describe land areas in the states outside of the original 13 colonies.

It is worth learning and using a coordinate system because it is a permanent way of describing locations. For example, older descriptions of mineral or fossil locations commonly refer to landmarks. Unfortunately, these old sites may now be difficult to find because road intersections, houses, small bridges, old trees, and railway lines have been moved or disappeared in response to ongoing development. A coordinate system allows a state to permanently keep track of the locations of abandoned oil wells, toxic waste sites, sealed mine shafts, and places hosting endangered species or breeding pairs. It allows geologists to describe important rock localities. And it allows hikers to precisely locate trailheads, remote camp sites, and other places worth returning to.

Universal Transverse Mercator (UTM) System

The UTM system, which is used throughout the world and in many global positioning system (GPS) units, is based on a grid of 60 north-south zones, each 6° wide; Figure 6.3A shows these zones in the U.S. The zones are numbered from west to east, beginning at the International Date Line; the zone number is indicated in the small print in the lower left corner of a USGS quadrangle

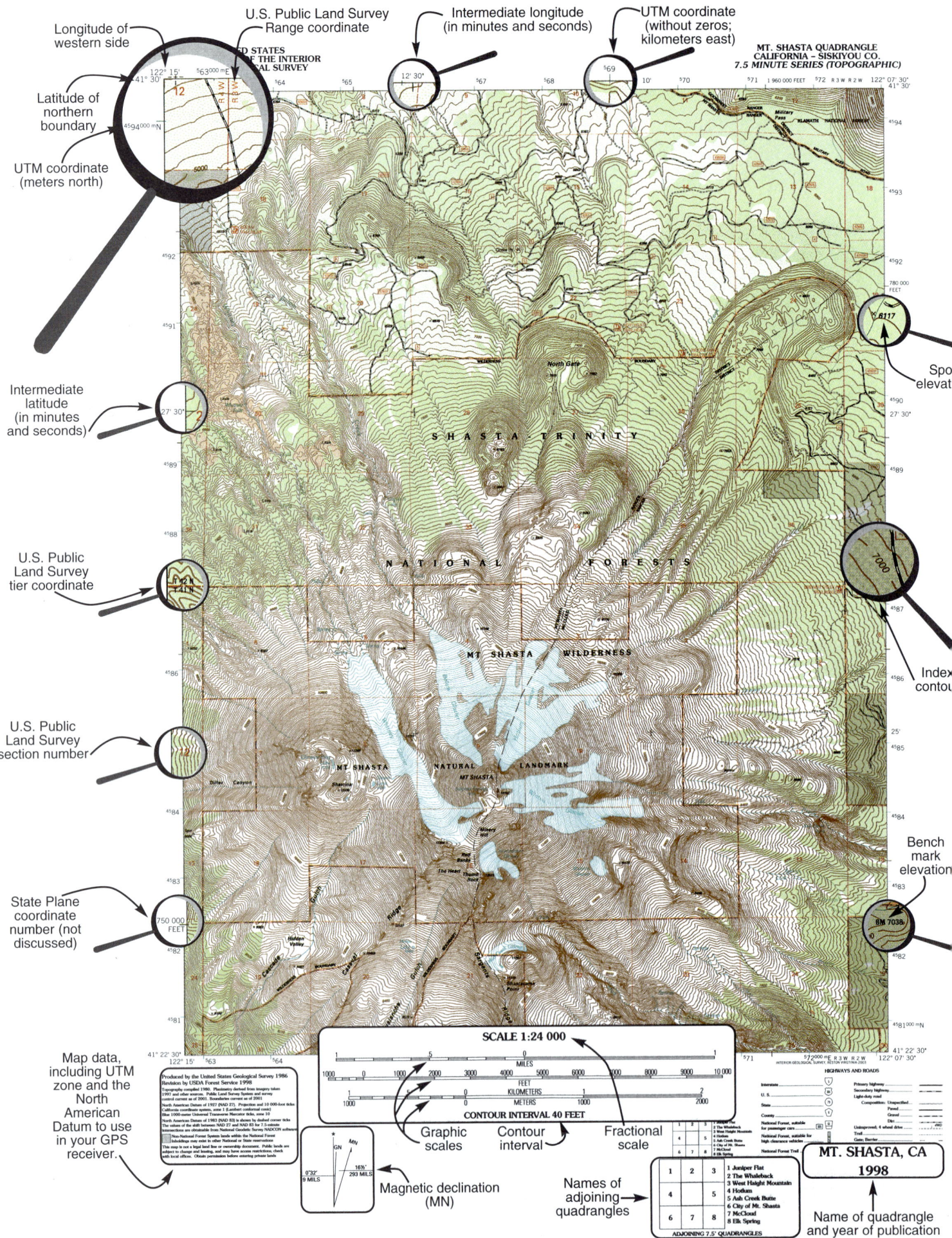

FIGURE 6.2
Reduced copy of the Mt. Shasta, California, 7½-minute quadrangle, with principal map features highlighted and magnified.

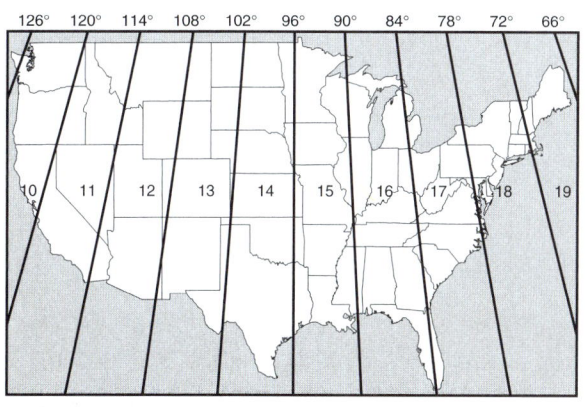
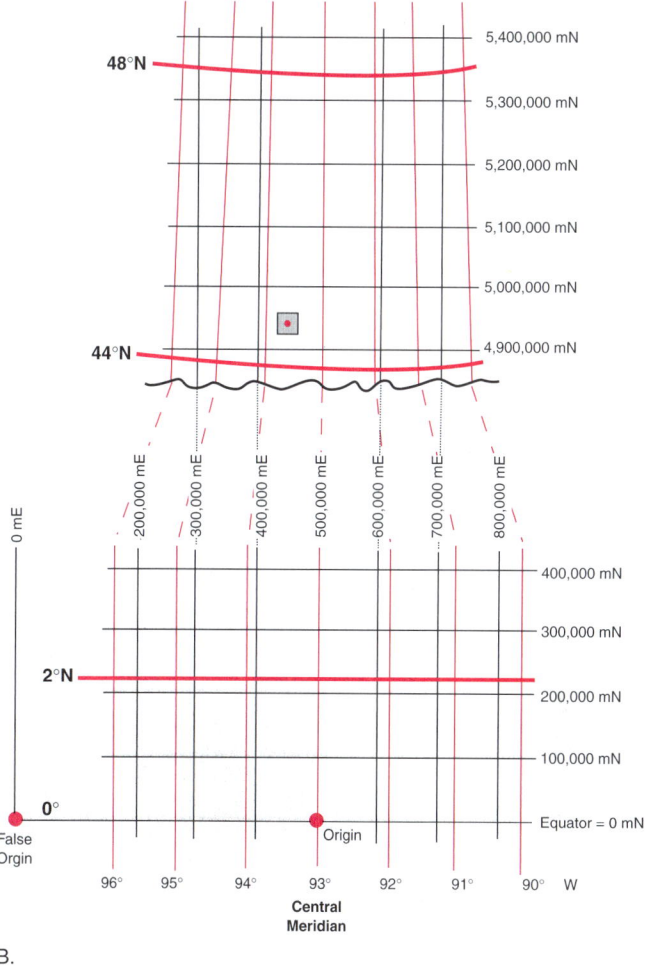

FIGURE 6.3
A. UTM-grid zones in the United States. B. Example showing two parts of UTM grid (black lines) in zone 15 superimposed on lines of latitude and longitude (red lines). Note that lines on the two grids are parallel near the equator but not at higher latitudes. This is because most latitude and longitude lines are projected as curves, whereas UTM lines are drawn as straight lines on that projection. The point in the shaded area is located in Figure 6.4.

map (Fig. 6.2). The grid system for each zone in the northern hemisphere has its origin at the intersection of the equator and its own central meridian, as shown for zone 15 (90° to 96°) in Figure 6.3B. A metric grid, with lines intersecting at right angles, is developed from this origin on a transverse-Mercator-type map projection. The grid is shown with black lines on newer USGS maps. Labels of east-west running lines are based on the number of meters north of the equator. Labels of north-south lines use a *false origin* of 0 mE, located 500,000 m west of the true origin; this makes it easier to locate features near zone boundaries.

Features are located by giving their UTM coordinates. UTM coordinates are shown in numbers of two sizes along the margins of USGS maps (Figs. 6.2, 6.4). The numbers are the distances in meters from the origins. For example, $^{49}43^{000}$ mN describes an east-west line 4,943,000 m (4,943 km) north of the equator. The east-west line one km (1000 m) to the north is $^{49}44^{000}$ mN; the 44 is larger for easy reference purposes. A location description is always given in the same sequence, as shown in Figure 6.4: north-south coordinate *(northings)*, east-west coordinate *(eastings)*, zone number, and hemisphere (north or south). Because north-south coordinates increase in value from bottom to top in the northern hemisphere, and east-west coordinate increase in value from left to right, the informal rule "read up-right" is a useful mnemonic device. An example of determining the UTM coordinates of a point between grid lines is given in Figure 6.4.

U.S. Public Land Survey System

In the western two-thirds of the United States, land is subdivided by the U.S. Public Land Survey System (commonly called the Township-Range System). The system was started in 1785, when the old Northwest Territory (Lake Superior region) was opened to homesteading, and has been used for ordinary and legal land descriptions since. This method subdivides land into 6- × 6-mile squares called *townships;* these are further subdivided into 1- × 1-mile squares called *sections.*

The starting point for subdivision is the intersection of selected latitude and longitude lines. The starting latitude is the **baseline,** and the starting longitude is the **principal meridian.** Baselines and principal meridians are established for a number of areas in the United States, such as the one shown in Figure 6.5A. Lines drawn six miles apart and parallel to the baseline form east-west rows called **tiers.** North-south lines parallel to the principal meridian and six miles apart form north-south columns called **ranges** (Fig. 6.5B). The squares formed by the intersection of tiers and ranges are called **townships.** Each township is approximately six miles square and has an area of about 36 square miles. Political townships, usually named after the largest town within the area at the

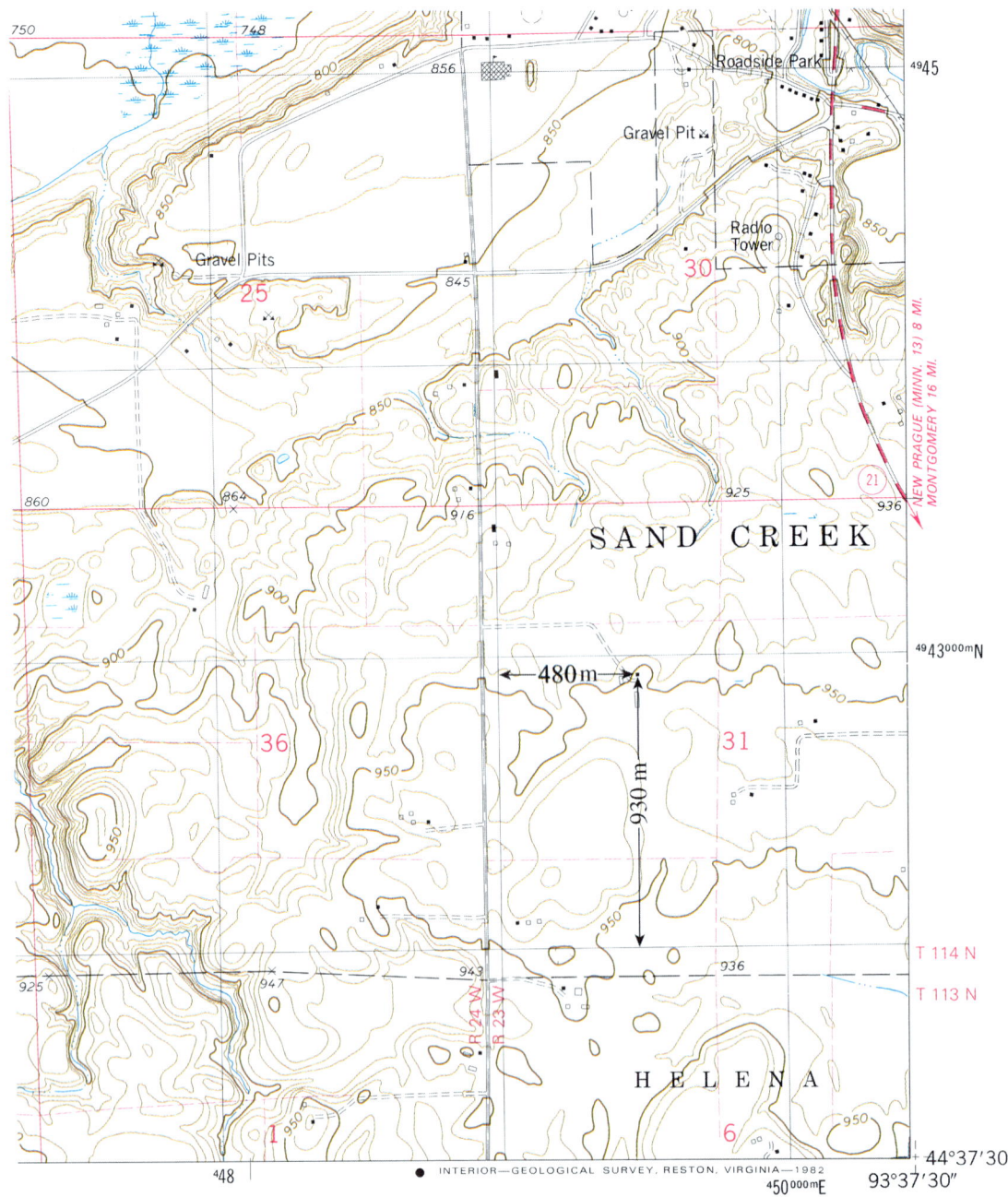

FIGURE 6.4

The UTM grid is shown with thin black lines. Along the side, ⁴⁹45 is shorthand for 4,945,000mN and ⁴⁹43⁰⁰⁰ᵐN for 4,943,000mN. Similarly, along the bottom, ⁴48 is shorthand for 448,000mE and ⁴⁵⁰⁰⁰⁰ᵐE for 450,000mE. On a full-sized map, the zone number is found in the lower left corner in the fine print. This map falls within zone 15.

Example: The house (small black square) falls within a 1,000-m square defined by grid lines 4,942,000mN (south side), 4,943,000mN (north), 449,000mE (west), and 450,000mE (east). Following the 'read up-right' rule, measure in millimeters the distance from the 4,942,000mN line to the house and then the total distance to the 4,943,000mN line. The house is located 39 mm out of a total 42 mm between grid lines. In percent, 39/42 = 0.93 or 93%. Since the grid distance represents 1,000 m, the house is located 0.93*1000 = 930 m above the southern line, or at 4,942,930mN. Similarly, the house is located 20mm/42mm or 48% of the way east of the 449,000mE line. This equals 449,480mE. The location of the house, to within a 10-m square, is formally given as: 4,942,930mN; 449,480mE; Zone 15; northern hemisphere.

time they were designated (for example, Baraboo Township), may or may not coincide with Public Land Survey townships.

Tiers and ranges are numbered by reference to the baseline and principal meridian (Fig. 6.5B). The first tier north of the baseline is Tier 1 North (abbreviated T1N); one in the fifth tier to the north is T5N, and so forth. Ranges are numbered to the east of the principal meridian (for example, R5E) and to the west (R2W). A Public Land Survey township (like the shaded one in Fig. 6.5B) is located using tier-range coordinates—T3S, R4E. *NOTE:* Tier is always written first; Range second.

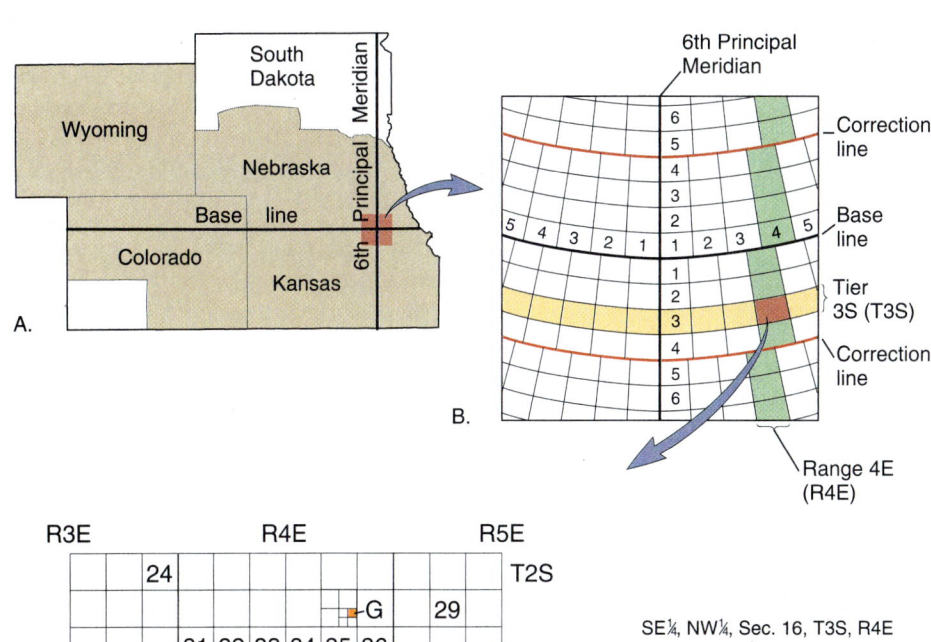

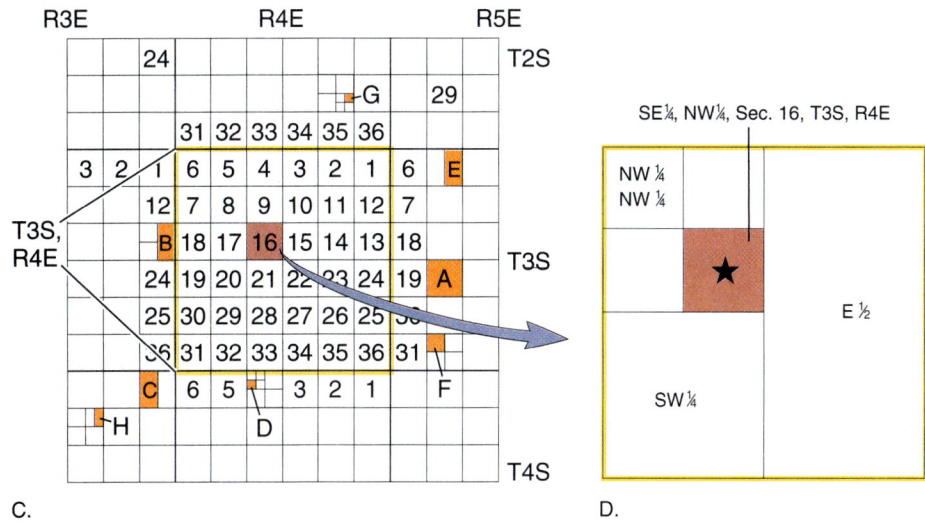

FIGURE 6.5
U.S. Public Land Survey subdivision, illustrated by successively smaller areas, A–D. A. Example of a baseline and principal meridian in the western United States. B. From a starting point at the intersection of a principal meridian and a baseline, 6-mile-wide tier and range bands subdivide land into 36-square-mile townships. C. Townships are subdivided into 36 1-square-mile sections. D. Sections can be divided into halves, quarters, eighths, or other fractions.

Because lines of longitude (meridians) converge toward the poles, it is impossible to maintain squares that are six miles on a side. Thus, a correction is made at every fourth tier line (labeled *correction line* on Fig. 6.5B), and new range lines six miles apart are established. The correction restores townships immediately north of the line to their proper size.

Each six-mile-square township is subdivided into thirty-six, 1- × 1-mile squares, called **sections**, which are numbered in a specific sequence (Fig. 6.5C). Each section consists of 640 **acres**. A section is subdivided into halves, quarters, eighths, sixteenths, and so on (Fig. 6.5D). A sixteenth of a section is 40 acres.

Points are located according to the smallest subdivision required. In Figure 6.5D, the star is located, to the nearest 40 acres, in the SE¼, NW¼, Sec. 16, T3S, R4E. *Locations are always written from the smallest unit to the largest, and tier is written before range.*

Section numbers and tier and range values are written in red on USGS topographic maps (see Fig. 6.4).

Elements of Maps

Scale

For a map to be usable, it must have a **scale**, so the user can tell the size of the area represented or the distance between various points on a map. Three types of scales are in common use.

A **ratio**, or **fractional scale**, shown at the bottom of Figure 6.2, is the ratio between a distance on a map and the actual distance on the ground. The ratio scale on Figure 6.2 is 1:24,000 (or 1/24,000), which means that one unit (for example, an inch) on the map equals 24,000 of the same units on the ground.

A **graphic scale** usually consists of a line or bar subdivided into divisions corresponding to a mile or kilometer (see Fig. 6.2). One mile or kilometer segment on the scale bar is commonly subdivided to allow more precise measurements of distance. The subdivided units are commonly placed to the left of zero on a scale bar, as in Figure 6.2. Thus, the total length of these scales is one unit longer than the numbers on the right end of the scales. This scale is helpful because it is readily visualized and will stay in true proportion if the map is enlarged or reduced. The graphic scale provides a convenient way of measuring distances on a map. To find the distance between two points on a map, lay a strip of paper between the points and make pencil marks on the strip adjacent to each point. Then lay the paper along the desired graphic scale at the bottom of the map to determine the distance.

A **verbal scale** is used to discuss a map but is rarely written on it. An example is "one inch on the map represents, or is proportional to, one mile on the ground" or, simply, "one inch to one mile." Because one mile equals 63,360 inches, the common fractional scale 1:62,500 corresponds closely to the verbal scale "one inch to one mile." Many U.S. maps, and essentially all foreign maps, use metric scales, making common fractional scales easily convertible to verbal scales. For example, for scales of 1:50,000, 1:100,000, and 1:250,000, one centimeter represents 0.5, 1.0, and 2.5 kilometers, respectively.

USGS 4° quadrangle maps are drawn at a fractional scale of 1:1,000,000; 2° quadrangles at 1:500,000; 1° at 1:250,000; 15' at 1:62,500 or 1:50,000; and 7½' at 1:24,000 or 1:25,000. Both graphic and fractional scales are shown at the bottom center of the map (see Fig. 6.2).

To show larger or smaller areas of the Earth's surface on conveniently sized maps, different scales must be used. For example, it may be possible to show a small city on a map where one inch on the map represents 12,000 inches (1000 ft) on the ground. This map would have a scale of 1:12,000. However, to show a mid-sized state, such as Indiana, on a map of similar size, the scale would have to be much smaller: one inch on the map to 500,000 inches (approximately eight miles) on the ground. In general, the larger the area shown on a map, the smaller the scale of the map.

Converting Among Scales

To convert from a verbal to a fractional scale:

1. Convert map and ground distances to the same units.
2. Write the verbal scale as the fraction:

$$\frac{\text{Distance on the map}}{\text{Distance on the ground}}$$

3. Divide both numerator and denominator by the value of the numerator:

$$\frac{\text{Distance on map / distance on map}}{\text{Distance on ground / distance on map}}$$

Example: Convert the following verbal scale to a fractional scale: 2.5 inches on the map represents 5000 feet on the ground.

1. Convert both map and ground distances to the same units, inches: $5000 \times 12'' = 60,000''$. The verbal scale is now 2.5 inches on the map represents 60,000 inches on the ground.
2. Write the verbal scale as the fraction:

$$\frac{2.5'' \text{ (distance on map)}}{60,000'' \text{ (distance on ground)}}$$

3. Divide the numerator and denominator by the value of the numerator:

$$\frac{2.5''/2.5''}{60,000''/2.5''} = \frac{1}{24,000} \text{ or } 1:24,000.$$

To convert from a fractional scale to a verbal scale:

1. Select convenient map and ground units to relate to each other (for example, inches and miles or centimeters and kilometers).
2. Express fractional scale using the map units (inches or centimeters).
3. Convert the denominator to the ground units (miles or kilometers).
4. Express verbally as "one inch [or one centimeter] equals X miles [or kilometers]."

Example: Convert a fractional scale of 1:62,500 to a verbal scale of one map inch equals x miles on the ground.

1. Units to be related are inches and miles.
2. $1:62,500 = 1''/62,500''$
3. Convert 62,500" into miles by dividing by the number of inches in one mile. One mile = 5280 feet and 1 foot = 12 inches. So, 1 mi = $5280' \times 12'' = 63,360''$. Working out the division:

$$\frac{62,500 \text{ inches}}{(63,360 \text{ inches per mi})} = 0.986 \text{ mi}$$

4. Expressed verbally, one inch on the map equals 0.986 mile on the ground.

Magnetic Declination

By convention, maps are usually drawn with north at the top. North on a map refers to **true geographic north.** At most places on the Earth, however, a compass needle does not point toward the geographic north pole but toward the **magnetic north pole.** The magnetic north pole is in the Canadian Arctic, but its exact position changes. For example, in 1955, it was located north of Prince of Wales Island near latitude 74° N, longitude 100° W; its 1994 location was on Ellef Ringnes Island at 78.3° N, 104.0° W.

The angular distance between true north and magnetic north is the **magnetic declination.** Because the location of the magnetic pole changes, the magnetic declination varies slightly with time at most locations. For very accurate work, these variations must be taken into account by adjusting the compass dial. Magnetic declination is shown at the bottom of most USGS maps by two arrows (see Fig. 6.2), one pointing to true north (commonly marked with a star, or T.N.) and one pointing toward magnetic north (commonly marked M.N.). The angular separation between them (the magnetic declination) also is given. When stating the magnetic declination of a map, it is always necessary to indicate whether the arrow pointing to the magnetic pole is east or west of the geographic pole. If it is east, the declination is stated as so many degrees east, for example, 2½° E. Most maps also have an arrow pointing toward **G.N.,** the location of the **grid north** direction for the Universal Transverse Mercator (UTM) grid system (see Fig. 6.2).

Symbols

Standardized symbols and colors are used on USGS maps to designate various features. Cultural features (those made by people) are generally drawn in black (purple in newly revised maps); forests or woods are shown in green (they are not always represented); blue is used for bodies of water; brown shows elevation (contours), some mining operations, and beaches or sand areas; red is used for the better roads and some land subdivision lines. See Figure 6.6 for symbols and Figure 6.2 for some examples.

Topographic Maps

A topographic map shows the size, shape, and distribution of landscape features; that is, the **topography,** or configuration of the land surface. It is a map to which a third dimension, elevation, has been added. Elevation differences can be shown by means of contour lines, shading or coloring, or short, closely spaced lines drawn to schematically indicate mountains or steep slopes. The emphasis here is on contour lines, because they provide the most precise means for depicting the third dimension. Figure 6.7 shows how the topography of the area sketched in the top diagram is depicted with contour lines in the bottom diagram. The section opener (p. 95) shows a contour map to which shading has been added.

Contour Lines

A contour line is a line on which all points have the same elevation. It is shown in brown on USGS topographic maps (see section opener, p. 95). The **elevation** or altitude of a point is the vertical distance between that point and a fixed **datum** (a point of reference). The datum from which most elevations are measured is mean or average sea level, which by definition has an elevation of zero. Figure 6.7 shows an area along a sea coast, with the sea at its average elevation of zero feet. Because the edge of the shore is everywhere at an elevation of zero feet, the shoreline coincides with the zero-foot contour. If sea level rose by 100 feet,

FIGURE 6.6

Standard symbols on USGS maps. Source: Data from the U.S. Geological Survey.

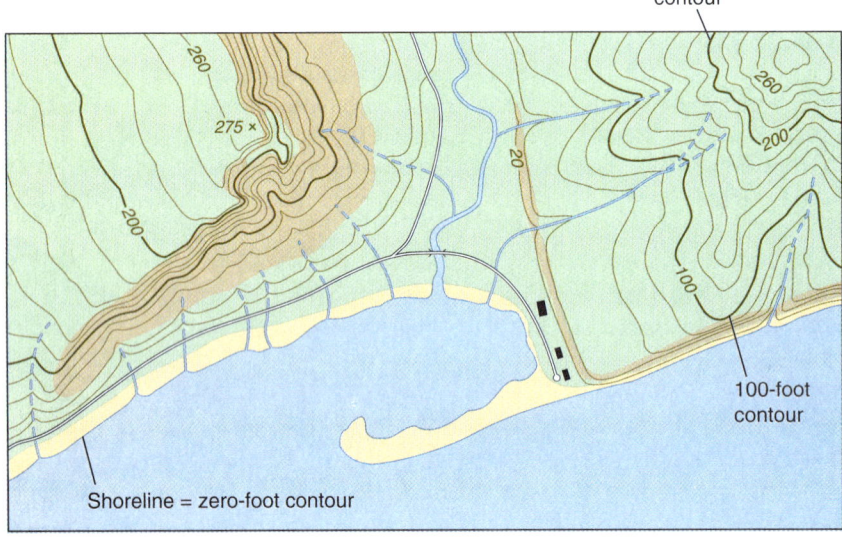

FIGURE 6.7
The area sketched in the top diagram is shown as a topographic map in the bottom diagram. Contour lines on the bottom diagram are drawn at intervals of 20 feet, starting with 0 at mean sea level. Note how contours bend upstream where they cross streams. Source: U.S. Geological Survey.

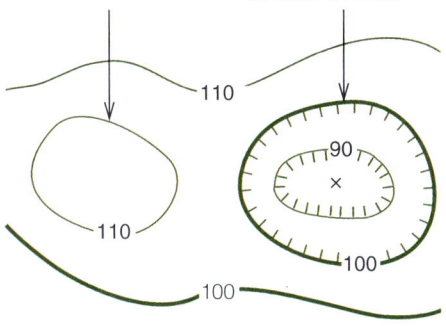

FIGURE 6.8
Elevations of normal and depression contours lying between two other contours. C. I. = 10 ft. The elevation of the closed contour on the left is 110 feet, the same as the contour upslope from it. The elevation of the outer *depression* contour, on the right, is the same as the *normal* contour downslope, namely 100 feet. The elevation of the inner depression contour is 10 feet lower, or 90 feet. The bottom of the depression is less than 90 feet, but more than 80 feet, because there is no 80-foot contour. The half-way elevation at X in the bottom is 85 ± 5 feet.

the shoreline would everywhere coincide with the 100-foot contour shown in Figure 6.7; if it rose 200 feet, it would coincide with the 200-foot contour.

Depression Contours

Depression contours are closed contours with hachures (short lines perpendicular to the contour line) pointing to lower elevations within a depression (Fig. 6.8). They generally encircle small depressions.

Characteristics of Contour Lines

The following characteristics of contour lines, most of which are illustrated in Figure 6.7, govern the construction and reading of contour maps:

1. Every point on the same contour line has the same elevation.
2. A contour line always rejoins or closes upon itself to form a loop, although this may or may not occur within the map area. Thus, if you walked along a contour, you would eventually get back to your starting point.
3. Contour lines never split.
4. Contour lines never cross one another; however, if there is a steep cliff, they may appear to overlap because they are superimposed on one another.
5. Slopes rise or descend at right angles to any contour line.
   - Evenly spaced contours indicate a uniform slope
   - Closely spaced contours indicate a steep slope
   - Widely spaced contours indicate a gentle slope
   - Unevenly spaced contours indicate a variable or irregular slope
6. Contours usually encircle a hilltop; if the hill falls within the map area, the high point will be inside the innermost contour (however, see discussion of depression contours).
7. Contour lines near the tops of ridges or bottoms of valleys always occur in pairs having the same elevation on either side of the ridge or valley.
8. Contours always bend upstream when they cross valleys. Because water runs downhill, this fact allows the rapid recognition of high and low areas on a contour map (Fig. 6.7).
9. If two adjacent contour lines have the same elevation, a change in slope occurs between them. For example, adjacent contours with the same elevation would be found on both sides of a valley bottom or ridge top.
10. **Depression contours** have the same elevations as the normal (unhachured) contours immediately downhill (Fig. 6.8).

Contour Interval

Contour lines are drawn on a map at evenly spaced intervals of elevation. The difference in elevation between two consecutive contours on the same slope is called the **contour interval** (abbreviated C. I.). It is a constant for a given map, unless otherwise stated, and is usually given at the bottom of the map just below the graphic scale (see Fig. 6.2).

The choice of contour interval depends on (1) the level of detail the topographer wishes to portray, (2) the scale of the map, and (3) the range in elevation or **relief** of the area to be mapped. Obviously, Florida and the Rocky Mountains cannot be mapped at the same scale with the same contour interval. Much of the Rocky Mountain area is mapped with a 100-foot interval, whereas a 2- or 5-foot interval is common for Florida.

Index Contour

As a general rule, every fifth contour, starting from sea level, is an **index contour,** which is drawn as a heavy line and labeled with its elevation (see Figs. 6.2 and 6.7). Contours between index contours are usually not labeled.

Reading Elevations

The elevations of points that fall on a contour line are the same as the elevation represented by that contour line. On most maps, only index contours are labeled. As you move uphill across unlabeled contours, keep track of the elevation by adding the value of the contour interval for every contour crossed. In Figure 6.9, moving from the 200' index contour to point X crosses two contours: 200' + 20' + 20' = 240' elevation. Similarly, you subtract contour intervals when moving downhill.

The elevation of a point that does not fall on a contour must be estimated. An estimate can be made by interpolation, assuming the slope between adjacent contours is uniform. For example, a point one-quarter of the way between contours with elevations of 200 and 220 feet (C. I. = 20 feet) would have an elevation of about 205 feet. However, slopes are often not uniform, so another approach is to give the *halfway elevation* between the two contours. A **halfway elevation** is the elevation halfway between the values of adjacent contours; thus, the elevation of a point between contours can be stated as the halfway elevation plus or minus one-half the contour interval. The halfway elevation is a shorthand way of saying the elevation is between two contours. Figure 6.9 provides examples.

As illustrated in Figure 6.8, a normal closed contour that lies between a higher and a lower contour always takes the same elevation as the higher one. A depression contour in the same situation takes the same elevation as the lower one.

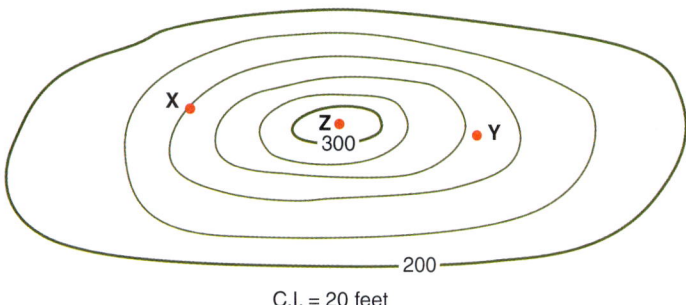

C.I. = 20 feet

FIGURE 6.9
Reading elevations from a contour map with contour interval of 20 feet. The elevation of X is 240 feet, because it falls on a contour with that elevation. Point Y falls between the 240- and 260-foot contours, so its elevation must be between those values. Its horizontal position is about three-quarters of the way between the two, and assuming a uniform slope, you might estimate an elevation of 255 feet. The half-way elevation method makes no assumptions about uniformity of slope. Point Y has a *half-way elevation* of 250 ± 10 feet: 250 is half-way between 240 and 260, and ± 10 indicates that point Y falls between 240 (250 – 10) and 260 (250 +10). Note that the error term (± 10) is found by dividing the contour interval by 2. What is the half-way elevation of point Z, at the top of the hill? (310 ± 10 feet).

Height and Relief

If someone asked you what your height was, you would say something like 5 feet 9 inches. This is the distance from the floor to the top of your head. You can also talk about the **height** of a hill, which is the difference in elevation between the top of the hill and the bottom.

A related but different term, **relief,** refers to the difference between the highest and lowest elevations in a given area. For example, in Winnebago County, Wisconsin, the highest elevation is about 920 feet, and the lowest is about 745 feet. Therefore, the relief of the county is 175 feet (920 – 745 = 175 ft). In Jefferson County, Colorado, immediately west of Denver, the highest and lowest elevations are approximately 11,700 feet and 5100 feet; the relief is 6600 feet. Relief is also used in a relative sense; a mountainous area has a high relief, whereas a plain has a low relief; Jefferson County has a high relief; Winnebago County has a low relief.

Figure 6.10 illustrates the differences between elevation, height, and relief.

Bench Marks and Spot Elevations

A **bench mark** is a point whose elevation and location have been precisely determined by government surveyors; its location is marked by a small brass plate. Bench marks are designated on maps by the symbol B. M. (Fig. 6.2). *Spot elevations* are marked by an "x" or are shown at many section corners, bridges, road intersections, hilltops, and the like (Fig. 6.4). Bench marks and spot elevations are used in conjunction with aerial photographs to construct topographic maps. Two aerial photos, taken from different points but overlapping the same area, will provide a three-dimensional view of the land surface when viewed through a stereoscopic viewer. By orienting the photos properly, two beams of light from different sources can be focused at any elevation. If the superimposed beams are moved around a hill, for example, they will trace a line at a precise elevation. The numerical value of this elevation can be determined from known elevations within the area (e.g., bench marks). Aerial photographs are discussed further in Chapter 7.

Making a Topographic Map

Given a number of elevations on a planimetric map (Fig. 6.11), a topographic map can be constructed as follows:

A. Select a contour interval that will illustrate the topography of the area. Choice of C. I. will depend on the total relief of the area. The C. I. used in Figure 6.11 is 10 feet. This means that contours will be drawn through all those points with elevations divisible by 10.

From the elevation values shown in Figure 6.11A, you can see that contours should be drawn representing elevations of 100, 110, 120, and 130 feet.

B. Mark the position of each contour line by interpolating between adjacent points of known elevation. The simplest way to do this is to assume a constant or even slope between points, then estimate the position between points at which the elevation represented by the contour line falls. For example, somewhere between the points in Figure 6.11B marked 122 and 136, there will be an elevation of 130 feet. There will also be an elevation of 130 feet between points marked 120 and 136 and between points marked 128 and 136. The locations of these 130-foot-elevation points shown on Figure 6.11B are rough estimates only; more information would be needed to locate them more accurately.

C. Start with any elevation, and connect points of the same elevation to make contour lines. Smooth out the lines, keep them reasonably parallel, and continue them to the edge of the map. Label each of the contour lines with the elevation that it represents (as in Fig. 6.11C), or label only index contours. Once you understand the process, you probably will be able to draw contour lines without first marking the points between known elevations, as outlined in step B.

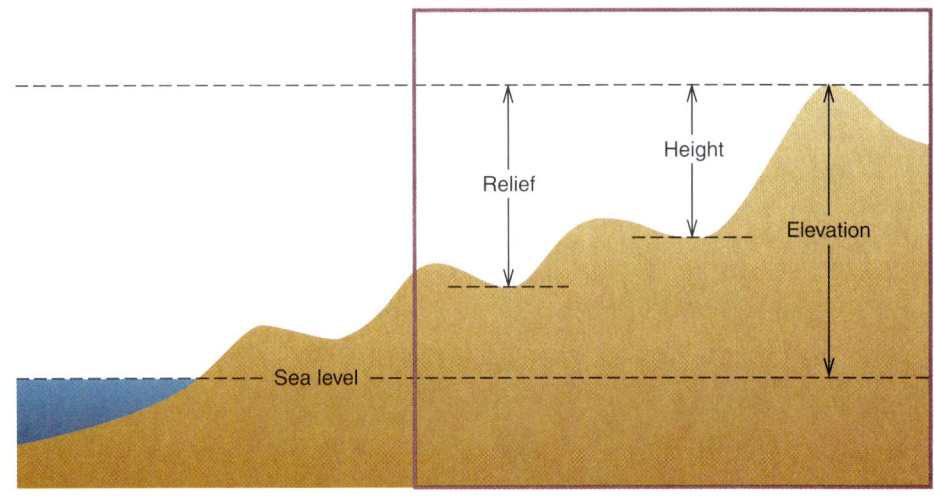

FIGURE 6.10
This profile view shows that the *elevation* of the hill on the right is measured from sea level, whereas its *height* is the difference in elevation between the top and bottom of the hill. *Relief* is the difference in elevation between the highest and lowest points *in a specified area*, such as the one that is outlined.

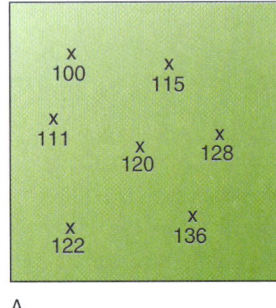 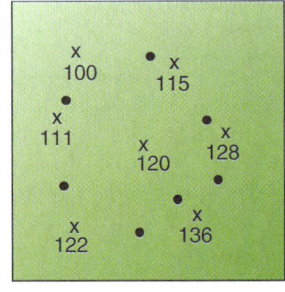 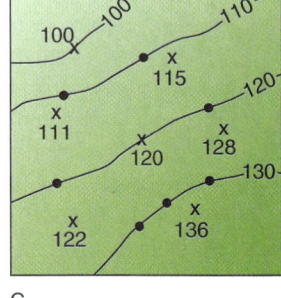

A. B. C.

FIGURE 6.11
Sketches illustrate how a contour map is made. A. Elevations are given for points (Xs) on map, and a contour interval is chosen (10 feet in this case). B. Points with elevations divisible by the contour interval are located and marked (dots). C. Smooth contour lines are drawn and labeled as appropriate.

Topographic Profiles

A topographic profile shows the shape of the land surface as it would appear in a cross section; it is like a side view of the land surface. Topographic profiles portray the shape of the land surface along a particular *line of profile*. They are useful for many practical purposes, such as planning roads, railroads, pipelines, canals, and the like, or for estimating the volume of material that will need to be excavated or filled during construction. Profiles are most easily made along straight lines, but they can also follow curved paths, such as a road or a stream.

A topographic profile is made from a contour map using the following procedure (Fig. 6.12):

A. Select the line or path along which the profile is to be made, such as line X–Y in Figure 6.12A.

B. Record the elevations along the line as shown in Figure 6.12B. To do this, lay the straight edge of a piece of paper along the line of profile. You can use the graph paper on which you will make your profile, or a separate piece of paper altogether. Mark on the paper the ends of the profile line and the exact place where each contour line meets the edge of the paper. If you are doing this on graph paper, use the top or bottom of the graph paper to make the marks. Label each mark on the paper with the elevation of the corresponding contour. Also mark the positions of any streams that cross the line of profile, because they will be low points on the profile.

C. Set up the graph on which the profile will be drawn (Fig. 6.12C). First note the differences in elevation between the highest and lowest points along the line of profile; this will determine the range of elevations on your profile. Label the vertical axis with a range of elevations that extends beyond the profile elevations and conveniently allows each contour to be graphed. In Figure 6.12C, the profile elevations range between 820 and 940 feet and are spanned by a vertical axis of 700 to

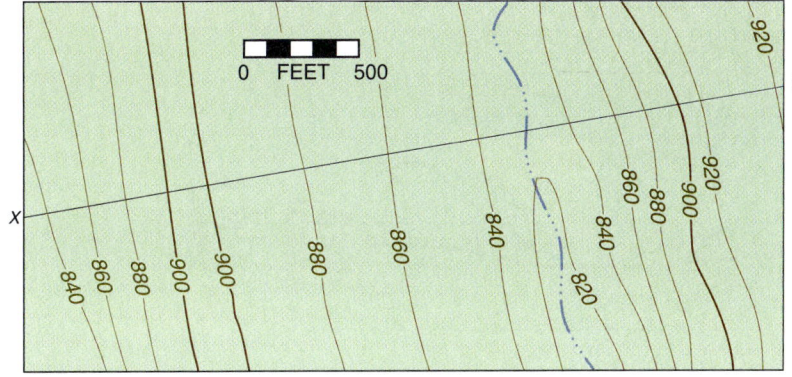

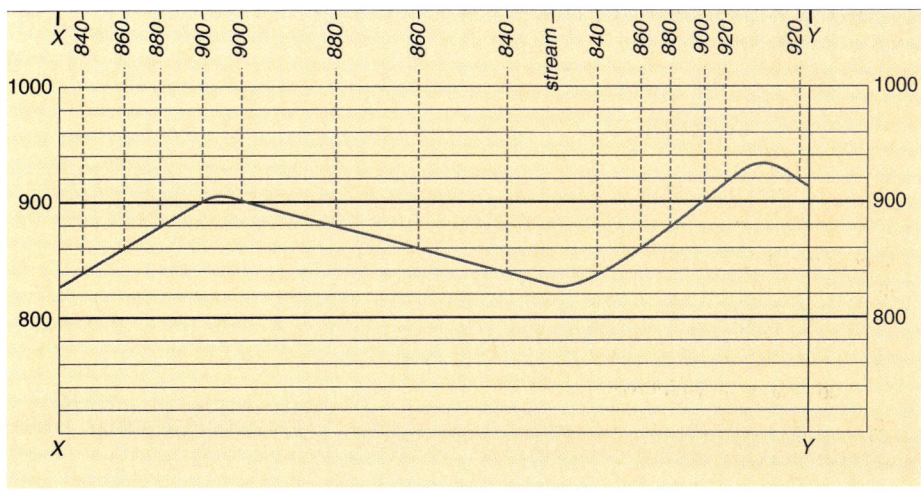

C.

FIGURE 6.12
Construction of a topographic profile. A. A line of profile (X–Y) is chosen. B. Intersections of contours and the stream are marked, and elevations are noted on paper laid along the profile line. C. A vertical scale is chosen, and the points from the previous step are transferred to the appropriate elevation. A smooth line is drawn to connect the points and complete the profile.

1000 feet. Horizontal lines on the vertical axis are 20 feet apart, which matches the contour interval and makes graphing simple. Commonly, as here, the vertical and horizontal scales are different. In Figure 6.12C, the horizontal scale is about 1" equals 800' (1:9600) whereas the vertical scale is 1" equals 160' (1:1920). If the scales were the same, the profile would look flat. Use of an expanded vertical scale highlights (exaggerates) topographic variations.

D. Transfer each of the marks made along the profile to the appropriate place on the graph paper (Fig. 6.12C). If you made marks on a separate piece of paper, place that paper along the bottom of your graph paper. Mark the ends of the profile on the graph paper. Then mark the contour and stream points on the graph paper at their appropriate elevations. This is done by going straight up from the mark on the paper (or, as illustrated here, down from the top of the graph paper if you made marks directly on it) to the horizontal line representing the same elevation; make a small dot on the paper at this point.

E. Connect the points on the graph paper with a smooth line representing the topography (Fig. 6.12C). When crossing a valley or a hilltop, there will be adjacent marks with the same elevation. Instead of connecting them with a straight line, draw your profile line so it goes up over a hilltop or down into a valley. In the case of a stream valley, the low point in the valley will be where the stream crosses the line of profile.

Vertical Exaggeration of Topographic Profiles

Profiles commonly are drawn with a vertical scale that is larger than the horizontal scale. This **vertical exaggeration** emphasizes topographic features that otherwise might not show up on the profile. The amount of vertical exaggeration is determined by the ratio of the horizontal map scale (for example, 1" to 1 mile) to the vertical scale on the profile (for example, 1/2" to 20').

To calculate the vertical exaggeration of a profile, first convert the horizontal

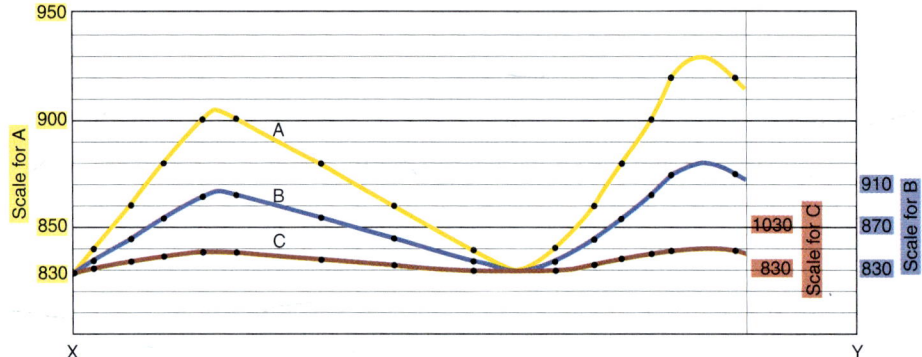

FIGURE 6.13
The profile from Figure 6.12 is shown using three different vertical scales. Assume a horizontal scale of 1 inch to 800 feet. In profile A, the vertical scale, shown in yellow on the left side of the profile, is 1 inch to 80 feet, so the vertical exaggeration is 800/80 = 10 times. In profile B, the vertical scale, shown in purple on the right side of the profile, is 1 inch to 160 feet, so the vertical exaggeration is 800/160 = 5 times. In profile C, the vertical scale, in red, is 1 inch to 800 feet, so the vertical exaggeration is 800/800 = 1 times—there is no vertical exaggeration.

scale and the vertical scale of the profile to the same units. For example,

*The horizontal scale is 1" to 1 mile,
which is the same as 1" = 5280'.
The vertical scale is 1/8" to 20',
which is the same as
1" = 160' (= 8 × 20').*

Next, divide the number of feet per inch in the horizontal scale by the number of feet per inch in the vertical scale:

$$\frac{1"\ horizontal}{1"\ vertical} = \frac{5280'}{160'} = 33$$

The vertical exaggeration is 33 times (33×). For example, the distance represent- ing a vertical difference in elevation of 25 feet on the profile would represent a horizontal distance of 25' × 33 = 825' on the horizontal scale.

Figure 6.13 shows the profile from Figure 6.12C exaggerated (A and B) and non-exaggerated (C). Note that an exaggeration of 5 × was chosen for the profile in Figure 6.12.

Gradient

Gradient represents the change in elevation over a specified distance and often is expressed as feet per mile or meters per kilometer. A gradient of 10 feet/mile means that the elevation of a given point is 10 feet higher than it is a mile away downhill. On a contour map, gradient is determined along a line or stream course by (1) using contour lines to determine the difference in elevation between two points, (2) using the horizontal scale to determine the distance between the same two points, and (3) dividing the vertical difference by the horizontal distance. For example, if the elevation along a stream changes 60 ft in a distance of 7.6 miles, the gradient is 7.9 feet/mile (60 feet divided by 7.6 miles). Note that in the case of a stream, the distance is measured along the stream itself; it is not the straight-line distance between two points (unless the stream is straight).

Hands-On Applications

Geology and other sciences depend on maps to convey basic information, because they provide a way to visualize large areas at a convenient scale. Contour maps are essential for recognizing and understanding the character and origin of many landforms. Geological data from a large area, when plotted on a map, may present a picture that could not be seen or understood from the perspective of a hilltop in the field. Although maps are commonly used to display such basic information as roads and political boundaries, they often play an essential role in furthering our understanding of how the Earth and other planets work. They also assist us in making land-use decisions that directly impact our quality of life.

Objectives

If you complete all the problems, you should be able to:

1. Define latitude and longitude.
2. Describe the boundaries of a quadrangle map in terms of latitude and longitude, and locate a point on a map using these coordinates.
3. Locate a point using the Universal Transverse Mercator (UTM) system.
4. Locate or describe a parcel of land using the U.S. Public Land Survey System, and give its area in acres.
5. Give the dimensions and area of a section and township (in miles and square miles).
6. Number the sections of a township if they are not already numbered on the map.
7. Determine the scale of a map and use it to measure distances.
8. Convert among verbal, fractional, and graphic scales.
9. Give the magnetic declination of a map (assuming it is printed on the map) and explain what it means.
10. Determine what the various symbols used on a map mean (symbols for streams, roads, houses, etc.).
11. Use a contour map to determine elevation, height, and relief.
12. Use the characteristics of contours to determine steepness of slope, direction of stream flow, and locations of hills and valleys.
13. Determine the contour interval of a map.
14. Starting with a map showing elevations, make a topographic map by drawing contours.
15. Construct a topographic profile and determine its vertical exaggeration.
16. Determine the gradient of a stream using a topographic map.

Problems

1. Convert the following fractional scales to verbal scales; show your calculations:
   a. 1:13,226. One inch to __1102__ feet.

   b. 1:88,000. One inch to __1.38̄__ mile(s).

   c. 1:125,000. One centimeter to __1.25__ kilometers.

2. Convert the following verbal scales to fractional scales; show your calculations:
   a. One inch to 2000 feet. $\frac{1\,in.}{(2000 \cdot 12)} \Rightarrow \frac{1\,in.}{24{,}000\,in}$ 1:24,000

   b. One inch to 4 miles. $4 \times 5280 \times 12 = \frac{1\,in}{253{,}440} = 1:253{,}440$

   c. One centimeter to 15 kilometers.   1:1,500,000
   $(15 \cdot 100 \cdot 1000)$

110 Part III Maps and Images

3. Use the UTM method to describe the location of the Radio Tower in Figure 6.4 to the nearest 10 m.

4. Use the Township-Range method to describe the locations of the areas labeled A, B, C, D, E, F, G, and H in Figure 6.5C.

5. One section (one square mile) contains 640 acres.
   a. What is the area, in acres, of a half-section? A quarter-section?

   b. What is the area, in acres, of A, B, C, D, E, F, G, and H in Figure 6.5C?

6. Using the topographic map in Figure 6.14:
   a. Determine the contour interval. 20 ft.

   b. Give the elevations of points 1, 2, 3, 4, 5, and 6 (write on the map). ✓

   c. Draw a topographic profile along the line A-B using the graph provided under the map.

   d. Use the graphical scale to determine the verbal horizontal scale (1 inch = X miles). Show your work.

   e. What is the verbal vertical scale of the profile you drew (1 inch = X feet)?

   f. What is the vertical exaggeration of the profile? Show your work.

   g. Determine the gradient (in feet per mile) of the stream between the points X and Y. Show your calculations.

7. Figure 6.15 is a map showing several streams and the elevations of selected points. Draw contours on the map using a *C. I.* of 20 feet. Label index contours and make them darker than intermediate contours. Hint: Stream valleys strongly affect the shape of contours, so study Figure 6.7 and rule 8 on page 104 before completing this map.

Use the map provided by your instructor for problems 8 through 27.
8. What is the name of the quadrangle?
   Fond du Lac

   What year was it published?
   1955

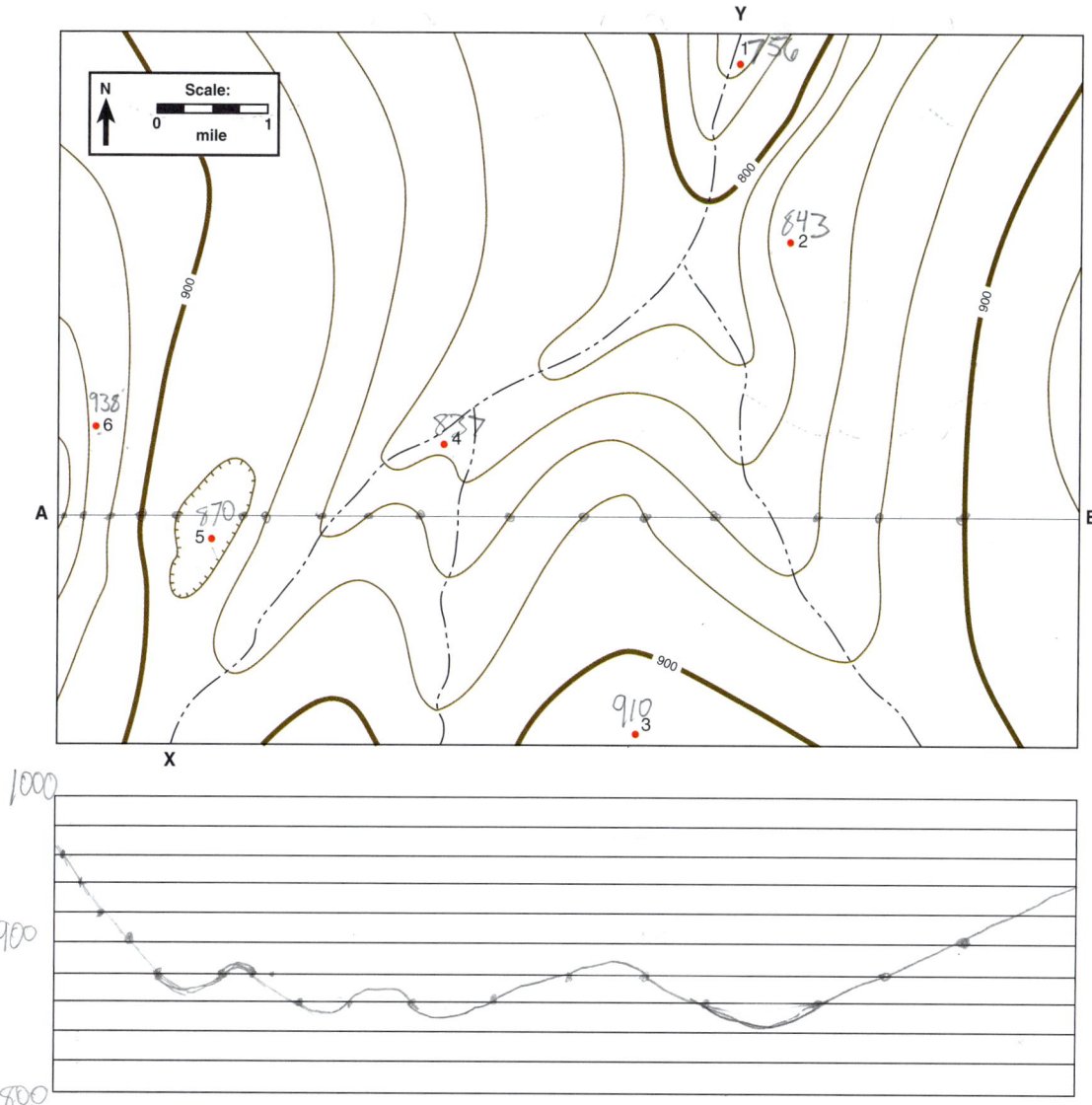

FIGURE 6.14
Topographic map for Problem 6. Note how the contours bend upstream when they cross drainages.

9. What is the name of the quadrangle to the east? Kiel

    The southwest? Waupun

    The north? Neenah

10. What is the latitude of the southern boundary? 43°45'-45'

11. What is the latitude of the northern boundary? 44°00'

12. What is the longitude of the eastern boundary? 88°15'

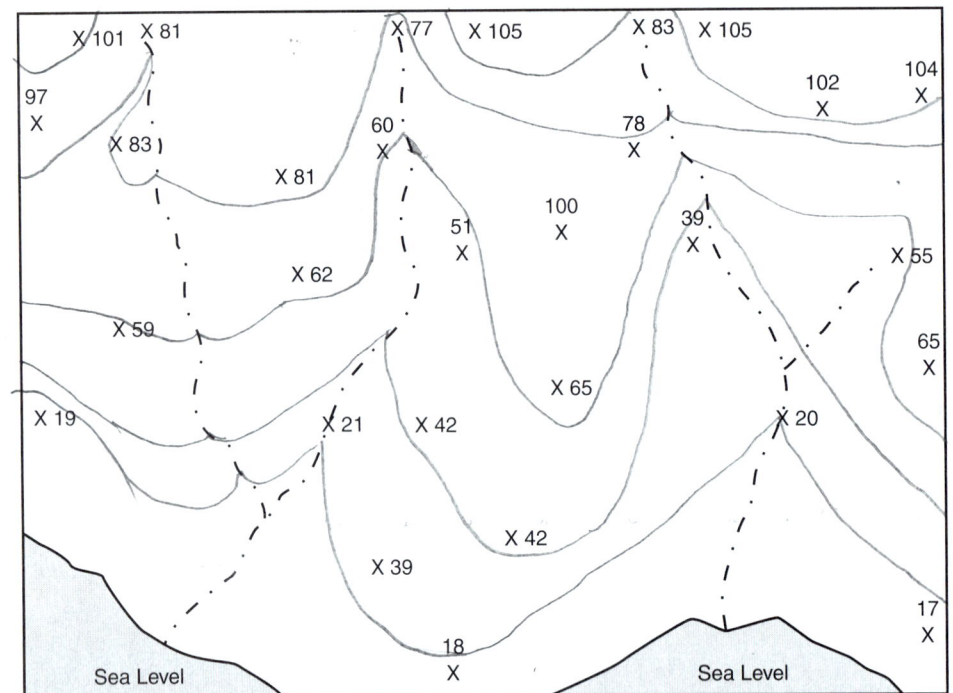

FIGURE 6.15
Xs mark locations with indicated elevations. Dot-dash lines are streams that flow into the ocean (gray). Draw a contour map using a C. I. = 20 feet.

13. What is the longitude of the western boundary? 88° 30'

14. What is the size of the quadrangle in angular units? 15'

15. Locate the feature designated by your instructor in terms of latitude and longitude, to the nearest minute. 43°55' N 86°20' W

16. Locate the feature designated by your instructor in terms of UTM coordinates, to the nearest 100 m.

17. If the map is subdivided by the Township-Range method, locate the feature designated by your instructor to the nearest 1/16th of a section.

18. What is the approximate size of the area designated by your instructor (in acres, if the map is subdivided by the Township-Range method; in square miles, if not)?

19. What is the fractional scale of the map? 1:62,500

20. Use the graphic scale to determine the distance in miles and kilometers between the features designated by your instructor. 3.9 miles, 6.3 km

21. What is the approximate verbal scale for this map in terms of inches and miles? 1 in = 1 mi

22. If you wanted to enlarge part of the map to a scale of 1 inch to 1000 feet, by what factor would it have to be enlarged? Explain your answer.

What would the enlargement factor be if you wanted a scale of 1 cm to 100 m?

23. What magnetic declination (in degrees) is indicated on the map, and for what year? 2°, 1955

24. What is the contour interval for this map? 10 ft.

25. What is the highest elevation within the area designated by your instructor?

    What is the lowest elevation in that area?

    What is the relief in that area?

    What is the height of the feature designated by your instructor in the same area?

26. Find the building designated by your instructor (black square), and give its elevation. 1020°

27. In what direction does the water flow in the stream designated by your instructor? East

28. Go to www.lib.utexas.edu/maps/national_parks.html (or link to it through www.mhhe.com/jones5e—see Preface). Select *Devils Tower National Monument, [Wyoming]* (shaded Relief Map) and answer the following:
    a. What is the elevation of Devils Tower?

    b. What is the contour interval of the map?

    c. What is the approximate height of Devils Tower?

    d. What is the top of Devils Tower like? Is it jagged, flat, dome-like?

    e. What does the dashed line that more or less circles Devils Tower appear to represent?

    f. Let's say you wanted to hike to the top of Devils Tower. Is it too steep? We can get the vertical distances from the contour interval, but unfortunately no scale is given on this map. Another source indicates that the maximum distance from the west to the east side of the tower top (the 5100-foot contour) is about 180 feet. You can see from the map that the horizontal distance between the 5100- and 4600-foot contours on the north side of Devils Tower is also about 180 feet. Thus, on the north side the elevation changes about 500 feet over a horizontal distance of 180 feet.
    What is the gradient (in vertical feet per 1 foot horizontal)?

g. What angle does this surface make with respect to the horizontal? Use the graph below to sketch the gradient you just got and either measure the angle with a protractor (less accurate) or use trigonometry to calculate the angle (more accurate). Label your graph!

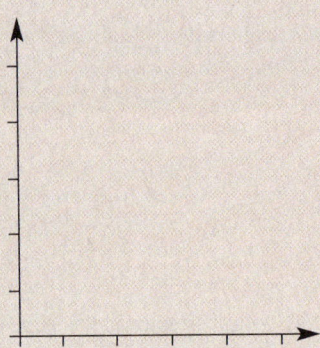

If you were on a roof pitched at this angle, you would find it very difficult to keep from slipping off. Thus, you would have to be a rock climber to scale Devils Tower.

Devils Tower is an interesting place. If you want to see what it looks like, try going to *http://den2-s11.aqd.nps.gov/grd/parks/deto/index.htm* (or link to it through *www.mhhe.com/jones5e*).

Climb the Tower: A web search reveals numerous sites dedicated to climbing Devil's Tower. It is a classic site for technical rock climbing. A National Park Service website *(www.nps.gov/deto/home.htm)* gives some information on historical climbs of the Tower (before the advent of modern equipment) as well as its geology (click on "Study the Tower").

Where is the magnetic north pole today? Magnetic north is always on the move. The Canadian Geologic Survey has set up a nice website *(www.geolab.nrcan.gc.ca/geomag/northpole_e.shtml)* showing the magnetic north pole's current and past positions on its journey through northern Canada and explaining why variations occur on daily and yearly time scales.

Topographic maps over the web: There are numerous sites that provide on-line access to U.S.G.S. topographic maps and that sell higher quality paper copies of these same maps.

- **Terraserver** *(http://terraserver-usa.com/)* allows you to see topographic maps and aerial photographs covering any place in the United States.
- **Sam Wormley's GIS Resources** *(www.edu-observatory.org/gis/gis.html)* lists site links that carry scanned U.S.G.S. topographic maps. Look under "DRG's Available Free Online"; DRG stands for "digital raster graphics."
- **United States Geological Survey Publications Page** *(www.usgs.gov/pubprod/)* lists publications and tells you how to purchase them. Included are links to many on-line retailers of U.S.G.S. maps (click on *Retail Sales Partners* to get to an alphabetical list).

In Greater Depth

29. Figure 6.16 is part of a Massachusetts road map. The distances between towns or intersections are given in miles.
    a. Determine the verbal and fractional scales.

    b. Draw graphic scales in both miles and kilometers.

    c. Use your scales to measure the direct-line distance between Pittsfield and Northampton.

    d. Use the mileages given on the map to determine the shortest road mileage between the two towns.

    e. You may want to compare your route and mileage with that determined from the *Driving Directions* link at the website *maps.yahoo.com/maps.*

Chapter 6 Topographic Maps 115

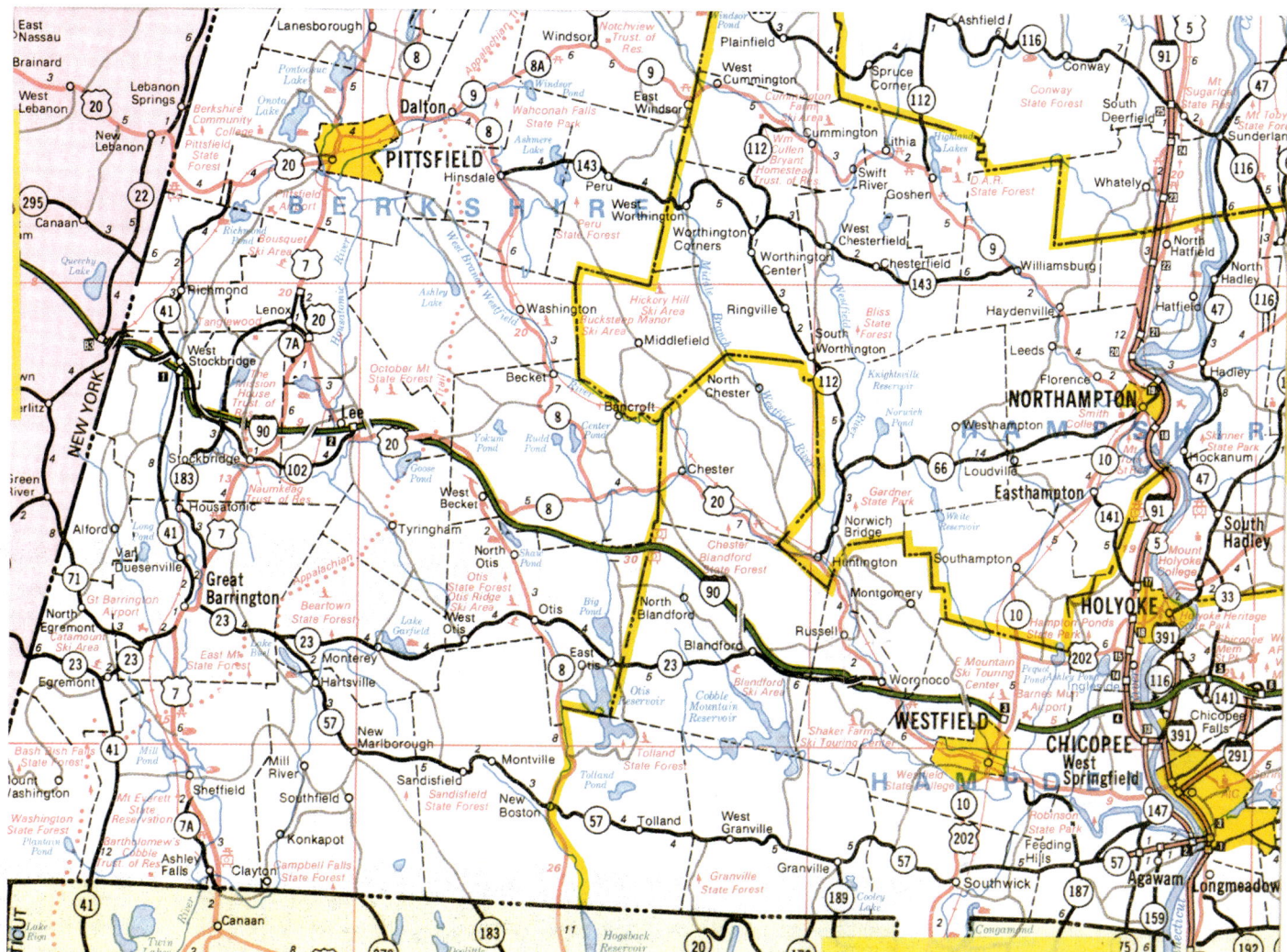

FIGURE 6.16
Portion of road map of Massachusetts for Problem 29.

CHAPTER 7

Aerial Photographs and Satellite Images

Materials Needed
- Pencil and eraser
- Ruler
- Calculator
- Stereoscope (provided by instructor)

Introduction

Aerial photographs and satellite images provide a lot of information about how the Earth works and about how we use the land. As you probably know, the view of a landscape looking down from an airplane is very different from that on the ground. You can see patterns and features that were not apparent before, and you see familiar objects from a new perspective. High-resolution aerial photographs taken from airplanes flying 12 km above the Earth have been used for years to help map rock units, geologic structures, topography, and land use. Satellites, because they fly hundreds or thousands of kilometers above a planet, provide a range of views—from planetary or regional views to quite detailed images covering small areas. Satellite images are not (yet) as high resolution as air photos, but as you will see they have a remarkably wide range of applications. The purpose of this lab is to introduce both air photos and satellite images so that you can profitably use them in subsequent lab exercises. You will learn to view air photos in 3-D and determine their scale; you will learn how to interpret the false colors frequently used in satellite images; and you will learn how to identify some common surface features in both air photos and satellite images.

Visible and Infrared Light

Visible light (the light we see every day) is only a small part of the **electromagnetic spectrum** that runs from X-rays through visible light and down to radio waves (Fig. 7.1). Each form of **electromagnetic radiation** travels through space in the form of a wave (Fig. 7.2), and different types of electromagnetic radiation are distinguished by their wavelengths (Fig. 7.1). **Wavelength** is the distance between successive wave crests (Fig. 7.2). A related term is **frequency,** which is the number of complete waves per second. The wave **amplitude** determines the intensity or brightness of the radiation.

The sun gives off ultraviolet, visible, and infrared radiation. The ozone layer blocks much of the ultraviolet light, so mostly visible and infrared energy hit the Earth's surface. We cannot see infrared wavelengths, but we have camera film and electronic detectors that can. Thus, **visible** and **infrared light** are the most useful for many Earth applications.

Incoming visible and infrared light is either reflected or absorbed when it hits the Earth. When white sunlight hits a plant, for example, mostly green and infrared wavelengths of light are reflected. This makes plants look green to us. If we could see infrared, plants would be a bright infrared color. The other wavelengths of sunlight are absorbed by plants; this warms them and fuels photosynthesis. All other objects on Earth similarly reflect or absorb different wavelengths of light. This not only gives them their characteristic colors, but also gives them distinctive "colors" within the infrared wavelengths. Because infrared wavelengths occupy a much larger part of the electromagnetic spectrum (Fig. 7.1), infrared images are useful for distinguishing and identifying particular rocks, soils, plants, ocean currents, and numerous other natural and artificial objects. Figure 7.3 shows how infrared wavelengths can be used to highlight rocks that are not distinguishable using visible wavelengths.

Electromagnetic radiation can also be emitted from an object if the object is hot enough. Stars, light bulbs, electric burners, and glowing embers all emit visible light

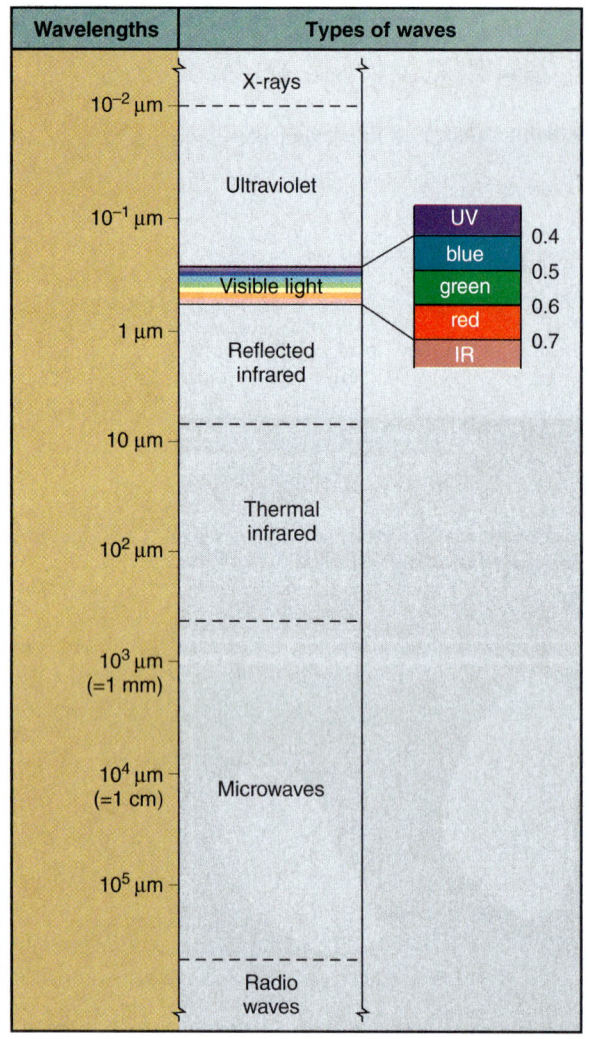

FIGURE 7.1
A portion of the electromagnetic spectrum. Wavelengths are in micrometers (10^{-6} meter or one millionth of a meter; symbol, μm); boundaries between types of waves are approximate.

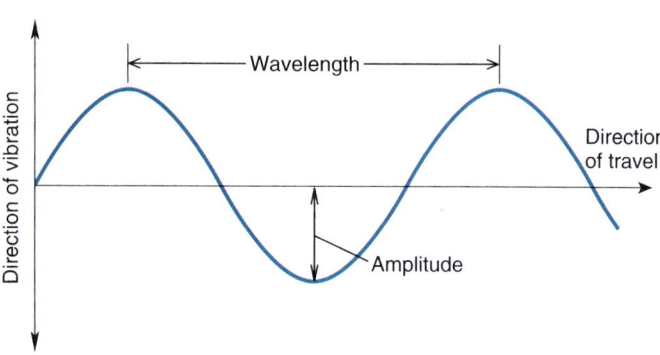

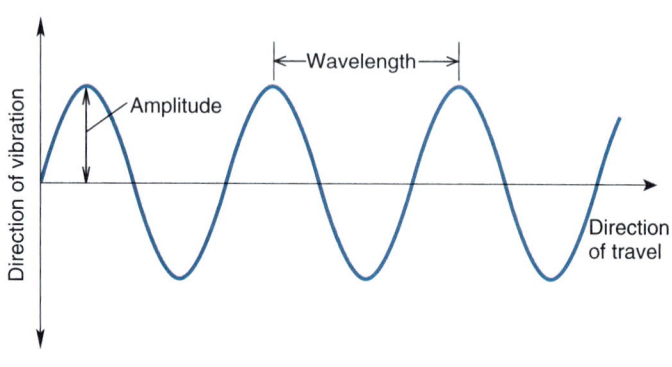

FIGURE 7.2
Examples of two waves. A. Wave with longer wavelength undergoes fewer complete vibrations per second, so has a lower frequency. B. Wave with shorter wavelength has higher frequency.

because they are hot. As objects cool, they emit more energy in the infrared spectrum. For earth observations, infrared light is broken into two components: **near** and **far infrared**. *Near infrared* has wavelengths closer to visible light and is also called **reflected infrared** because, like visible light, it is useful as a reflected form of radiation. *Far infrared* is also called **thermal infrared** because these longer wavelengths allow such phenomena as volcanic eruptions and forest fires to stand out clearly in an image.

Aerial Photographs

Most aerial photographs are taken from airplanes, although some of the more spectacular ones were taken by astronauts from space. Air photos are taken using a camera and film. Types of film include black-and-white, infrared black-and-white, natural color, and infrared color. Various filters can be used to block out a portion of the spectrum; for example, visible light is often cut out when using infrared films. As will be discussed in more detail in the section "Satellite Images," infrared films record images not in true colors, but in **false colors.** For example, healthy green plants appear red, not green, and clear, clean water appears black, not blue. Examples of black-and-white and false-color air photos are given in Figure 7.4.

Because of their low distortion, the most useful for extracting data are **vertical photographs,** those taken with the camera pointing straight down. **Oblique photographs,** taken at other angles, are excellent for illustrations, but not for mapping because features become distorted. Black-and-white photos are the most common, and such photographs of most of the United States have been available since the early 1950s. The U.S. Geological Survey provides high-altitude (40,000 feet), cloud-free, black-and-white and color-infrared photographs of the 48 conterminous states at scales of 1:80,000 and 1:58,000, respectively. It also offers high-quality color-infrared photos taken from 20,000 feet (scale 1:40,000); black-and-white photos can be made from these if desired.

Vertical photos are taken at regular intervals as an airplane flies along a

A. Visible light image   0   10 km

B. False color infrared   0   10 km

FIGURE 7.3
ASTER satellite images of the Escondida copper-gold-silver open-pit mine in the Atacama Desert of Chile. A. Image created using largely visible wavelengths of light. B. Image of the same area created using false-color infrared. The infrared image reveals a complex geology of varied rock types surrounding this major ore deposit, a geology that is not easily appreciated using visible light alone.

predetermined flight line at a specific altitude. The intervals are such that photographs along the flight line overlap each other by about 60 percent, and adjacent flight lines overlap by about 30 percent. A common photograph size is a square 23 cm (approximately 9 inches) on a side.

Scale

The average scale of a photograph or satellite image can be determined if you know the distance between two points on the image. From a map, you can use its scale and a ruler to get the distance between two points. For example, say a map with a scale of 1:50,000 has two road intersections 25 mm apart. This scale means that 1 mm on the map equals 50,000 mm on the ground. Thus, 25 mm equals 25 × 50,000 = 1,250,000 mm = 1,250 m between the two road intersections.

You then use a ruler to measure the distance between the same points on the photograph or satellite image. In this case, the distance is 31 mm. Thus, 31 mm on the image equals 1,250 m on the ground. To express this as a scale, be sure the units are the same and divide by the image distance:

$$31 \text{ mm} = 1{,}250 \text{ m} = 1{,}250{,}000 \text{ mm}$$
$$31 \text{ mm}/31 = 1{,}250{,}000 \text{ mm}/31$$
$$1 \text{ mm} \approx 40{,}000 \text{ mm, or the photo scale} \approx$$
$$1{:}40{,}000$$

Distortion

The *approximately equals* symbol (≈) is used in the scale equation because, unless the terrain is perfectly flat, the scale of most aerial photographs varies from one part of the photo to another. This *distortion* is greatest near the edges of the photo and most severe where relief is high. When compared to points at intermediate elevations, points at high elevations are shifted away from the center of the photo, and points at low elevations are shifted toward the center. In addition, the flying altitude varies as hills and valleys are crossed. The altitude above hills or mountains is less than above valleys, so the photo scale is larger over hills and mountains and smaller over valleys. (Larger scale means that the quotient of the fractional scale is a larger number; the fractional scale 1:50,000, with quotient 0.00002, is a larger scale than 1:62,500, with quotient 0.000016.)

Stereoscopic Viewing

Overlapping aerial photographs can be viewed with a **stereoscope** to see the image *stereoscopically*, that is, in three dimensions (Fig. 7.5). The slight difference in perspective of the two photos resembles the slightly different perspectives from which our two eyes see objects; our brains process this information to give us three-dimensional viewing.

To use a stereoscope:

1. Place the photos so that identical points on each photo are directly below each lens of the stereoscope. The photos in this manual are already in position.

2. Select an obvious point on the photos and place the stereoscope over them. Adjust the interpupillary distance of the stereoscope for your eyes (Fig. 7.5).

3. As you look through the stereoscope, let your right eye look at the right-hand photo and your left eye look at the left-hand photo. If the three-dimensional scene doesn't pop out at you, rotate the stereoscope slightly about a vertical axis. The stereoscope is properly aligned when an imaginary line connecting the centers of the two lenses is parallel to a line connecting equivalent points on the two photos. If you are having difficulty seeing three dimensions, draw a light line on the lab manual photos between equivalent points to guide you.

You may be able to view a stereo-pair (a pair of overlapping aerial photographs) stereoscopically without a stereoscope. Use the same approach recommended to see those strange-looking images in the comic section of your Sunday newspaper. Put the

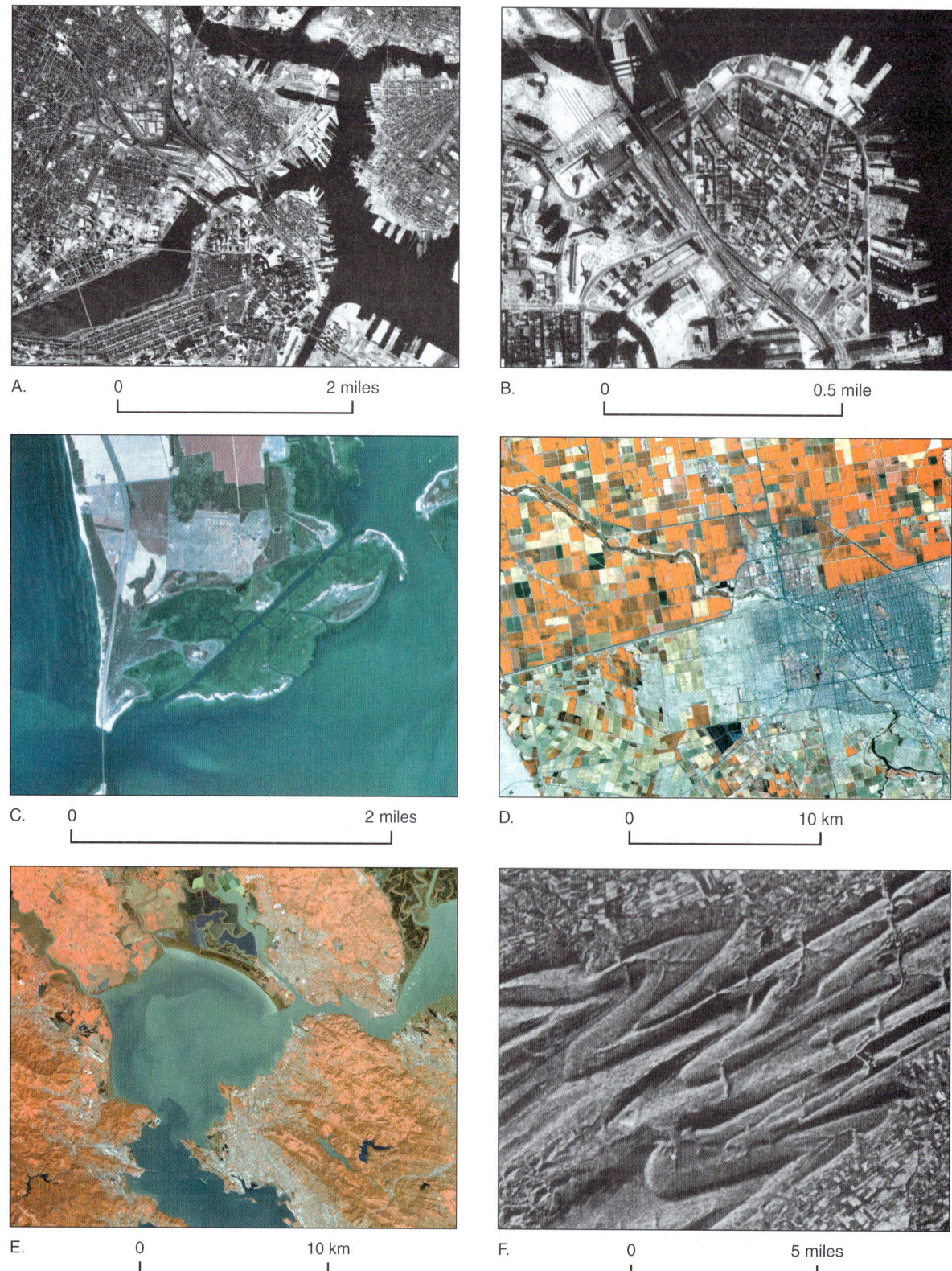

FIGURE 7.4
Examples of different aerial and satellite images. A. High-altitude, black-and-white photograph of Boston, Massachusetts (scale 1:80,000). B. Low-altitude black-and-white photograph of Boston (scale 1:20,000). C. High-altitude, color-infrared photograph of Cape Charles, Virginia (scale 1:58,000). D. False-color visible and near-infrared ASTER image of U.S.-Mexico border in California showing, on the U.S. side, lush farm fields (bright red) due to irrigation from the Colorado River (scale 1:315,000). E. False-color visible and near-infrared ASTER image of the northern San Francisco Bay showing sediment-rich water (light blue) carried in from the north and east by rivers, and clear water carried in from the south by rising tides (scale 1:617,000). F. Side-Looking Airborne Radar (SLAR) image of Appalachian Mountains, central Pennsylvania (scale 1:250,000).

two photographic images close to your face with your nose on the line separating them. Stare straight ahead while gradually moving them away until the three-dimensional image pops into focus.

The height of features is exaggerated three or four times in the typical stereoscopic view. One of the most misleading effects of this is that slopes are also exaggerated. For example, a 15° slope appears to be about 40°, and a 30° slope appears to be about 60°. This will not cause you problems in this class, but it is something that you should realize.

Comparison with Topographic Maps

Figure 7.6 is a stereographic pair of aerial photos of a volcanic cone near Idaho Falls, Idaho, and Figure 7.7 is a topographic map of this, plus a second cone. Look at the photos with a stereoscope and compare them with the map. Note how the cone in the stereo pair is exaggerated vertically

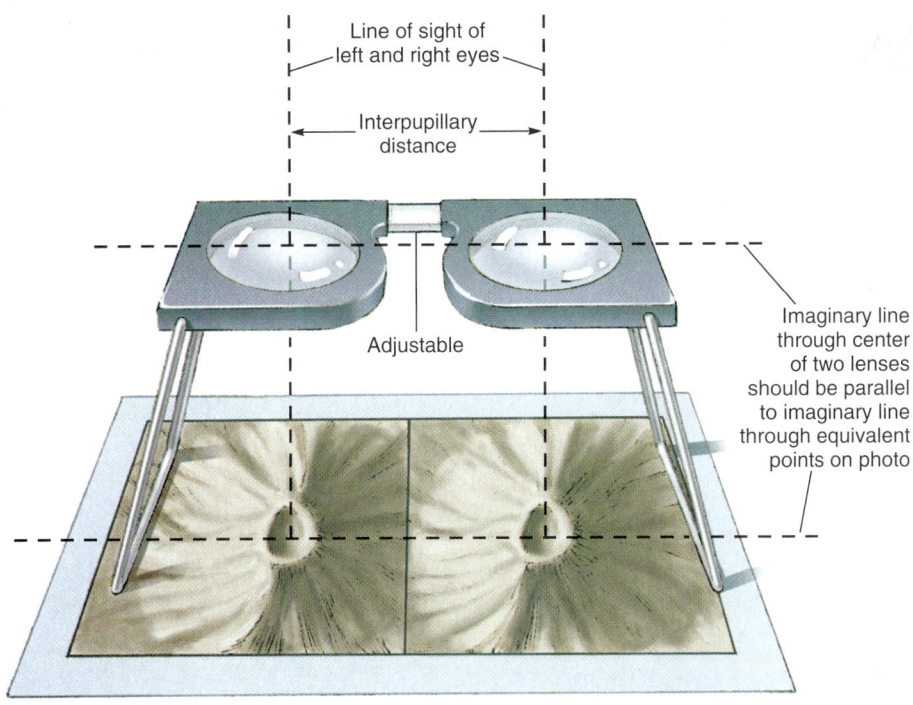

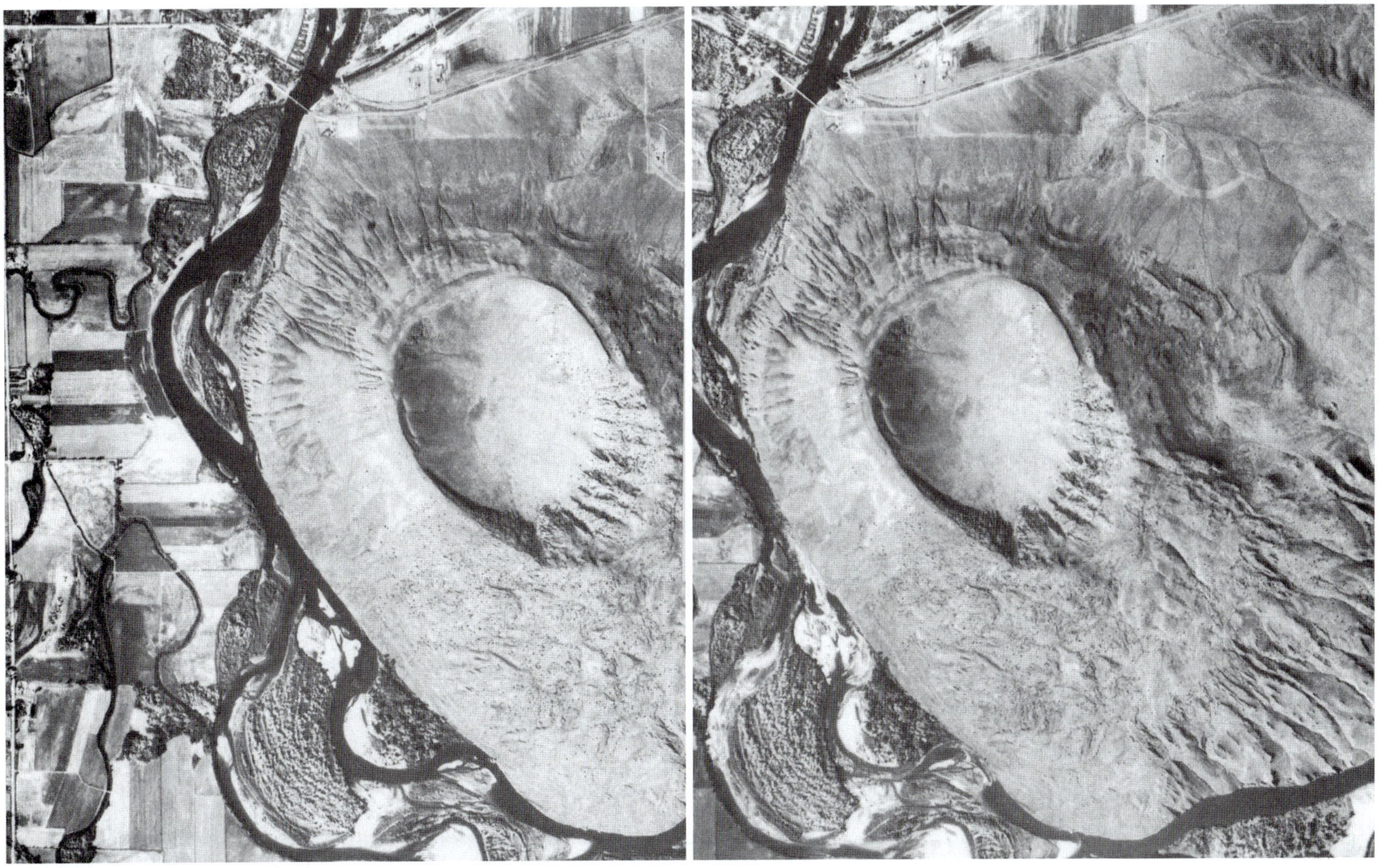

FIGURE 7.5
Stereoscope in position for three-dimensional viewing of a stereopair of aerial photographs.

FIGURE 7.6
Stereographic pair of one of the Menan Buttes near Idaho Falls, Idaho. The Menan Buttes are volcanic cones made of basalt ash. Normally, basalt cones are made of walnut-size cinders, but here, rising basaltic magma encountered abundant groundwater in the gravel along the Snake River and erupted explosively, forming fine ash-size particles.

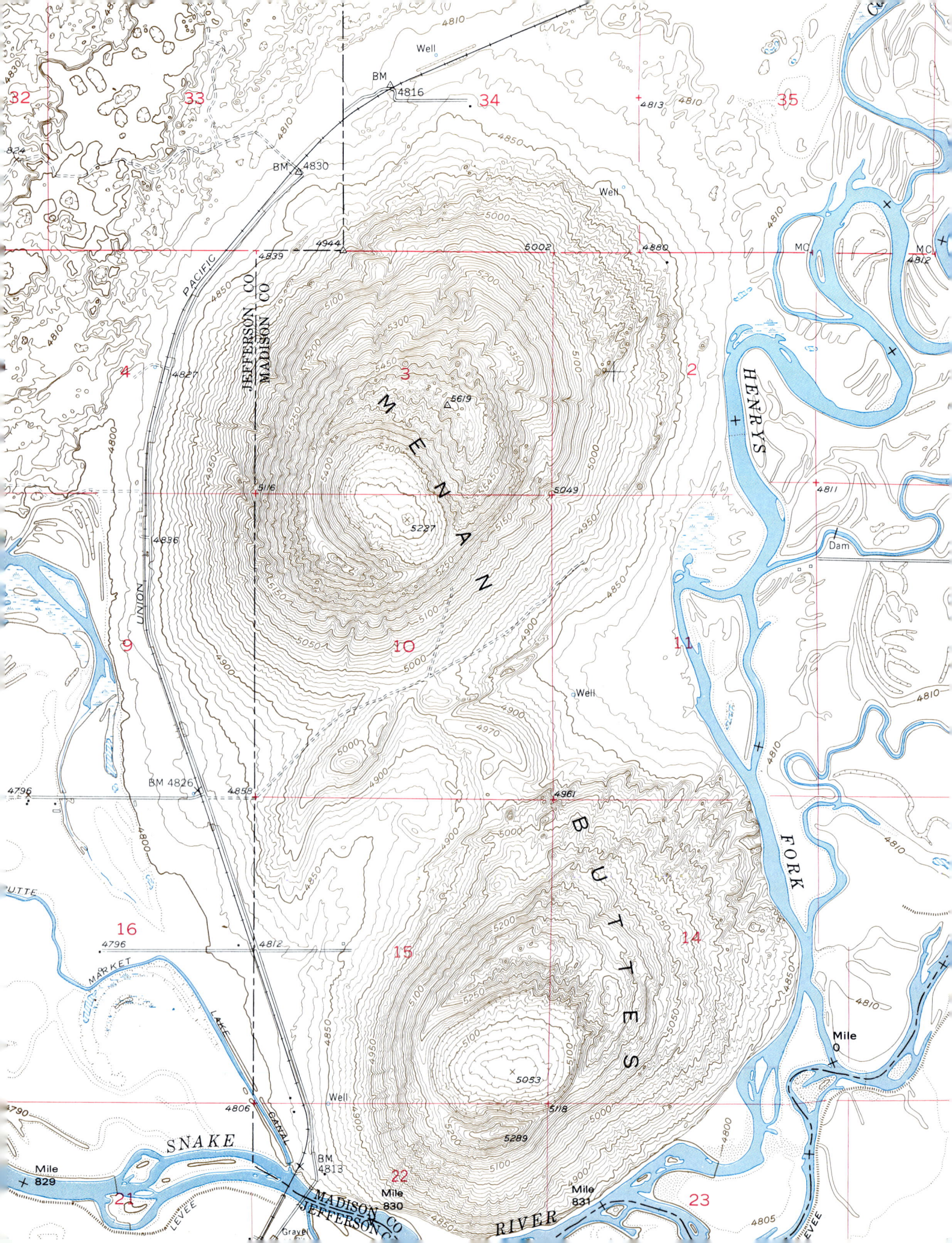

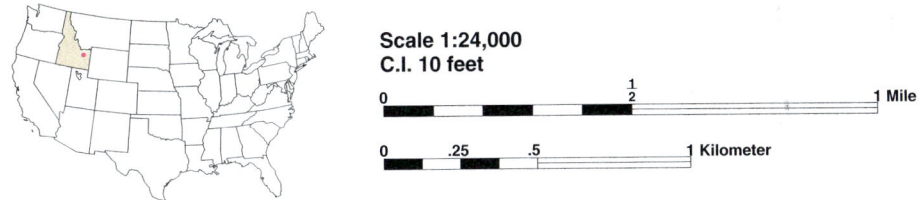

FIGURE 7.7
(Page 122) Portion of topographic map of Menan Buttes from Menan Buttes, Idaho, 7½ minute quadrangle.

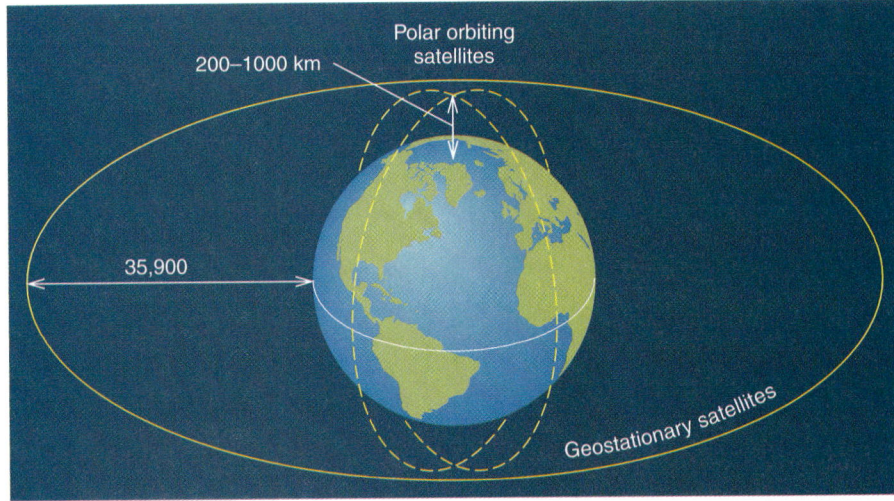

FIGURE 7.8
Polar and geostationary orbits of satellites. *Landsat* satellites are in polar orbit.
Source: Data from Ray Harris, *Satellite Remote Sensing: An Introduction,* 1987 Routledge & Kegan Paul, London, & New York.

and how pronounced the crater in the center appears.

Satellite Images

Instead of using film, satellites use a detector or a series of detectors to *scan* the Earth's surface. A single detector can focus on a certain minimum area of the Earth's surface. For example, a high-resolution detector can cover an area of about 15 meters (50 feet) square. This means that all light and color variations within a given 15-meter square are averaged out into a single reading on the detector. A satellite scans across a portion of the Earth using its 15-meter (or whatever value) squares to build up a complete image that may cover many tens, hundreds, or thousands of kilometers. Each minimum square used to create the whole image is called a **pixel.** The whole process is analogous to a newspaper photograph or a TV image: if you look very closely you see that an image of a person's face, for example, does not reveal details like tiny hairs or pores but instead only pixels of color and/or shading. Because a string of color information from each pixel must be transmitted back to Earth, satellites that focus on large-scale features, such as the weather, tend to have lower resolution (one pixel covers a larger area) than those intended to study geologic and land-use problems.

Each detector on a satellite measures the intensity of electromagnetic radiation of a specific wavelength. Color film, by contrast, collects information from the entire spectrum of visible light. To overcome this limitation, satellites carry an array of detectors, each designed to sample a different useful wavelength within the visible and infrared spectrum. Because satellites do not get all visible wavelengths and because the infrared wavelengths do not correspond to any visible wavelengths, satellite images are created by assigning red, green, and blue colors to particular visible and infrared wavelengths. This creates what is known as a **false-color image.** In general, because plants reflect so much infrared light, red on a false-color image corresponds to plants. Because clear water reflects no infrared light, it tends to be black. Bare rock, soil, snow, clouds, and many human-made structures frequently end up with colors that are close to natural. *However,* because colors are assigned by whomever is processing the satellite data, read the figure caption before following the above rules of thumb!

Most remote sensing from satellites is done from one of two types of orbits, polar or geostationary (Fig. 7.8). **Polar orbits** take satellites over or near the north and south poles at altitudes of 200 to 1000 km. As the satellite orbits, the Earth spins below it, so that, with time, the satellite will pass over all or most of the Earth. Satellites designed to study the Earth's surface fly in polar orbits. **Geostationary orbits** (Fig. 7.8) are much higher (35,900 km) and follow the equator. The orbital velocity is such that the satellite remains in the same apparent position above the Earth. Communication and weather satellites use geostationary orbits.

The most useful information for geologists has been gathered by the *Landsat* satellites in polar orbits. *Landsat 1* was launched in 1972; *Landsat 7* was launched in 1999. *Landsats 1* through *5* carry a *Multispectral Scanner (MSS)* capable of scanning a 185-km-wide swath in four separate wavelength bands (or spectra), corresponding to green, red, and two *reflected infrared* bands. *Landsat 7* carries an Enhanced

FIGURE 7.9
This infrared *Landsat* image of Great Salt Lake, Utah, was created by projecting four wavelength bands (green, red, and two near-infrared) through filters and combining them to create a false-color image. Reds (the false color assigned to infrared) signify growing vegetation: the more vigorous and dense the growth, the more intense the red. With decreasing vigor and density, pale red is replaced by shades of white, green, or blue that reflect the type and moisture content of the underlying soil or rock. Water is black if clear, but becomes pale blue with an increase in sediment load. Very shallow water takes on the color of the bottom sediment. Rocks and most human-made features have tones that are approximately natural: asphalt roads are dark, concrete roads are light, and buildings are various colors.

Thematic Mapper with an eight-band scanner that includes a band for heat or *thermal infrared* (10.4 to 12.5 μm). Figure 7.9 is an excellent example of a *Landsat* image. Another satellite instrument of particular interest to geoscientists is called ASTER. It is equipped with a variety of visible, near infrared, and thermal infrared detectors and is designed to provide a wide variety of data relating to geology, volcanic and other natural hazards, water supplies, changes in the glacial ice caps, urban growth, desert growth, and land use (Fig. 7.4D, E).

Radar Images

Satellites produce radar images by beaming microwaves onto the ground and recording the signals that bounce back. This is the essence of the radar gun familiar to you speeders! Unlike visible and infrared light, microwaves pass through clouds, making radar techniques useful day or night, rain or shine. Because microwaves are beamed down at an angle to the ground, the land surfaces inclined toward the satellite will reflect more energy and appear brighter in the image. If inclined away, they will appear dark. The overall effect is similar to the light and shadows produced when sunlight hits the ground at an angle. However, the microwave beams can be projected from unnatural angles and thus can cast "shadows" in unnatural directions. Such images can at first appear inverted (ridges look like valleys and vice versa) (Fig. 7.4F). In addition to large-scale topography, radar images are sensitive to the roughness of a surface. A rough landscape has more surfaces inclined toward the satellite and thus appears brighter than a smooth landscape, which allows most microwaves to simply bounce away. Several Space Shuttle-based radar missions have been deployed to develop further applications for radar imagery and to create a comprehensive topographic map of the Earth's surface. One spectacular success is the Magellan Project to Venus, which created a complete surface feature map of this cloud-covered planet.

Hands-On Applications

Like maps, aerial photos and satellite images provide a way to view the Earth at a convenient scale. Unlike maps, they are direct images and convey firsthand information, not interpreted information. Geologists use aerial photos to record data, help interpret relations between bodies of rock, and generally guide them in the field. Careful study of air photos and satellite images, in conjunction with field work, enables hypotheses to be formulated and tested, and aerial images thus are an integral part of many aspects of geological science.

Objectives

If you complete all the problems, you should be able to:

1. Identify common features on an aerial photograph or a false-color satellite image.
2. Determine the scale of an aerial photograph or satellite image if you know the actual distance between two points on the photo.
3. View a pair of suitable aerial photos stereoscopically (some may be unable to do this).
4. Use a sequence of satellite images to monitor a volcanic eruption.

Problems

1. To help you get started using aerial photos, refer to Figures 7.6 and 7.7 and answer the following:
    a. Which of the Menan Buttes is shown in Figure 7.6? How do you know?

    b. Draw an arrow pointing north on Figure 7.6.
    c. Outline on Figure 7.7 the area covered by the right-hand aerial photograph in Figure 7.6.
    d. From what direction was the sunlight coming in Figure 7.6? How do you know?

    e. Determine the approximate scale of Figure 7.6; show your work. Circle or describe the features on the map and the photo that you used for your measurements.

    f. The Menan Buttes formed from the accumulation of basaltic ash during a small but unusually violent volcanic eruption. In what direction was the wind blowing when the southern Menan Butte formed? The northern butte? Explain your answer.

2. To illustrate how familiar things appear from the perspective of an airplane, identify the labeled features in Figure 7.10. Use a stereoscope!

    A.

    B.

    C.

    D.

    E.

    F.

    G.

    H.

    I.

125

126 Part III Maps and Images

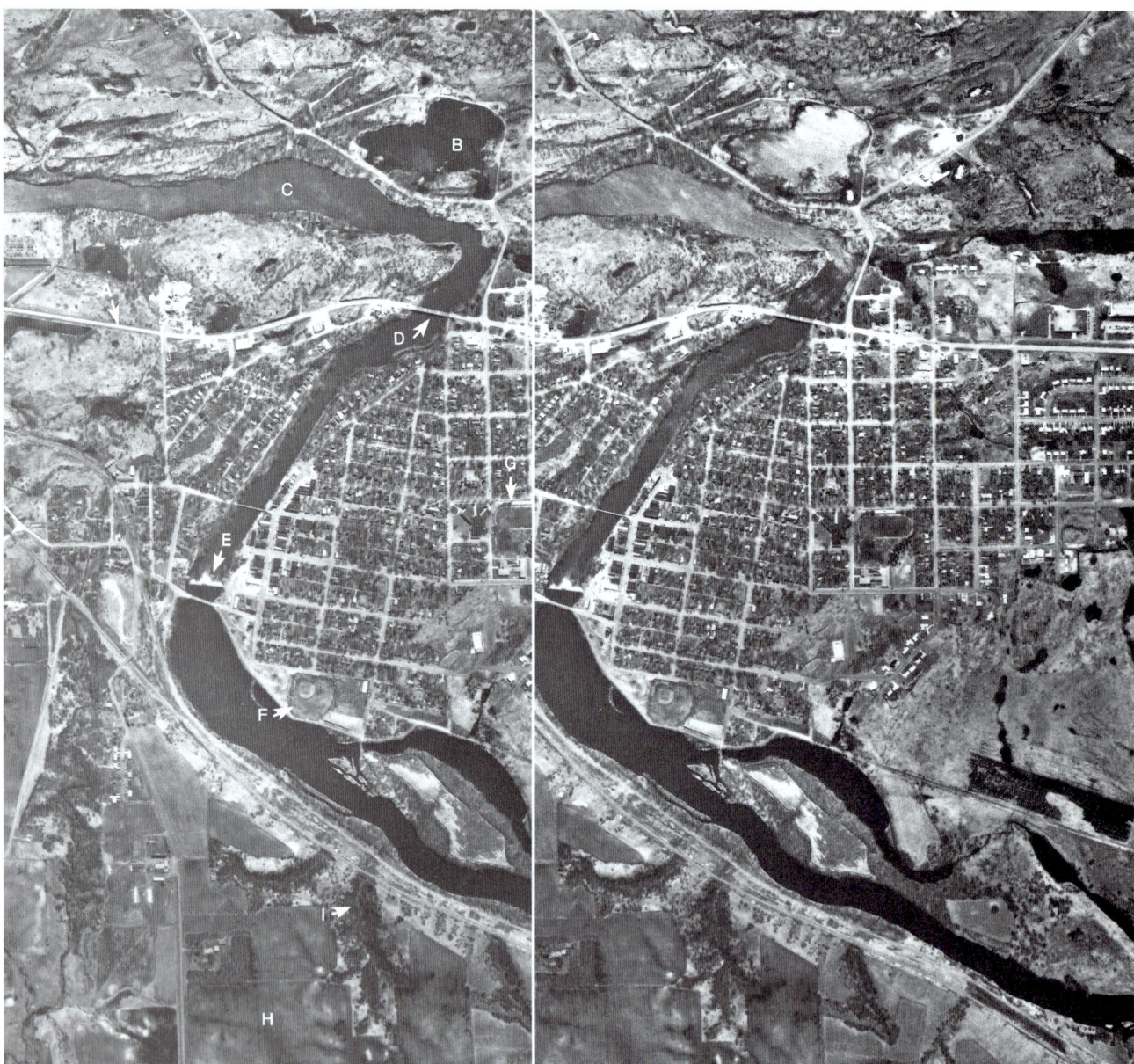

FIGURE 7.10
Aerial photographs (stereographic pair) of Granite Falls, Minnesota (scale 1:17,996), for use with Problem 2.

Which way is north? (Hint: Use the shadows. The photographs were taken about noon in late April.)

Which direction is the water flowing at point E?

3. To help you learn to interpret satellite images, refer to Figure 7.9, a composite, false-color, *Landsat* image of the Great Salt Lake, Utah, and answer the following (north is toward the top of the image):
   a. In what part of the area is vegetation the most abundant? How do you know?

b. On the east side of the image, in the red area, there is a discontinuous series of white areas trending north-northwest. What is the white?

Some white patches can also be seen in the red areas on the south side of the image. What are they?

c. Two large areas of white occur southwest and west of the lake, and some smaller ones on the north. Knowing the name of the lake and realizing that it has been larger in the past, what might these white areas be?

d. Much of the land in the center and western part of the image, including the islands, is a greenish tan color. What would you expect to find in those areas?

e. Locate Salt Lake City on the southeast end of the Great Salt Lake. Interstate 80, a light gray line on the image, goes directly west from downtown Salt Lake City to the lake, where the road bends southwest. Immediately southeast of the bend, you can see an evaporating pond with a shape that points west. What might the evaporating pond be used for?

f. The lake is very dark blue on the south end, and light blue on the north end and to the west of the small island in the southwest. The colors are separated by straight lines, an indication that the cause is human activity. The line across the middle of the lake marks the position of the causeway for the Southern Pacific Railway, and levees block the area in the southwest. The lighter blue colors could reflect shallower water depths, more sediment in the water column, or different water chemistry. Given that most fresh water enters the darker southern half, suggest a possible reason for the bluer colors behind the artificial barriers.

g. The largest island (Antelope Island), in the southeastern part of the lake, is approximately 25 km long. What is the approximate scale of the image?

h. The world's largest open-pit mine, at Bingham Canyon, Utah, appears on this image. It is the light blue area near the south edge of the image; Antelope Island "points" at it. It is principally a copper mine, but when copper prices are low, recovery of minor, but more valuable, elements such as gold and silver enable the mine to continue operating. What are the length and width of the area disturbed by this mining operation?

4. a. **Radar Images of Landforms on Venus.** Go to *http://nssdc.gsfc.nasa.gov/planetary/magellan.html* (or link to it through *www.mhhe.com/jones5e*), click on *Magellan Images of Venus* near the bottom of the page, and examine the following radar images:

   (1) Page 1: Click on *Sapas Mons volcano.* Younger flows tend to partially cover older flows. Examine the image for flows that appear to be largely complete (some flows appear to have originated on the side of the volcano, not the central crater) and therefore are the youngest. Are the younger flows smoother or rougher than the older flows they bury?

   (2) Page 1: Click on *Mona Lisa crater.* Around the crater is the typical Venusian surface. Is it smooth or rough? Note the many fractures. How does the roughness of the sides of the volcano compare to inside the crater?

   (3) Page 5 (hit left arrow at bottom of page 1): Click on *Sapas Mons volcano (3-D color press release version).* The radar data can be processed to show perspective views of planetary surfaces. Here you see the Sapas Mons volcano in the foreground, with its lava flows going across smoother surfaces. Based on what you've seen so far, what type of feature is the hill directly behind Sapas Mons? What is the vertical exaggeration of this image (see image caption)?

   Despite being similar in size and composition to the Earth, Venus shows many geologic features that have no parallels on our planet. To see several interesting craters, hit the left arrow at the bottom of page 5 to go to page 4 and check out the *3-D perspective view of the 'crater farm' on Venus.* The orange colors selected for these images are based on photographs taken on the surface by the Soviet "Venera" lander. Hunt through the rest of the page to see fracture zones, pancake domes, and other exotic Venusian landscapes.

   b. **Comparison of Different Types of Images.** Next go to *http://satftp.soest.hawaii.edu:80/space/hawaii/index.html* (or link to it through *www.mhhe.com/jones5e*—see Preface), and click on *Remote Image Navigator.* Then click on *Maui,* then *Images from Space;* there are three links to three types of images: Shuttle photographs, *Landsat,* and SIR-C images.

   (1) To see how these images differ, choose each type and click on the outlined area that best covers the island of Maui. Describe each and indicate what kinds of things that image shows better than the others.

   (2) Next, from the bottom of the page, choose the button *Go to Virtual Field Trips,* click on *Big Island,* then *"We now have six ways to enjoy Kilauea Volcano, virtually"* and *"Take a tour along Chain of Craters Road"* and select location 5 in yellow. This will give you an image of the Royal Gardens area, which was overrun by basaltic lava flows emanating from the Pu`u `O`o crater in the early 1980s. You may want to revisit *Go to Virtual Field Trips* for a tour of the Big Island or one of the other Hawaiian Islands.

   c. **Excellent Radar Topographic Map.** Go to *www.winona.msus.edu/geology/imagearchive/remoteimages/remoteimages.html* (or link to it through *www.mhhe.com/jones5e*—see Preface), and select the *Wyoming state relief map from radar imagery.* Mountain ranges show very well on this map. Use a state road map of Wyoming that gives the names of mountain ranges and locate the following in terms of latitude and longitude:

   (1) Bighorn Mountains   (5) Laramie Mountains

   (2) Absaroka Mountains   (6) Black Hills

   (3) Teton Mountains   (7) Medicine Bow Mountains

   (4) Wind River Mountains

TABLE 7.1
G.O.E.S. Data, Mount St. Helens, 1980

Time of Image	Maximum Distance from Mount St. Helens	Travel Time

In Greater Depth

5. Figure 7.11 shows Geostationary Operational Environmental Satellite (GOES) images taken from 35,900 km above the Earth following the May 18, 1980, eruption of Mount St. Helens. The times (Pacific Daylight Time) at which the images were taken are given, and the state boundaries are superimposed on the images.

    a. Transfer the boundaries of the ash cloud from each image onto the map in Figure 7.12 using the state boundaries as guidelines. Label each ash-cloud boundary with the time of the image. Write the time of each image in Table 7.1.

    b. Using Figure 7.12, measure the *maximum* distance from Mount St. Helens to the leading edge of the ash cloud for each image, and enter your results in Table 7.1.

    c. Calculate the travel time represented in each image by subtracting the time of the eruption (8:32) from each image-time, and enter the result in Table 7.1.

    d. Prepare a graph of the data in Table 7.1, using Figure 7.13, by plotting the travel time on the y (vertical) axis and distance traveled, in kilometers, on the x (horizontal) axis for each image. Connect the data points. When the points are connected, do they form a straight line, a smooth curve, or an irregular line?

    e. The velocity at which the ash plume moved is represented by the slope of the lines connecting the points; the steeper the slope, the slower the velocity. What does the graph tell you about the wind velocity—was it constant, did it gradually increase or decrease, or was it variable?

    f. The velocity at which the ash plume traveled can be calculated by dividing the distance traveled during a particular time interval by the time interval. For example, if the ash cloud traveled 425 km in 200 minutes, velocity, v = 425 km/200 min = 2.13 km/min. Calculate the overall, or *average,* velocity between 8:32 and 16:15. Give the answer in (1) km/min, (2) km/hour (multiply [1] by 60), and (3) miles/hour (1 km = 0.62 miles).

    g. In what general direction was the prevailing wind (see Fig. 7.11)?

    Was the direction the same along the entire path, or did it change; if so, how?

    h. Based on the color-shading of the ash plume during the period of the images, predict where the thickest deposits of airfall ash were likely to have accumulated, and outline them on Figure 7.12.

130   Part III   Maps and Images

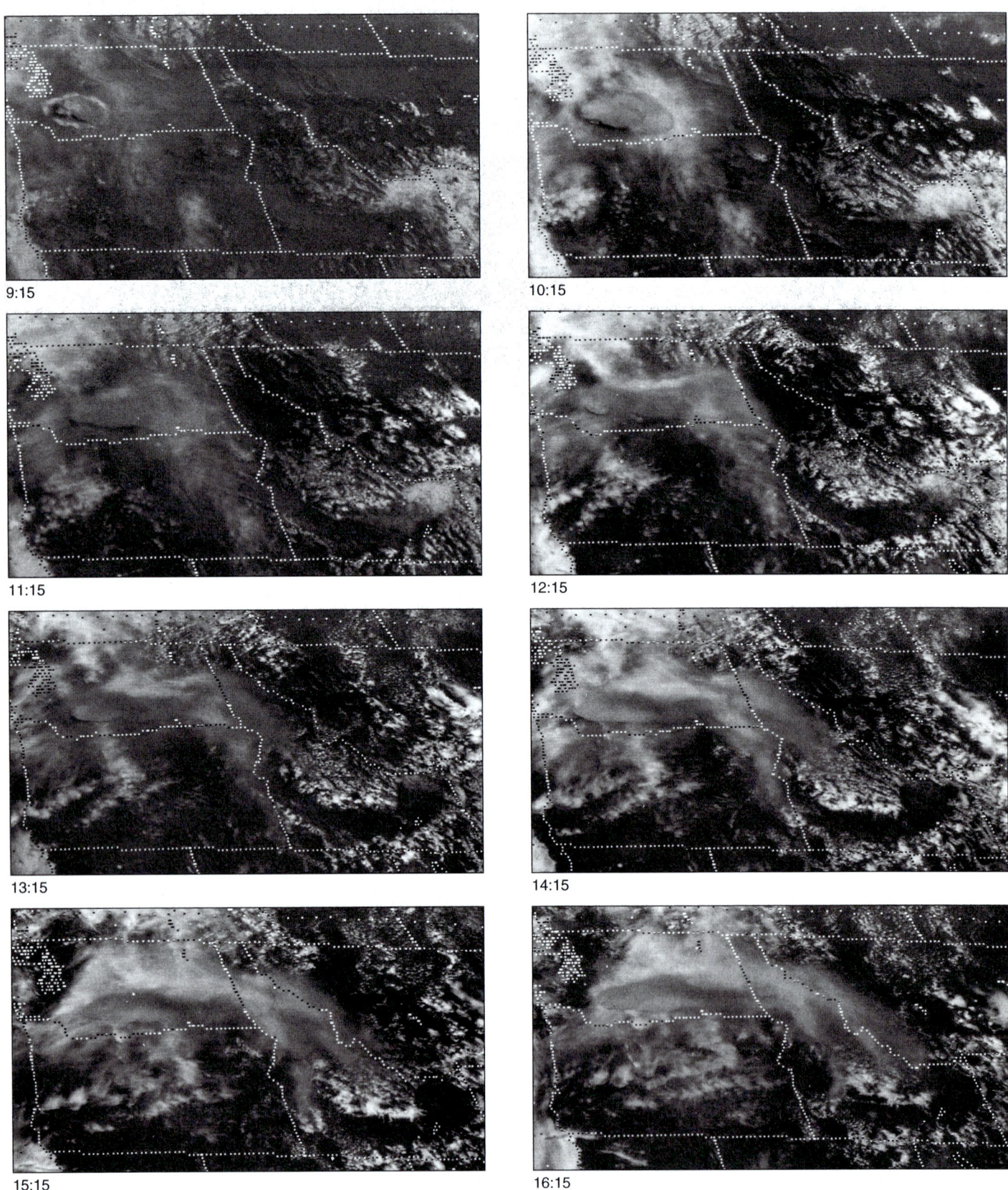

FIGURE 7.11
Satellite images taken by GOES from a geostationary orbit illustrate the movement of the ash cloud following the May 18, 1980, Mount St. Helens eruption. The ring in the first image is a shock wave. The ash cloud is inside the ring. As the original scientists examining these images had to do, do your best to distinguish between the expanding ash cloud and the normal water vapor clouds that were present on the day Mount St. Helens erupted. For use with Problem 5.

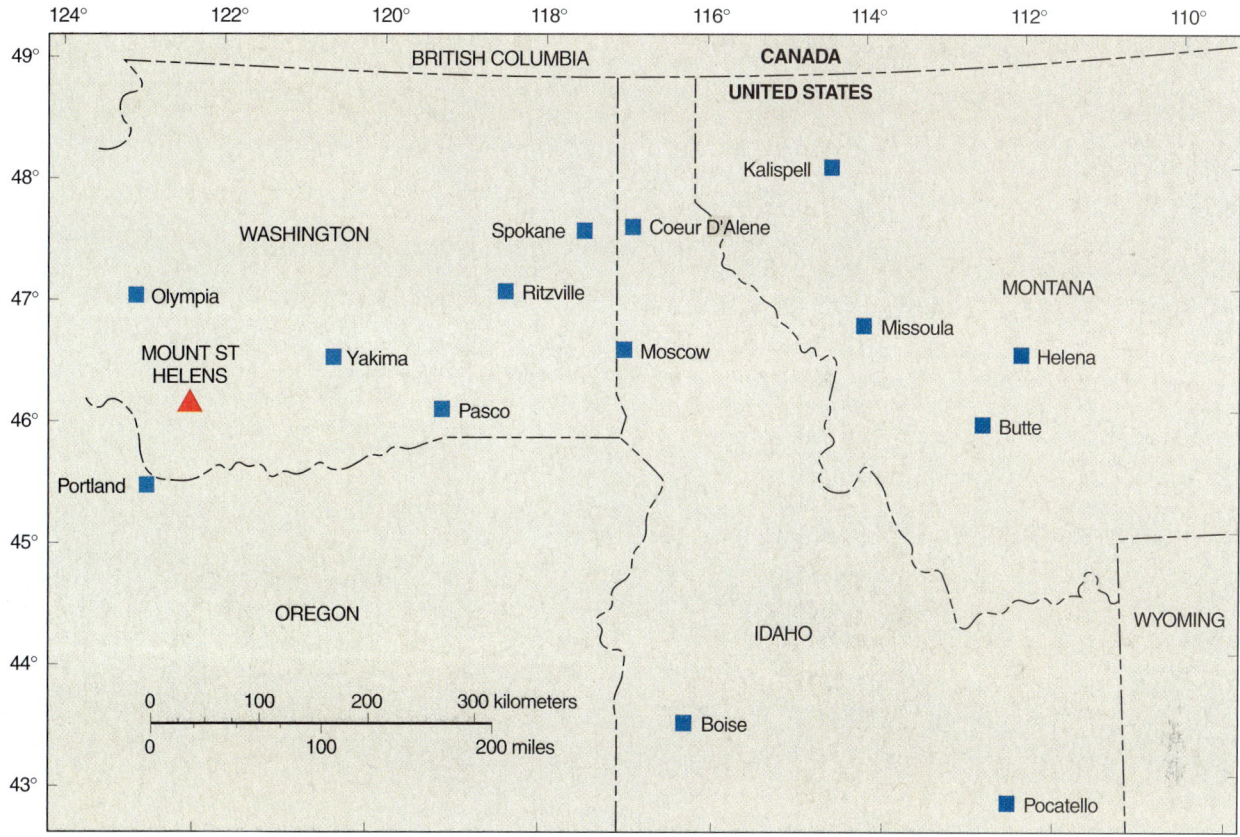

FIGURE 7.12
Outline map of area affected by ash from Mount St. Helens. For use with Problem 5.
Source: U.S. Geological Survey.

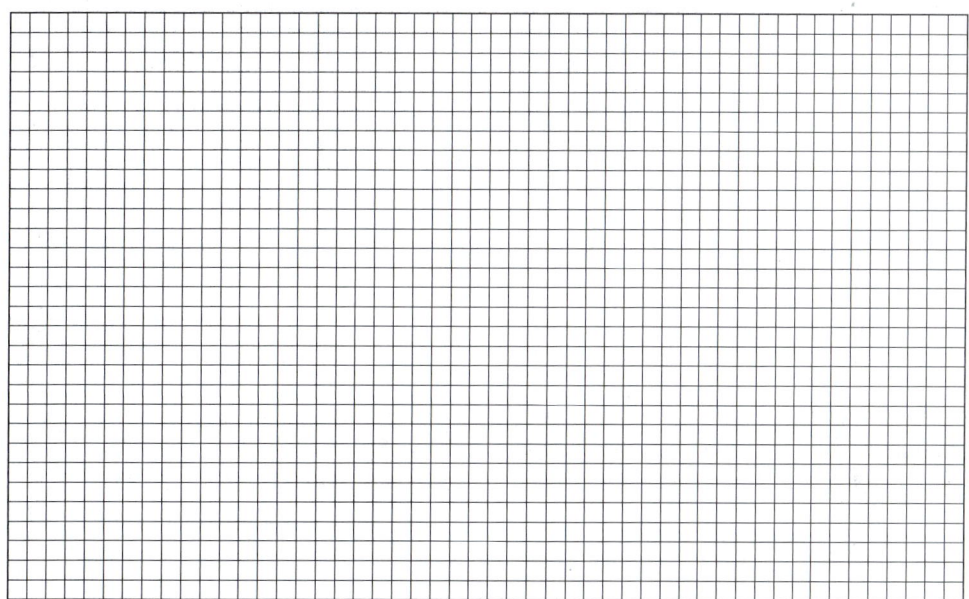

FIGURE 7.13
Graph for use with Problem 5.

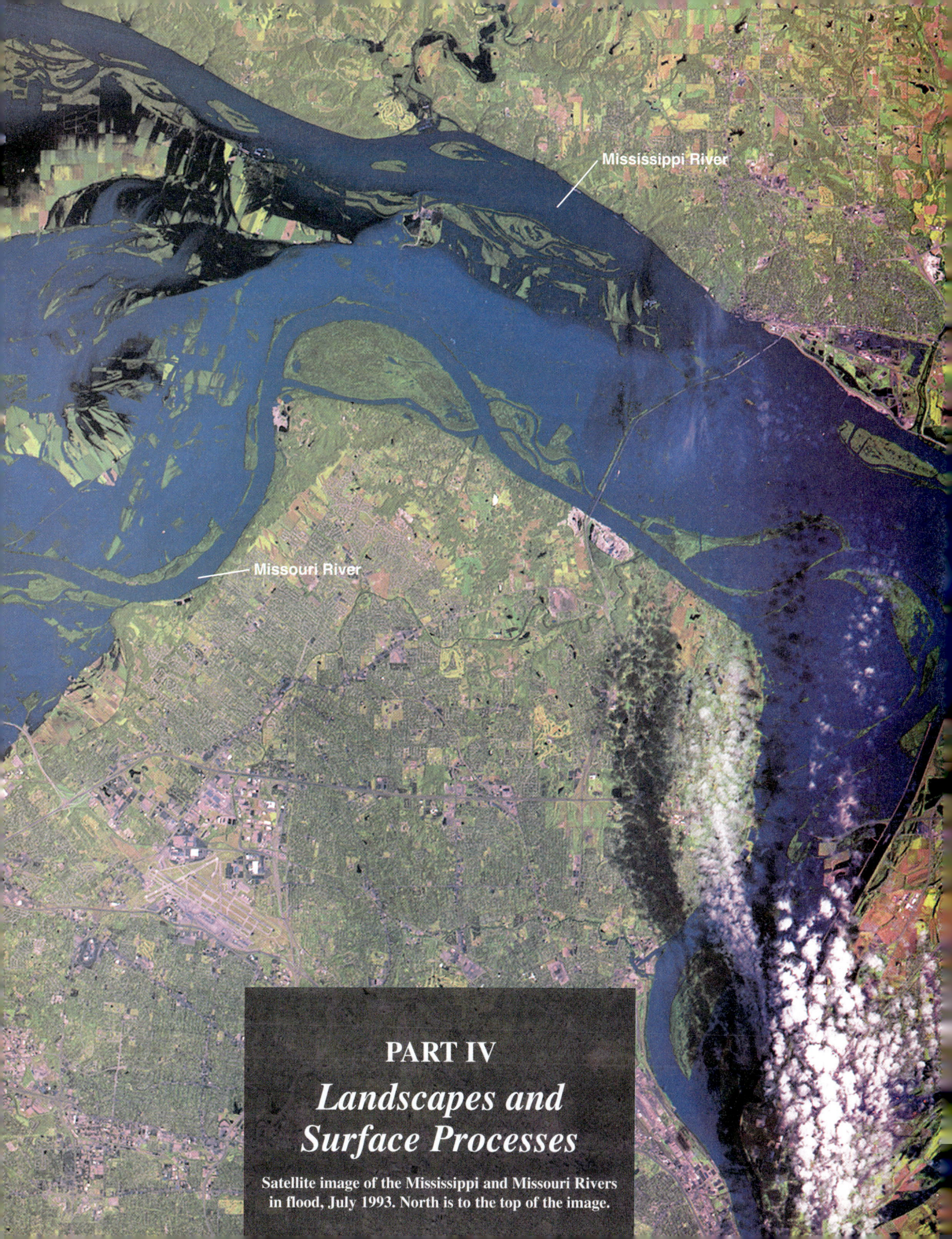

PART IV
Landscapes and Surface Processes

Satellite image of the Mississippi and Missouri Rivers in flood, July 1993. North is to the top of the image.

CHAPTER 8

Streams and Humid-Climate Landscapes

Materials Needed
- Pencil and eraser
- Colored pencils
- Calculator
- Ruler

Introduction

Running water is the most important force shaping the Earth's surface. Even desert landscapes are dominated by drainage systems carved by infrequent rainstorms. Thus, many familiar landforms owe their origin to streams. In addition, many cities are located on rivers and depend on them for drinking water, freight transport, and recreation. In this lab you will learn how to determine the size, shape, and runoff characteristics of a drainage basin and to recognize and interpret the common erosional and depositional features formed by running water. You will see how these features depend on the underlying geology, climate, and changes in elevation relative to base level and time. You will also learn to use data from gaging stations to predict the frequency, size, and extent of major floods using data available on the World Wide Web.

The first step in erosion by running water is weathering. Weathering causes rocks at the surface to disintegrate and decompose. Because rainfall supports vegetation and vegetation holds a layer of soil in place over unweathered rock, humid landscapes tend to have relatively rounded ridges, valleys, and slopes. Despite these stabilizing influences, loose sediment and soil eventually move downhill by mass wasting or water erosion. Once they reach a stream, the sediments are carried farther downslope by fast currents. This lab focuses on the landscapes produced by streams in humid regions—those with annual precipitation of more than 50 cm (Fig. 8.1).

Runoff and Drainage Basins

Streams are part of the hydrologic cycle, as shown in Figure 8.2. They carry water precipitated on the surface back to the oceans as surface runoff. Some of the precipitation returns to the atmosphere by evaporation and transpiration (emission through the leaves of plants), and some soaks into the ground and is carried away below the surface (the *infiltration loss* in the equation below). Water that enters the streams and makes up the *runoff* comes from direct overland flow and from water that moved underground before being discharged into the streams. Surface runoff, the runoff due to overland flow, can be expressed as:

surface runoff = precipitation – infiltration loss – evaporation – transpiration.

Anything affecting one of these terms affects surface runoff, so the percentage of precipitation that naturally runs off varies considerably worldwide. In Arizona, where precipitation is about 25 cm/year, evaporation and infiltration are high, and runoff is very low, typically 2 cm/year or less. In contrast, precipitation in Alabama is about 125 cm/year, and, because infiltration and evaporation are low, runoff is about 75 cm/year. Human activity also affects runoff. For example, runoff increases when the land surface is paved so that rain doesn't soak into the ground, or when vegetation is removed so that the amount of transpiration is reduced.

When water runs off a surface, it flows downhill and into a stream. The stream is part of a drainage network in which smaller streams feed larger streams. As shown in Figure 8.3, each stream, no matter how

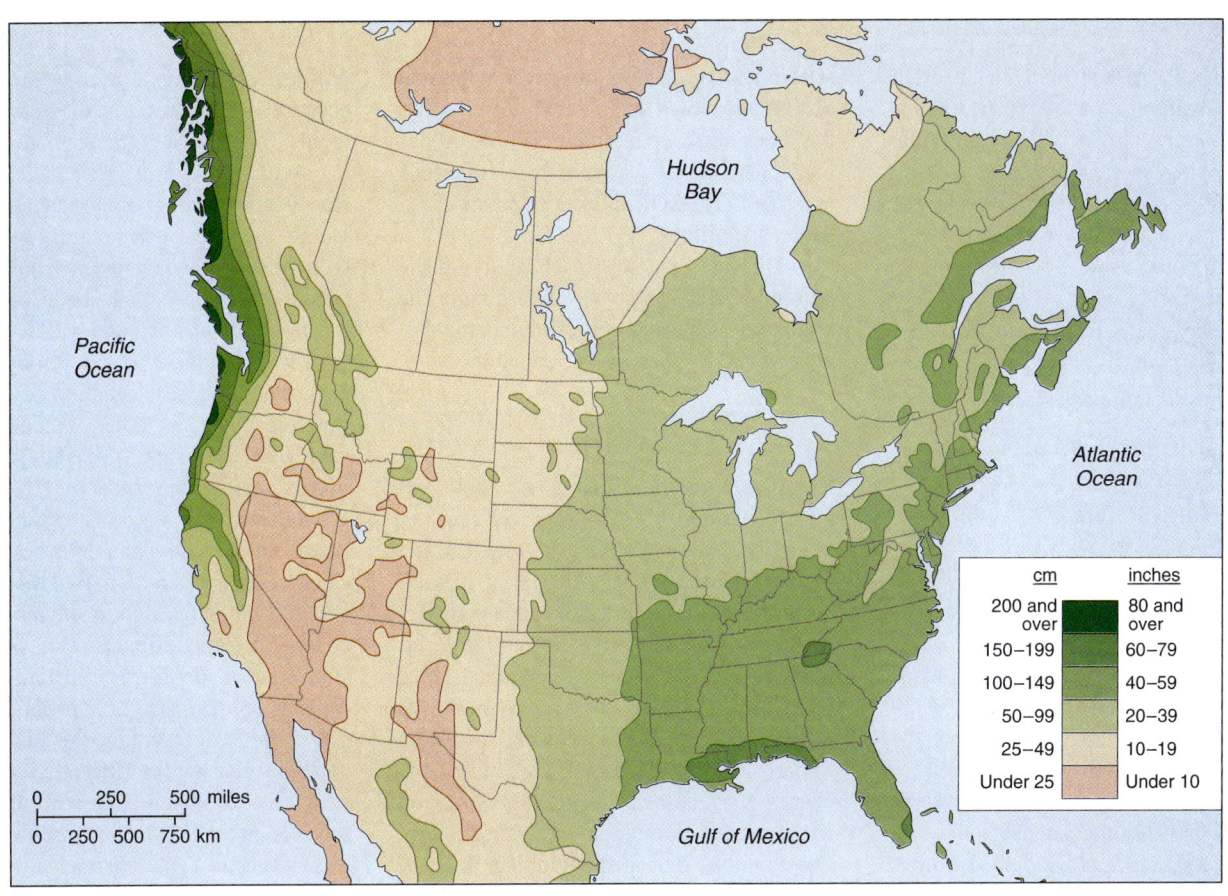

FIGURE 8.1
Average annual precipitation in the United States and Canada. Areas with more than 50 cm per year are considered humid. From *Geosystems,* 2nd edition by Robert W. Christopherson. Copyright © 1994. Reprinted by permission of Prentice-Hall, Inc. Upper Saddle River, N.J.

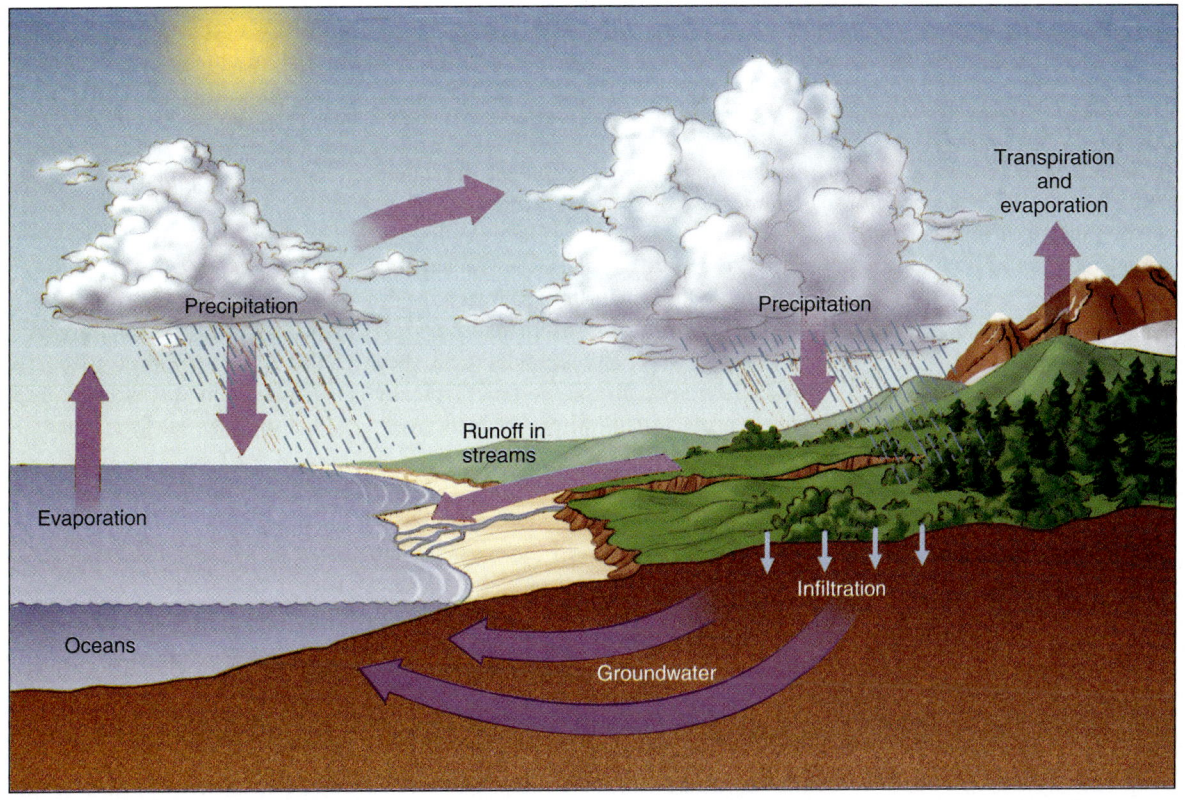

Water resources: Oceans and inland seas, 97.22%; glaciers, 2.15%; ground water, 0.62%; lakes, rivers, and streams, 0.01%

FIGURE 8.2
The hydrologic cycle and relative volumes of Earth's water resources. This chapter emphasizes runoff.

large or small, has an area that it drains called a **drainage basin** or **watershed.** Drainage basins are separated by **divides,** which are higher points between basins. Rain falling on one side of a divide goes downhill into one drainage basin; that falling on the other side goes into another drainage basin.

Using Clayborn Creek in Figure 8.3 as an example, divides can be located, and the drainage basin outlined, as follows:

1. Locate the stream in question (Clayborn Creek) and its tributaries (branches) and note which way each is flowing (i.e., which way is downhill). Note how contours crossing valleys "point" uphill.

2. Locate other streams on the map and note which direction they are flowing. For example, in Section 31 (red number 31 on Figure 8.3) a tributary of Clayborn Creek flows south-southeast from the vicinity of Inspiration Point. Another stream in the northeast corner of Section 31 flows north. Both streams flow away from a divide into their own drainage basins.

3. Find the highest point on a line between the ends of the two streams. That point will lie on a divide and, therefore, on the margin of the drainage basin of Clayborn Creek. In this case, the elevation is about 1440 feet.

4. The divide here is a well-defined ridge. In this case, a road more or less follows the ridge on the north side. By following the ridge, the drainage basin of Clayborn Creek can be outlined (the dashed line).

Stream Channels and Valleys

Figure 8.4 shows a contour map of a stream and a topographic profile drawn along its length. The **longitudinal profile** illustrates how the **gradient** (see Chapter 6) gradually decreases from the **head** to the **mouth** of the stream. The mouth is the **base level** for a stream: it limits the depth to which that stream can erode. Over time, a stream adjusts its channel and longitudinal profile in response to changes in discharge, base level, and erodibility of the rock or sediment over which it flows. Ideally, the adjustments lead to a near *balance* between erosion and deposition along the course of a stream and produce a smooth longitudinal profile, as shown in Figure 8.4. A stream that does not have a smooth profile erodes or deposits so as to attain one: waterfalls and rapids are eroded, lakes or ponds along streams are filled.

The size of a stream channel and the volume of water increase downstream. The volume of water per unit of time is the **discharge** and is given by this equation.

$$discharge = velocity \times cross\text{-}sectional\ area\ of\ channel$$

Common units for discharge are cubic meters per second (m^3/sec) or cubic feet per second (ft^3/sec); for velocity, meters per second (m/sec) or feet per second (ft/sec); and for cross-sectional area, square meters (m^2) or square feet (ft^2). As discharge increases, so do all of its components. Thus, during flooding, velocity and channel size increase as the volume of water increases. Stream velocity, channel area, and discharge are recorded at *gaging stations* on many streams throughout the United States. For example, during the summer flood of 1993 (see Part IV opening image, p. 133), the gaging station on the Mississippi River at St. Louis, Missouri, recorded a peak discharge of 1,070,000 ft^3/sec on August 1. This is more than eight times the average August discharge of 133,000 ft^3/sec.

Features of Streams and Their Valleys

Streams vary, from turbulent mountain streams rushing down narrow valleys to great rivers with wide valleys flowing across a nearly flat landscape. As streams vary, so do their characteristic features.

Streams with steep gradients tend to erode downward more rapidly than they erode laterally. Therefore, they typically have narrow valleys with V-shaped cross-profiles. Longitudinal profiles are irregular because of the presence of waterfalls and rapids. Figure 8.5 illustrates these features.

With decreasing gradient, lateral erosion becomes more important, and broad valleys develop. Such streams have a variety of distinctive features, as illustrated in Figure 8.6. The actual stream channel is much narrower than the valley, most of which is occupied by the **floodplain,** the area that could be submerged during a flood. Just adjacent to the channel are **natural levees,** low ridges formed by sandy sediments rapidly deposited by flood waters. **Backswamps** develop in low areas on the floodplain behind natural levees. A river channel is not straight, but meanders or winds about in the floodplain; a bend in the channel is a **meander.** Erosion on the outside of a meander forms a **cutbank,** and deposition on the inside of a meander forms a **point bar.** This combined erosion and deposition causes the meanders to move across the floodplain. The **meander belt** is the zone in the floodplain within which meanders occur. The channel may take a shortcut across a meander loop to form a **cutoff,** or abandon the loop altogether to form an **oxbow lake.** Where floodplains are wide and natural levees are high, tributary streams may flow in the floodplain for long distances before joining the main river; such tributaries are known as **yazoo streams. Stream terraces** are step-like benches above the level of the present-day floodplain. They represent the remnants of preexisting floodplains or valley floors.

Floods and Recurrence Intervals

A flood occurs when a stream overflows its channel. The size of a flood, as measured by maximum discharge, or by **stage** (elevation of water surface), varies from year to

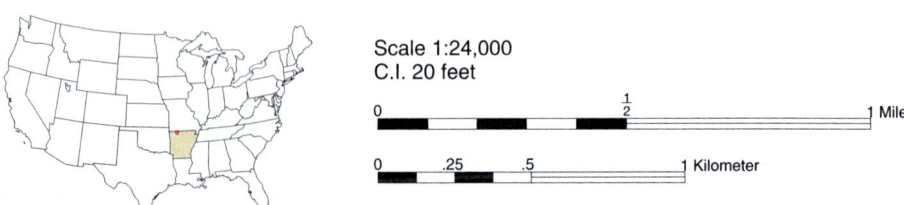

Scale 1:24,000
C.I. 20 feet

FIGURE 8.3

(Page 137) The drainage basin of Clayborn Creek and its tributaries is outlined on this portion of the Beaver, Arkansas/Missouri, quadrangle. Note that the divides connect the highest elevations between adjacent drainage basins. Water falling within the drainage basin moves downhill toward Clayborn Creek and eventually makes its way into the White River.

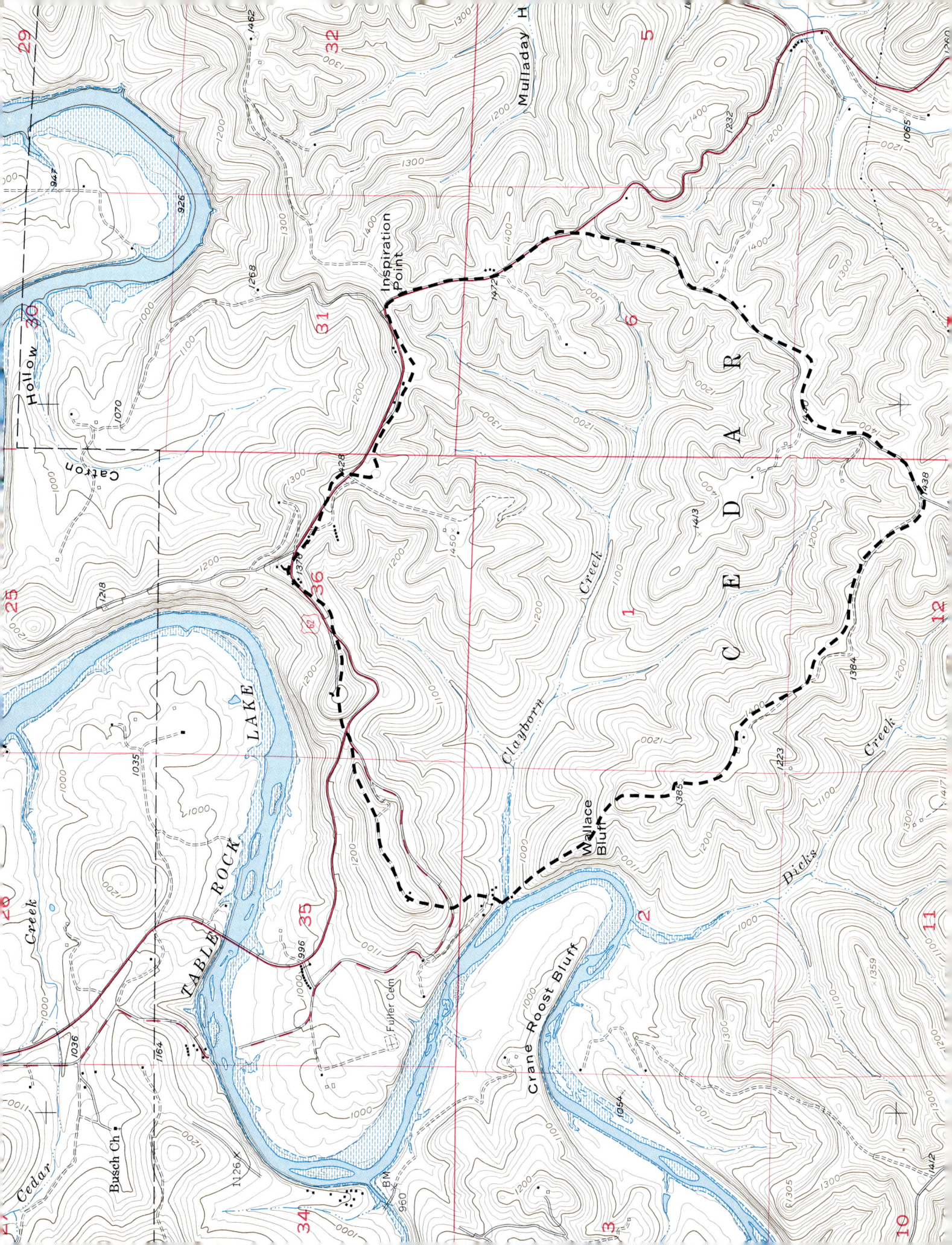

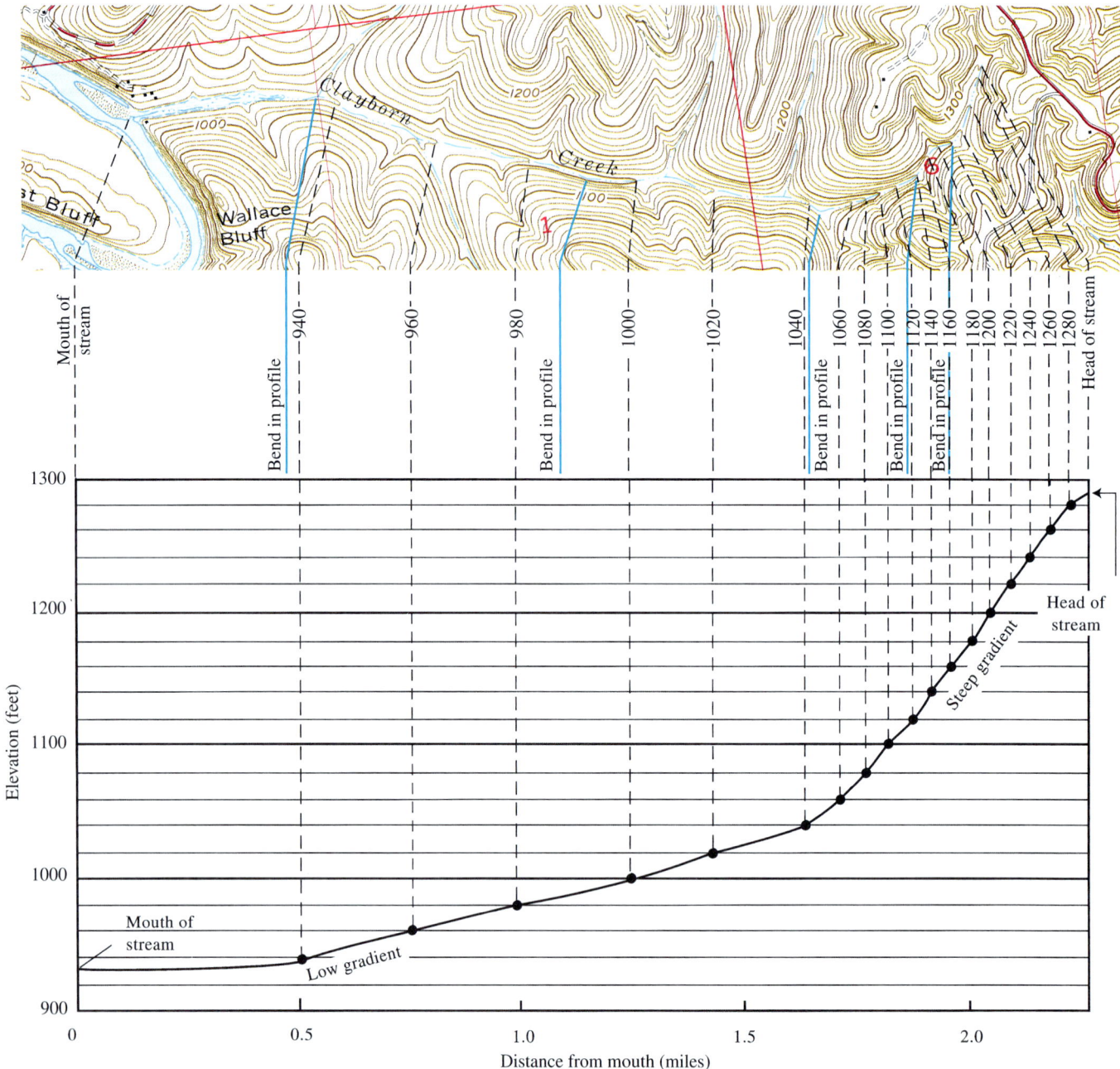

FIGURE 8.4

The longitudinal profile along Clayborn Creek was drawn by marking, on the top of the profile paper, points where contours cross the stream and labeling these points with their elevations. The profile follows the bends in the stream, not the straight-line distance between the head and the mouth. An easy way to follow along the stream when marking the contour-crossing points is to hold the paper down with a pencil where the stream bends, then rotate the paper to follow along the next stream segment. The locations of the bends are shown here for illustrative purposes only. Note that the gradient decreases from the head to the mouth of the stream. The vertical exaggeration is 15.9 times. (The horizontal scale is 1:22,850, and the vertical scale is 1 inch to 120 feet.)

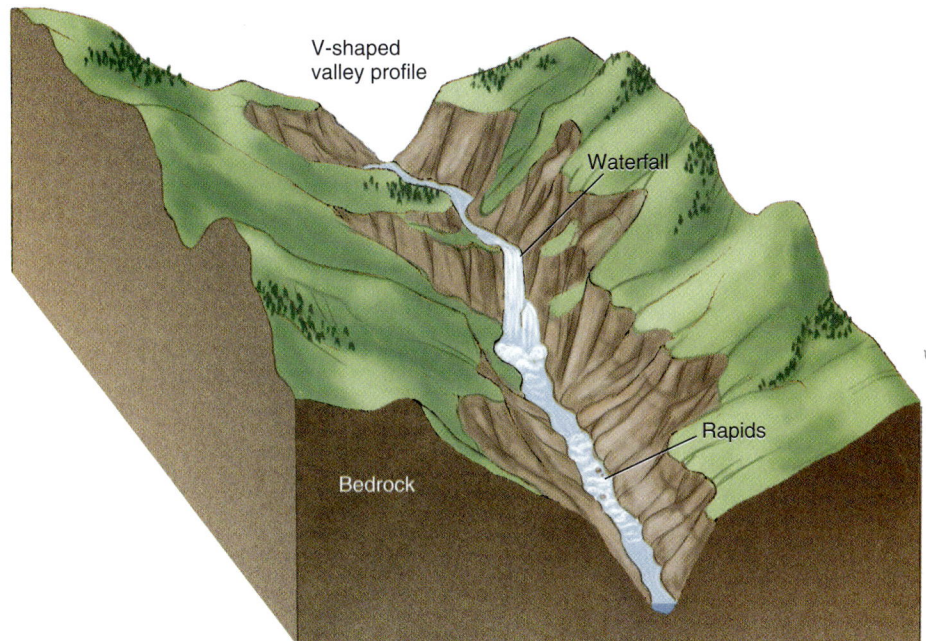

FIGURE 8.5
Features of a stream with a steep gradient.

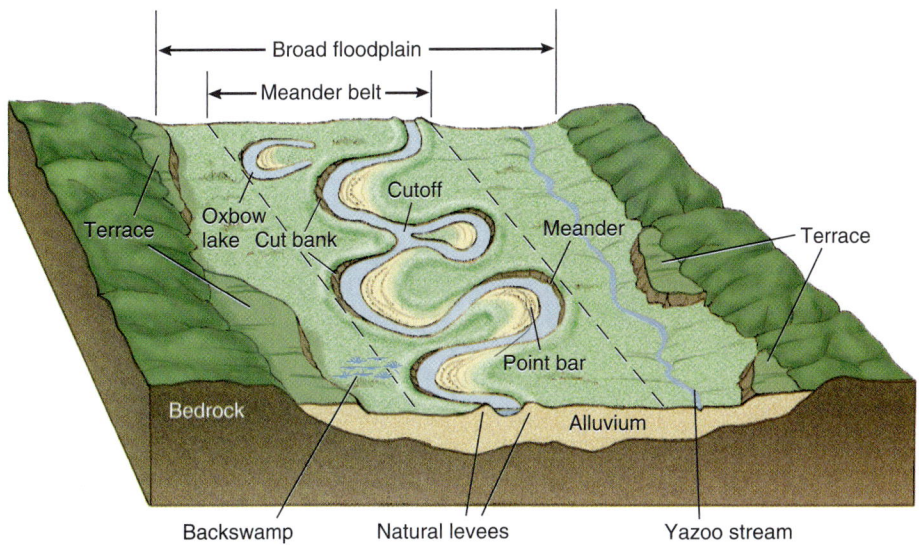

FIGURE 8.6
Features of a stream with a gentle gradient.

year. By analyzing the frequency of floods of various sizes, a *recurrence interval* can be developed for a river at a particular locality. The **recurrence interval,** usually measured in years, is the average interval between floods of a particular size. Thus, *on average,* a 100-year flood will recur at intervals of 100 years. That does not mean that a flood that size could not occur two years in a row; it means that the chance of it occurring in *any* year is 1 in 100. A *flood-frequency curve* plots discharge, or in some cases, stage, against recurrence interval, as illustrated in Figure 8.7.

Floodplain zoning is based on recurrence intervals. If planners know what stage a stream will reach in a 50-year flood, they can determine what parts of a floodplain are likely to be flooded about once every 50 years. They can then zone accordingly, deciding which areas are better for homes, businesses, and hospitals and which for parks and recreation. The width of a 50-year floodplain is determined from gaging-station records and flood-frequency curves. However, this width may be uncertain if a gaging station was in place too few years or if there is no station on the stream in question.

Landscapes in Humid Areas

Drainage Patterns

Drainage patterns reflect the characteristics of the underlying rocky materials (Fig. 8.8). Landscapes developed on flat-lying, homogeneous rock or unconsolidated sediments will naturally develop a **dendritic** (branching like a tree) drainage network. Water running off a volcano or uplifted dome will create a **radial drainage pattern** of streams flowing from a central point. If the underlying rock has been cut by fractures, these are more easily eroded and cause drainages to form a **rectangular drainage pattern.** In mountainous areas where rock layers have been folded and eroded to expose long resistant ridges, most streams will run along the valleys to join a larger river that has cut across the ridges. This forms a **trellis drainage pattern** (Fig. 8.8).

Evolution of Stream Systems

Stream valleys change with time in ways that are fairly predictable *if* the controlling factors remain constant. The following characteristics can describe the evolution of a single, long-lived drainage system or the features of a single river moving from the headwaters to the midsection to the lowlands, especially if the headwaters are in the mountains.

Early-stage streams show steep gradients down to a base level created by a lake, larger river, or ocean. High-gradient streams cut deep into the landscape and form V-shaped valleys with narrow floodplains (Fig. 8.5). Waterfalls or rapids frequently occur. The surrounding landscape shows high hills, deep valleys, and prominent drainage divides.

With time, erosion does three things: it cuts farther back into the hills in a process called **headward erosion** (Fig. 8.9); it brings the elevation of most of the stream channel nearer to base level, thus reducing

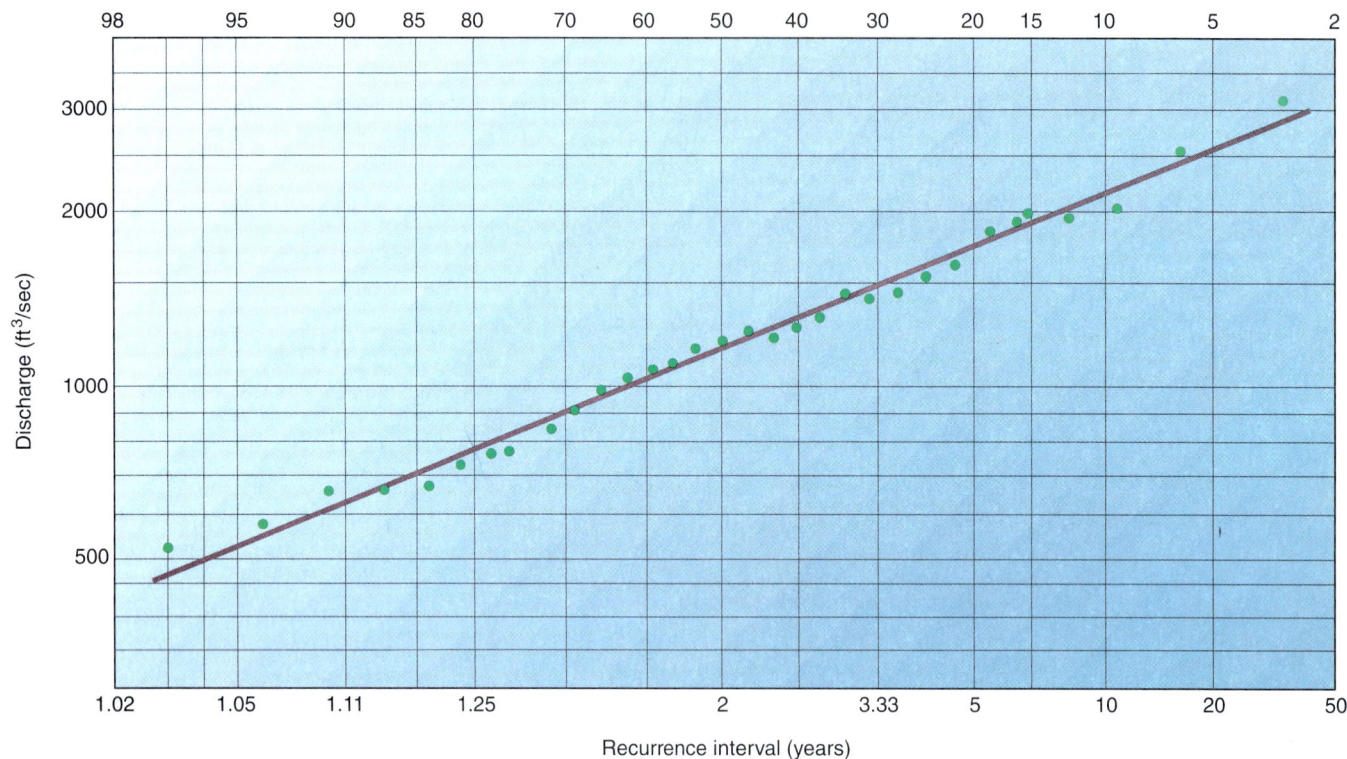

FIGURE 8.7
Flood-frequency curve for Rock Creek near Red Lodge, Montana, 1932 to 1963, using maximum annual discharge data from the USGS. From the *Recurrence Interval (RI)* scale (bottom), it can be predicted that, on average, a discharge of 2600 ft³/sec will be attained every 20 years. Using the *Percent Probability of Recurrence (P)* scale on top, the same thing can be said in a different way: there is a 5% probability that a discharge of 2600 ft³/sec will be reached in a given year. $P = 100/RI$.

the river gradient; and it reduces the elevation and relief of the surrounding landscape. As a result, **middle-stage streams** are characterized by longer drainage systems with more tributaries, moderate gradients, and increasingly rare waterfalls and rapids. In addition, lower gradients reduce the importance of downcutting while increasing the importance of lateral (side-to-side) erosion. Lateral erosion creates meandering channels, wider flood plains, and broad, flat valleys within otherwise hilly landscapes.

With still more erosion, middle-stage streams become **late-stage streams,** which feature low gradients and extensively developed meander belts characterized by wide floodplains and extensive systems of oxbow lakes, backswamps, and occasional yazoo rivers (Fig. 8.6). While low hills may bound one or both sides of a late-stage floodplain, the surrounding landscape is overall fairly flat.

Headward erosion puts streams in adjacent drainage basins in competition with each other. Clayborn Creek (Fig. 8.3), for example, is trying to expand its area while at the same time the surrounding creeks are trying to expand their areas. If erosion rates on either side of a divide are the same, headward erosion stalls, the position of the divide stays about the same, and the landscape merely loses elevation. If erosion rates are higher on one side, headward erosion causes the faster-eroding stream to cut into the drainage basin of the other stream as it wears down the landscape. Eventually, the dominant stream can snare one or more tributaries of the other stream, thus **beheading** it in a process known as **stream capture** or **stream piracy** (Fig. 8.9).

Early geologists formulated the idea of stages in stream development in terms of a grand *erosion cycle*. However, this idea is complicated by the fact that the controlling factors of landscape development are rarely constant over the many millions of years it takes to go from early- to late-stage streams. A change in climate can increase runoff,

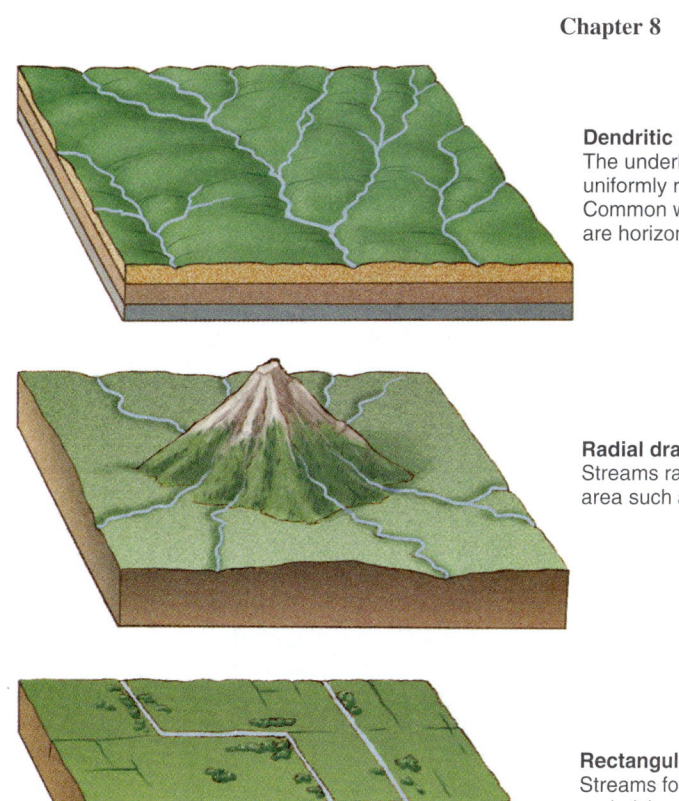

Dendritic drainage pattern.
The underlying rock or sediment is uniformly resistant to erosion. Common where sedimentary layers are horizontal.

Radial drainage pattern.
Streams radiate outward from a high area such as a volcano or dome.

Rectangular drainage pattern.
Streams follow fractures or faults in underlying rocks. Right-angle intersections are common.

Trellis drainage pattern.
Most streams follow valleys that parallel inclined layers of rock.

FIGURE 8.8
Four common drainage patterns.

thus causing a middle-stage stream to cut into its meander belt sediments and leave river terraces. Regional uplift of a stream or a drop in base level would also cause erosion by downcutting. The uplift of headwater tributaries or an increase in glacial activity could increase the amount of sediment coming into the middle stages of a river system and thus fill the valley with meander belt sediments. A rise in sea level or lowering of the land surface could also increase deposition of meander belt sediments. Additionally, a rim of resistant rock can temporarily create a secondary base level at a relatively high elevation within a river system. This causes meanders to develop upstream. Once the resistant rocks are breached by erosion, the river will cut into its old meander belt sediments and create terraces. Overall, many river systems reflect a complex mixture of present and past conditions.

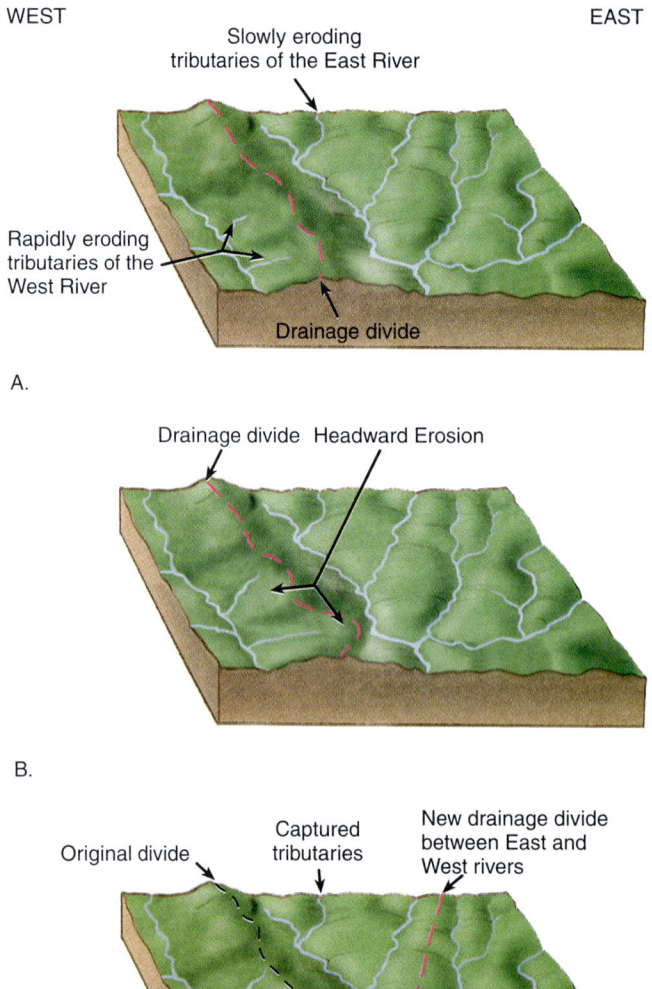

FIGURE 8.9

Headward erosion and stream piracy. A. Erosion is faster on the west side of the drainage divide than on the east. B. This enables several stream tributaries to cut into the divide and lengthen their drainages in a process called headward erosion. C. Stream piracy occurs when headward erosion cuts into a tributary from another stream system, thereby capturing its water flow.

Hands-On Applications

Geologists have approached the study of streams and associated landscapes in several ways. Early approaches were descriptive systems of classification of natural phenomena, which assumed cause-and-effect relations among various features. Working models were constructed to help understand rivers and river systems. Eventually, enough numerical data became available from both natural and model systems to develop mathematical models, and quantitative predictions could be made. The problems in this lab illustrate both qualitative and quantitative approaches.

Objectives

If you complete all the problems, you should be able to:

1. Identify the following features on a map or aerial photograph: floodplain, meander, meander belt, backswamp, oxbow lake, cut bank, point bar, natural levee, cutoff.

2. Determine whether the characteristics of a stream or landscape are those of early, middle, or late stage.

3. Identify dendritic, radial, trellis, and rectangular drainage patterns, and use them to predict the general structure of the underlying rocks.

4. Use data from a gaging station to calculate the *percent probability of recurrence* or the *recurrence interval* of a particular discharge or river stage and to construct a flood-frequency curve.

5. Outline a drainage basin on a topographic map, estimate its area, and calculate its annual runoff from precipitation, evaporation, transpiration, and infiltration data.

6. Draw both cross and longitudinal profiles of a stream valley.

Problems

1. Figures 8.10, 8.11, and 8.12 illustrate segments of three major U.S. rivers that show different characteristics.

    a. Examine the maps and complete Table 8.1. Note that the maps are at different scales. The purpose of this problem is to recognize common stream features on a map and to learn which stream features are typically associated with one another. When calculating gradients for Table 8.1, measure the distances along the stream's actual course, as demonstrated in Figure 8.4.

    b. Levees are often too low to show up on contour maps. On which river(s) would you expect to find natural levees?

    c. Levees are marked on the Mississippi River map. Do you think these are natural or made by people? What is your evidence?

    d. Because a river normally occupies the lowest points within its valley, small streams generally flow in a fairly direct path to join the main river. Carefully examine the somewhat complicated drainages formed by the numerous bayous (these are both permanent and intermittent streams) west of the Mississippi. Do they flow toward or away from the main river channel? Why do they flow this way?

e. Levees can be most obvious during a flood. Examine the satellite image of the confluence of the Missouri and Mississippi rivers just north of St. Louis, Missouri (Part IV opening image, p. 133). St. Louis covers the whole area south of the rivers and west of the clouds. The Mississippi comes in from the northwest corner and the Missouri comes in just south of the Mississippi. Which river shows beautiful examples of levees?

Where is there a good example of a cutoff?

Where is there an artificial canal roughly half the width of the nearby river?

The curved landforms east of the clouds reflect current oxbow lakes, curved ridges from old point bars, and filled-in oxbow lakes. These are typical of meandering river landscapes.

TABLE 8.1
Characteristics of Three Stream Valleys

Stream Characteristics	Missouri River	Arkansas River	Mississippi River
Are there meanders?			
backswamps?			
oxbow lakes?			
cutoffs?			
River gradient within map area (in feet per mile)	(See footnote 1)		(See footnote 2)
Approximate width of floodplain (from one side to the other)			
Sketch a valley profile. Mark channel position in profile and show approximate position of floodplain.			
Importance of downcutting (high, medium, low)			
Importance of lateral erosion (high, medium, low)			
Development stage (early, middle, late)			

[1] A supplementary contour with an elevation of 750 feet crosses the river 1.5 miles below the bottom edge of the map. Use this information, together with what can be observed, to calculate the gradient.

[2] The 100-foot contour crosses the river 74 miles above the top edge of the map. Use this, along with what is observed on the map, to calculate the gradient.

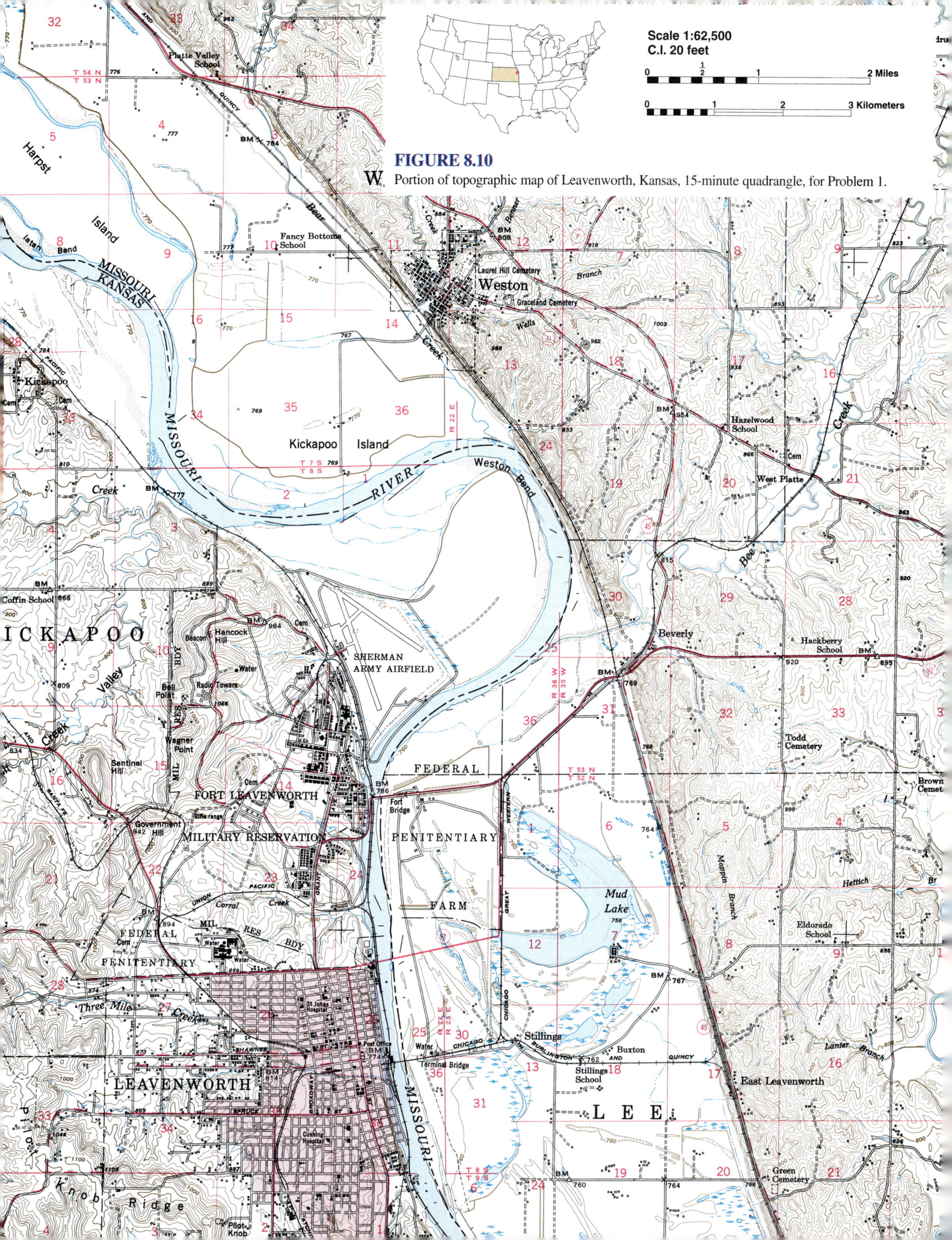

FIGURE 8.10 Portion of topographic map of Leavenworth, Kansas, 15-minute quadrangle, for Problem 1.

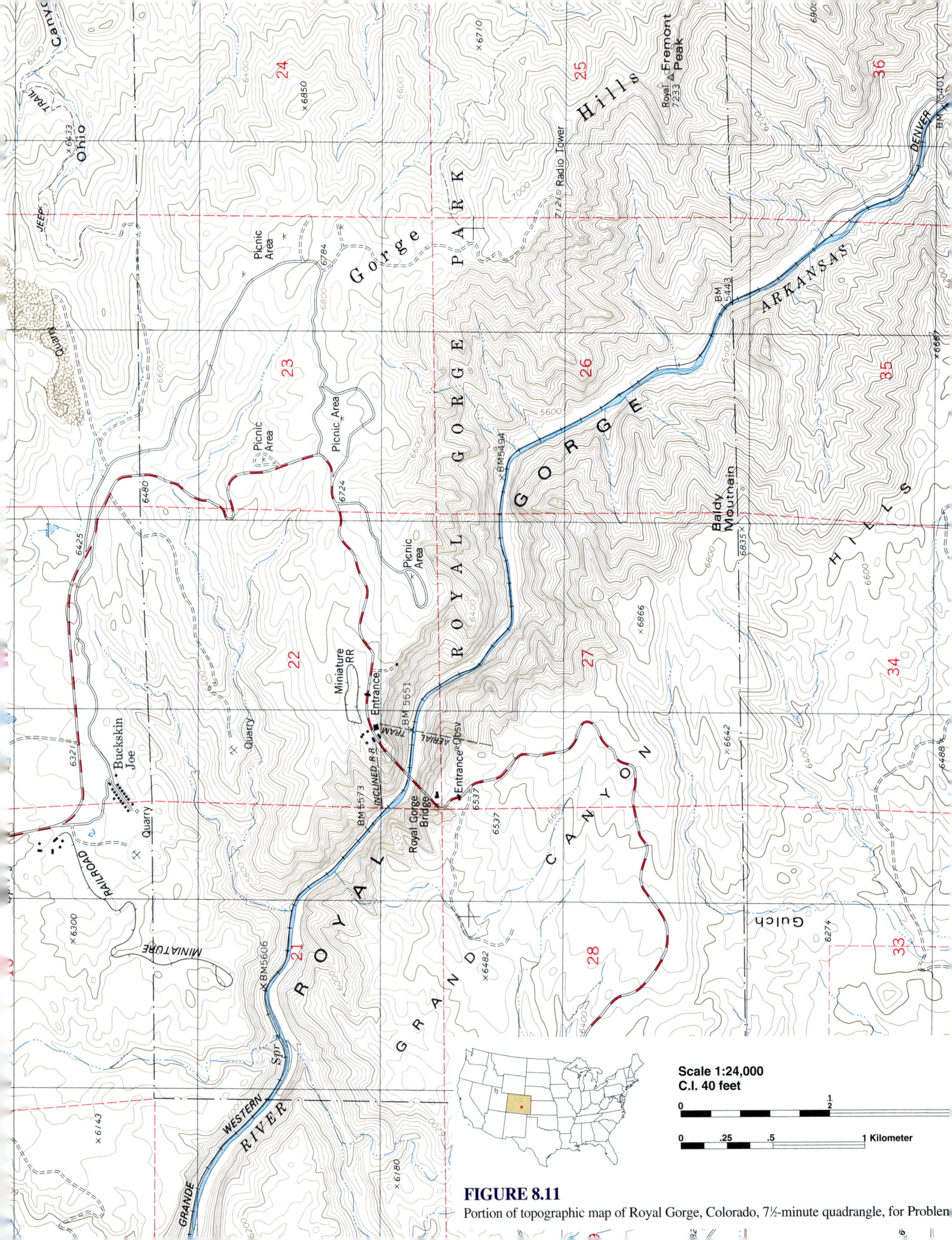

FIGURE 8.11
Portion of topographic map of Royal Gorge, Colorado, 7½-minute quadrangle, for Problem

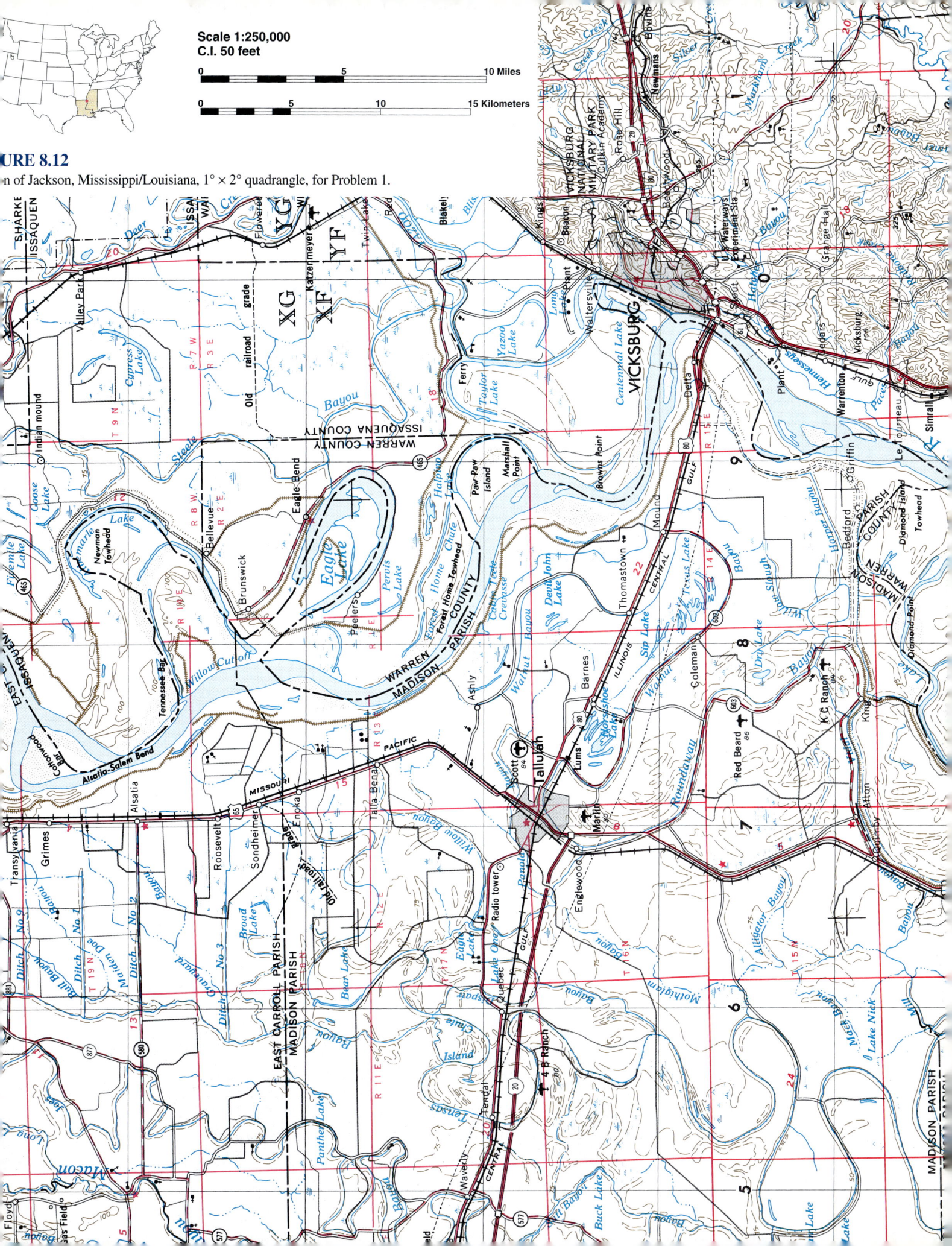

FIGURE 8.12
Portion of Jackson, Mississippi/Louisiana, 1° × 2° quadrangle, for Problem 1.

2. Refer to Figure 8.13, a portion of the Kaaterskill, New York, quadrangle. This area, in the northeast corner of the Catskill Mountains of southeastern New York, shows a contrast in topography between the east and west sides of the map. The western part of the area is underlain by sedimentary rocks with nearly horizontal layers, and the eastern part is underlain by sedimentary rocks with steeply inclined layers.

   a. How do the drainage patterns in the eastern and western parts of the area differ?

   b. Are rocks in the eastern or western part of the area more resistant to erosion? How do you know?

   c. Which of the two creeks, Schoharie Creek or Plattekill Creek (between its head and West Saugerties), has the steepest gradient?

   Which do you think is eroding headward most actively, and why?

   What is the likely future for the headwater tributaries of Schoharie Creek?

   d. Kaaterskill Creek has committed stream piracy and "beheaded" Gooseberry Creek. Use a colored pencil to indicate on the map the probable course of Gooseberry Creek before this tyrannous event.

RE 8.13 of topographic map of Kaaterskill, New York, quadrangle, for Problem 2.

3. Figure 8.14 is a map showing the Minnesota River at Jordan, Minnesota. The *gaging station* at the bridge over the Minnesota River north of Jordan is used to monitor the *stage* or level of the river as well as discharge and water quality. The stage here is the number of feet above the gaging-station elevation of 690'. Table 8.2 shows the maximum stage reached during selected years from 1935 to 2003. A relative magnitude, m, has been assigned to each stage and is recorded in column 3 of Table 8.2. The highest maximum stage was reached in 1965, so its magnitude, m, is 1; the lowest maximum stage was attained in 1935 (not shown in the table) and its magnitude is 69 (the table summarizes 69 years of records).

   a. Construct a **flood-frequency curve** as follows:

   (1) Calculate the *percent probability of recurrence, P,* for each of the events listed in Table 8.2, and record it in column 4 of that table. P is the percent probability that a stage of a particular height will recur—be reached or exceeded—in a given year. It is determined by:

   $$P = 100 \times m/(n + 1)$$

   where n is the number of years for which records were kept, and m is the relative magnitude of the event. For example, for a stage that ranked 30th in magnitude for a period of 69 years, $P = 100 \times 30/(69 + 1) = 42.9\%$.

   (2) Plot the data in Table 8.2 on the graph paper in Figure 8.15. Be sure to use the scale on the top of the graph, labeled *Percent probability of recurrence (P),* when plotting your numbers. Then draw a best-fit straight line through the data points to complete the flood-frequency curve. The graph paper is called probability paper, because it can be used to determine the probability, or likelihood, that the event plotted on it will occur. In the example above, there is a 42.9% probability that a stage equal to that reached during the $m = 30$ event will be reached in any given year.

   b. The **recurrence interval,** or average *time* interval between similar-sized floods, is another way of describing recurrence. The recurrence interval, RI, is given by:

   $$RI = (n + 1)/m$$

   and $RI = 100/P$. Values for the recurrence interval are shown on the lower part of the graph in Figure 8.15. Note that a 42.9% probability of recurrence also means that a stage of that value would be reached, on average, about every 2.4 years. The probability of a 50-year flood (one with a recurrence interval of 50 years) in any given year is 0.02, or 2%. Thus, there is always a probability of a 50-year flood, even if one had occurred just the previous year. Conversely, there is no guarantee that a 50-year flood will occur every 50 years.

   (1) Major flooding occurred in the spring of 2001, when the maximum stage reached 33.11 feet. Use your flood-frequency curve and best-fit straight line to determine what the recurrence interval was for this flood at this location.

   c. Use the flood-frequency curve you constructed (Fig. 8.14) to determine the maximum stages for a 100-year flood (one with a recurrence interval of 100 years). From the stage values, calculate the elevation of this flood (by adding stage to 690'), and using a colored pencil, outline the 100-year floodplain on the topographic map. Note that the obvious broad valley is not the current floodplain, but was carved when much more water was coming down the river than at present. This took place about 10,000 years ago during the melting of continental glaciers and the draining of a vast lake (Glacial Lake Agassiz—see Fig. 10.6).

   The floodplain you have outlined (or similar ones based on different recurrence intervals) can be used by government agencies to plan the use of floodplains or by insurance companies to set their flood insurance rates. For example, two-year floodplains should not be used for housing or other permanent types of development and flood-insurance rates would be prohibitively expensive. People building in areas of the floodplain with longer recurrence intervals should know they can expect to be flooded sometime and should be prepared to take that risk.

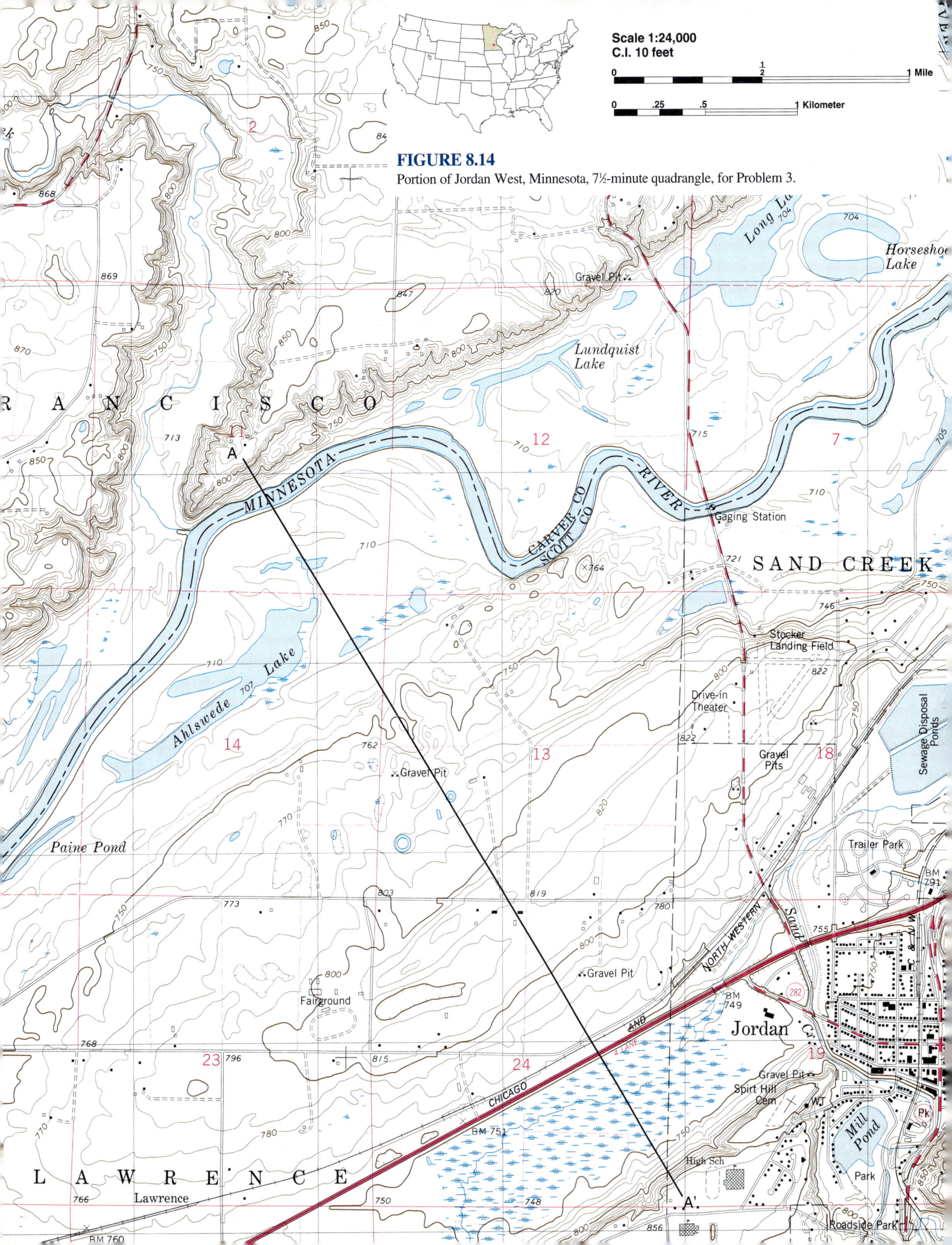

FIGURE 8.14
Portion of Jordan West, Minnesota, 7½-minute quadrangle, for Problem 3.

TABLE 8.2
Selected Data From the Gaging Station on the Minnesota River Near Jordan, Minnesota, 1935–2003

Year	Maximum Stage, feet	Magnitude, M	P
1936	22.43	30	
1942	14.36	57	
1946	18.22	48	
1952	28.31	6	
1958	12.30	63	
1959	8.37	67	
1964	16.15	54	
1965	34.37	1	
1969	32.85	4	
1975	23.77	23	
1976	11.19	65	
1984	27.54	8	
1985	25.05	17	
1986	26.30	9	
1990	20.23	38	
1993	33.52	2	
2003	19.10	44	

Source: USGS

4. **Watershed Data on the Web.** You can investigate local flood risks, or see whether your favorite river is likely to have too little (or too much!) water for a planned canoe trip, by going to *http://water.usgs.gov/* (or link to it through *www.mhhe.com/jones5e*—see Preface). There is a lot to explore on this website, but for now click on *NWISWeb Water Data* (NWIS stands for National Water Information System) and then click on *Real-time* to see a national map of stream gaging stations. Although this exercise focuses on a particular gaging station, the steps below work for any station and your instructor may select a local stream for you to work on. Your station may lack data for certain questions.

   a. At the top, under "Geographic Area," select *Pennsylvania*. Click on *Statewide Streamflow Table*, and under "Group table by" change the tab from *Major River Basin* to *County* and hit *Go*. Under Allegheny County, find (don't click!) "Monongahela River at Elizabeth, PA." What is the stream flow (discharge) in cubic feet per second? How does this compare to the long-term median flow for this date?

   b. Click on the link for this station, *03075070*. Is the flow to this station entirely natural or is it affected by such factors as hydroelectric plant discharges or other dam and lock controls?

   c. Scroll down to the graph of discharge. From the graph, what are highest and lowest discharges over the past week? How does the weekly flow compare overall to the median discharges plotted on the graph?

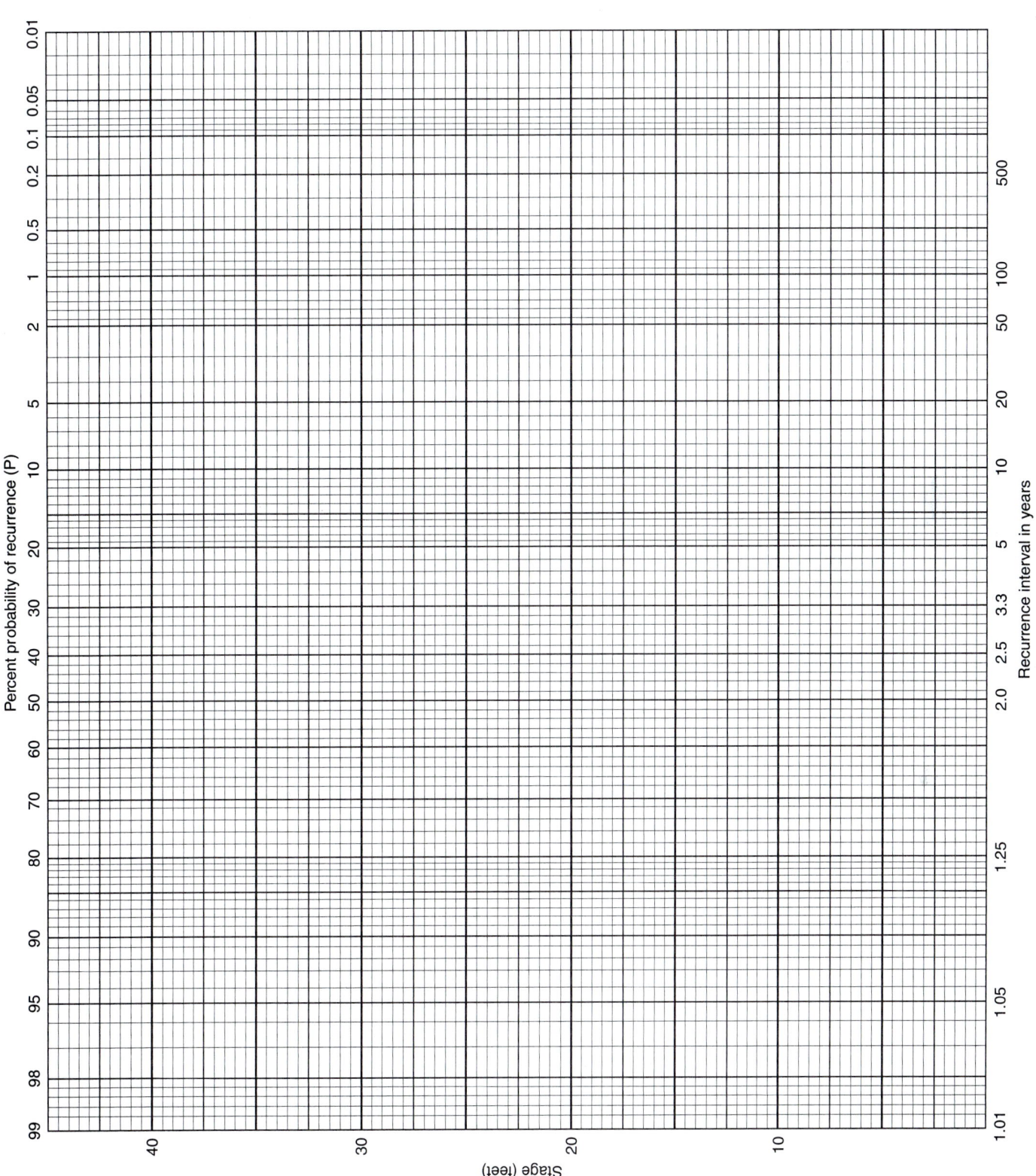

FIGURE 8.15
Graph paper for Problem 3.

d. Back at the top, change "Available data for this site" from *Real-time* to *Station site map*. Describe the location of the gaging station relative to the nearest major city (use the *Zoom In* or *Zoom Out* functions under the maps).

e. At the top, change "Available data for this site" from *Station site map* to *Surface Water: Daily Streamflow*. Under "Retrieve data from", put "1995-01-01 to today's date" (use the proper data format). Ensure the *Graphs of data* button is selected. Hit *submit*. In which year was the discharge the highest? The lowest?

The tick marks on the time axis should divide the year into quarters, each three months long. Does the discharge vary consistently by season? If so, when are the highest discharges?

f. To construct a flood-frequency curve and calculate recurrence intervals, you need to find the maximum flood for each year back through time. Under "Available data for this site" go to *Surface water: Peak streamflow*. Under "Output formats" click on *"Table."* This gives the required numbers. How many years make up the record for this station?

Compared to the 67-year record used in Problem 3, would you expect the data from this station to do a better or worse job of constraining the magnitude of a 100-year flood event? Why?

Unless directed to do more by your instructor, finish this problem off by listing the year in which the peak discharge was the highest and the year in which it was the lowest.

To find out about the water quality of this river, change the "Available data for this site" to *Water-Quality: Discrete samples*. Under "Retrieve Water quality samples for Selected Sites," check *Inventory of water-quality data* and, after changing the selection box from *For printing* to *With retrieval*, hit *Submit*. You will see a table listing the various aspects of water quality available for this site. By checking one or more boxes to the left of the table and clicking on *Submit,* you can see the actual data.

In Greater Depth

5. Figure 8.16 is a portion of the Urne, Wisconsin, quadrangle to be used for the following problems.
    a. On a separate piece of paper, draw a longitudinal profile from the mouth of Deer Creek to its head at the pond near the center of Section 3. Find the length of the stream (or the profile) by following all the curves and measuring the actual distance along the channel; mark and label those points where a contour crosses the stream. This can be done most easily by using the edge of a piece of paper. Use a vertical scale of 1 inch to 100 feet for the profile.

    What is the vertical exaggeration? Show your calculations.

    In what part of the stream is the gradient greatest? How do you know?

FIGURE 8.16
Portion of Urne, Wisconsin, 7½-minute quadrangle, for Problem 5.

What is the shape of the profile? Is it concave or convex upward? Is it smooth, or are there abrupt changes in elevation such as might be caused by a waterfall?

Does the shape of the profile suggest that the stream has or has not attained a near balance between erosion and deposition, and how did you tell?

b. Outline the stream channels and drainage basin of Deer Creek and its unnamed tributaries. The boundary of a drainage basin is a drainage divide, the line of separation between adjacent drainage basins. It can be a ridge, summit, or tract of high ground. Water falling within the drainage basin will flow toward the mouth of the stream; water falling outside the drainage will flow into another stream system.

c. Estimate the area of the drainage basin in square miles by adding the number of whole sections and fractions of sections that fall between the divides (one section = 1 square mile).

Using your estimate, calculate the area of the drainage basin in square feet (1 mile = 5280 feet). Show your work.

d. The average annual precipitation for this area is 29.1 inches (2.4 feet). Calculate the average volume of water that falls on the drainage basin per year in cubic feet (volume = area × annual precipitation). Show your work.

e. The average volume of water that runs off and flows to the mouth of Deer Creek per year is the average annual discharge of Deer Creek. Calculate the average annual discharge, assuming the following different conditions:

(1) Infiltration loss, evaporation, and transpiration are all zero, and no groundwater is contributed from outside the drainage basin.

(2) Infiltration loss = 2 inches per year, evaporation + transpiration = 15 inches per year, and no groundwater comes from outside the drainage basin. Show your calculations.

The Urne quadrangle is in the unglaciated or driftless area of Wisconsin. Note that little flat or gently sloping land can be found in the Urne quadrangle, except in the floodplain of the Mississippi River. This is one indication that the area has not been glaciated and that streams have been the principal eroding agents. If glaciers had covered the area, the topography would be more subdued: hilltops would have been eroded, valleys would have been wholly or partly filled with glacial drift, and surface drainage would not be as well developed.

CHAPTER 9

Groundwater and Groundwater-Influenced Landscapes

Materials Needed
- Pencil and eraser
- Calculator
- Ruler

Introduction

Water beneath the Earth's surface that occupies open spaces in rock or sediment is called **groundwater.** If the spaces are sufficiently connected, it can flow from one place to another. This enables us to pump water from a well without it immediately running dry, but it also means that we can contaminate important freshwater resources through careless waste disposal. In carbonate rocks, naturally acidic water flowing between grains and in cracks dissolves the rock and enlarges the open spaces. These openings can grow in size from a few millimeters up to huge cave systems.

Groundwater infiltrates from the surface and is part of the water cycle, as illustrated in Chapter 8. Typically, surface water percolates downward through unconsolidated soil and sediment until it enters a zone in which all of the interconnected spaces between sediment grains are filled with water, as illustrated in Figure 9.1. Near the surface, in the **unsaturated zone,** or **zone of aeration,** these **pore spaces** are mostly filled with air. Below the **water table,** interconnected pore spaces are filled with water in the **zone of saturation.** In humid regions, the elevation of the water table generally rises below hills and falls below valleys; its shape is commonly a subdued replica of the land surface.

Problems for this chapter illustrate how the shape of a water table can be determined in areas where lakes are abundant, how the flow direction of groundwater can be determined and used to predict the movement of pollutants, how water-bearing zones differ, and how groundwater can shape the surface topography of areas underlain by carbonate rocks.

Occurrence and Movement of Groundwater

Porosity

The **porosity,** or percentage of open (pore) space in the rock or sediment, determines the volume of water in the zone of saturation. In most rocks or sediments, the only pore spaces are those between adjoining particles or the narrow openings along fractures (cracks). In carbonate rocks and some lava flows, however, there may be large open spaces, and water may actually flow as underground rivers.

Permeability and Flow

The rate (velocity) at which groundwater flows is determined by **permeability** (ease of conducting water), **effective porosity** (the percent of open space through which the groundwater actually flows), and **hydraulic gradient** (for this lab, the slope down which the groundwater flows). Highly permeable material includes rock or sediment with large, interconnected pore spaces; low permeability materials have poorly interconnected pores. Groundwater flows fastest moving down a steep slope through a highly permeable material like sand or gravel. It moves much more slowly though an impermeable material like shale, especially when there is a low hydraulic gradient. The rate at which water flows through all such porous materials is expressed by the following equation derived from Darcy's law:

$$v = \left(\frac{K}{P}\right)\left(\frac{h}{l}\right) \qquad \text{Eq. 9.1}$$

where v is the velocity of groundwater flow (also known as the *seepage velocity*); K is the **hydraulic conductivity,** which depends on the nature of the medium and is directly related to permeability; P is the effective porosity; h is the **head,** which in this

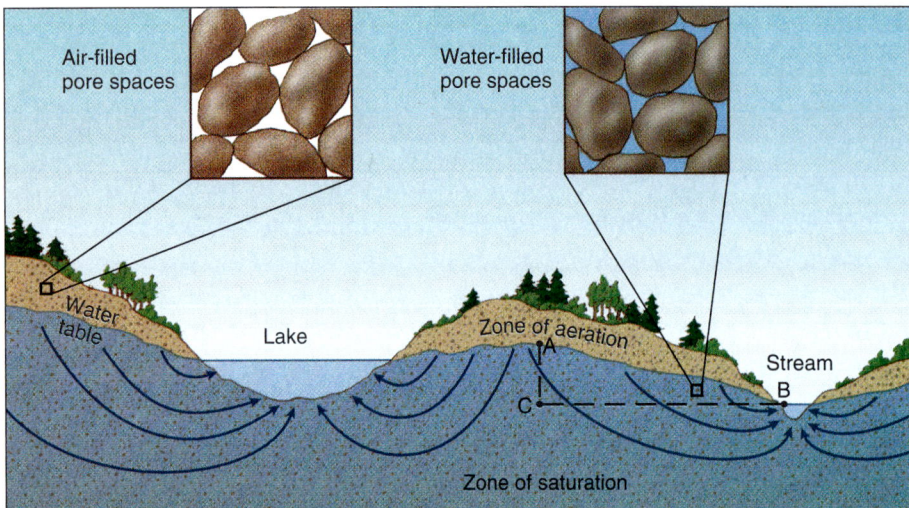

FIGURE 9.1
Cross section of an unconfined aquifer illustrating the zones of aeration and saturation, the water table, and groundwater flow paths. The exploded views show that air fills the pores in the zone of aeration while water fills the pores in the zone of saturation. Water falling on the land surface flows straight down through the zone of aeration. Once in the zone of saturation, it flows toward nearby rivers and lakes. Groundwater builds up to higher elevations away from where it exits to rivers, lakes, and springs, thus building a hydraulic gradient to drive groundwater flow. This combination of sideways flow plus more water coming in from above (from rain) creates the curved paths. Water flowing from A to B has a hydraulic gradient equal to the head (A to C) divided by the horizontal flow distance (C to B).

Example

If, in Figure 9.1, the head *(A-C)* is 75 m, the horizontal distance traveled *(C-B)* is 1.2 km, the hydraulic conductivity of the sediment through which the water moved is 3.0×10^{-3} cm/sec, and the effective porosity is 20 percent, calculate the velocity of the water discharging into the stream in cm/sec.

Because the answer should be in cm/sec, first convert *A-C* and *C-B* to centimeters: 75 m = 75 m × 100 cm/m = 7500 cm = 7.5×10^3 cm; 1.2 km = 1.2 km × 1000 m/km × 100 cm/m = 120,000 cm = 1.2×10^5 cm. Then use Equation 9.1, where $A-C = h$, $C-B = l$, the hydraulic conductivity is K, effective porosity is P, and the velocity is v:

$$v = (K/P)(h/l)$$
$$v = (3 \times 10^{-3} \text{ cm/sec}/0.2)(7.5 \times 10^3 \text{ cm}/1.2 \times 10^5 \text{ cm})$$
$$v = 9.4 \times 10^{-4} \text{ cm/sec}$$

What would the velocity be in centimeters per day? First determine the number of seconds in a day. One day has 24 hours, each hour has 60 minutes, and each minute has 60 seconds. Therefore, there are 24 hours/day × 60 minutes/hour × 60 sec/min = 86,400 sec/day = 8.64×10^4 sec/day. The velocity per day is, $v = 9.4 \times 10^{-4}$ cm/sec × 8.64×10^4 sec/day = 81 cm/day.

lab is taken as the difference in elevation between the top and bottom of the flow path; and *l* is the horizontal distance over which the groundwater flows. The hydraulic gradient is *h/l*. Figure 9.1 shows a common situation in a humid area in which water flows underground, following a curved path, and discharges into a stream or lake. Water going from point A to point B, Figure 9.1, drops a distance *A-C* and travels a horizontal distance *C-B*: *A-C* is the head, *h*, and *C-B* is the length, *l*, in Equation 9.1.

Aquifers

An **aquifer** is a porous and permeable sediment or rock from which a useful amount of groundwater can be obtained. Sand and gravel, sandstone, limestone, dolomite, basalt flows, and some *fractured* granites and metamorphic rocks are good aquifers.

An **unconfined aquifer,** like the one in Figure 9.1, is one in which movement of water is unrestricted, and the aquifer is recharged by water percolating downward from the overlying surface. Sand and gravel make excellent unconfined aquifers.

A **confined aquifer** is overlain by, or sandwiched between, comparatively impermeable **confining layers,** as shown in Figure 9.2. Common confining layers are clay and shale. Confined aquifers are recharged where they are exposed at the surface *(the recharge area),* and by whatever water can leak through the confining beds.

Springs

Springs occur where groundwater comes to the surface. This may happen where fractures, faults (fractures on which movement occurs), caves, or a contact between a permeable layer and underlying impermeable layer intersect the surface.

Human Use

We make use of groundwater by drilling wells and pumping it to the surface. We also contaminate groundwater by carelessly disposing of contaminants in a way that allows them to enter the groundwater system.

Wells

A well is a hole dug or drilled into the ground for the purpose of obtaining groundwater (Fig. 9.3). Wells may penetrate unconfined or confined aquifers. In either case,

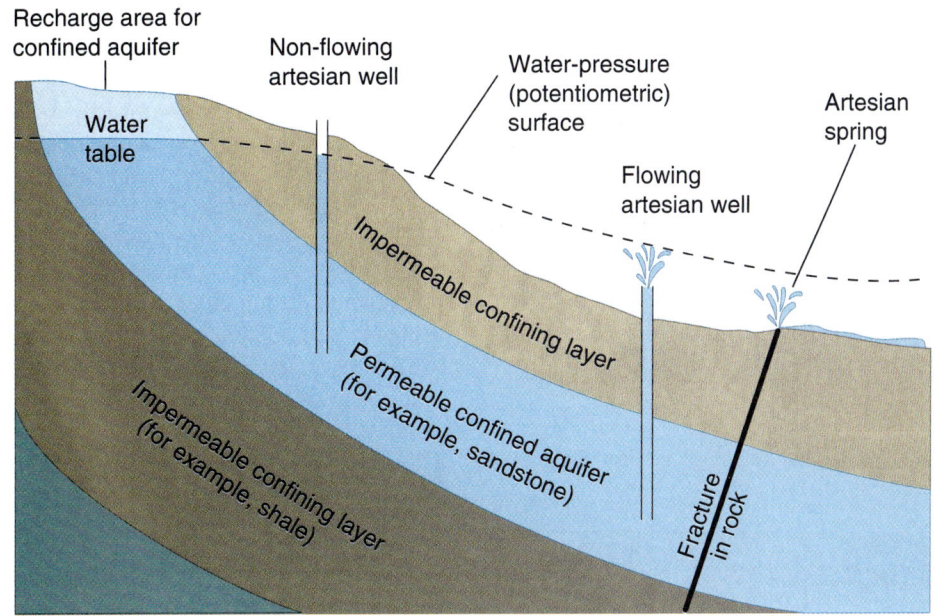

FIGURE 9.2
Cross section illustrating a confined aquifer system with recharge area, water-pressure surface, and flowing and non-flowing artesian wells.

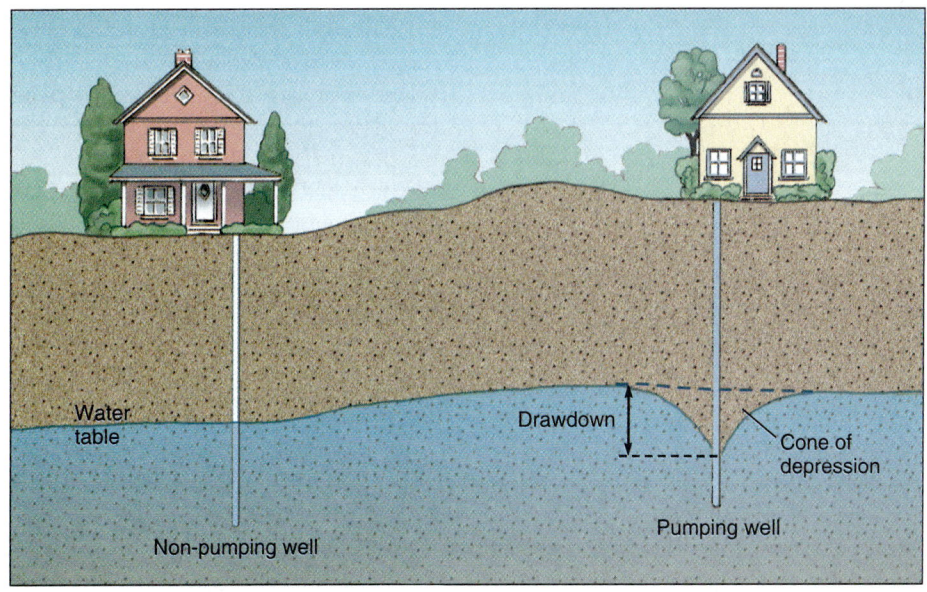

FIGURE 9.3
Cross section illustrating a non-pumping and pumping well in a simple unconfined aquifer. Pumping causes the water table to be drawn downward, forming a cone of depression.

water from the aquifer flows toward the well, an area of low relative pressure. If the aquifer is unconfined, the hole will fill with water up to the level of the water table (Fig. 9.3).

Confined aquifers have their recharge areas at higher elevations than most of the aquifer (Fig. 9.2). This pressurizes the confined water in the aquifer. If you drill a well into a confined aquifer, water generally rises above the level of the aquifer, forming an **artesian well** (Fig. 9.2). The level to which water in a well rises above the confining layer defines the **potentiometric** or **water-pressure surface** (Fig. 9.2). An **artesian spring** forms where water moves up along a fracture (Fig. 9.2).

Water is extracted from most wells by pumping. Pumping causes the level of the water and the pressure in the well to drop, allowing more water to flow into the well. If the aquifer is unconfined, continuous pumping causes a **cone of depression** to develop on the water table around the well, as shown in Figure 9.3. This also occurs when you use a straw to quickly suck the flavoring from a crushed-ice slurpee.

If the amount of water withdrawn from an unconfined aquifer exceeds the amount added by recharge, the water table throughout the region will drop. Excessive withdrawal from confined aquifers will lower the potentiometric surface.

In extreme cases, extraction of water from pore spaces in the aquifer may cause the pore spaces to collapse. As a result, the aquifer may be permanently damaged. In some cases, porosity in the aquifer or in confining beds decreases so much that the land surface subsides. For example, so much water has been withdrawn from the lake sediment on which Mexico City is built that the opera house has settled 3 m, and parts of the city have subsided as much as 8 m.

Fluids with Different Densities

Lighter, less dense fluids tend to float on heavier, more dense fluids, whether in a mixing bowl in the kitchen or in a permeable rock below the surface. Four natural fluids, listed in order of increasing density, are natural gas, oil, fresh water, and salt water—if all four are present, gas will be on the top, salt water on the bottom.

Salt water fills the pore spaces in sediment below the oceans, and some continental aquifers contain saline water. In coastal areas, saline and fresh groundwater come in contact, as shown in Figure 9.4. Because salt water is more dense, it underlies fresh water in the groundwater system. Active removal of fresh water by pumping will change the position of the salt water interface and may cause infiltration of salt water into some wells, a problem known as saltwater encroachment.

Oil and natural gas, being insoluble in water and lighter than either saline or fresh groundwater, rise through permeable rocks until they reach an impermeable barrier. If that barrier cannot be passed, the gas and/or oil is trapped, and an oil or gas pool will form. As shown in a simple example in Figure 9.5, when water, oil, and gas are all

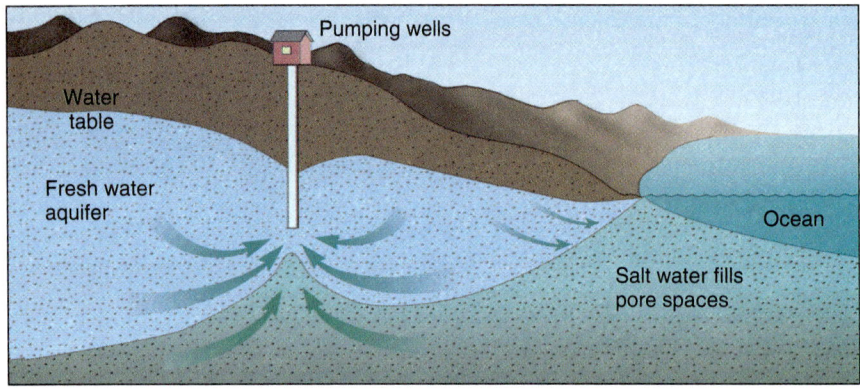

FIGURE 9.4
Cross section illustrating saltwater encroachment caused by excessive withdrawal of fresh water.

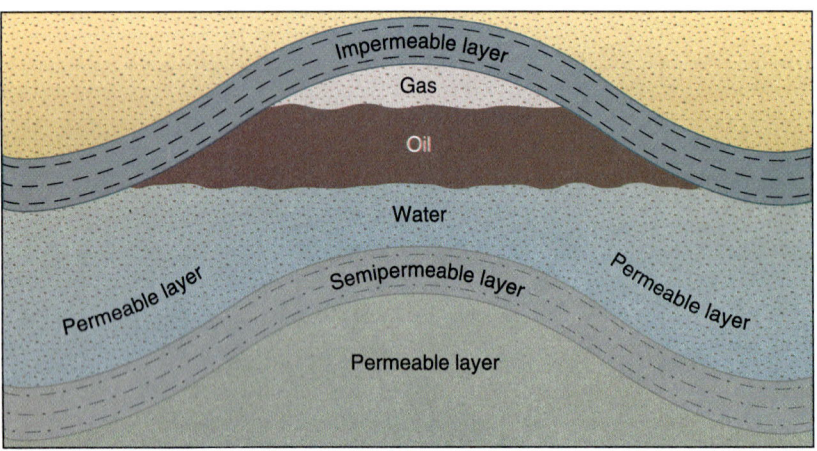

FIGURE 9.5
Oil and gas rise in permeable rocks until they encounter an impermeable layer and are trapped, as illustrated in this cross section.

present, gas, being the lightest, rises to the top, while water sinks to the bottom and oil settles in the middle.

Contamination

Contamination of groundwater can be a severe problem because it is generally not detected until it affects a well or enters a stream. Once contaminated, an aquifer is difficult and expensive to purify. Figure 9.6 illustrates some of the ways in which groundwater can be contaminated by human activities. Many contaminants are somewhat soluble in water, allowing them to disperse and making it difficult or impossible to recover them. In some cases, it is possible to dispose of such materials by injecting them into deep, permeable rocks containing dense salt water that is isolated from freshwater aquifers. Insoluble contaminants may segregate within the system on the basis of density. For example, waste oil disposed of at the surface would percolate down to the top of the water table, but would not mix with underlying fresh water. If the oil does not migrate very far laterally, it may be possible to reclaim it.

Groundwater in Carbonate Rocks

Although most of the world's groundwater occurs in porous rock, it can also be present in fractures or even underground streams. This is particularly true of groundwater in carbonate rocks because carbonate minerals—especially calcite—dissolve in naturally acidic groundwater. Groundwater is slightly acidic because it contains dissolved carbon dioxide derived from the atmosphere or organic acids released by the decay of organic matter in soil. Water combined with dissolved carbon dioxide makes carbonic acid (H_2CO_3), which then acts on carbonate minerals. The overall reaction is:

$$H_2O + CO_2 + CaCO_3 = Ca^{+2} + 2HCO_3^{-1}$$

As a result, calcium (Ca^{+2}) and bicarbonate (HCO_3^{-1}) ions are carried away in groundwater, and cavities form in carbonate rocks.

Caves, Disappearing Streams, and Sinkholes

The largest subsurface cavities are **caves,** which commonly develop as groundwater

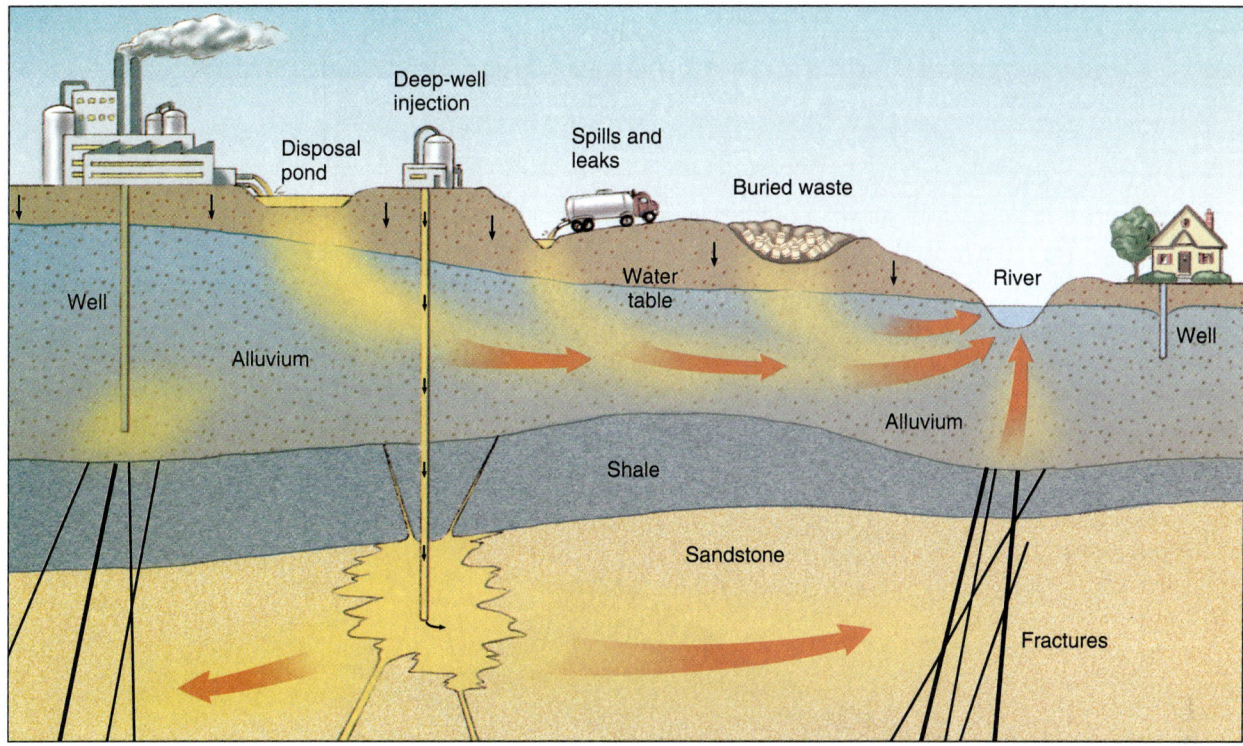

FIGURE 9.6
This cross section illustrates a number of ways in which groundwater can be contaminated by human activity.
Source: Data from USGS Circ. 875.

flows along sedimentary layers or fractures. Where dissolution occurs along horizontal surfaces, such as sedimentary layers, caves have dendritic or braided patterns when viewed on a map. Where fractures control water flow, a more regular pattern is common, with chambers intersecting at consistent angles. Most caves form at or below the water table, but they can also form above it. Caves serve as channels for ground-water flow and act as underground rivers. **Disappearing streams** are surface streams that plunge underground to become underground rivers. **Sinkholes** are depressions formed by the dissolution of carbonate rock beneath the overlying soil or by the collapse of portions of near-surface caves.

Karst Topography

Areas underlain by carbonate rocks in which dissolution has been active develop a distinctive surface appearance known as **karst topography.** Figure 9.7 shows features of karst topography in various stages of development. In the early stages, scattered sinkholes appear, and the surface drainage is disrupted as some streams disappear underground. Continued cave collapse leads to the presence of valleys, which form as sinkholes enlarge and join. Eventually, all that may remain are haystack-shaped hills consisting of rock that has not dissolved. It has been estimated that 20 percent of the Earth's land surface and 40 percent of the U.S. east of Oklahoma show some manifestation of karst topography. In particular, Kentucky, southern Indiana, and much of Florida are famous for their sinkholes and karst topography.

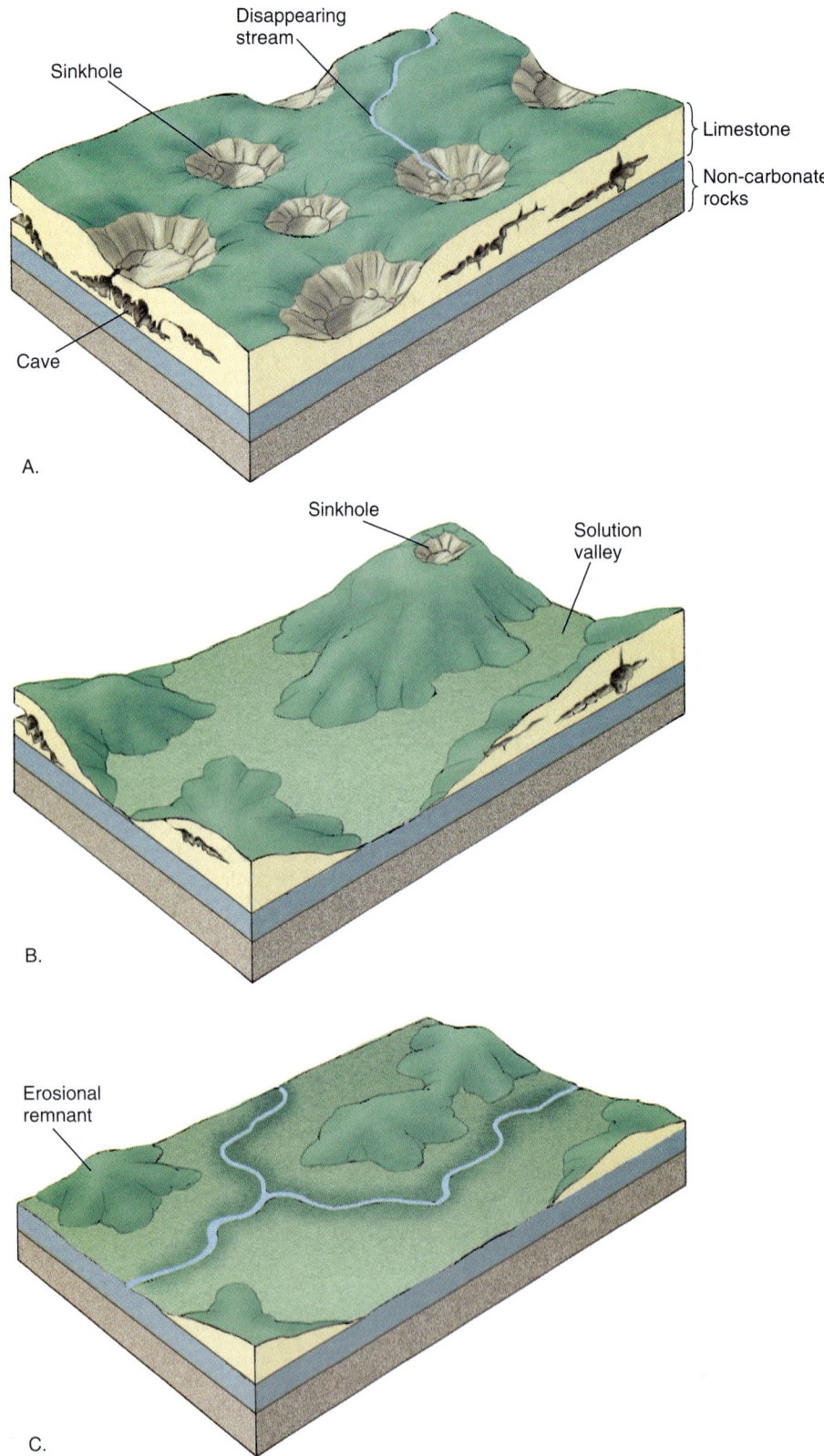

FIGURE 9.7
The evolution of karst topography. A. Scattered sinkholes and disappearing streams characterize the early stage. B. Continued dissolution and collapse of caves leads to the formation of valleys. C. Eventually, all that remains are scattered limestone hills.

Hands-On Applications

Because groundwater and the rock or sediment through which it moves are not directly visible, our ability to understand and predict its behavior is strongly dependent on models or hypotheses. These models can be tested by various means. Where carbonate rocks have been dissolved by groundwater, direct observations can be made in caves and at large springs. In most areas, however, data are obtained from wells, which are samples of a very small volume. In this lab, you will learn some of the methods used by hydrogeologists to interpret and predict the movement of groundwater, and its relation to the rock or sediment that contains or confines it.

Objectives

If you complete all the problems, you should be able to:

1. Use water-table contours and flow lines to predict groundwater flow.
2. Use water-level elevations, whether in lakes or wells, to contour the water table or draw it on a cross section.
3. Determine a hydraulic gradient.
4. Recognize distinctive karst topography and the following features on a topographic map: sinkhole, disappearing stream, solution valley.
5. Explain how water wells in confined and unconfined aquifers differ.
6. Use water-table and water-pressure-surface information to determine how deep a well must be drilled to encounter water from confined and unconfined aquifers.
7. Predict how fluids of different densities would behave in a groundwater system.

Problems

1. Figure 9.8 is part of the Lakeside, Nebraska, quadrangle, and is part of the Sand Hills area of western Nebraska, an area composed of old sand dunes that now are mostly stabilized by vegetation. The dunes themselves have irregular surfaces with many small depressions. This makes for a complex, "busy" contour pattern. Some of the low areas between the dunes contain small lakes. Because the sand is so permeable, these lakes form where the water table intersects the surface, and their elevation, printed on most lakes, is that of the water table.

    a. Draw a contour map of the water table using the lake elevations in the same way you used spot elevations to make a topographic map in Chapter 6, Problem 7. Use a contour interval of 5 feet and start with the east side of the map by drawing the 3900-foot contour.

    b. In what direction does the water table slope, and what is the average gradient in feet per mile?

    c. Assuming the gradient just determined is the hydraulic gradient for groundwater entering the system at Ellsworth and discharging near Lakeside, calculate the velocity, in cm/sec, at which it would flow between these points. Assume a *hydraulic conductivity, K,* of 1×10^{-3} cm/sec and an *effective porosity, P,* of 20%. Hint: Use Equation 9.1, but first convert your hydraulic gradient from units of feet/mile to feet/foot; once this is done, the hydraulic gradient will be a unitless number.

d. If the distance between Ellsworth and Lakeside is 7 miles (11.3 km = 1.13×10^5 cm), how long would it take for the groundwater to make this journey?

e. Note the windmill in the SW1/4, Sec.22, T24N, R44W (southeast of Lakeside) and its map symbol; the surface elevation is also given. Many other windmills are shown on the map by symbol only. The windmills are connected to water wells and used to pump groundwater to fill water tanks for cattle.

Locate the windmill in the NW1/4, Sec.19, T24N, R43W. How high must the water be raised by the pump to fill the stock tank? Explain your answer.

2. Mammoth Cave, in Kentucky, is one of the best-known caves in the United States—and the largest, with more than 560 km (348 miles) of passageways. The entrance to the cave is near the northeast corner of the topographic map in Figure 9.9, where red lines mark roads. Three areas with different topography are evident in this map. In the northeast, a series of ridges, some with nearly flat tops, is dissected by valleys without streams. In the northwest, the valleys are not as deep, but they do contain streams. Both areas extend south to a sharp topographic break, known as (but not labeled on the map) the Dripping Springs Escarpment, which is just north of the Louisville and Nashville Road shown on the map. The third area, south of the escarpment, is characterized by numerous depressions.

The area is underlain by nearly horizontal layers of sedimentary rock of two kinds: (1) sandstone (with some interbedded shale), which is essentially insoluble in groundwater; and (2) limestone, which is soluble in groundwater. The sandstone is on top and at higher elevations than the limestone.

a. Carefully study the distribution of depression contours on the map (a hand lens or magnifying glass may help). Make a simple outline around the large area in which depression contours are generally *not* found. Overall, are depression contours more likely to be found in areas with elevations higher than about 700 feet or lower? In which of the two types of rock in the area, sandstone or limestone, are depressions more likely to occur?

b. Why are there no streams in the northeast part of the map?

c. What evidence suggests that there once were streams in the northeast? Hint: Note the pattern of the valleys.

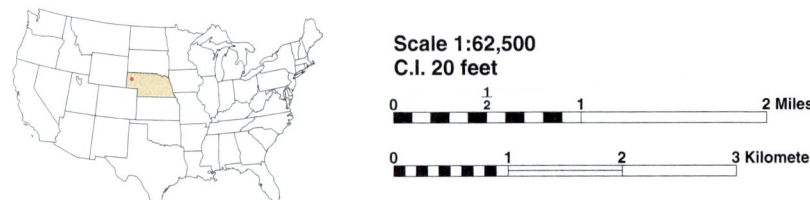

FIGURE 9.8
(Page 165) Portion of Lakeside, Nebraska, 15-minute quadrangle, for Problem 1.

d. Why are there streams in the northwest?

e. In what direction does Gardner Creek (southeast corner of map) flow? What happens to it?

3. Figure 9.10 shows the flow pattern of groundwater in the Mammoth Cave area (red lines), contours of the water-pressure or potentiometric surface (called water-level contours in the Legend), and divides separating groundwater basins and sub-basins (black dashed and dotted lines, respectively).

   a. What relation is there between the flow lines and the contours (that is, are they parallel, perpendicular, or is there no relation)?

   b. The black dot on the northwest side of Cave City marks the location of a cyanide spill that occurred along Interstate Highway 65 (road not shown on either Figure 9.9 or 9.10) in 1980. The spill was cleaned up quickly, but what if it had not been? Trace the likely groundwater flow path on the map and explain how you drew your path. Hint: Do not cross groundwater basin boundaries (dashed lines). Sub-basins drain one sub-basin into the next.

   Would the cyanide have found its way into the underground streams of the main part of the Mammoth Cave system? Explain.

   Could the cyanide have reached the Green River? If so, circle the entry point on the map.

   c. A creek disappears at point X on the map (about two map inches to the west-southwest of Park City). Dye placed in the creek just before it disappears can be used to determine where the water reappears on the surface.

      (1) Predict where the dye would reappear at the surface, and mark those points on the map.

      (2) What is the hydraulic gradient in feet per mile between the point where the creek disappears and the first place its waters reappear? Hint: Use the water-level contours to calculate h and the map scale to measure l. Show your calculations.

      (3) If the dye took 152 hours to travel the distance in (2), what was its average velocity in feet/hour? Show your calculations.

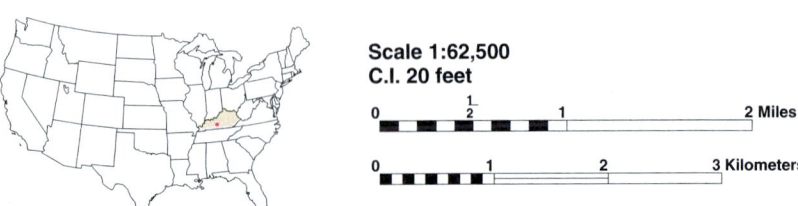

FIGURE 9.9
(Page 167) Portion of Mammoth Cave, Kentucky, 15-minute quadrangle, for Problem 2.

168   Part IV   Landscapes and Surface Processes

4. **Groundwater Data.** Go to *http://water.usgs.gov/local_offices.html* (or link to it through *www.mhhe.com/jones5e*—see Preface), and select *Wisconsin* (try other states if you wish, but the path to the well data may be different than described here) by clicking on it in the list of states. Select *Water Data* from the Navigation Menu, then choose *Ground-Water Network,* then select *Current Water Table Hydrographs* and choose one of the wells, such as *MT-0007* in Marinette County (northeastern Wisconsin). Your instructor may want to modify this problem to fit data from a local well, which may vary considerably from the example well used here.

   a. How deep is the well?

      What is the elevation at the top of the well, at the land surface?

   b. What type of rock or sediment makes up the aquifer?

   c. What is the current depth to water in the well?

   d. How much has the depth to water varied in the past year? During what season were water levels the highest?

   e. For how many years has a record been kept of the depth to water in the well?

      What is the highest level for the period of record? The lowest?

      What is the range in depth to water in the well for the period of record?

   Other groundwater-related websites of interest include the National Park Service sites with links to parks with caves *(www2.nature.nps.gov/geology/tour/caves.htm)* and hot springs *(www2.nature.nps.gov/geology/tour/hotsprin.htm).* To see excellent images of cave formations, try The Virtual Cave at *www.goodearthgraphics.com/virtcave.html.*

In Greater Depth

5. This problem illustrates how cross sections help us understand groundwater flow and contamination. Refer to the top of Figure 9.11, a map with three types of contours. Regular surface contours are shown in brown. The top of an important sandstone aquifer is contoured with red lines. The water-pressure surface (potentiometric surface) is shown in black. The stream (in blue) flows throughout the year and is recharged from both surface flow and groundwater.

   a. Complete the cross section at the bottom of Figure 9.11. The section is from points A to A' on the map. It shows the surface topography along that line and the unconsolidated sediment that covers most of the area. Although not shown on the cross section, the bedrock beneath the unconsolidated sediment is shale, and below the shale is the sandstone aquifer. To complete the cross section, draw the top of the sandstone aquifer and the water-pressure surface. Do this by using the intersections between the contours and line A-A' in the same way you would to draw a topographic profile. Label everything!

   b. Shallow wells in the unconsolidated sediment at points X and Y fill with water to within 10 feet of the surface. Using this information, mark the positions of X and Y on the cross section and sketch the probable position of the water table in the unconsolidated sediment. Hint: consult Figure 9.1.

      If you were to build a house where the boundary between the shale and unconsolidated sediment hits the surface, what problem might you encounter with the basement?

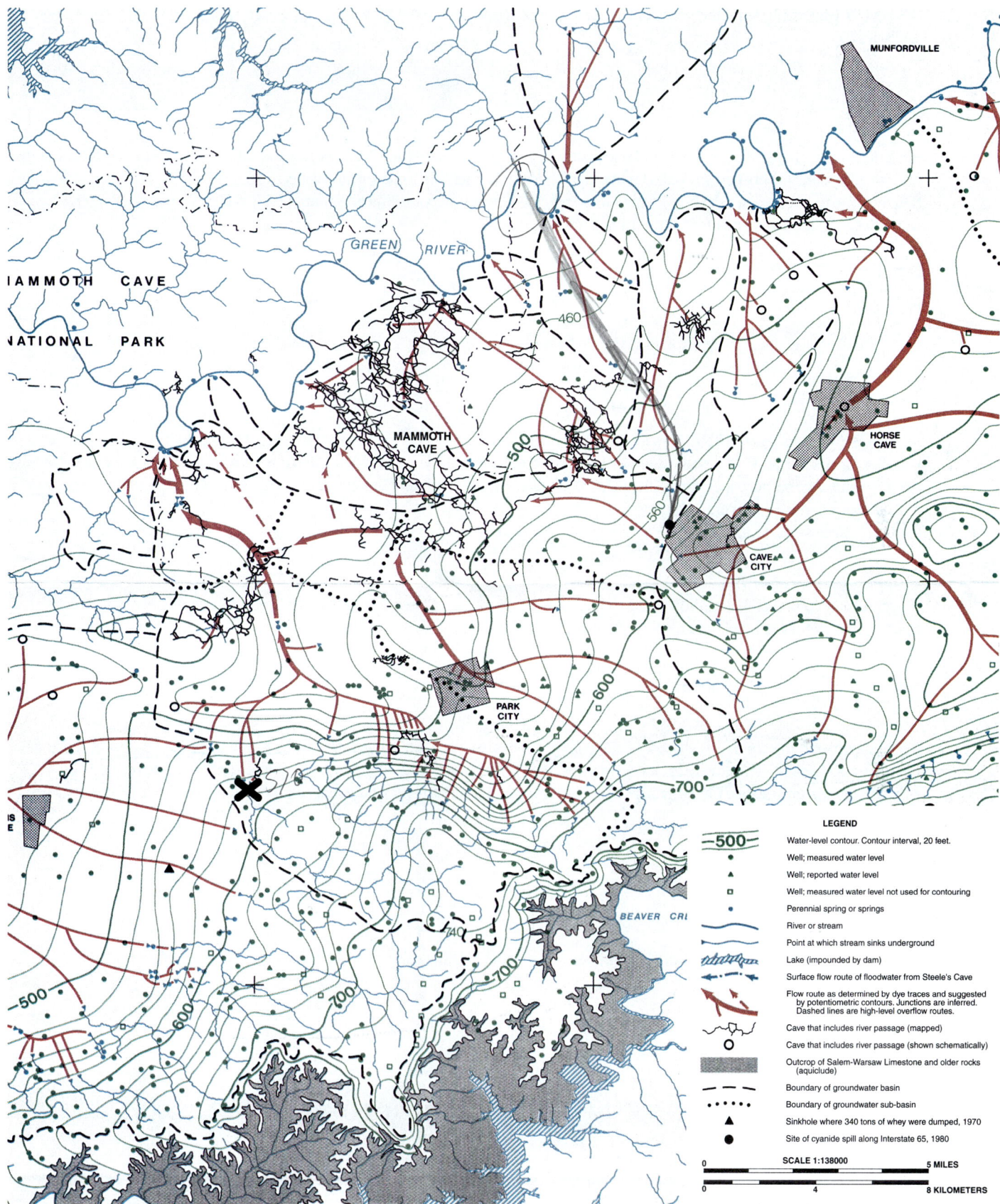

FIGURE 9.10
Portion of *Groundwater Basins in the Mammoth Cave Region, Kentucky,* map for Problem 3. *Quinlan and Ray,* 1981, revised 1989.

c. Deep wells were drilled into the sandstone at X, Y, and Z. How deep would the wells be at each spot to intersect the top of the sandstone?

X: Y: Z:

Assume that the shale bedrock below the unconsolidated sediment and above the sandstone aquifer is impermeable, meaning that no connection exists between the water in the sandstone aquifer and the near-surface water. Also assume the wells are cased (lined so water cannot enter the hole) where they pass through the unconsolidated sediment. How high (to what elevation) would water from the sandstone aquifer rise in each of the three wells?

X: Y: Z:

Are these deeper wells flowing artesian wells, non-flowing artesian wells, or non-artesian wells? Explain.

d. In the late 1940s and early 1950s, a small chemical plant was located near point X on the map. One of their waste by-products was an orange chemical. At first the engineers disposed of it by drilling a shallow well into the unconsolidated sediment, but they were soon caught because the orange chemical reappeared. Where do you think it reappeared? Mark the general location on the map and explain your answer.

The chemical company then drilled a 150-foot disposal well in the same spot. At first they made a slightly heavier-than-water mixture by adding a heavy liquid to the orange chemical, but after a few years, that fluid was also detected. How do you think it was detected?

Next, they mixed the chemical with oil to make a lighter-than-water fluid. This time the waste stayed hidden until well after the engineers had retired. Then, in 1985, a day-care center drilled a water well into the lower aquifer only to discover oily orange stuff in their water. Indicate on the profile where the waste migrated to and where, approximately, the day-care center well was located.

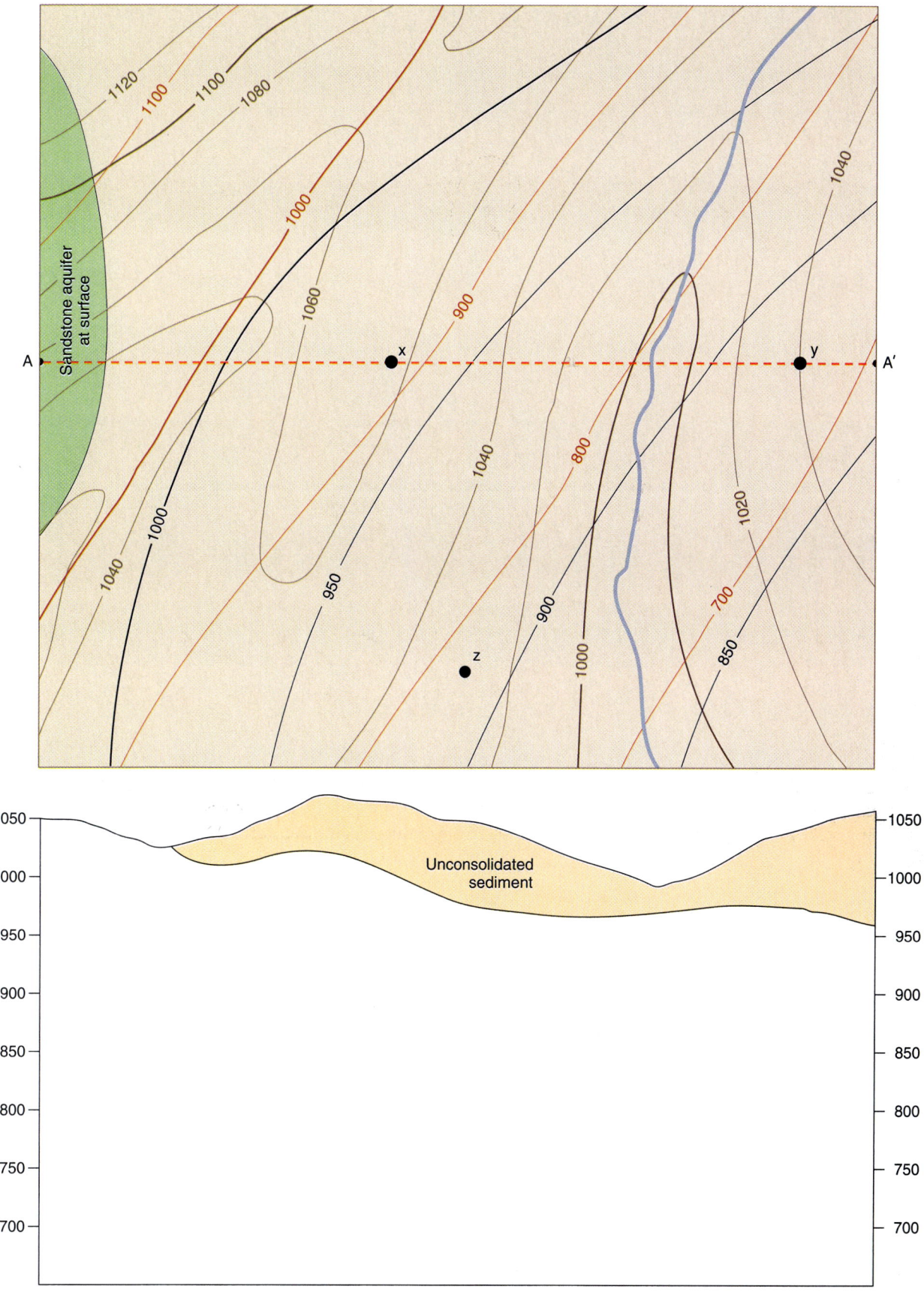

FIGURE 9.11
Top: Map for Problem 5 shows surface contours in brown, contours on the top of a sandstone aquifer in red, and contours of the water-pressure surface of the sandstone aquifer in black. *Bottom:* Cross section along line A-A'.

CHAPTER 10

Glaciation

Materials Needed
- Pencil and eraser
- Colored pencils
- Ruler
- Calculator
- Stereoscope (provided by instructor)

Introduction

Glaciers are slow-moving, thick masses of ice that form when low temperatures and sufficient snowfall allow winter snow and ice to survive the summer. Where glaciers occur, they dramatically shape the landscape by their ability to erode, transport, and deposit huge quantities of sediment. Glaciers are currently found on many tall mountains around the world as well as across most of Antarctica and Greenland. In the geologically recent past (before about 10,000 years ago), vast ice sheets also covered much of Canada and the northern parts of the U.S., Europe, and Russia. In this lab, you will learn to recognize and analyze on aerial photographs and topographic maps erosional and depositional features of glaciers. You will also learn how measurements taken over a period of years can be used to monitor the activity of a glacier in response to changes in climate.

There are two main types of glaciers: *alpine* and *continental*. **Alpine glaciers** occur in mountains. They typically begin at high elevations and flow down preexisting stream valleys to lower elevations, where they melt, as shown in Figure 10.1. As ice moves downhill, it erodes the valley and imparts a character to it that is unmistakable: a glaciated valley is straighter, deeper, and wider than the original stream valley, and has a U-shaped, rather than V-shaped, cross profile. Should the ice melt and the glacier recede from the valley, evidence of its presence is etched into the landscape.

Continental glaciers are thick, broad ice sheets that spread out to cover virtually the entire landscape (Fig.10.2). They may reach thicknesses of 4000 m or more, and they are not confined by valley walls. Present-day continental glaciers cover most of Antarctica and Greenland, but during the Pleistocene Epoch of the Quaternary Period, approximately 2 million to 10,000 years ago, vast areas of the northern continents were alternately covered and exposed by the repeated waxing and waning of the great ice sheets.

Material eroded by glaciers is transported by the moving ice and by meltwater. Deposition occurs when the ice melts or the running water slows. The depositional features of alpine glaciers, formed in valleys or at valley mouths, are distinctive when young. However, because they are in areas of active erosion, they largely disappear within a few tens of thousands of years, geologically a very short period of time. Continental glacier deposits, on the other hand, may be several hundred meters thick and impart a distinctive character to the landscape that remains for many tens of thousands of years. Some glacier deposits have been converted to sedimentary rock and have survived for more than two billion years.

Formation, Movement, and Mass Balance

As snow accumulates over years at high elevations or high latitudes, the lower parts of the snow mass compact and eventually recrystallize to ice. When thick enough (about 60 m), the ice begins to flow under its own weight. Alpine glaciers flow downhill until they melt or hit the ocean. Continental glaciers flow outward from where they are thickest, usually areas of highest snow accumulation, and continue until they cross into warmer climates or hit an ocean. In both cases, a great deal of rock is frozen into the ice where it is in contact with the ground. As the ice moves, it picks up this rocky debris and carries it along.

Figure 10.3 shows a longitudinal (lengthwise) cross section of an alpine glacier. Snow accumulates in the **zone of ac-**

FIGURE 10.1
Oblique aerial photograph of a retreating alpine glacier near Mt. Spurr, Alaska, taken in late summer. The glacier starts in a cirque and flows down the valley to its terminus, which is marked by a prominent end moraine. The sides are marked by lateral moraines, and medial moraines occur within the glacier. The snow line is at the elevation where the medial moraines disappear; it approximates the equilibrium line.

FIGURE 10.2
Oblique aerial photograph of Antarctica, illustrating the vast sweep of a continental glacier. Only here and there do mountains poke through the thick ice.

cumulation. As old snow is buried, it moves downward and is converted to ice. Eventually it flows downhill until finally at lower, warmer elevations, the surface ice melts away in the **zone of wastage** (or **ablation**). The boundary between these two zones is the **equilibrium line.** Continental glaciers also have these zones.

The **mass balance** of a glacier is the annual difference between the mass of glacier ice added to the zone of accumulation and that lost from the zone of ablation. In perfect equilibrium, the mass balance is zero and the front edge of a glacier neither advances or retreats. A **positive mass balance** means an excess of accumulation over loss, which means that the front of the glacier will **advance** farther down the valley. A **negative mass balance** means more melting than accumulation and thus that the glacier front will **retreat** up the valley. In both cases, gravity ensures that the ice *within* the glacier is always moving forward. Whether a glacier front advances or retreats depends on the rate of forward movement of ice within the glacier, which is driven by accumulation, compared to the rate at which summer melting removes ice at the front edge.

Alpine glaciers are very sensitive to climate change. Many glaciers in Europe and western North America advanced substantially during the Little Ice Age, a cold spell lasting from about 1350 to 1850, and most alpine glaciers around the world are retreating in response to global warming over the past few decades. Continental glaciers also respond to climate change, but over the longer timescales of centuries and millenia. During the Pleistocene Epoch (1.8 million to 10,000 years ago), climatic fluctuations caused continental glaciers to advance and retreat as many as 21 times. Clearly, the study of glaciers is important to our understanding of the causes and effects of past and future climate change.

Erosional Landforms

A mountainous area that is or has been occupied by alpine glaciers has distinctive erosional landforms, as shown in Figure 10.4.

A **cirque** is an amphitheater-shaped depression high on a mountain, which either is or was filled with glacial ice (Fig. 10.3). If it does not contain ice, the bottom may be the site of a bedrock-basin lake, a **tarn.** The upper end, or headwall, of a cirque is very steep, and erosion by frost and glacial action causes it to migrate headward with time. *Cirque glaciers* are no larger than the cirque itself, but for many *valley glaciers,* the cirque is simply the head of the glacier; cirque and valley glaciers are types of alpine glaciers.

Headward erosion of cirques from opposite sides of a ridge may eventually cause them to join and carve a low point, or *pass,* in the ridge called a **col.** The intersection of three or more cirques may result in the formation of an isolated high peak called a **horn.**

Valley glaciers reshape preexisting stream valleys with V-shaped cross profiles into **U-shaped valleys.** (Fig. 10.4) Depressions left in the upper part of such valleys may be occupied by several small lakes connected by a stream; when viewed from the air or on a map, they look like a string of beads and are called **paternoster lakes.** The divides or ridges between adjacent glaciated valleys are commonly sharp, jagged features known as **arêtes** (a rets´). Small tributary glaciers are unable to erode as deeply as the large glaciers they join. As a result, the junction of such former glaciers

Chapter 10 Glaciation 175

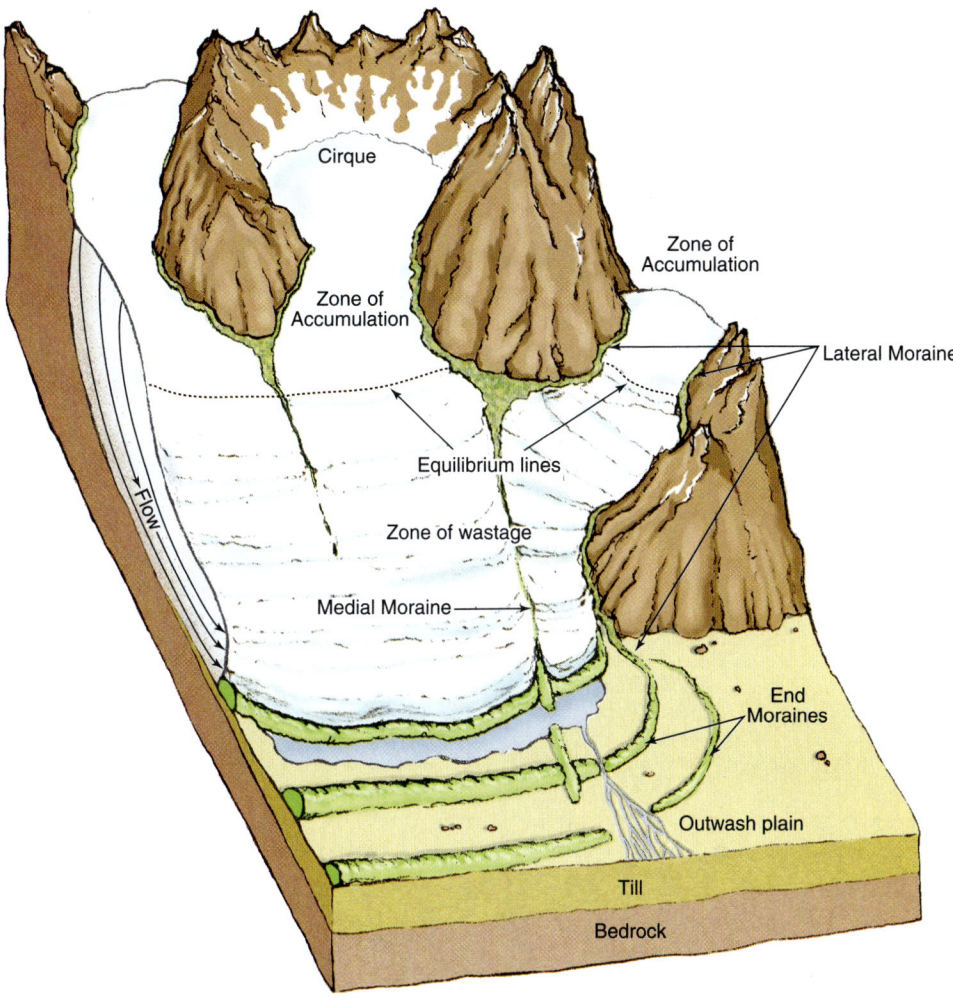

FIGURE 10.3
Glacier cross section illustrates how ice flows from the zone of accumulation, past the equilibrium line, to the zone of wastage. An end moraine forms during a pause in the retreat of the front edge, which allows flowing ice to bring sediment forward to a single stationary line. A cirque, lateral and medial moraines, and a braided outwash stream are also shown.

FIGURE 10.4
Oblique aerial photograph of the eastern side of the Juneau icefield, British Columbia. The U-shaped valley in the foreground was carved by an alpine glacier. Cirque and small valley glaciers are still present in the background. Other features characteristic of alpine glaciation are labeled.

is marked by a **hanging valley,** a U-shaped valley at a higher elevation than the one it joins, commonly with a waterfall. **Fiords** (or *fjords*), which are narrow, steep-sided inlets of the sea, are submerged valleys once occupied by valley glaciers.

Erosion by continental glaciers generally smooths and rounds topography by shaving down hills. Because distinctive erosional features like those formed by alpine glaciers are not left behind, past continental glaciations are best known by their deposits.

Depositional Landforms

Glaciers pick up huge volumes of loose sediment ranging in size from clay to massive boulders and transport it to the zone of wastage (Fig. 10.3). There it can be dropped straight from the melting ice or transported some distance by running meltwater or wind.

The material deposited in association with a glacier is called **glacial drift.** Two kinds of drift are **till** (unsorted, unstratified debris deposited directly from ice) and **stratified drift** (sorted and stratified debris deposited from glacial meltwater).

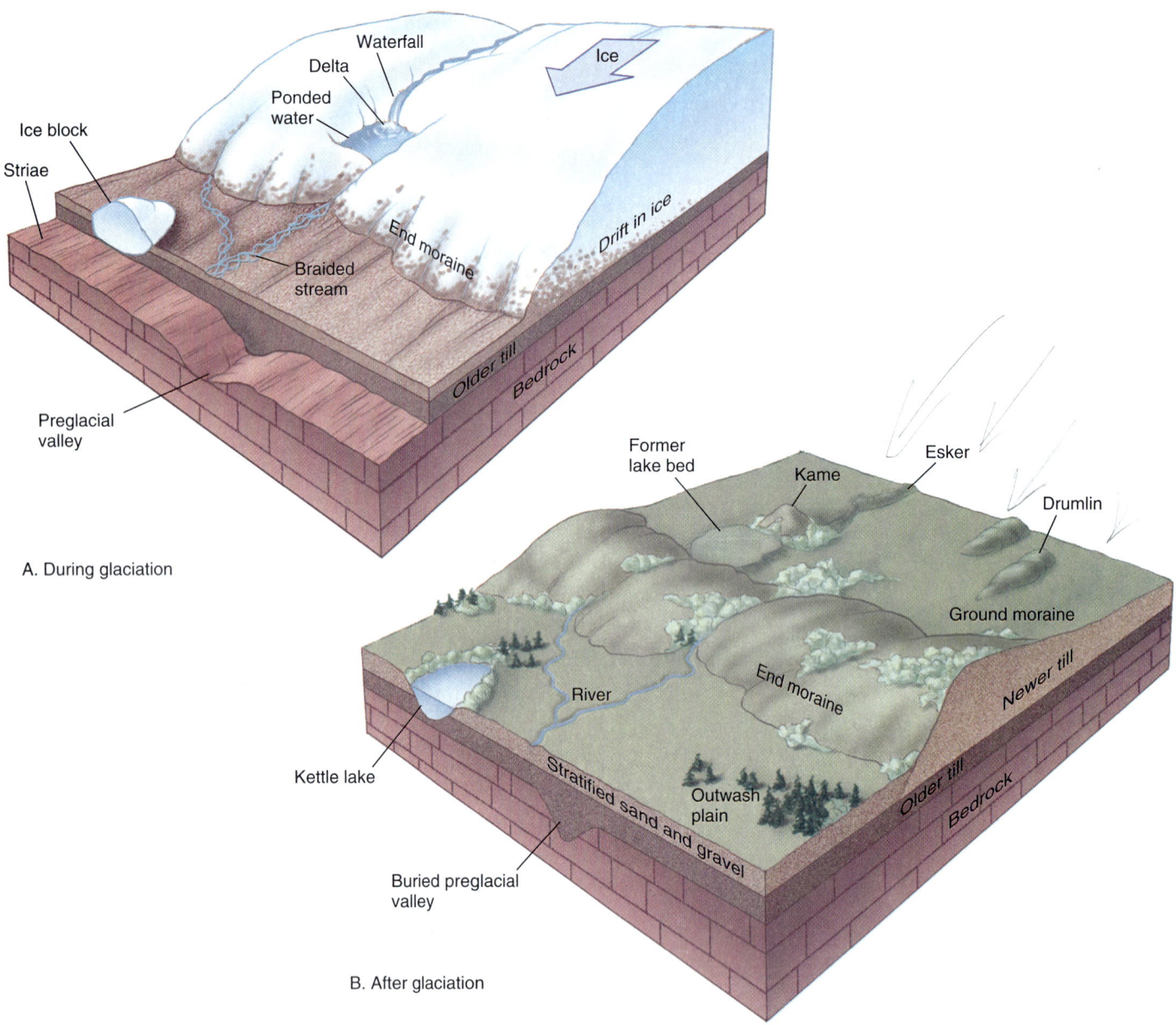

FIGURE 10.5
A. Terminus of continental glacier. B. Same area several thousands of years after retreat. Bedrock is solid rock that underlies soil and loose sediment.

Moraines are landforms composed mostly of till that form on or within a glacier, or are left behind when the ice melts. Figures 10.1 and 10.3 illustrate moraines on alpine glaciers that are named on the basis of their position. **Lateral moraines** are low ridges that form on each side of a glacier largely from rocks falling from valley walls. A **medial moraine** is a ridge that forms in the middle of a glacier when two valley glaciers merge and combine lateral moraines (Fig. 10.3). **End mo-** **raines** are ridges that form when a glacier achieves equilibrium for a period of time before retreating: the front edge remains stationary while the ice conveyor continues to bring sediment to the zone of ablation (Fig. 10.3). A glacial advance will destroy an end moraine. Continental glaciers form long, curved end moraines that reflect the main lobes of ancient retreating ice sheets. **Ground moraine** is the uneven blanket of till between the other moraines.

Stratified drift deposits are most prominent at the end of the glacier, where they consist of **outwash**—sand and gravel washed out of the glacier by running water. Outwash deposited in a valley forms a **valley train.** An **outwash plain** forms where braided meltwater streams deposit sediment over a wide area. Outwash plains are common features in areas of continental glaciation, as illustrated in Figure 10.5, but also form where alpine glaciers flow out of a valley and spread out (Fig. 10.1). Some

FIGURE 10.6
Maximum extent of Glacial Lake Agassiz (purple area). Blue areas within Glacial Lake Agassiz are present-day lakes. Arrow indicates drainage of Glacial River Warren; the valley is now occupied by the Minnesota River.
Source: Data from illustration by J.A. Elson, in Mayer-Oakes, *Life, Land and Water,* 1967. University of Manitoba Press, Winnipeg, Manitoba, Canada.

outwash plains, formed during glacial retreat, contain abundant, undrained depressions called **kettles** or, if filled with water, **kettle lakes,** which form by the melting of buried blocks of ice (Fig. 10.5). Deposits of stratified drift that form beneath or within a glacier include **eskers** and **kames,** as shown in Figure 10.5. An esker is a long, narrow, winding ridge formed by deposition from a stream flowing within or at the base of the ice. A common type of kame is a steep-sided mound formed where meltwater flowed into a depression or hole in the ice. Eskers and kames are common features of continental glaciation in some areas, but are usually destroyed by erosion in alpine settings.

Figure 10.5 also illustrates **drumlins,** which are streamlined, elongated hills that are steeper on one end. The steeper end faces the direction from which the ice advanced. Typical drumlins are 400 to 800 m long and 8 to 60 m high. They form in swarms near the outer edge of continental glaciers and appear to have been shaped from glacial drift by the advancing ice.

The effects of continental glaciation extend beyond the ice itself. Abundant meltwater may pond to form large **proglacial lakes** on the perimeter of the melting glacier, as illustrated in Figure 10.6. Sedimentation in such lakes results in a flat surface, recognizable long after the water has drained away. In some lakes, drainage occurred in stages, and *former shorelines* are left to record these stages.

Rivers draining these lakes may carve large valleys, which are later occupied by much smaller rivers. Such *underfit* streams are obviously too small to have carved the valley they now occupy. An example is the present-day Minnesota River, which occupies a broad valley cut by Glacial River Warren as it drained Glacial Lake Agassiz (Figs. 10.6 and 8.14).

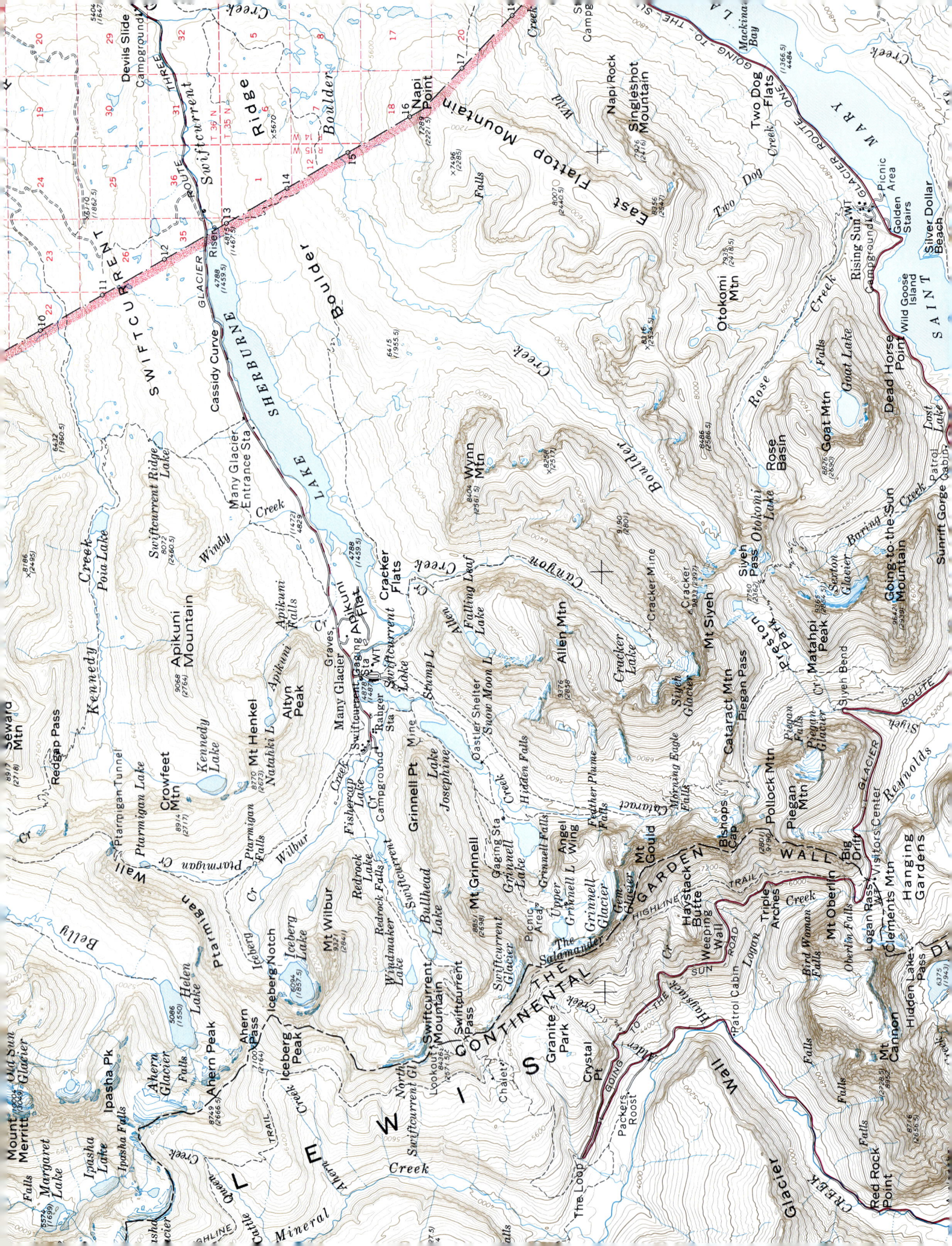

Hands-On Applications

Continental glaciers have only recently (geologically speaking) retreated from vast land areas in the northern hemisphere, and there is a possibility that they could return within ten or twenty thousand years. An alternative possibility is that, with global warming, existing glaciers in Antarctica and Greenland could melt. Either occurrence would have a devastating impact on humans, and it is clearly in our best interests to know as much as possible about glaciers. In this lab, you will learn some of the scientific methods geologists use to study active glaciers, and how it is possible to reconstruct the history of past glaciers.

Objectives

If you complete all the problems, you should be able to:

1. Determine from a photograph or historical measurements whether a glacier has advanced or retreated in the recent past.
2. Determine the flow direction of glacial ice from a map, photograph, or historical measurements, and given the necessary data, calculate the rate of movement or the distance moved during a given time span.
3. Recognize the following alpine-glacier erosional features in a photograph or on a topographic map: cirque, tarn, col, horn, U-shaped valley, paternoster lakes, arête, hanging valley, fiord.
4. Recognize the following depositional features of valley and continental glaciers in a photograph or on a topographic map: end moraine, lateral moraine, medial moraine, outwash plain, kettle or kettle lake, esker, kame, drumlin.
5. Interpret a map of glacier deposits.

Problems

1. Glacier National Park, in northern Montana, is part of the Canadian–U.S. Waterton-Glacier International Peace Park. It sits astride the continental divide, which separates drainage to the Pacific from that to the Gulf of Mexico or to Hudson Bay. Present-day glaciers are small and occupy cirques or well-shaded areas. These particular glaciers, though always small, were probably at their maximum during the middle of the 19th century. At that time, there were some 150 small glaciers in the park area, about three times as many as at present. During the period from about 1920 to the mid 1940s, wastage of glaciers was extensive; from that time to the present, most glaciers have retreated somewhat, although some have advanced. During the last major continental glaciation in North America, about 20,000 years ago, alpine glaciers were much more extensive in the park and occupied valleys tens of miles long. The spectacular erosional features that remain in the park reflect both earlier and recent glaciation. Figure 10.7 is a topographic map of one of the most visited parts of the park and includes the Many Glacier area.

    Locate examples of the following features of alpine glaciation. Use a geographic name to describe the location (e.g., an example of a cirque glacier is Grinnell Glacier).

    a. Cirque.

    b. Col.

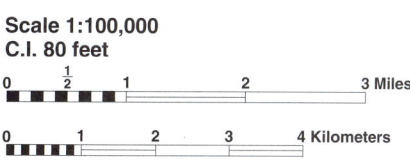

FIGURE 10.7

(Page 178) Portion of topographic map of Glacier National Park, Montana.

180 Part IV Landscapes and Surface Processes

    c. Horn.

    d. U-shaped valley.

    e. Arête.

    f. Hanging valley.

    g. Tarn.

    h. Paternoster lakes.

2. Refer to the aerial photographs of North Crillon Glacier (Fig. 10.8) and to the map of part of Glacier Bay National Park, southeastern Alaska (Fig. 10.9). North Crillon Glacier also appears on the map.

    a. Look at Figure 10.8 with a stereoscope. North is marked.

        (1) Look at the small white glacier north of North Crillon Glacier. What geological term describes the valley occupied by the small glacier compared to the larger valley hosting the larger glacier?

            What evidence suggests that the small glacier was once thicker? While you are looking, note that the small glacier forms an *ice fall* near the top of the photo.

        (2) Examine the valley wall north of North Crillon Glacier. Draw a line indicating the height to which the glacier ice may have once extended. Did the North Crillon Glacier fill its valley up to the level of the small glacier, or is it more likely that the small glacier made another ice fall as it entered the larger valley?

        (3) Examine the terminus (end) of the North Crillon Glacier. Why is a big piece missing from its north end?

            Although you cannot see a meltwater stream, there is strong evidence that summer meltwater transports a great deal of sediment along the southwestern edge of the glacier. What evidence can you see?

        (4) What are the dark streaks on the upper end of North Crillon Glacier?

            Approximately how many are there?

            Assuming the above number is correct, infer how many tributary glaciers flow into North Crillon Glacier.

        (5) Describe why the glacier is darkest near its terminus.

FIGURE 10.8
Stereophotographs of North Crillon Glacier, Glacier Bay National Park (see Fig. 10.9 for location). Scale is 1:40,000.

Part IV Landscapes and Surface Processes

(6) Note the cracks, or *crevasses,* near the terminus. What is their orientation relative to the direction of ice flow?

Did they form from tension (pulling apart) or compression (pushing together)? Explain your answer.

b. As you can see in the map (Fig. 10.9), North and South Crillon glaciers flow southwest, bend 90° as they enter a northwest-trending valley, and then go either northwest or southeast. Incidentally, the northwest-trending valley marks the position of one of the major geologic faults of southeastern Alaska. Earthquakes along the Fairweather Fault occasionally jar the area.

(1) Starting with the ridge between North and South Crillon glaciers, trace the drainage divide that defines the drainage basin of the North Crillon Glacier (Chapter 8 explains drainage basins). Look for the sharp-edged arêtes and horns to help guide your way; parts of the divide are snow-covered.

Approximately how many tributary glaciers, large and small, can you spot flowing into the North Crillon Glacier? Ignore the smallest "buds." Your answer probably will not match what you got from the air photos because closely spaced medial moraines tend to merge as additional glaciers squeeze into the flow.

(2) Cite evidence to support or refute the suggestion that glaciers once flowed down the creeks north of Lituya Bay (i.e., Fourmile, Eagle, and Portage creeks).

(3) What evidence suggests that Lituya, North Crillon, and South Crillon glaciers were more extensive in the past?

Based on this evidence, mark their probable former extent on the map.

What type of glacial feature is Lituya Bay?

What type of glacial feature is at least partly responsible for the shallow water and major features at the mouth of Lituya Bay and for the ridge curving around the southern end of Crillon Lake?

c. The brown stippled areas on the glaciers in Figure 10.9 represent rocky sediment on the surface of the ice. Find and label representative lateral, medial, and end moraines.

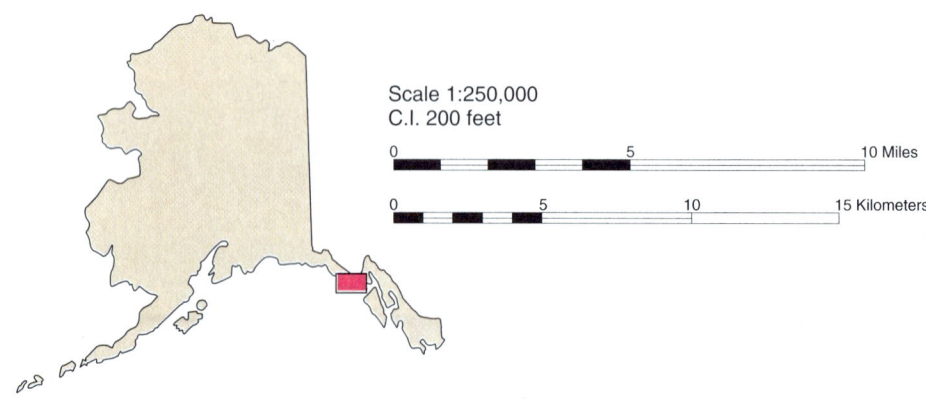

FIGURE 10.9
(Page 183) Portion of Mt. Fairweather, Alaska/Canada, 1° × 2° quadrangle, including part of Glacier Bay National Park.

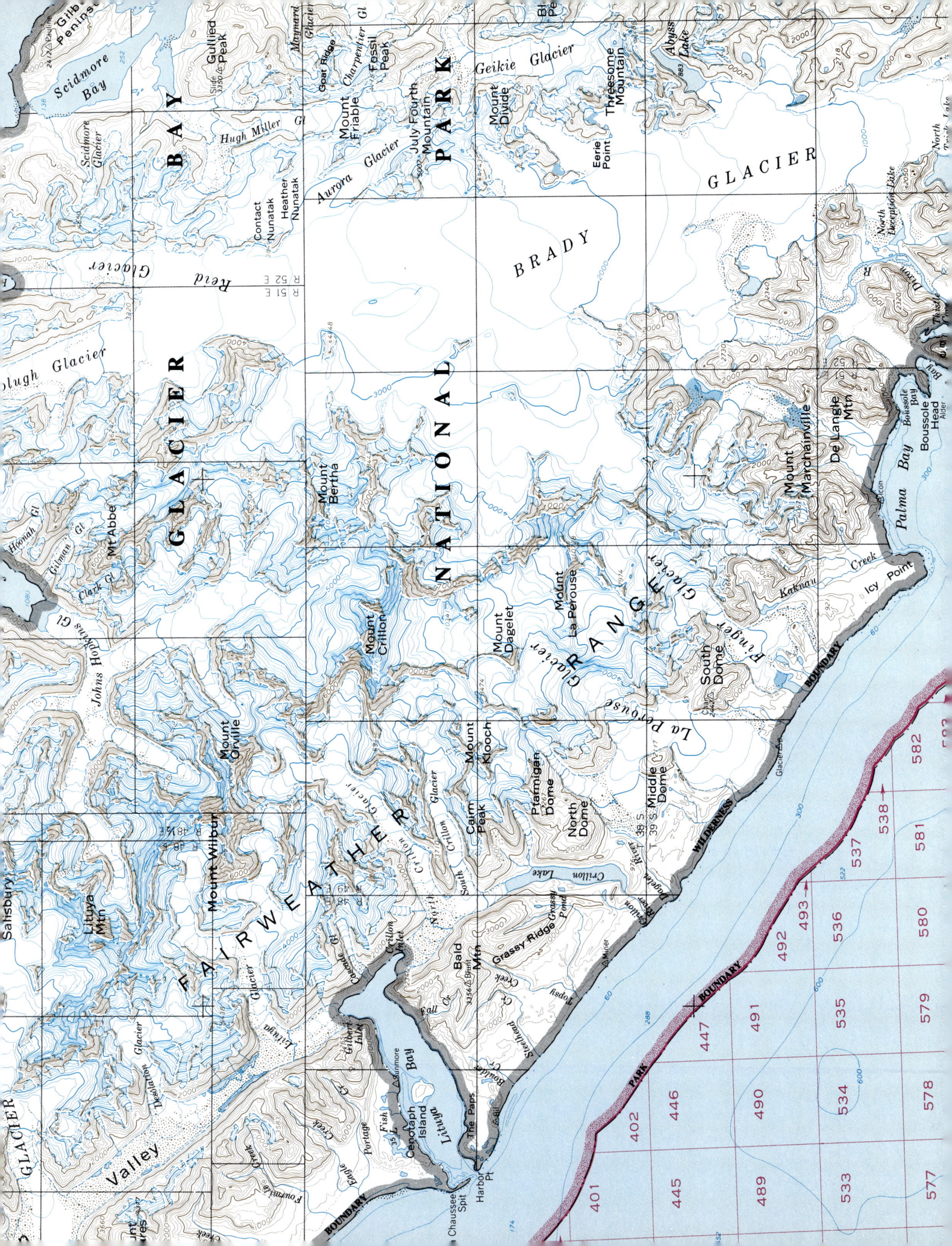

184   Part IV   Landscapes and Surface Processes

3. Figure 10.10 is a portion of the Glacial Map of the United States East of the Rocky Mountains. It shows the kinds of glacial deposits present at the surface. If you were to go to Indiana, for example, and stand on one of the belts marked in green, you would know by consulting the map explanation on the left that you were standing on an end moraine of Wisconsin age. Because this map shows only surface deposits and features, it displays only the most *recent* glacial event to have affected individual areas. Two major periods of continental glaciation, both within the Pleistocene Epoch (see Figure 13.7), are well represented on this map. Most of the map shows, in shades of green, deposits of Wisconsin age, because this is the most recent glaciation. The pink areas in southern and southwestern Illinois were not affected by Wisconsin-age glaciers, but were affected by glaciers of the earlier Illinoian age. The position of the ice front shifted back and forth considerably during glacial times, but each glacial age ended with total retreat.

   a. What glacial feature is represented by the dark green, curved bands, some of which are named?

   Was the front of the ice advancing, retreating, or staying more or less in place as these features were being formed?

   What glacial feature, shown in light green, occurs between the dark green bands in Illinois, Indiana, and Ohio?

   Locate the dark green named bands in eastern Indiana. The one on the northeast is labeled *Fort Wayne,* the one on the southwest *Union City.* Which is older, the Fort Wayne or the Union City? How can you tell?

   b. What do the gray areas surrounding especially lakes Huron and Erie represent?

   What do the bright yellow areas represent? Hint: Examine the branching patterns seen especially in Indiana.

   Today the Great Lakes drain to the east via the St. Lawrence Seaway out to the Atlantic. As the ice sheets melted, did the water always drain *only* east? What evidence supports your answers?

   c. The familiar lakes, such as Lake Michigan, are shown in light blue on the map, but other lakes, such as Lake Wauponsee or Lake Watseka (southwest of Chicago), may be less familiar. What are these lakes?

   Explain how and when they came to be, and what evidence for them you might find in the landscape today.

4. All of the northern Midwest except for the southwestern part of Wisconsin and a small part of adjacent Illinois was covered by continental glaciers several times during the Pleistocene Epoch (Fig. 10.10). The aerial photograph in Figure 10.11 includes the boundary between glaciated and unglaciated areas. The red area on the east is hilly and partially forested. The area on the west, with the circular irrigated fields, is a flatter terrain that is mostly cropland.

   a. Which side has been glaciated, east or west, and how do you know? Hint: Consult both Figure 10.10 and Figure 10.11.

   b. What type of glacial landform marks the former outer edge of the continental glacier?

   c. What kind of glacial sediment would you expect to find in the eastern part of the area?

   d. What type of glacial landform occurs in the western part of the area?

   e. What kinds of glacial sediment make up this type of landform?

5. Figure 10.12 represents an area of southern Michigan just south of Jackson and located in one of the yellow areas of Figure 10.10. The Jackson area consists mainly of glacial deposits associated with stagnant (nonmoving) ice. These are mostly deposits from meltwater that flowed on, in, or at the base of ice that later melted away.

   a. How did the several lakes—such as Mud, Crispell, Skiff, and so forth—probably form?

   b. What type of glacial feature is Blue Ridge, and how did it form?

   c. What type of glacial feature is Prospect Hill, and how might it have formed?

   d. Newer, 1:24,000-scale maps of the north half of this area show crossed-shovel symbols in several places. What resource do you think is being extracted?

   Circle several areas in the northern half of the map where you think comparable resources might be located.

FIGURE 10.10
Portion of glacial map of the United States east of the Rocky Mountains

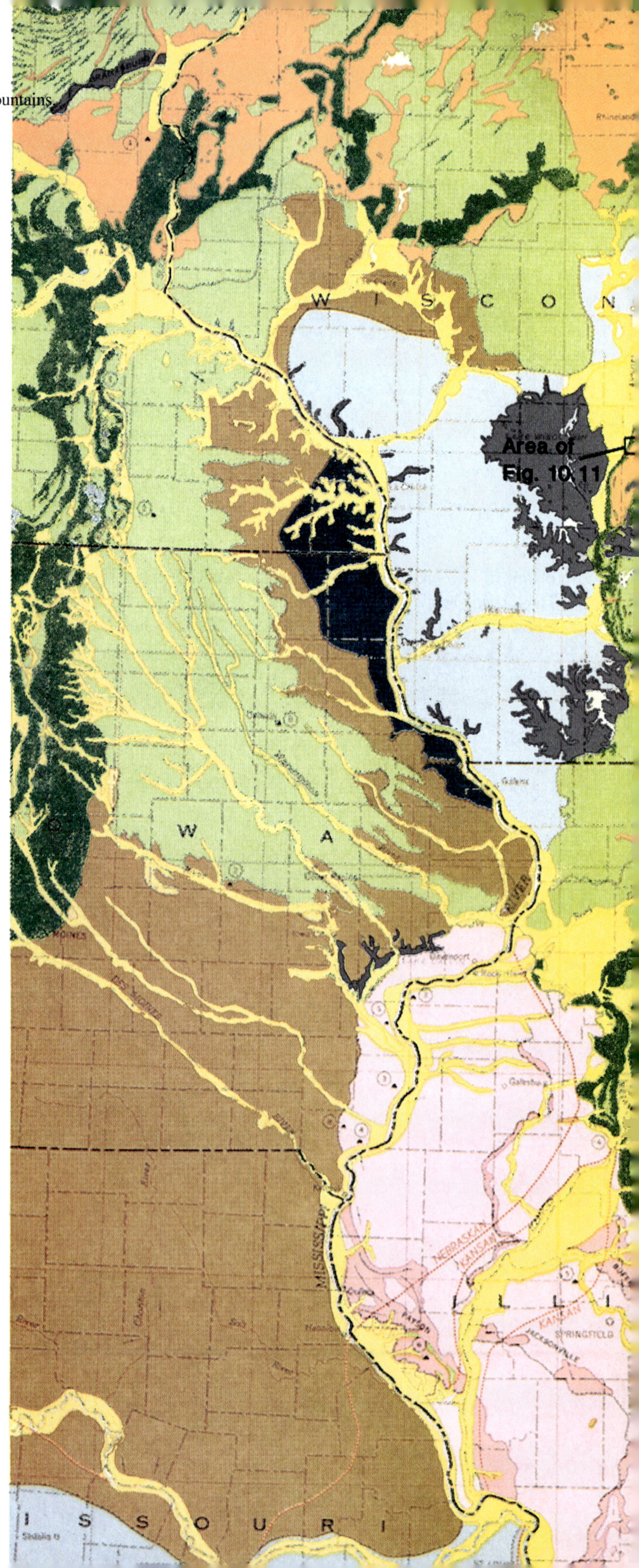

EXPLANATION
AREAL COLORS

Lacustrine sediments
Mainly sand, silt and clay, deposited in lakes dammed by glacier ice or by outwash sediments. Includes areas of till or bedrock. Labels show the names of lakes except ancient Great Lakes.

Outwash sediments
Mainly sand and gravel, deposited by proglacial streams. Includes some alluvium; in Kansas, Nebraska, South Dakota, North Dakota, and Montana, includes alluvium with little or no outwash. Arrow shows direction of flow.

Ice-contact stratified drift
Mainly sand and gravel, occurring as kames, eskers, kame terraces, pitted plains and collapsed stratified drift.

End moraines of Wisconsin age
Mostly till. Local names are lettered on or beside the moraines.

Drift, other than end moraines and outwash, of Wisconsin age
Chiefly till. Includes many areas of bedrock and older drift.

End moraines of Illinoian age
Mostly till. Local names are lettered on or beside moraines.

Drift, other than end moraines and outwash, of Illinoian age
Chiefly till. Includes many areas of bedrock.

Drift of Kansan age
Chiefly till. Includes large areas of bedrock.

Drift of Nebraskan age
Patches of till and scattered erratics.

Area not glaciated

- Streamline features
- Striation direction
- Strandlines

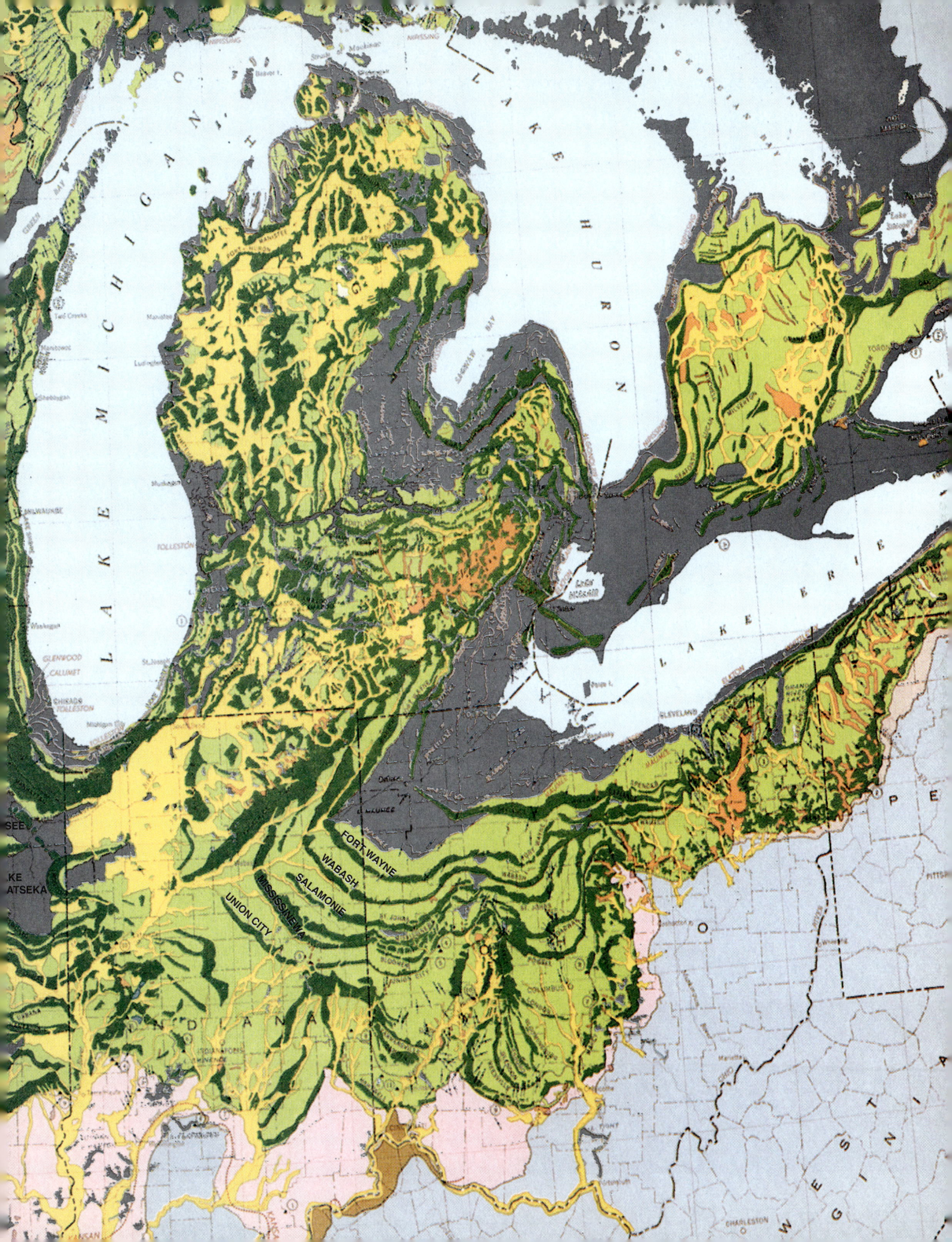

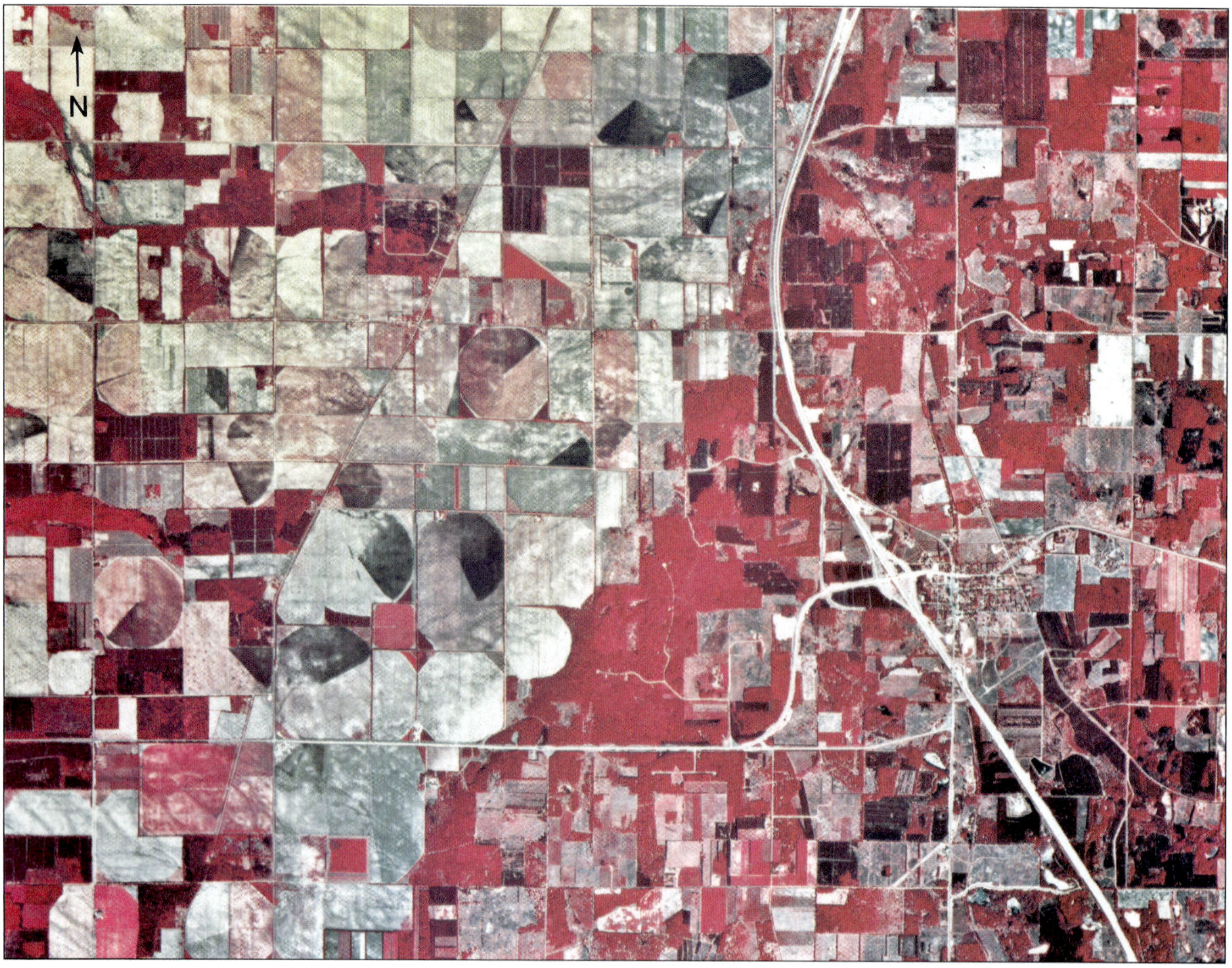

188 Part IV Landscapes and Surface Processes

FIGURE 10.11
High-altitude, color-infrared aerial photograph of the Coloma, Wisconsin, area (scale 1:58,000) for Problem 4. The circles are irrigated fields. Red indicates growing vegetation. Greens and lighter shades reflect land with little vegetation. Photo location is shown in Figure 10.10.

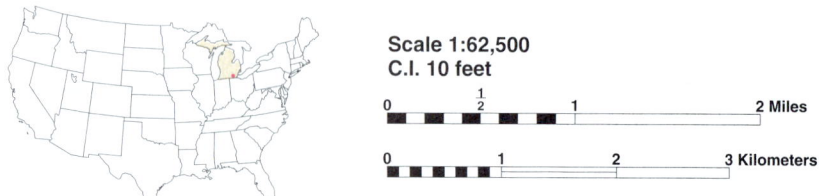

FIGURE 10.12
(Page 189) Portion of Jackson, Michigan, 15-minute quadrangle topographic map.

190 Part IV Landscapes and Surface Processes

6. **Glaciers on the Web:** Go to *http://nsidc.org/glaciers* (or link to it through *www.mhhe.com/jones5e*—see Preface) and click on *General Information,* then *Q&A.*

   a. Click on *What is a glacier?* What percentage of the world's land area is occupied by glaciers today?

   What percentage was occupied during the last Ice Age?

   b. Click on *Where are glaciers located?* What is the total area covered by glaciers around the world?

   What *percent* of this is in Antarctica? Show your work.

   In the United States?

   In Africa?

   Does Iceland have more or less ice than Greenland?

   c. Click on *What types of glaciers are there?* What are the differences between an *ice sheet, ice cap,* and an *ice field?*

   d. Click on *Do glaciers affect people?* What essential resource do some glaciers provide people?

   Click on *Are glaciers dangerous?* What are the main hazards for people living near glaciers?

   e. Click on *How do glaciers reflect climate change?* How can glaciers give us information regarding the history of climate over the past several hundred thousand years?

   What might they be telling us about climate change over the past century?

   Under *Why do glaciers move?* you can see a striking example of glacial retreat.

Another site of interest is the U.S. Geological Survey site *http://ak.water.usgs.gov/glaciology,* which describes the "benchmark" glacier program. This program monitors climate, glacier motion, mass balance, geometry, and stream runoff at three glaciers. Many other interesting sites can be reached through links from this site.

In Greater Depth

7. The huge Glacial Lake Agassiz (Fig. 10.6) once drained south out the Minnesota River. A portion of this river is shown in Figure 8.14 as part of Problem 3 in Chapter 8. If you worked this problem, you learned that the 100-year flood plain is located roughly at the level of the 730-foot contour.

   a. Use Figure 10.13 to create a topographic profile along the line A-A' shown in Figure 8.14.

   b. Locate and mark (on Fig. 10.13) the position of the floodplain for the *modern* Minnesota River.

   c. Do you think the Minnesota River carried the same average flow as today? Mark reasonable limits for the ancient floodplain on the profile you drew on Figure 10.13.

A   A'

FIGURE 10.13

8. Figure 10.14 is a map that summarizes pre-1970 data for Grinnell Glacier, a cirque glacier in Glacier National Park (west center of Fig. 10.7).

   a. The arrows or vectors shown on the map summarize the total movement of a number of rocks exposed on the surface of the glacier, as indicated in the *Explanation*. The lengths of the vectors depend on both the time over which observations were made and the *rate* at which the ice moved. What if we wanted to know whether the rate varied from year to year, or whether it varied from place to place in the glacier? Table 10.1 contains the data from which the length of the vectors was determined, and it shows the average annual rate of movement for one of the rocks.

      (1) First, fill in the table by calculating the average annual rate of movement for all the rocks.

      (2) Then separate the rates into four categories ranging from slowest to fastest (for example, if the difference between the slowest and fastest rate is 40 feet/year, and the slowest rate is 25 feet/year, the categories would be 25 to 34 feet/year, 35 to 44 feet/year, 45 to 54 feet/year, and 55 to 65 feet/year). List your categories next to Table 10.1.

Part IV Landscapes and Surface Processes

TABLE 10.1
Movement of Marked Rocks on Grinnell Glacier
Source: Data from A. Johnson, *USGS Professional Paper 1180*, 1980.

Rock Number	Period	Movement (feet)	
		Total	Annual average
47-1	1947–57	380	38
47-2	1947–53	215	
50-1	1950–64	530	
50-2	1950–69	700	
52-1	1952–62	485	
52-2	1952–56	210	
52-3	1952–59	285	
58-1	1958–69	385	
59-1	1959–64	170	
59-2	1959–68	355	
59-3	1959–69	355	
59-4	1959–65	275	
59-5	1959–66	320	
63-1	1963–66	110	
63-2	1963–69	190	
64-1	1964–69	185	
65-1	1965–69	170	
65-2	1965–69	180	
65-3	1965–69	185	
66-1	1966–68	80	

(3) Using colored pencils or an equivalent, highlight on the map those vectors representing the highest rates in red, second highest in yellow, third in green, and slowest in blue.

(4) Does the ice seem to move more rapidly in some parts of the glacier? If so, give a possible explanation.

b. It is clear from Figure 10.14 that Grinnell Glacier has receded since 1887, when it was first described by the George B. Grinnell party (see sequence of "edge of glacier" lines). However, the position of the terminus of such a small glacier over a short time period may or may not reflect the true mass balance of the glacier. The terminus position can be strongly influenced by rate of ice movement and by bedrock topography. Another way to look at changes is by periodic observations of ice-surface elevations, such as those reported in Table 10.2.

The data in Table 10.2 are mean (average) elevations of five segments of the line B-B' shown on the map. The segments are designated with the distance from plane table benchmark 6425, located near the intersection of lines A-A' and B-B' on Figure 10.14. For example, the column headed 100–500' lists the mean elevations of the segment of line B-B' between 100 and 500 feet south-southwest of the point labeled 6425.

(1) Use these data to construct a graph on Figure 10.15 that illustrates how the average elevation of the ice surface in just the 500- to 1000-ft interval has changed with time. Plot mean elevation on the vertical *y*-axis and the date (year) on the horizontal *x*-axis. Draw lines between adjacent data points to emphasize the changes. Although these data represent one small area of the glacier, changes elsewhere on the glacier appear to be similar, so you can assume your graph represents the glacier as a whole.

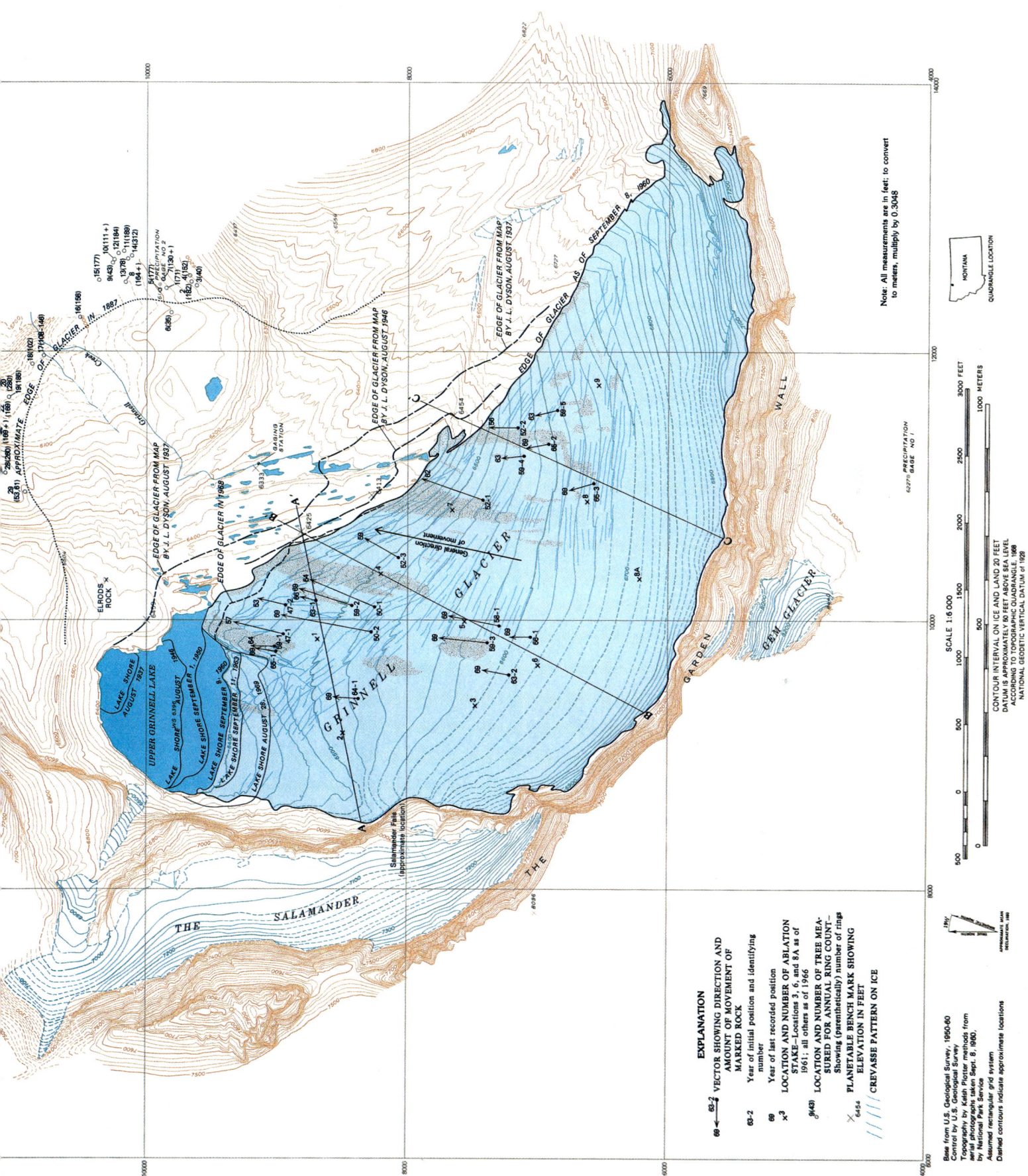

FIGURE 10.14
Map of Grinnell Glacier, Glacier National Park, Montana.

(2) How has the elevation of the glacier changed during this short time interval?

(3) Is this generally consistent with the changes in position of the terminus shown on the map?

(4) Does this imply that Grinnell Glacier has advanced or retreated during this period?

(5) Has the *rate* at which elevation changed been consistent throughout the period? If not, indicate when it changed, and identify those time periods when it was generally consistent. When considering rate changes, look at rates established over periods of five or more years, not year-to-year changes. (Hint: Rate is expressed graphically by the slope of the line.)

c. An interesting website showing photographs of Grinnell Glacier at various times between 1910 and 1997, and an aerial photograph showing the boundaries of the glacier in 1850, 1937, 1968, and 1993, is at *http://www.nrmsc.usgs.gov/research/grinnell.htm.*

TABLE 10.2
Mean Ice-Surface Elevations of Segments of Profile B–B', Figure 10.13
Source: Data from A. Johnson, *USGS Professional Paper 1180*, 1980.

Line Segment	100–500'	500–1000'	1000–1500'	1500–2000'	2000–2500'
Date		Mean elevation of line segment			
1950 Sept. 14	6,460.1	6,523.3	6,564.8	—	—
1951	—	—	—	—	—
1952 Aug. 22	6,460.3	6,522.6	6,563.8	6,604.8	—
1953 Sept. 4	6,458.4	6,519.5	—	—	—
1954 Sept. 27	6,459.5	6,522.0	6,564.4	—	—
1955 Sept. 6	6,460.6	6,521.8	6,563.9	—	—
1956 Aug. 30	6,461.7	6,521.6	6,563.8	6,604.6	6,659.9
1957 Sept. 10	6,456.6	6,517.9	6,560.6	6,600.9	6,654.9
1958 Sept. 15	6,446.4	6,509.8	6,551.6	6,591.2	6,642.7
1959 Sept. 12	6,449.9	6,513.6	6,555.7	6,597.1	6,649.5
1960 Sept. 6	6,449.9	6,514.1	6,555.5	6,594.7	6,646.7
1961	—	—	—	—	—
1962 Sept. 2	6,442.8	6,506.4	6,547.4	6,596.9	6,640.7
1963 Sept. 12	6,437.5	6,499.2	6,541.8	6,582.4	6,634.4
1964 Sept. 15	6,436.5	6,499.1	6,541.1	6,581.4	—
1965 Sept. 6	6,436.4	6,499.6	6,540.9	6,580.2	6,635.0
1966 Aug. 17	6,437.6	6,501.5	6,543.1	6,582.3	6,637.7
1967	—	—	—	—	—
1968 Aug. 28	6,432.1	6,498.8	6,539.5	6,578.8	6,633.9
1969 Aug. 27	6,428.2	6,493.3	6,534.2	6,574.2	6,628.3

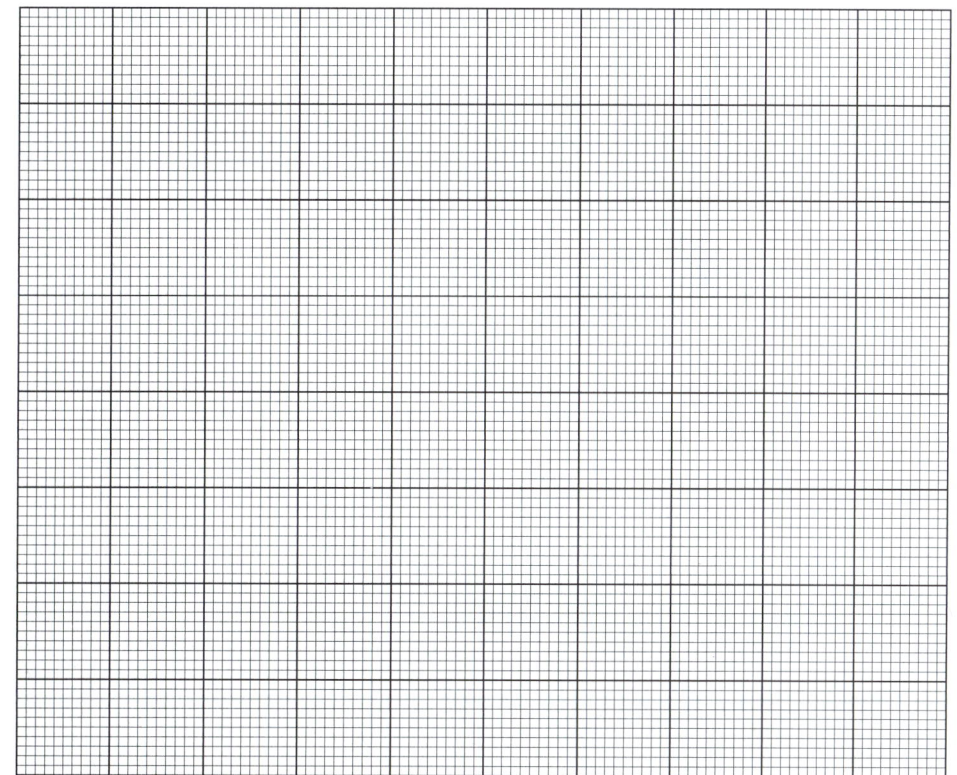

FIGURE 10.15

Graph paper for Problem 7b.

CHAPTER 11

Sea Coasts

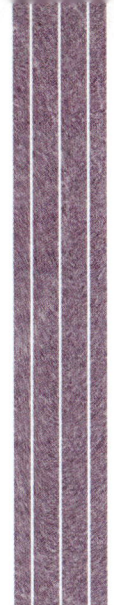

Materials Needed
- Pencil and eraser
- Ruler
- Stereoscope (provided by instructor)

Introduction

The land along the coasts is subject to relentless pounding by waves. These waves move a great deal of sand and other sediment that can maintain beaches, make them disappear, move them inland, or push them across the mouths of bays and harbors. Where waves hit coastal cliffs, they focus their energy on the irregularities that jut into the sea. These promontories gradually wear away and collapse into the sea, and the sediment they produce is moved along to help fill in nearby bays and estuaries. Since over 75 percent of the U.S. population lives near the seacoasts, it is worth understanding how waves, longshore currents, and changes in sea level affect the coasts. In this lab, you will learn to recognize common coastal features on maps and in air photos. You will learn to use them to interpret the processes that have affected, are affecting, and will affect a given coast.

Waves, generated by wind blowing across the water, play the greatest role in shaping the coast. Figure 11.1 illustrates a wave and shows how its passage causes movement of water. Note that while the crest of a wave travels across the water, the water itself simply moves in a circular path. Water motion decreases to zero at the **wave base,** a depth approximately equal to half the wavelength.

As waves approach the shore, they begin to interact with the bottom, their shapes change, and the pattern of water movement changes, as shown in Figure 11.2. When the water depth becomes less than wave base, waves begin to erode and transport fine-grained bottom sediment. In the shallow **surf zone,** the wave shape breaks down, and water motion becomes turbulent; it is here that most work is accomplished. Sand and pebbles are thrown into suspension by the turbulent surf and are in near-constant motion.

Waves nearly always approach a coast with their crests at an acute angle to the shoreline. As the waves begin to interact with the bottom, they slow down and gradually bend, or **refract,** and become more parallel to the coast, as shown in Figure 11.3. In spite of refraction, however, the waves still hit the shore at an angle. Although most of the wave energy is directed at the beach, some is parallel to the beach. This produces a **longshore current,** which flows along the coast in the direction in which it is being pushed by the waves (Fig.11.3). Sediment in the surf zone is carried by the current in a process known as **longshore drift.**

Wind energy is thus transferred through wave action to the surf zone, where erosion, transportation, and deposition combine to modify the coastline. We look first at common landform features that result from coastal erosion, then turn to those features formed by deposition.

Erosion

All coasts undergo erosion, but the intensity differs among coasts. The evidence for active erosion is best seen on rocky coasts. Figure 11.4 shows waves being refracted by the shallows around a **headland,** which is a rocky point of land between bays. Notice how the waves bend to focus their destructive energy directly on the headland. The concentration of erosive activity just above and below sea level creates steep **wave-cut cliffs.** These are progressively undercut and collapse into the water. In time, this process creates a gently sloping erosional surface just below sea level called a **wave-cut platform.** More resistant portions of rock temporarily left standing offshore on the wave-cut platform are **sea stacks.**

Deposition

While headlands erode because wave energy focuses on them, deposition takes place in the bays because wave energy dissipates there. Deposition results in the formation of a **beach,** commonly made of sand, pebbles, or cobbles eroded from the headlands, as well as material carried to the coast by rivers. In time, the combination of erosion and deposition reduces the irregularities of the coastline.

Longshore drift aids in straightening out coastlines. When a longshore current

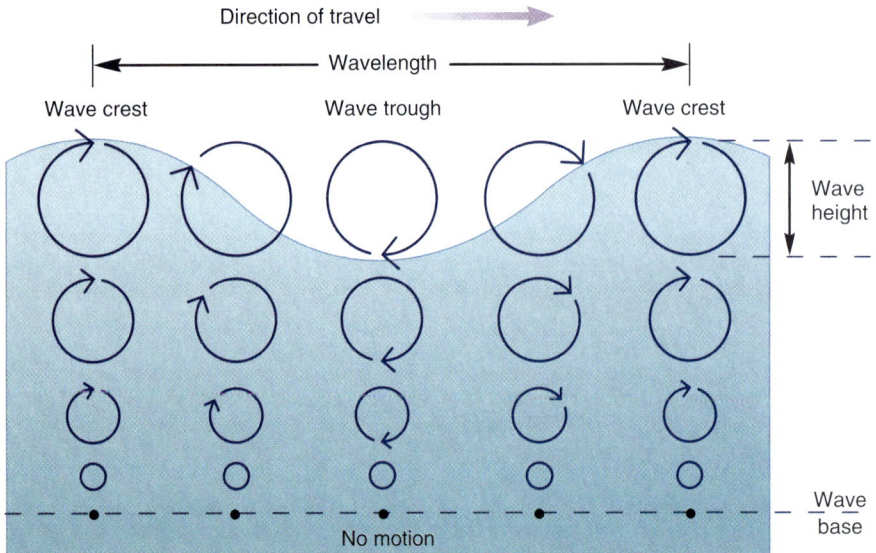

FIGURE 11.1
Cross section of wave, showing how water motion decreases to zero at wave base (a depth approximately equal to half the wavelength).

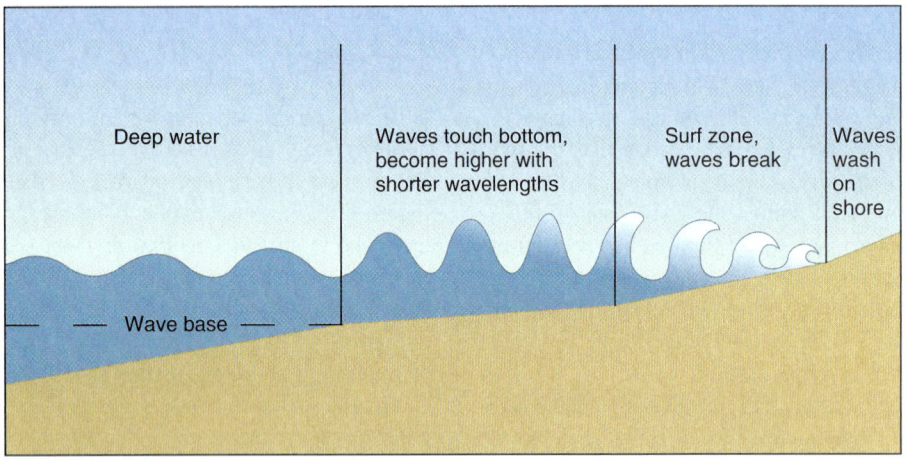

FIGURE 11.2
When waves enter water shallower than the wave base, they change form until, in the surf zone, they break, and water motion becomes turbulent.

passes a bend in the coast and enters deeper water, it slows, and some of its load is deposited. A common feature formed by this process is a **spit,** an emergent ridge of sand growing in the direction of longshore drift, as shown on Figure 11.5. A spit extending completely across the mouth of a bay is a **baymouth bar. Tombolos** are sand ridges connecting islands to the shore (Fig. 11.5). They develop because the island blocks and refracts the waves, interrupting the longshore current. The refracted waves push the sand behind the island where it is deposited.

Rivers provide most of the sediment for beaches and longshore drift. If longshore currents are strong, most of the sediment at the mouth of a river is swept away. If longshore currents are weak, or if the sediment load of a river is excessive, the sediment is deposited at the mouth of the river as a **delta** (Fig. 11.5).

Low-lying coasts, such as the southeastern U.S. coast from New York to Mexico, may have long, narrow, sandbar islands paralleling the coast. These **barrier islands** (Fig. 11.6) are 2 to 5 km wide and 10 to 100 km long. They are separated from the mainland by a *lagoon* some 3 to 30 km wide. Barrier islands slowly migrate landward due to big storm waves washing sand from the ocean side to the lagoon side. This process creates **overwash fans,** lobes of sand that breach the vegetated dunes and extend toward the lagoon. Gaps between islands, known as **tidal inlets,** develop strong tidal currents as the tide rises and falls. Sediment transported by these currents is deposited as **tidal deltas** both landward and seaward of the inlets. Longshore currents often push spits into tidal inlets. These spits generally curve toward the lagoon due to interactions between waves, longshore currents, and tidal currents. As the spit closes off one side of the inlet, tidal currents erode the other side to keep the inlet open. This causes tidal inlets to migrate along barrier islands in the direction of longshore drift. Both wave action and inlet migration can dramatically change a barrier island within even a lifetime. Despite this, several major cities (including Atlantic City, New Jersey; Miami Beach, Florida; and Galveston, Texas) and hordes of resort developments are located on barrier islands.

Human Interaction

People living along coasts, and especially on barrier islands, commonly do not understand the power and complexity of the natural processes of coastal erosion and deposition. If their part of the beach is eroding, they may try to protect their property from encroaching sea waves by erecting a concrete **seawall.** Unfortunately, reflection of wave energy off this seawall further enhances erosion of the beach, and a narrower beach allows waves to more forcefully pound the seawall. Eventually the seawall fails and a larger, uglier, more expensive seawall must be built.

Another approach is to build **groins,** which are low walls built perpendicular to the beach to interrupt the longshore current (Fig. 11.7). Although sand is deposited on the upstream side of the groins, it is eroded away on the downstream side. Thus, some property owners have a wider beach, others have a narrower beach, and everyone gets to look at ugly groins cutting up the beach. Hard structures should in general be avoided for beach protection because they have unintended negative consequences, are expensive to maintain, and are politically impossible to remove once in place.

Another popular beach-saving strategy is to dredge sand from a tidal inlet or off-

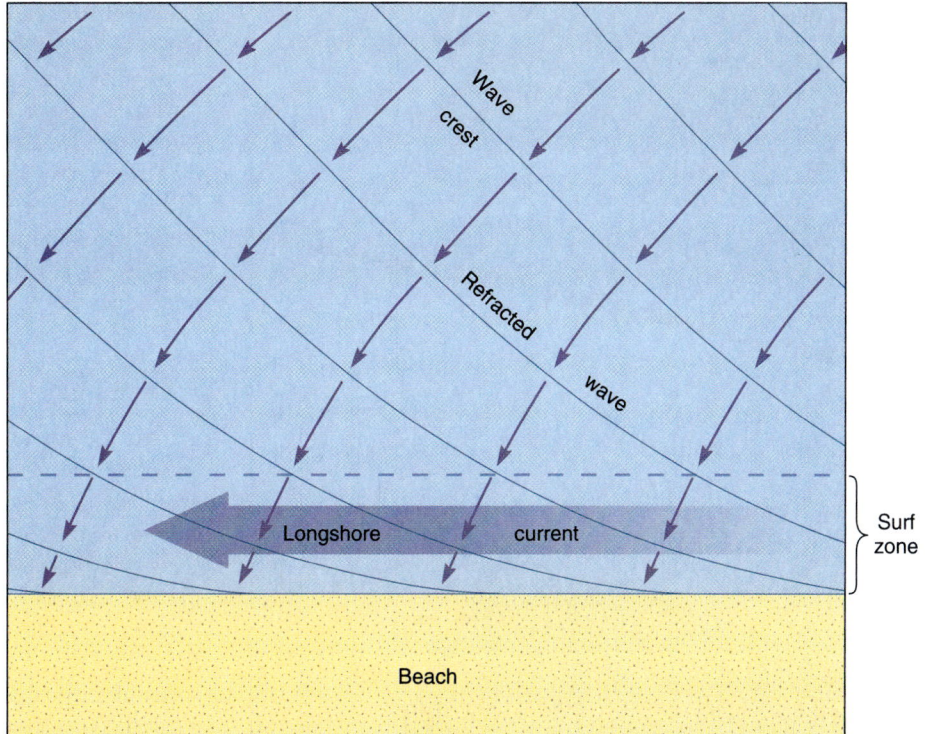

FIGURE 11.3
Map view showing refraction (bending) of waves as they approach shore. Solid lines represent wave crests, and arrows indicate wave movement. As the water is pushed against the shore, a longshore current develops in the surf zone. The longshore current carries sediment with it in the surf zone, a process known as longshore drift.

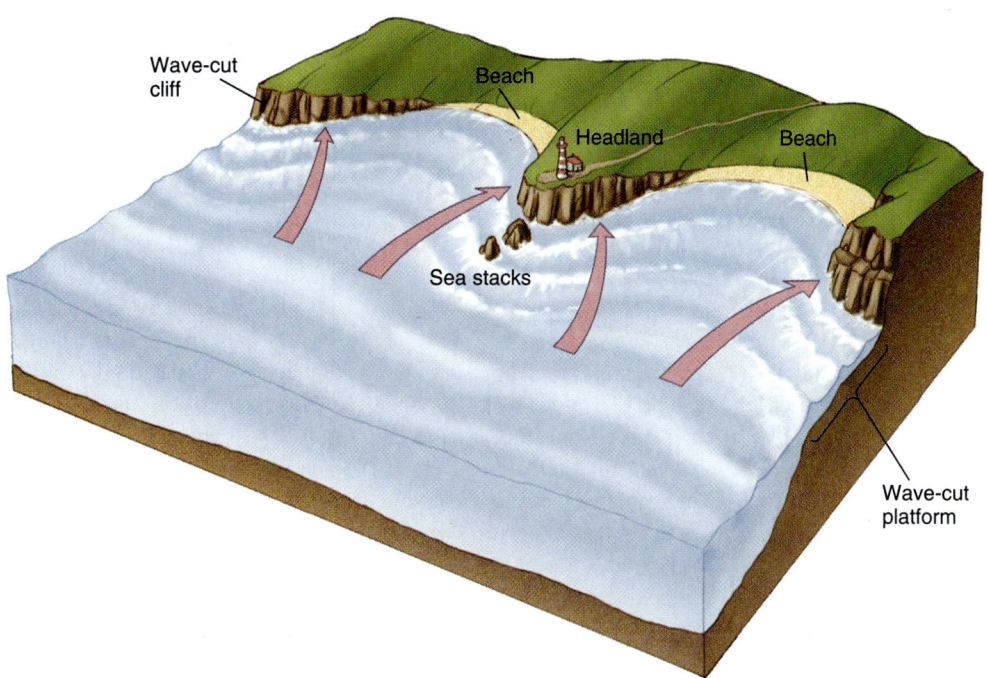

FIGURE 11.4
Waves approaching an irregular coastline are refracted so that energy is concentrated at headlands. Erosion results in the formation of wave-cut cliffs, sea stacks, and a wave-cut platform.

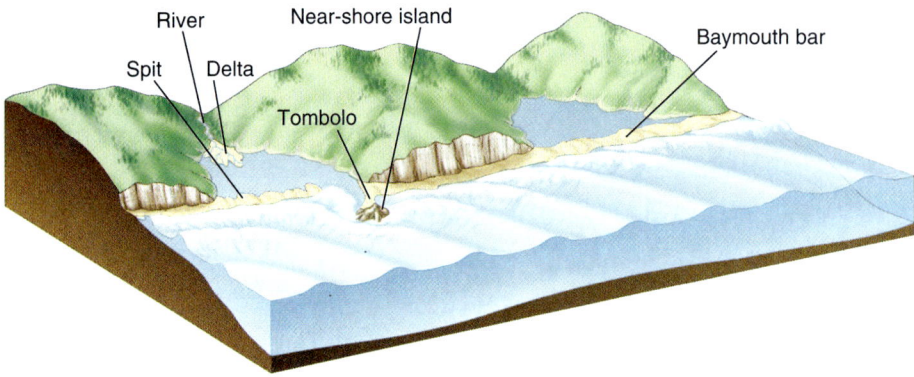

FIGURE 11.5
Longshore drift is from left to right, as indicated by the spit. Wave refraction around the near-shore island causes the tombolo to form.

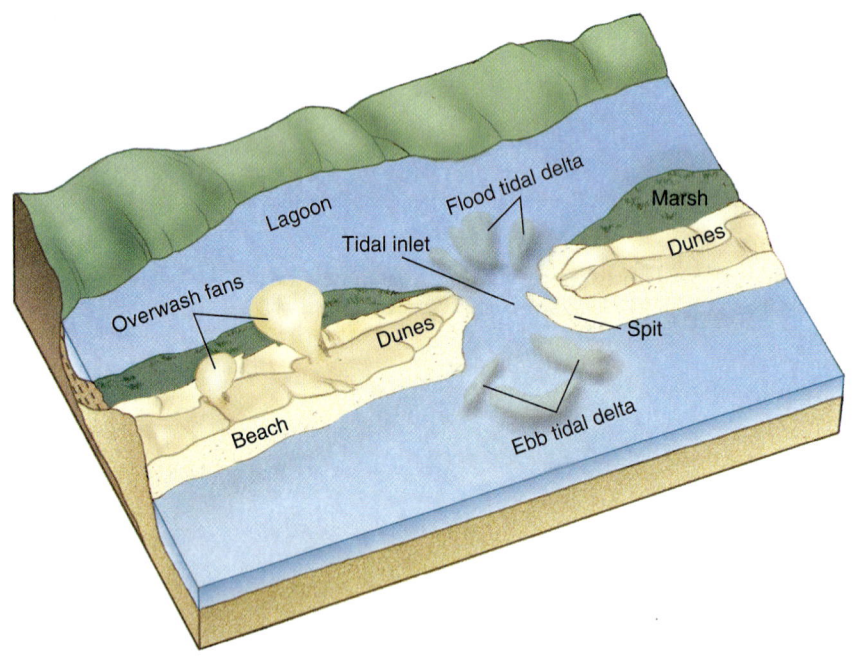

FIGURE 11.6
Features of barrier islands include beach, dunes, overwash fans, marsh, lagoon, tidal inlet, and tidal deltas (usually submerged).

shore area and, at a cost of tens of millions of taxpayer dollars, pump it onto the beach. Of course, the processes that cause the beach to naturally move have not changed, and the precious sand from such "renourishment" projects frequently disappears within a few short years. It is difficult to reconcile the natural mobility of the beach with immobile buildings and fixed property lines.

Submergent Coasts

The position of shorelines has fluctuated throughout geologic time. Over the past three million years, dramatic changes in the volume of ice stored in the polar ice caps have caused sea level around the world to fluctuate by over 100 meters (>325 feet). There are many examples around the world of coasts that were drowned (submerged) by the most recent sea level rise, which started some 20,000 years ago as the great ice sheets began melting. Shorelines can also move in response to tectonic and other forces that cause the edges of continents to rise and fall relative to sea level.

The characteristics of **submergent coasts** depend on the nature of the landscape before submergence. If the topography was rugged, the submerged coastline is irregular, with such features as **estuaries** (drowned river valleys) or **fiords** (drowned glacial valleys) and islands. If the topography had low relief, such as the east coast of the United States, the submerged coastline may have large estuaries, extensive tidal flats, salt marshes, and barrier islands.

Emergent Coasts

Despite the recent sea level rise, tectonic and other forces have uplifted many coasts, especially along the tectonically active Pacific margins, to create **emergent coastlines. Marine terraces** are their most distinctive feature (Fig. 11.8). Many marine terraces represent normal wave-cut platforms that were later uplifted by brief episodes of tectonic activity. Some were also cut 125,000 years ago when global sea level was about 2 meters higher than today or at other times over the past few hundred thousand years when sea level was similarly high.

Recent and Future Sea Level Rise

The Earth's climate has warmed by about 0.6°C (1°F) over the past 100 years. This warming and its likely continuation will cause global sea level to rise. A rise of 20 cm over the past 100 years has already resulted from the accelerated melting of mountain glaciers and continental ice sheets and the warming and consequent thermal expansion of the oceans. Unfortunately, even such small increases in sea level are likely to significantly increase coastal erosion and flooding. Along coasts with sea cliffs, a higher sea level allows waves to carry more of their energy over the beach and directly into the sea cliffs. With barrier island beaches, a raised sea level allows more storm waves to breach the vegetated sand dunes and wash sand into the lagoons (Fig. 11.6). The net result is accelerated erosion of both coastal sea cliffs and beaches. In addition, storm surges during hurricanes and other major storms are more likely to flood low-lying areas and destroy property. Global warming affects much more than just the weather!

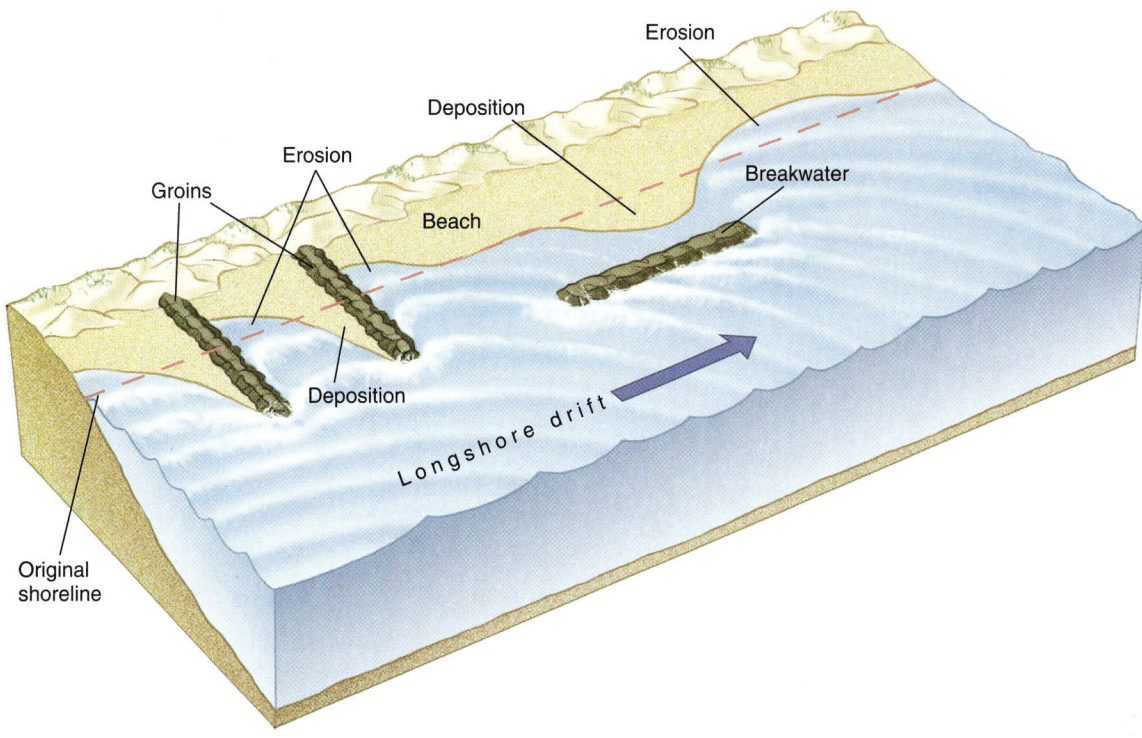

FIGURE 11.7
Groins and breakwaters interrupt longshore drift, causing deposition *and* erosion.

FIGURE 11.8
The gently sloping surface near Pacific Valley, California, was a submerged wave-cut platform that has been uplifted to form a marine terrace. Old sea stacks are visible on the terrace. Although now vegetated, such terraces often preserve well-rounded beach gravel and marine shells (clams, snails, etc.) as evidence of their former submergence.

Hands-On Applications

Coastlines are among the most dynamic environments in which we live, so it is important that we thoroughly understand coastal processes. Geologists learn to use the past and the present to predict the future. This lab demonstrates how careful observation of past results and present processes, together with the application of general principles, leads to an understanding of the coastal environment.

Objectives

If you complete all the problems, you should be able to:

1. Illustrate the refraction of waves as they strike a smooth or irregular coastline.
2. Explain the origin of longshore currents and longshore drift.
3. Explain how wave refraction may determine where erosion and deposition take place along a coast.
4. Identify the following erosional features on a map or in a photograph: headland, wave-cut cliff, wave-cut platform, sea stack.
5. Identify the following depositional features on a map or in a photograph: beach, spit, baymouth bar, tombolo, delta, barrier island, and associated features such as lagoon, tidal inlet, tidal delta, tidal flat.
6. Determine the dominant direction of the longshore current from a map or photograph showing features such as spits or groins.
7. Based on a topographic map, identify a coast as rocky or sandy, and use evidence on the map to determine whether erosion or deposition is the dominant process.
8. Recognize submergent and emergent coasts and their distinctive features, such as estuaries, fiords, and marine terraces.

Problems

1. Figure 11.9 illustrates the fragility of barrier islands with a sequence of aerial photographs. Shown in the photographs is part of Matagorda Island, a barrier island off the Texas coast in the Gulf of Mexico. Hint: Figures 11.5 and 11.6 may help you answer the following questions.

    a. Photo A, from 1943, shows a tidal inlet called Greens Bayou. Note the road to the left (southwest) of Greens Bayou. What is the minimum width of Greens Bayou? (The fractional scale is 1:10,200.)

    Based on the orientation of the waves in Photo A, determine the direction of the longshore current (right to left or left to right). What, if any, evidence is there to indicate whether this is the usual direction of longshore current?

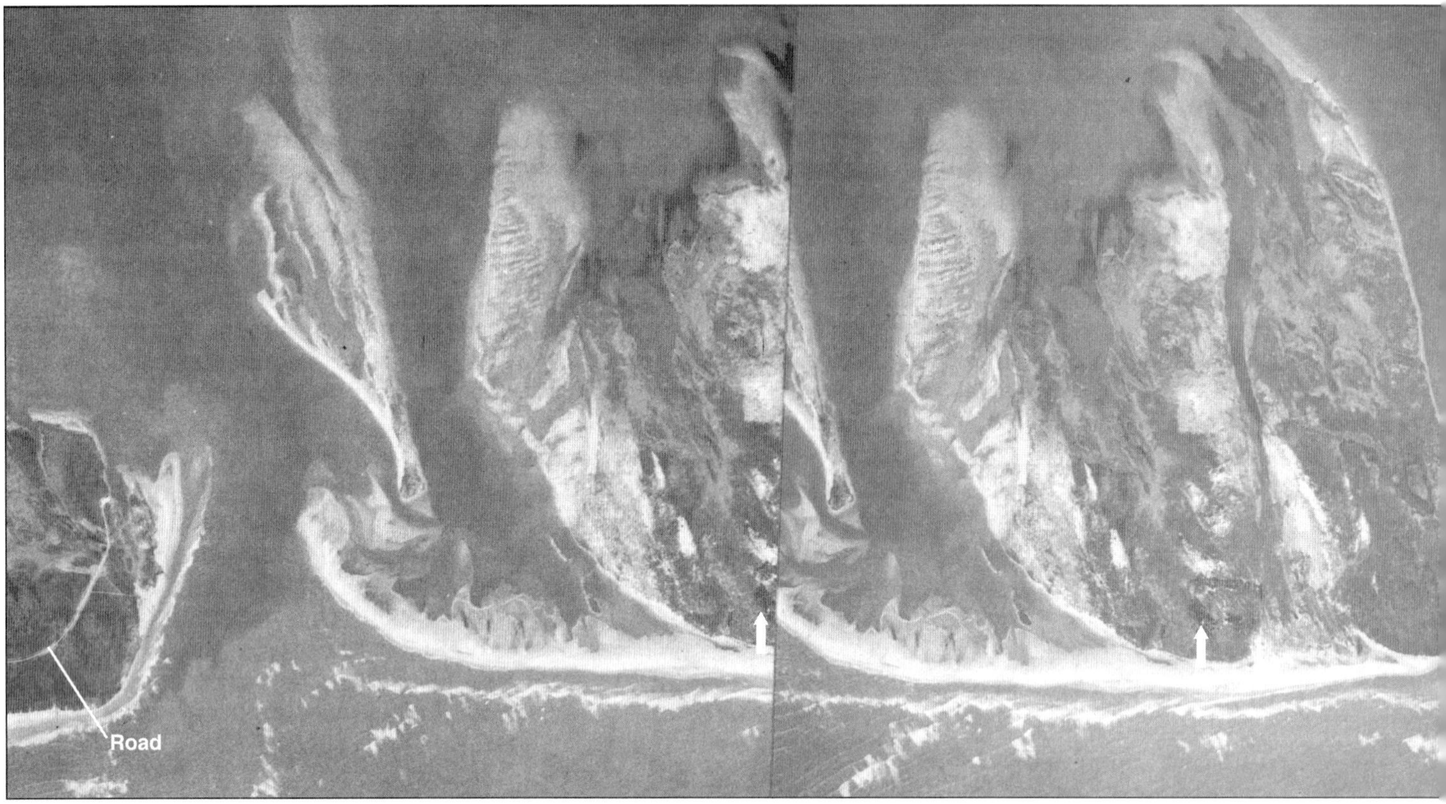

Road

A. 1943, scale 1:10,200

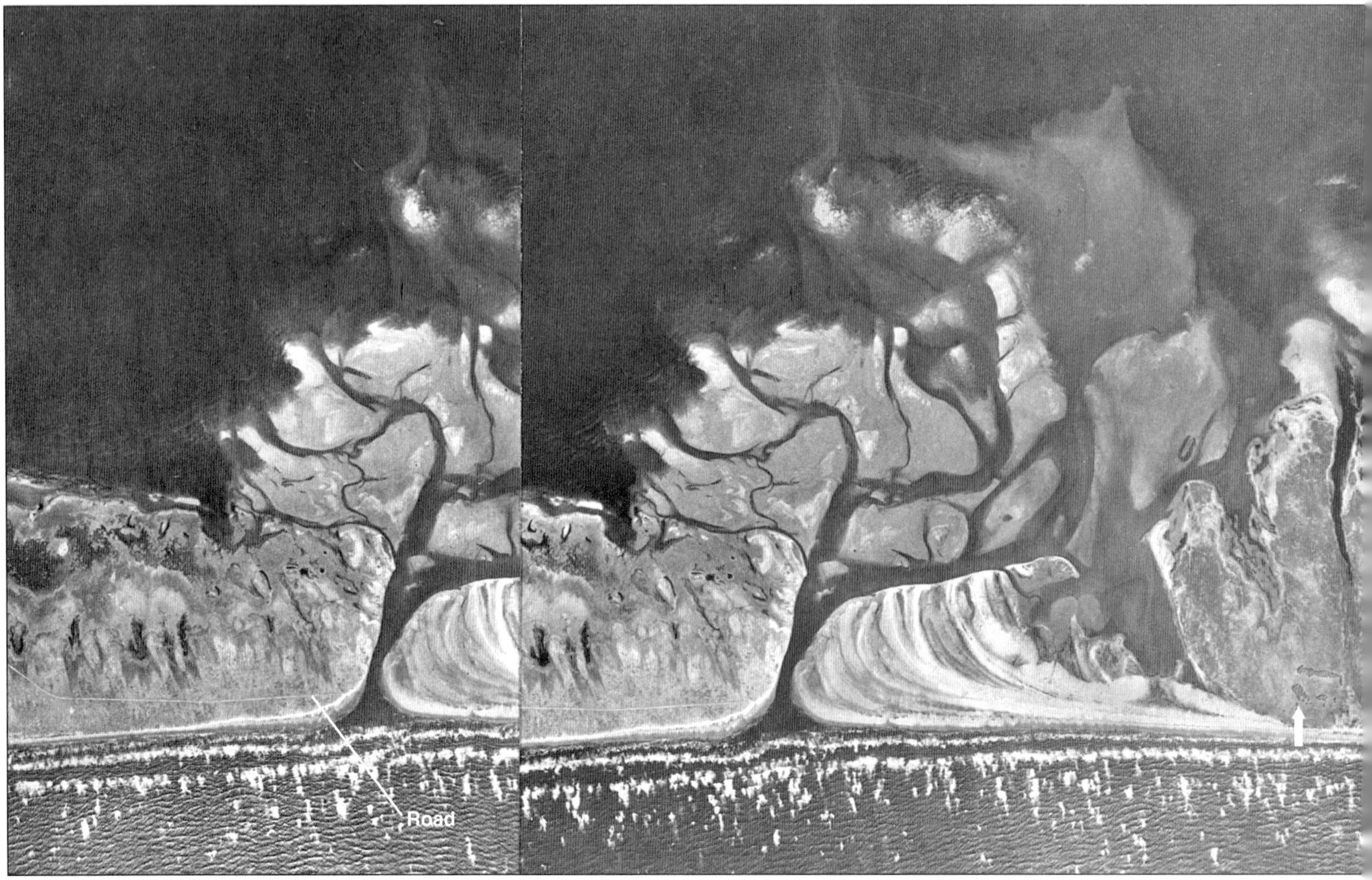

Road

B. 1957, scale 1:25,400

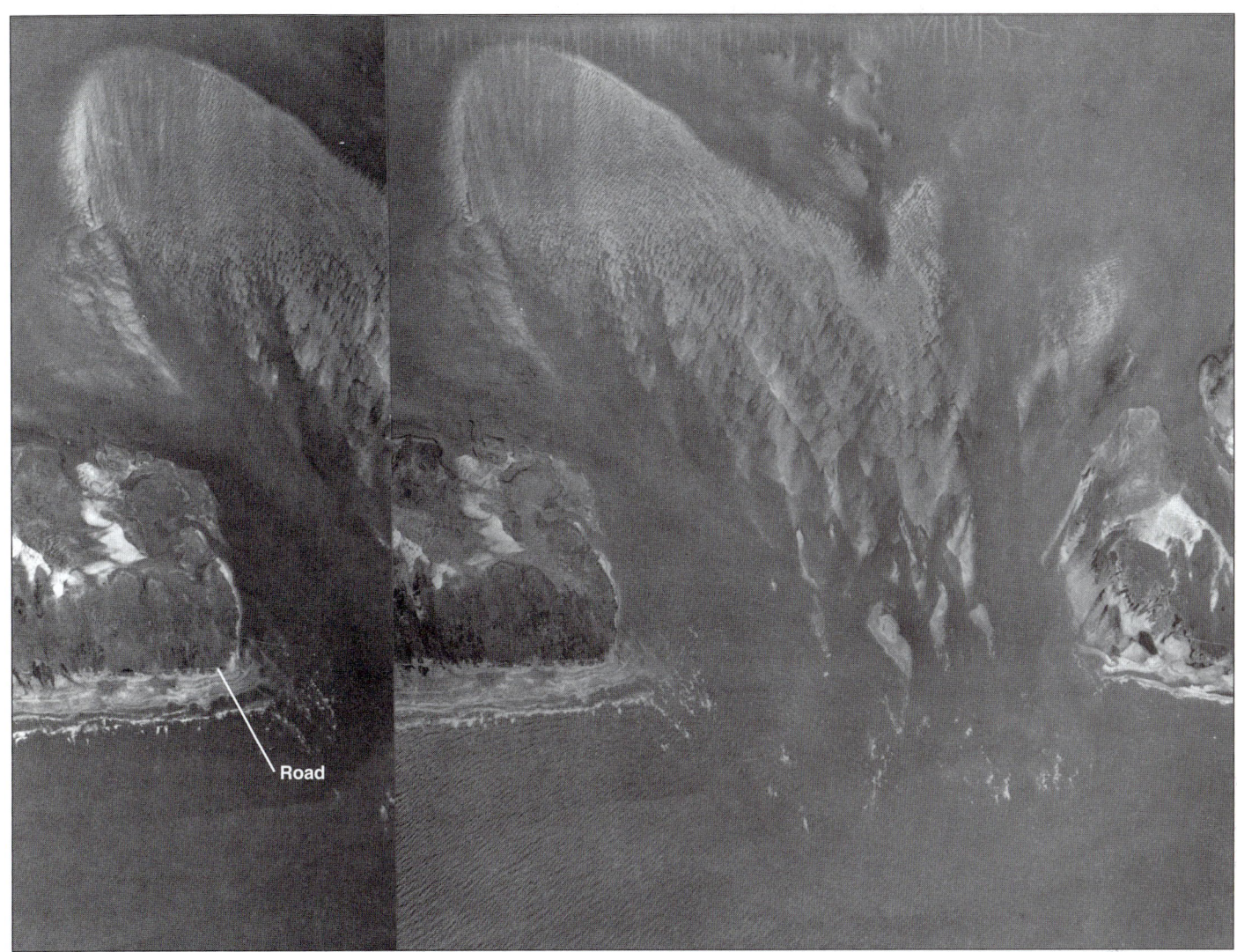

C. 1961, scale 1:18,500

FIGURE 11.9
Sequence of three sets of stereophotographs of Greens Bayou, Matagorda Island, Texas, taken in: A. 1943 (scale 1:10,200); B. 1957 (scale 1:25,400); and C. 1961, after Hurricane Carla (scale 1:18,500). The white arrows mark an identical feature on A and B.

b. Photo B shows the same area in 1957. The coastal road and white arrow should help you to orient yourself with respect to Photo A.

What is the minimum width of Greens Bayou now? Measure the width just landward of the coastal road. The fractional scale is 1:25,400.

Where have erosion and deposition occurred?

You have seen how the width of a tidal inlet can change, which is of some concern to boaters. How about the position of the tidal inlet? If you owned property on a barrier island, you would be more concerned about a moving inlet destroying your yard and home. The white arrows point at the same spot (probably a clump of vegetation) in photos A and B. Measure the distance in a line parallel to the coast between this clump of vegetation and the far side of the tidal inlet for each photo. Remember, the scales are different in A and B.

Distance in 1943:

Distance in 1957:

How far has the left-hand side of the inlet moved over these 14 years?

What is the origin of the set of curved features located to the right of the inlet in 1957?

As the tidal inlet has narrowed, the current velocities resulting from the tides moving in and out through the inlet have dropped. How might this be related to the large shallow-water sand deposit behind the inlet? What is this sand deposit called?

c. The Texas coast is frequently hit by hurricanes; 46 struck during the twentieth century. On September 14, 1961, one of the largest hurricanes in recorded history, Hurricane Carla, hit Matagorda Island. The hurricane virtually destroyed the Matagorda Island Air Force Base and considerable erosion occurred along the island. Photo C shows Greens Bayou in 1961 after Carla struck. The coastal road is barely visible just behind the beach.

What is the minimum width of Greens Bayou in Photo C? The fractional scale is 1:18,500.

Comparing the width of the tidal inlet and the position of the coastal road with the earlier photos, describe the major changes that have taken place as a result of Carla.

Is a barrier island a good place to use your retirement savings to build a condo?

2. Figure 11.10 is a portion of the Cayucos, California, 1:62,500 quadrangle, which is about 15 miles northwest of San Luis Obispo.
   a. Where would you go to see a rocky beach? A sandy beach? (Hint: See the table of topographic map symbols in Chapter 6.)

   What coastal landform is represented by the small circular land areas just offshore 2 to 6 km northwest of Morro Beach (for example, Whale Rock)?

FIGURE 11.10
Portion of the Cayucos, California, 15-minute quadrangle, for Problem 2.

b. What is the landform on the west side of Morro Bay, labeled Morro Bay State Park?

Morro Rock is shown as an island on older maps. What landform now connects Morro Rock to the mainland?

What happens to Morro Creek as it approaches the coast? Explain. How might this situation change during a period of abundant rainfall?

c. What does the pattern of dots in Morro Bay indicate? (Hint: See the table of topographic map symbols in Chapter 6.)

What depositional feature is the swampy area on the east side of Morro Bay?

What does the future hold for Morro Bay?

d. In what direction is the longshore drift in the area between the town of Morro Beach and the bottom of the map, as indicated by (1) the breakwater and (2) two different natural landforms (specify each)? What is the evidence?

3. Figure 11.11 is a portion of the Boothbay, Maine, 1:62,500 quadrangle. This area, about 40 miles northeast of Portland, is made of metamorphic rocks whose resistant layers intersect the surface in generally north-south bands. The striking topography and extremely irregular coastline result from differential erosion of these rocks. Blue contours show water depth in feet.

   a. Is this a rocky coast or one dominated by sandy beaches? Cite evidence for your answer.

   b. Is erosion or deposition the dominant process along most of the coast? Cite evidence for your answer in terms of landforms and map symbols. Consider both on-land and offshore topography.

   c. Is this coast emergent or submergent? Cite evidence.

   d. Use geographic names to cite examples of the following features: wave-cut platform (identified by the symbol for "rock or coral reef"—see Fig. 6.6), sea stack (identified by the symbol for "rock, bare or awash"—see Fig. 6.6), headland, wave-cut cliff, estuary. Consider subaerial and submarine topography when searching for examples.

FIGURE 11.11
Portion of Boothbay, Maine, 15-minute quadrangle, for Problem 3.

e. If you have completed the chapter on glaciers, you may recognize landforms suggesting glacial erosion. In fact, continental ice sheets covered this area from roughly 25,000 to 11,000 years ago. During this time so much water was locked away in continental ice sheets that sea level fell approximately 120 meters and exposed the entire area seen in Figure 11.11. Identify one aspect of the topography suggesting that the area was once glaciated. What about the topography indicates the ice flow direction?

f. If climatic and sea-level conditions remain similar to those of the present, predict how this coastline might look in ten or twenty thousand years. Consider the effects of erosion and deposition by both marine and terrestrial processes. Sketch a possible new shoreline on the map, and clearly label those areas removed by erosion or built by deposition.

4. Go to *http://ceres.ca.gov/ceres/calweb/coastal/geography.html* (or link to it through *www.mhhe.com/jones5e*—see Preface), and answer the following questions about marine terraces and beaches along the California coast.

   a. Marine terraces

   What are marine terraces and where along the California coast are they seen?

   How are marine terraces formed?

   In some places, up to twenty marine terraces are present. What combination of events occurred that allowed so many wave-cut platforms to form, then rise above sea level to form terraces?

   b. Beaches

   From where does the sand in California beaches come?

   What is *littoral drift?*

   How is sand permanently lost from the beach system?

   How have humans altered the natural beach processes and what effect might this have on the system?

In Greater Depth

5. Figure 11.12 is a portion of the Redondo Beach, California, quadrangle from 1951. Figure 11.13 is a 1990 stereopair of the same region. The area is about 25 miles south of Los Angeles proper.

   a. Is the coast rocky or sandy in this area? Cite your evidence.

   b. Draw a topographic profile between A and A' using the graph paper of Figure 11.14. Note that the contour intervals on land and offshore are different, and that on land, every *fourth* line is an index contour. Use every contour when preparing your profile or you will not see that the profile is steplike, with areas of gentler slope separated by areas with steeper slopes.

   c. Refer to your topographic profile. How did the gently sloped area immediately seaward of the coastal cliff form, and what landform name is given to such a feature?

   d. How do you think the other gently sloped areas on your topographic profile formed?

   How many of these areas are there, including the one just off the coast? Number them for reference purposes; assign "1" to the one just off the coast, "2" to the first one inland, and so forth.

   What is the approximate change in elevation between each of the flat areas? For example, between 1 and 2, 2 and 3, and so forth. Measure elevation differences between the top and bottom of each area with steep slope.

   e. Look at the stereopair in Figure 11.13. These photos were taken in 1990. Note the amount of development since 1951, and where it has occurred.

   Is erosion or deposition likely to be the dominant process along the coast? Circle those areas that ought to pay the highest rates for natural hazards insurance, and explain your answer.

   f. What does the topographic profile along A-A' suggest about the manner in which emergence took place? That is, was it gradual, episodic, or can't you tell? Explain. (Be careful; this is tricky, because two variables must be considered.)

   g. Assign relative ages to the flat areas on the profile. That is, which appears to be the youngest, which the oldest, and where in the sequence do the others fit? (Hint: Which have been subjected to the least on-land erosion, which the most, and which are intermediate?)

FIGURE 11.12
Portion of the 1951 Redondo Beach, California, 7½-minute quadrangle, for Problem 5.

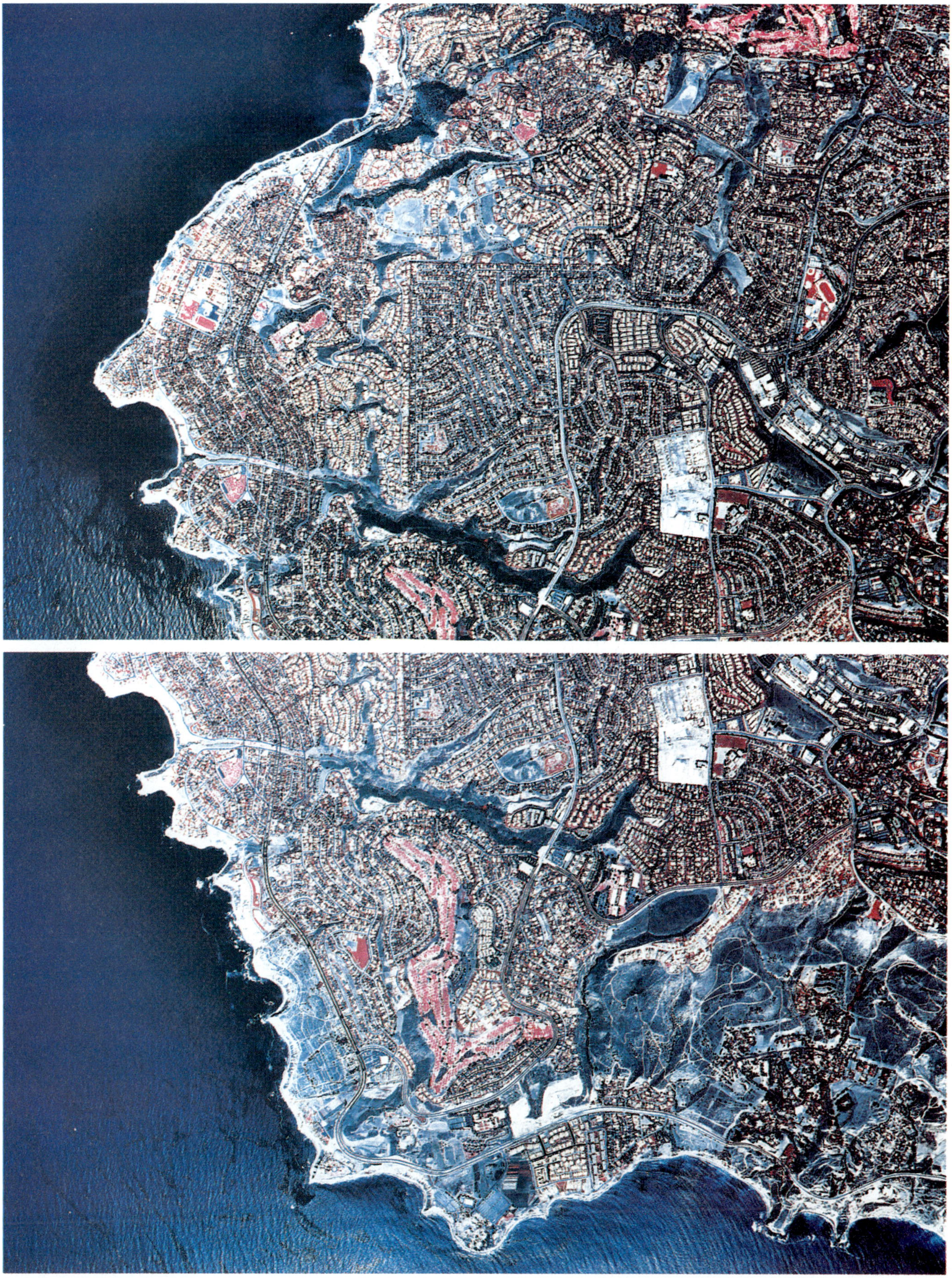

FIGURE 11.13
Infrared stereopair of Palo Verdes Point area, California, 1990. Scale is 1:40,000. North is to the top. Compare with Figure 11.12 to see population changes from 1951 to 1990. Large red areas are golf courses.

FIGURE 11.14
Graph paper for a topographic profile, Problem 5.

CHAPTER 12

Arid-Climate Landscapes

Materials Needed
- Pencil and eraser
- Colored pencils
- Metric ruler
- Calculator
- Stereoscope (provided by instructor)

Introduction

The bold, stark landscapes of arid and semiarid lands contrast with the rounded, more subtle forms of humid regions. Although running water is the principal agent of erosion in warm arid regions, it commonly is present only after infrequent cloudbursts. Flash floods remove the products of mechanical weathering and mass wasting and leave behind steep banks and cliffs. Plateaus, mesas, and buttes develop from deep erosion of horizontal layers of resistant bedrock. As erosion wears down mountains, alluvial fans form at canyon mouths and, with time, gently sloping erosional and depositional surfaces extend outward from the mountain front. Where strong winds blow and sand is available, dunes of several types may form. In this lab, you will learn to recognize and analyze common landforms that develop in warm arid climates, such as those found in the U.S. southwest. Arid landscapes also form in cold regions (e.g., the north slopes of Alaska and the dry valleys of Antarctica), but we do not consider them here.

Warm arid and semiarid regions (precipitation, <50 cm/year; see Fig. 8.1) are characterized by sparse vegetation, little water, thin soils, frequent strong winds, and sharp, angular landforms. Mechanical weathering predominates, and sedimentary particles tend to be coarser than in humid regions. Slopes typically are steeper, a requirement for the removal of coarser particles. Even rock such as limestone, which is readily attacked by chemical weathering in humid climates, forms steep cliffs in dry climates.

In spite of the aridity, water is the principal agent of erosion in most deserts. Rain commonly comes as downpours that quickly fill streams and cause flash floods. These floods cause extensive erosion, and sediment is quickly transported down normally dry stream beds and out into valleys before it is deposited. When sediment-choked streams spill out into a valley, they spread out to form numerous interwoven shallow channels called **braided streams** (Fig. 12.1). These braided streams rapidly deposit their sediment to form deposits of sand and gravel. Individual channels have steep, easily eroded banks. Each new flood tends to rework the older sediment and may carve a new main channel among the many secondary ones.

Wind is also an important agent of erosion and transportation in deserts, but it does not move as much material as water does. In fact, large, wind-carved rock landforms are unusual. Sand dunes, however, are common, and demonstrate that wind is an active agent of sediment transport and deposition.

Erosional Landforms

While most landforms in arid and semiarid regions are shaped by running water, the streams in arid regions are different from those in humid climates. Most streams are **intermittent,** carrying water only part of the year. Many, whether intermittent or **perennial,** have braided channels (Fig. 12.1). Deep canyons in dry climates may have near-vertical walls, and box-like cross profiles.

Where rivers traverse and cut into horizontal layers of sedimentary or volcanic rocks, *cliff-and-bench* topography, like that in Figure 12.2, develops. The more resistant layers of sandstone, limestone, or volcanic rock form steep cliffs where erosion has cut through them; flat benches develop on their upper surfaces. Less resistant layers of shale or soft volcanic rock form gentler slopes between the benches. The cliff-and-bench topography is developed in a spectacular way on the *Colorado Plateau* of the western United States. This region, in eastern Utah, western Colorado, northwestern New Mexico, and northern Arizona, has sixteen national parks or monuments featuring this scenery, including Grand Canyon and Canyonlands.

Plateaus are comparatively flat upland areas that form in regions with nearly horizontal rock layers. Figure 12.3 illustrates several common features developed by the

215

FIGURE 12.1
Upstream view of a dry stream bed in Death Valley National Monument, California, with braided channels and easily eroded banks.

FIGURE 12.2
Cliff-and-bench topography in Canyonlands National Park, Utah, along the Green River.

dissection of a plateau. As stream valleys enlarge by lateral and headward erosion, parts of the plateau may be cut off to form mesas. A **mesa** (table, in Spanish) is a generally flat-topped area, bounded by cliffs, that is wider (commonly by quite a bit) than it is high. **Buttes** are smaller, flat-topped landforms isolated by erosion. They are more or less as wide as they are high. **Monuments** (or spires) are slender features that are much higher than they are wide.

Where the bedrock does not consist of horizontal layers of unequal resistance, different erosional landforms develop. In the *Basin and Range* area of the southwestern United States, rugged mountain ranges are separated by wide valleys with flat floors. The mountains were uplifted along steep faults. Although some faults are still active and are readily discernible, others have been obscured by erosion and deposition. Figure 12.4 shows a profile from a mountain into the adjoining valley. A sharp break in slope at the edge of the mountain marks the *mountain front*. The gentle valleyward slope from the mountain front is the **piedmont**. It consists of two parts, the *bajada* and the *pediment*. The bajada is formed by deposition, as discussed in the next section. A **pediment** is a gently sloping, erosional surface. It cuts across bedrock, and typically has a thin veneer of sediment on its surface. It is generally thought that a pediment develops as the steep slope at the mountain front retreats under the attack of erosion. The erosional debris is transported across

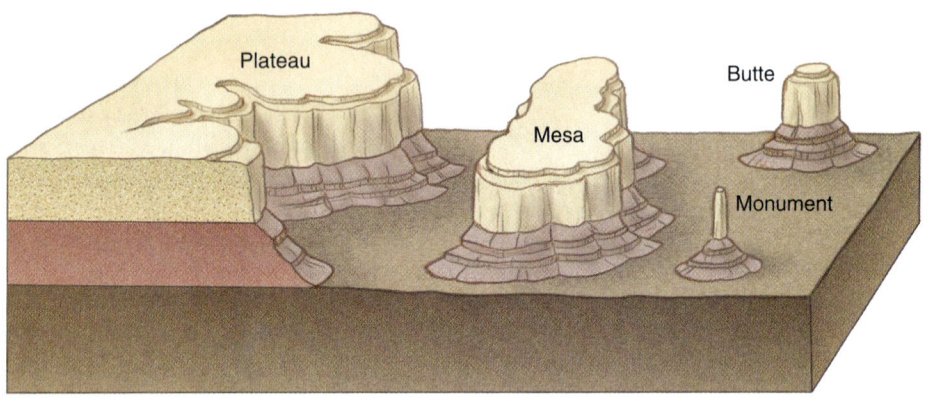

FIGURE 12.3
Characteristic erosional landforms in arid or semiarid areas with horizontal rock layers. Retreat of a plateau leaves mesas, buttes, and monuments as remnants.

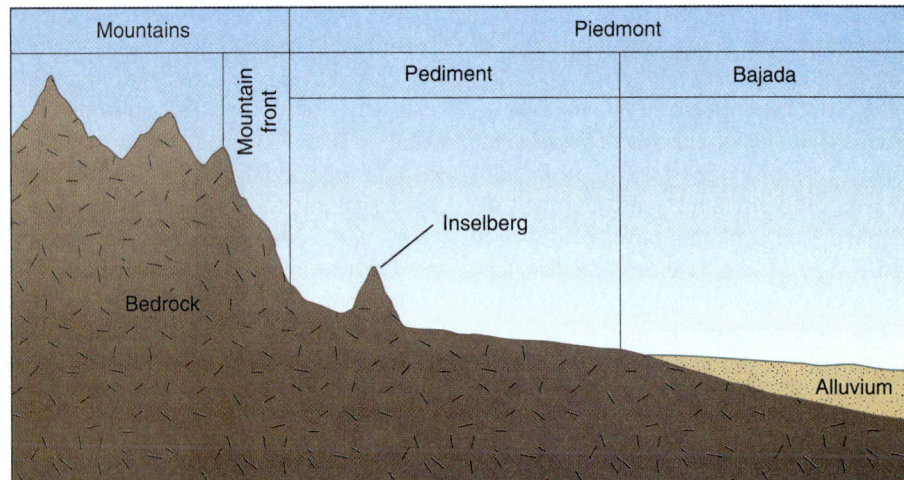

FIGURE 12.4
Profile from mountain into valley, showing the mountain front and the piedmont, with pediment, inselberg, and bajada.

FIGURE 12.5
Two alluvial fans occur along the base of the Black Mountains in Death Valley National Monument, California. There are two wet playa lakes in the foreground. The white areas are evaporite minerals that were deposited as more extensive playas dried up.

FIGURE 12.6
The dark area in front of the background mountains is a piedmont. The upper part of the dark area is the pediment and the lower part is the bajada, but you would need to be on the ground to map the boundary between them. White areas in the middle of the valley are evaporite deposits. Panamint Range, Death Valley National Monument, California.

the pediment to the bajada. Erosional remnants, or **inselbergs,** form isolated hills on the pediment.

Depositional Landforms

Water Deposits

Most sediment in arid lands is transported by intermittent streams. In mountainous areas, these streams cut and occupy rock-walled channels, and sediment transport is confined within the rock walls. As the canyons empty into a valley, the stream is no longer confined; it spreads out, losing its capacity to carry sediment. The resulting series of braided stream deposits at the canyon mouth eventually builds up a large **alluvial fan,** like those shown in Figure 12.5. With time, alluvial fans from adjacent canyons coalesce to form a **bajada,** the gently sloping apron of sediment along the mountain front in Figure 12.6. If a pediment is present, the bajada lies basin-ward of the pediment, and together, they form the piedmont at the base of the mountain (see Fig. 12.4).

When an intense rainstorm occurs in arid areas, water runs out of the highlands, down canyons, and across the piedmont. If a pediment exists, much of the water washes across it to the bajada. As it crosses the bajada, a substantial amount of water may soak into the alluvium and become part of the groundwater. But, if it is a real downpour, much water flows out into the valley. If there is no through-going drainage, water may pond to form temporary lakes called **playas** (Fig. 12.5). Evaporation of water from playas results in the deposition of evaporite minerals, such as gypsum or halite.

Wind Deposits

The most obvious wind deposits are sand dunes. They develop from the accumulation of sand-size grains that have been transported by the wind. Grains move by rolling and hopping, and rarely get more than a meter above the surface; many are set in motion by impact from other grains.

Initial deposition generally is caused by an obstacle or irregularity on the surface. Wind velocity decreases after it blows over an obstacle, and some of the sand is deposited on the downwind side; continued deposition produces a dune. Figure 12.7 is a cross section of a dune, showing how a steep, downwind face develops as sand rolls and hops up the windward side and tumbles and slips down the leeward side, forming cross-strata (see Fig. 4.4). As this process continues, the dune migrates downwind.

Differently shaped dunes develop under different conditions. The principal controls are wind velocity, vegetation, and sand supply.

Transverse dunes (Fig. 12.8A) are wave-like ridges of sand perpendicular to the predominant wind direction. They form in areas of abundant sand supply, scarce vegetation, and moderate, unidirectional winds. They are found not only in deserts, but also commonly behind beaches and on barrier islands.

Longitudinal dunes (Fig. 12.8B) are long, narrow ridges of sand parallel to the prevailing wind direction. They form where sand supply and vegetation are meager and winds are strong. Although they parallel the general wind direction, the slip face varies from one side to the other along the axis of the dune, reflecting variations in wind direction.

Parabolic dunes (Fig. 12.8C) are crescent shaped, with the steep slip face on the convex side so that the horns point upwind. They typically develop where vegetation is available to anchor the horns. In many, the area in front of the dune, between the horns, is a **blowout,** a small depression excavated by the wind.

Barchan dunes (Fig. 12.8D) are crescent shaped, like parabolic dunes, but the steep slip face is on the concave side so that the horns of the crescent taper and point downwind. Barchans develop best on barren desert floors where prevailing wind direction is constant, vegetation is scarce,

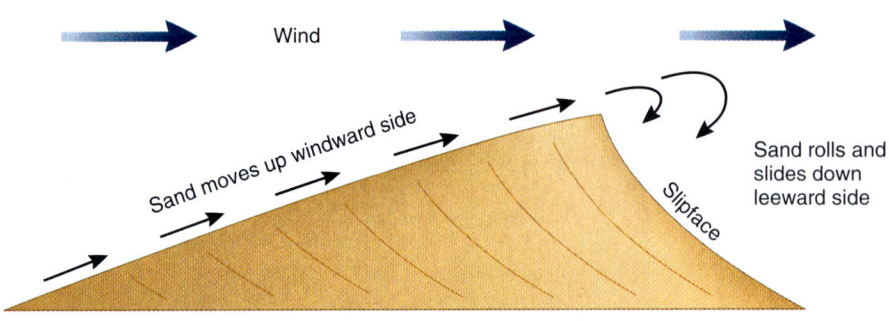

FIGURE 12.7
Cross section of a dune, showing sand moving up the gentle windward slope and falling off the steeper leeward slope.

and sand supply is low. They may occur singly or in groups, where they may join to form more complex shapes.

Dune fields may become inactive as conditions change. Vegetation, in particular, acts to stabilize dunes. For example, you can see the crescent shapes of vegetated dunes in the Sand Hills region of Nebraska (Fig. 9.8). However, if patches of vegetation die or are killed by overgrazing or excessive human activity, blowouts develop and active dunes may again migrate over the surrounding landscape.

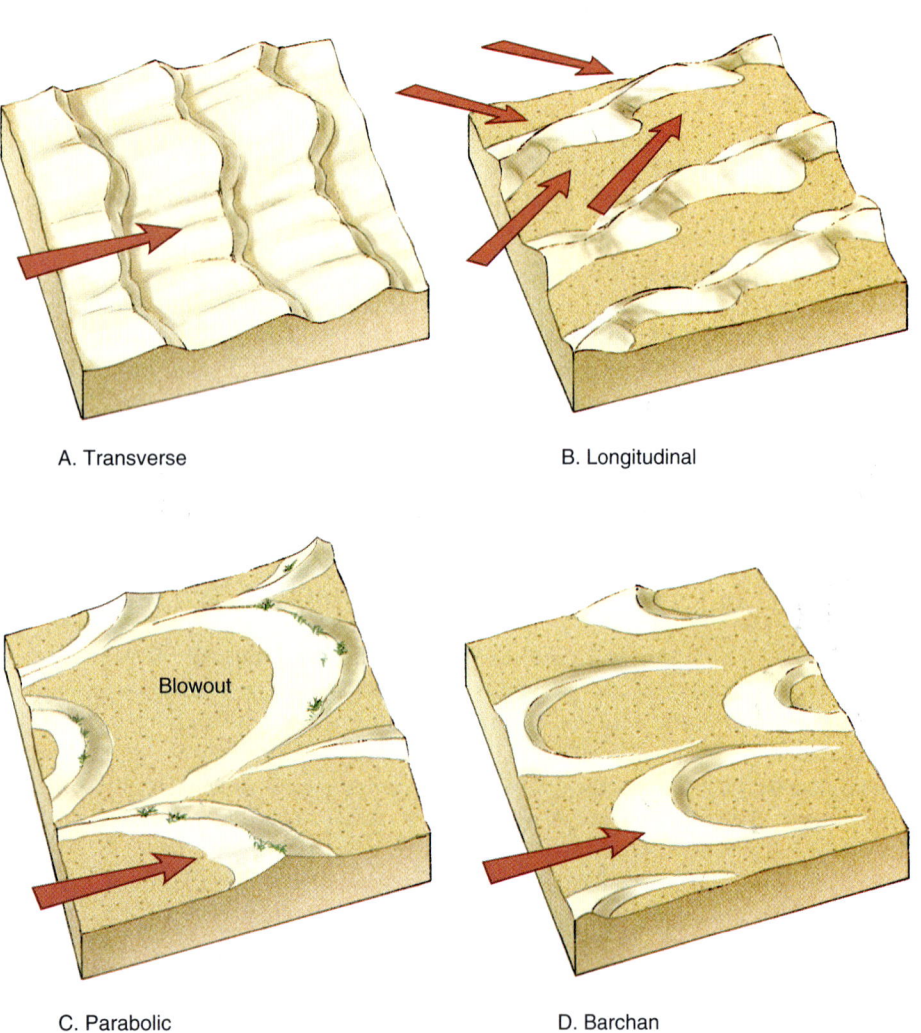

FIGURE 12.8
A. Transverse dunes; B. Longitudinal dunes; C. Parabolic dunes; D. Barchan dunes.
Source: Data from U.S. Geological Survey.

Hands-On Applications

Desert environments are very fragile, and as the human population increases, more and more people live in, or interact with, deserts. To live successfully in arid regions, the geologic processes operating there must be understood. In this lab, you will learn how careful observation and measurement of desert landforms elicits a variety of questions regarding the processes that formed them. The answers to the questions are hypotheses, some of which can be tested using available evidence, others of which require field evidence.

Objectives

If you complete all the problems, you should be able to:

1. Recognize on a map or in a photograph the following desert landforms and explain how they develop: plateau, mesa, butte, monument.

2. Distinguish pediment, inselberg, bajada, alluvial fan, and playa in a photograph or on a map and explain how they form.

3. Recognize transverse, longitudinal, parabolic, and barchan dunes in a photograph or on a map and describe the conditions favoring their development.

4. Recognize a braided stream and use its presence to interpret the nature of the sediment in the streambed.

Problems

1. Figure 12.9 is a stereographic composite of aerial photographs of the Stovepipe Wells area of Death Valley National Monument, California. The buildings near the center of the photograph mark the small community of Stovepipe Wells. Its elevation is 0 feet—sea level—and the highest elevation in the photograph is about 2500 feet. North is to the left in the photographs.

   a. What are the prominent landform features extending from the mountain front, and how did they form?

   b. What happens to the single stream channel as it leaves the mountain canyon?

   Formulate a hypothesis to explain your observation, and suggest a way to test it.

   c. Rock or sediment exposed at the surface in Death Valley (and many other places) becomes coated with a very thin, dark-colored layer of clay, iron, and manganese minerals called rock varnish. The longer a rock surface is exposed without being eroded, the thicker and darker it becomes. The mountain-front features in Figure 12.9 have dark, light, and intermediate gray areas. Examine the photos with a stereoscope, noting the positions and relative elevations of the different colored areas.

   (1) What are the relative ages of these areas?

   (2) What seems to be happening at present in each of the three differently colored areas? For example, are some areas active channels, and others abandoned channels? Is deposition occurring in some areas, and erosion in others?

   d. Using a blue pencil, color the presently active channels.

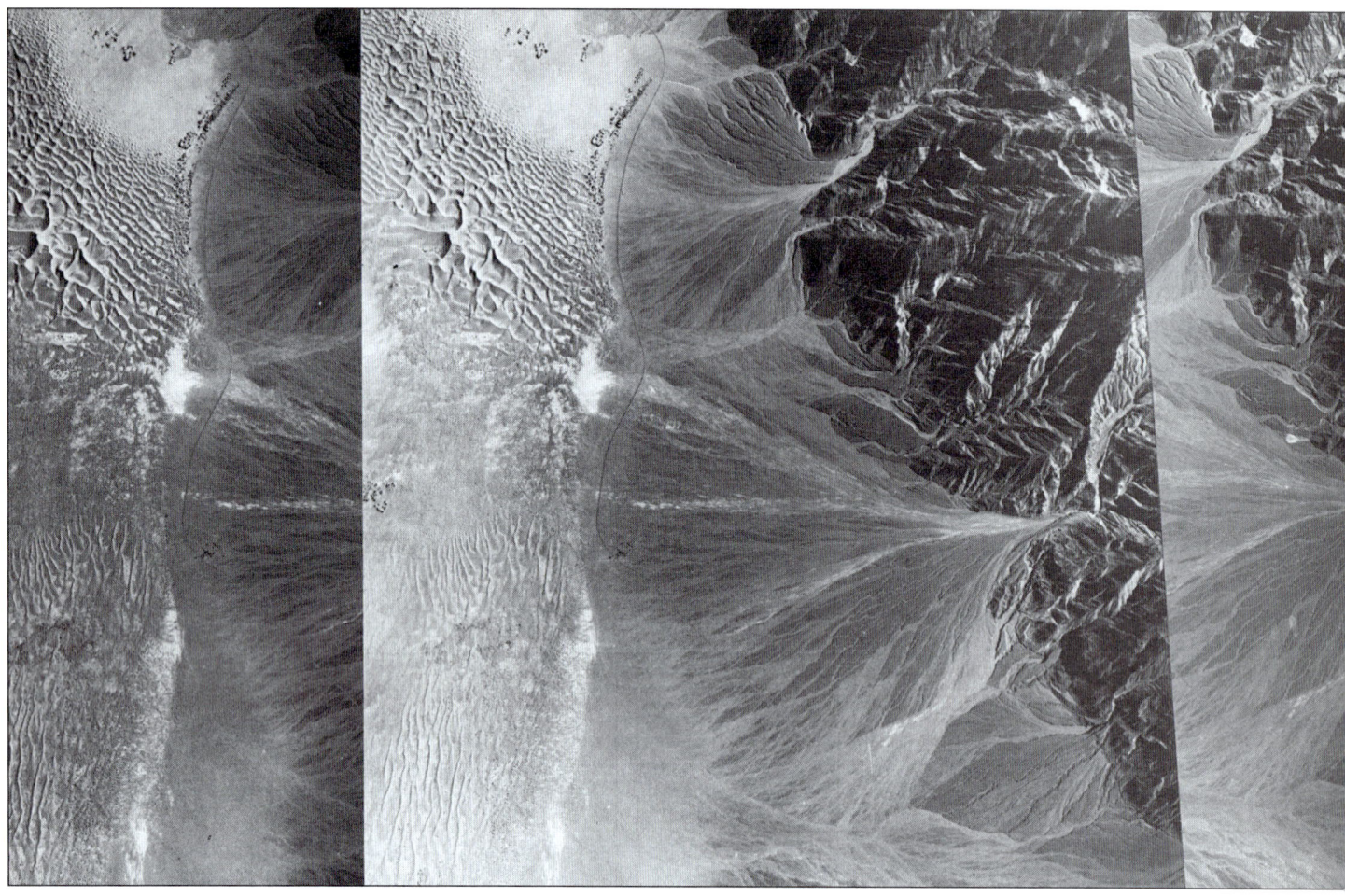

FIGURE 12.9
Stereophotographs of the Stovepipe Wells area, Death Valley, California, for Problem 1. Scale 1:65,000; photos taken November 11, 1948. North is to the left.

2. Figure 12.10 is a portion of the Bennetts Well, California, 1:62,500 quadrangle. This is the heart of Death Valley, about 30 miles south-southeast of Stovepipe Wells (see Fig. 12.9). The lowest elevation in the United States, 282 feet below sea level, occurs in this quadrangle. Because Death Valley is below sea level, all of the drainage in this area flows into the valley. No streams flow out. The Amargosa River flows north—when it flows. Rainfall averages about five centimeters per year, but it commonly comes in big storms after years of virtually no rain. Most erosion occurs during storms.

    a. Most valleys form as erosion removes loose material and streams transport it out of the valley. But that cannot happen here. Death Valley formed when a series of faults developed in response to crustal stretching across eastern California and Nevada. Movement along one of these faults caused (and still causes) the floor of Death Valley to tilt down toward the fault. At the same time, this fault movement causes the mountain range on the other side of the fault to rise. Along one side of the valley, the fault forms a series of curved surfaces that run deep under the valley floor.

    On which side of the valley is this fault? Cite at least two pieces of evidence to back your choice.

    b. What name is given to the landform between the valley and the mountain front on the west side of the map?

    What is the gradient of this feature, in feet/mile, between BM-242 (on the road) and the 19 in Section 19 to the west?

3. Figure 12.11 is a portion of the Antelope Peak, Arizona, quadrangle. U.S. Interstate 8 crosses the area on its way from San Diego, California, to Tucson, Arizona. The area is within the *Basin and Range* province, but the basin-bounding faults have long since been covered by alluvium. The piedmont is extensive and occupies a large part of the map area. The rocks exposed in the mountains in this quadrangle are mostly metamorphic and igneous rocks of Precambrian age.

   a. How does the map-view shape of the mountain front differ from the mountain front in Figure 12.10?

   b. Determine the gradient in feet/mile along a line extending from the southwest corner of Sec. 26 (near center of map) to the southwest corner of Sec. 6 (northeast corner of map). How does the slope of the piedmont surface compare with the slope of the piedmont in Figure 12.10?

   How does the general contour pattern on the piedmont differ from the contour pattern in Figure 12.10?

   c. Note that the piedmont area north of the interstate is punctuated with a few hills to the northeast of the mountain front. How did these hills form?

   What part of the piedmont (see Fig. 12.4) lies between these hills and the mountain front? Explain.

   d. Based on the differences you have observed between the Antelope Peak and Death Valley areas, which area appears to have been subjected to erosion for the longest period without rejuvenation (for example, by tectonic activity or change in base level)?

   e. Note the water wells in the northeast part of the map. In what part of the piedmont are they likely to be located? Explain.

   What is the probable source for the groundwater tapped by the wells? Explain.

   f. Locate the canyon just south of Indian Butte. What landform has developed at its mouth?

   The canyon has a box-like cross profile, with steep walls and a rather flat floor. One intermittent stream is shown on the canyon floor, yet the upstream bends in the contours indicate several potential channels. With a blue pencil, sketch as many possible channels as you can.
   What pattern is shown by these channels, and what does that suggest about the nature of the sediment on the floor of the canyon?

   Directly across the canyon from Indian Butte, a wedge-shaped area slopes to the east from a hilltop with elevation 2258 feet. How do you think this feature formed?

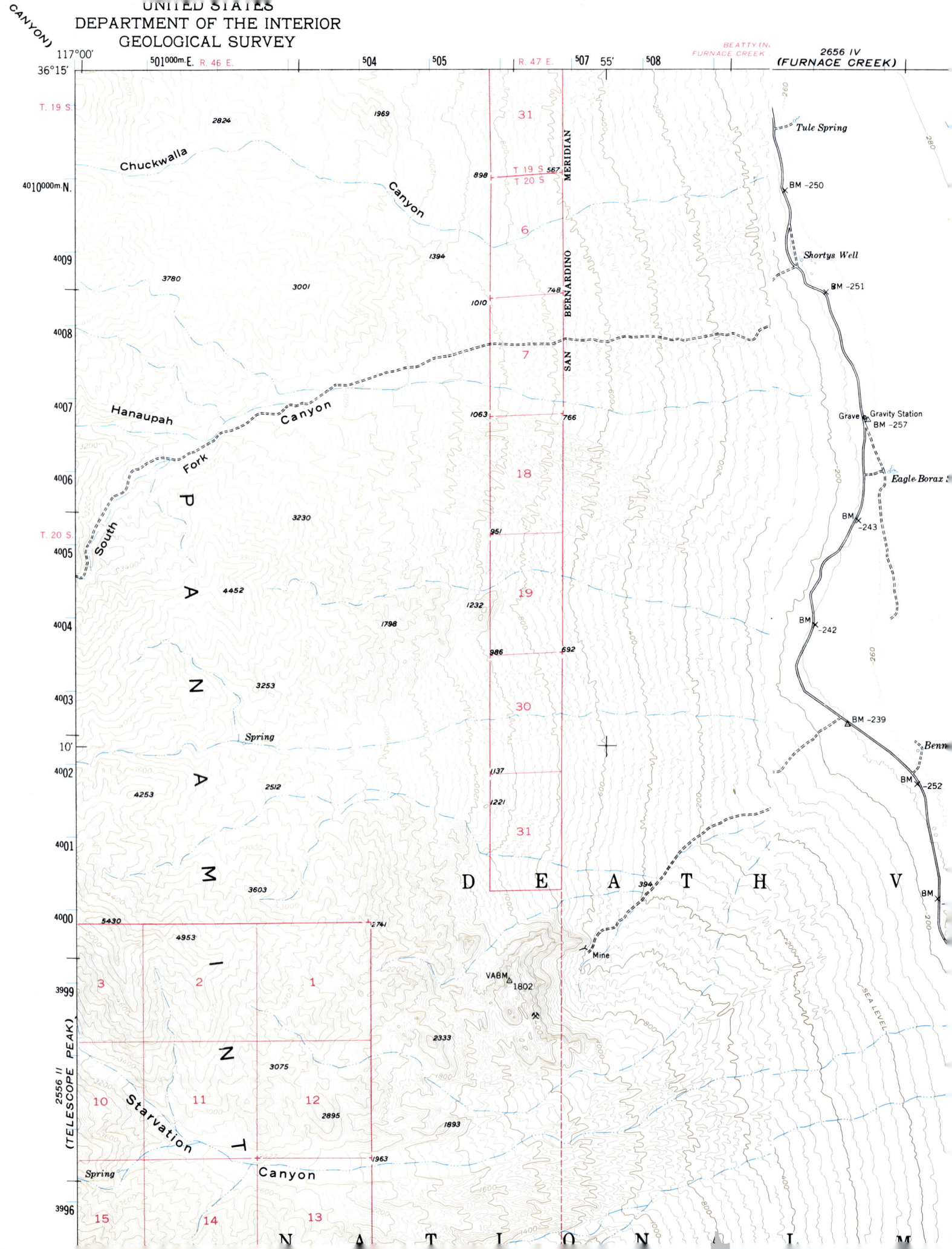

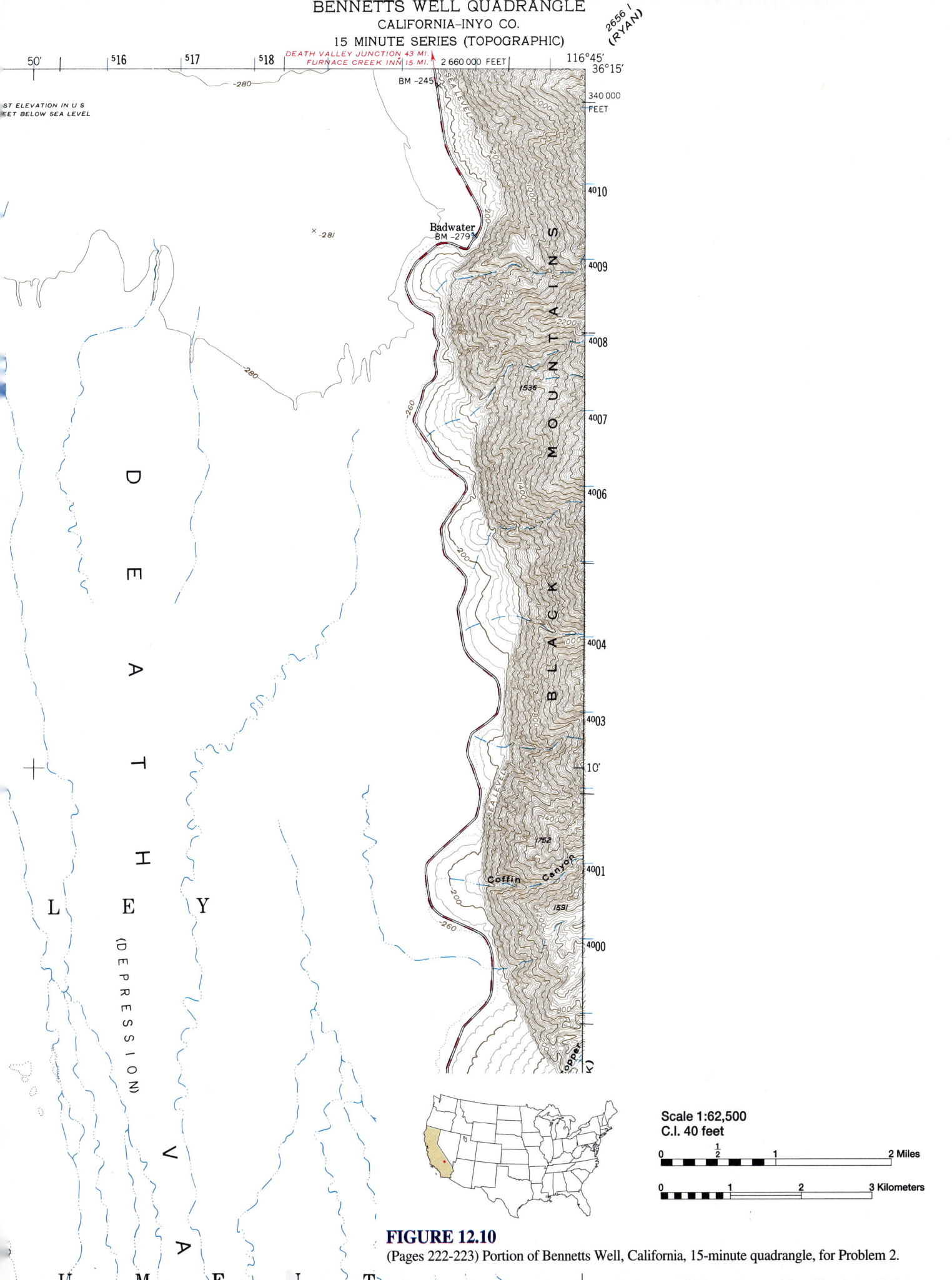

FIGURE 12.10
(Pages 222-223) Portion of Bennetts Well, California, 15-minute quadrangle, for Problem 2.

224 Part IV Landscapes and Surface Processes

4. Figure 12.12 is a portion of the Upheaval Dome, Utah, 1:62,500 quadrangle. It is within Canyonlands National Park, an area with beautiful canyons cut by the Colorado and Green Rivers (see Fig. 12.2). Reddish Paleozoic and Mesozoic sedimentary rocks are carved into a series of cliffs and benches that gradually descend to the two rivers.

   a. Some sedimentary rocks are highly resistant to erosion while others erode quite easily. The landforms in Figure 12.12 directly reflect the distribution of resistant and easily eroded rock layers. Based on the topography, are the rock layers in this area roughly horizontal, strongly folded, or tilted? Explain.

   b. Locate *The Neck* on the road just northeast of Grays Pasture. The Neck connects the plateau to the north to an area known as *Island in the Sky*. Island in the Sky is the Y-shaped, relatively flat area that includes Grays Pasture, Grand View Point, and the unnamed area to the west.
   What is the approximate width of The Neck?

   With the head of Taylor Canyon to the west (only the word Canyon appears on the map), and the South Fork (of Shafer Canyon) to the east, what does the future hold in store for The Neck?

   Should your prediction come true, what kind of landform will the Island-in-the-Sky area become?

   c. Extending to the east of Grays Pasture are three, small, rather flat areas bounded by steep cliffs and connected by "necks"; the high point on the easternmost area is marked with its elevation of 5932 feet. What landform name would be given to these areas if the necks were to be severed by erosion? Assume that their base elevation is 4500 feet.

   d. Washer Woman is just southeast of Grays Pasture. What type of landform is just south of the word "Woman"?

   e. At some point in the future, the resistant rocks that make the high, flat areas of Grand View and Grays Pasture will completely erode away. Scan the map for signs that another set of resistant layers may support a future plateau. At what elevation will this plateau form (using modern elevations) and what about the topography led you to pick this level?

5. Figure 12.13 is a stereopair showing sand dunes on the southwest side of Salton Sea, south of Salton City, California, in the Imperial Valley.

   a. What type of dune is most prominent? How did you distinguish this type from the others?

   b. What is the prevailing wind direction (north is to the right)?

   c. These are active, migrating dunes. A comparison of the positions of selected, well-defined dunes in the years 1956 and 1963 shows just how active these dunes are. The distance each dune moved during this 7-year period is given in Table 12.1. The table also shows the height of the slip face of the dune, which is a measure of dune size.
   Prepare a graph on Figure 12.14 showing distance moved during the 7-year period on the x-axis, and the height of the slip face on the y-axis. Is there a relation between distance moved and size of dune, and if so, what is it?

   Give a possible explanation for this relation.

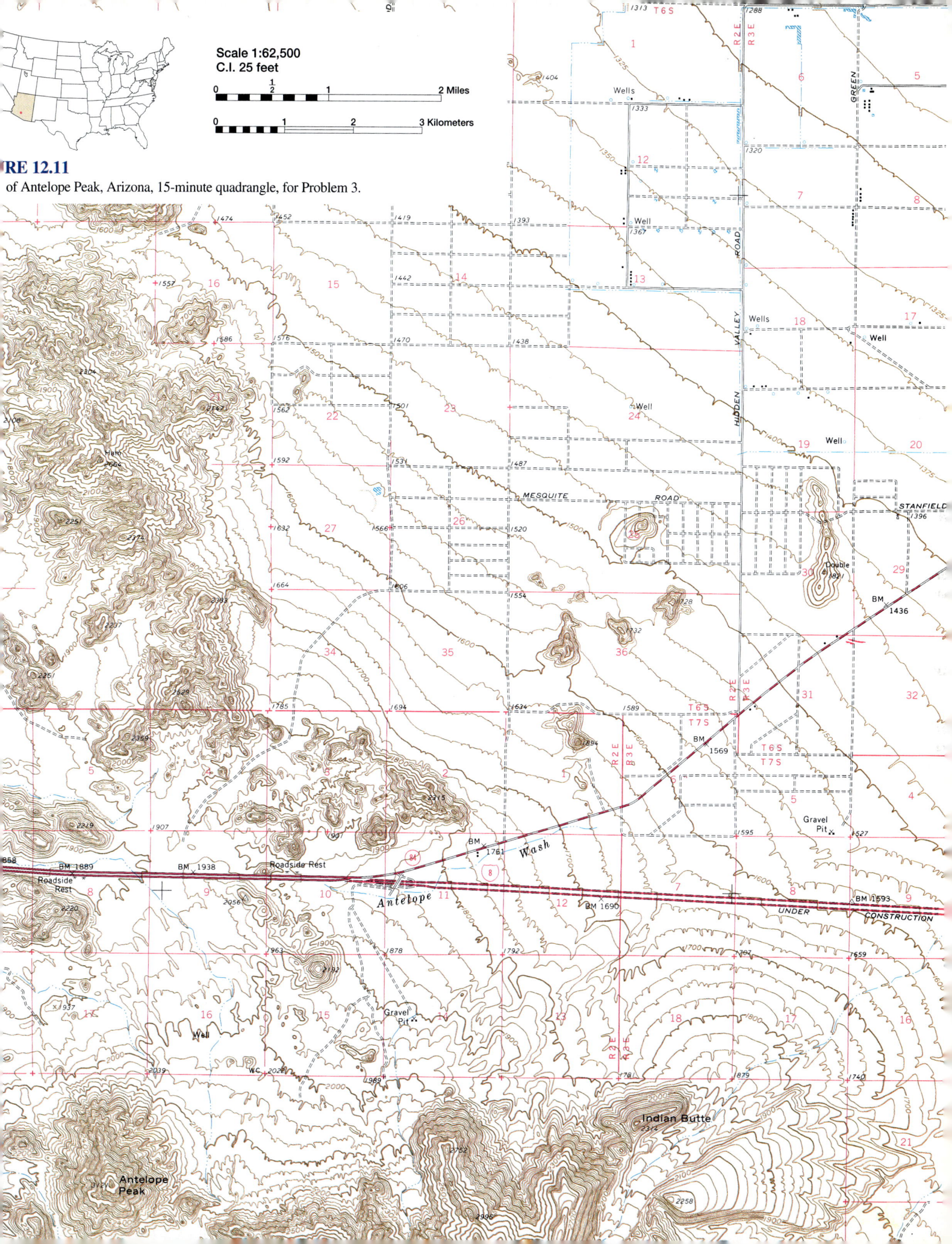

FIGURE 12.11 of Antelope Peak, Arizona, 15-minute quadrangle, for Problem 3.

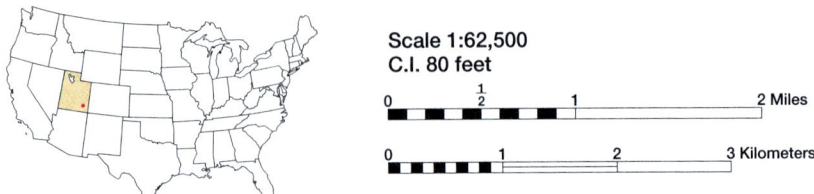

FIGURE 12.12
(Page 227) Portion of Upheaval Dome, Utah, 15-minute quadrangle, for Problem 4.

d. Dune number 9 in Figure 12.13 moved in the direction N75°E during the 7-year period. Using the data in Table 12.1, determine the rate, in meters per year, at which it moved. Show your work.

What is the distance on the air photo in centimeters in a N75°E direction from the top of the slip face in the center of dune 9 to the road?

Convert this photo distance to actual ground distance in meters (scale of photo is 1:20,000). Show your work.

Assuming the dune has continued to move at the same rate, how far has it moved since 1959, the date the photographs were taken? Show your work.

In what year did (or will) the dune cross the road?

228 Part IV Landscapes and Surface Processes

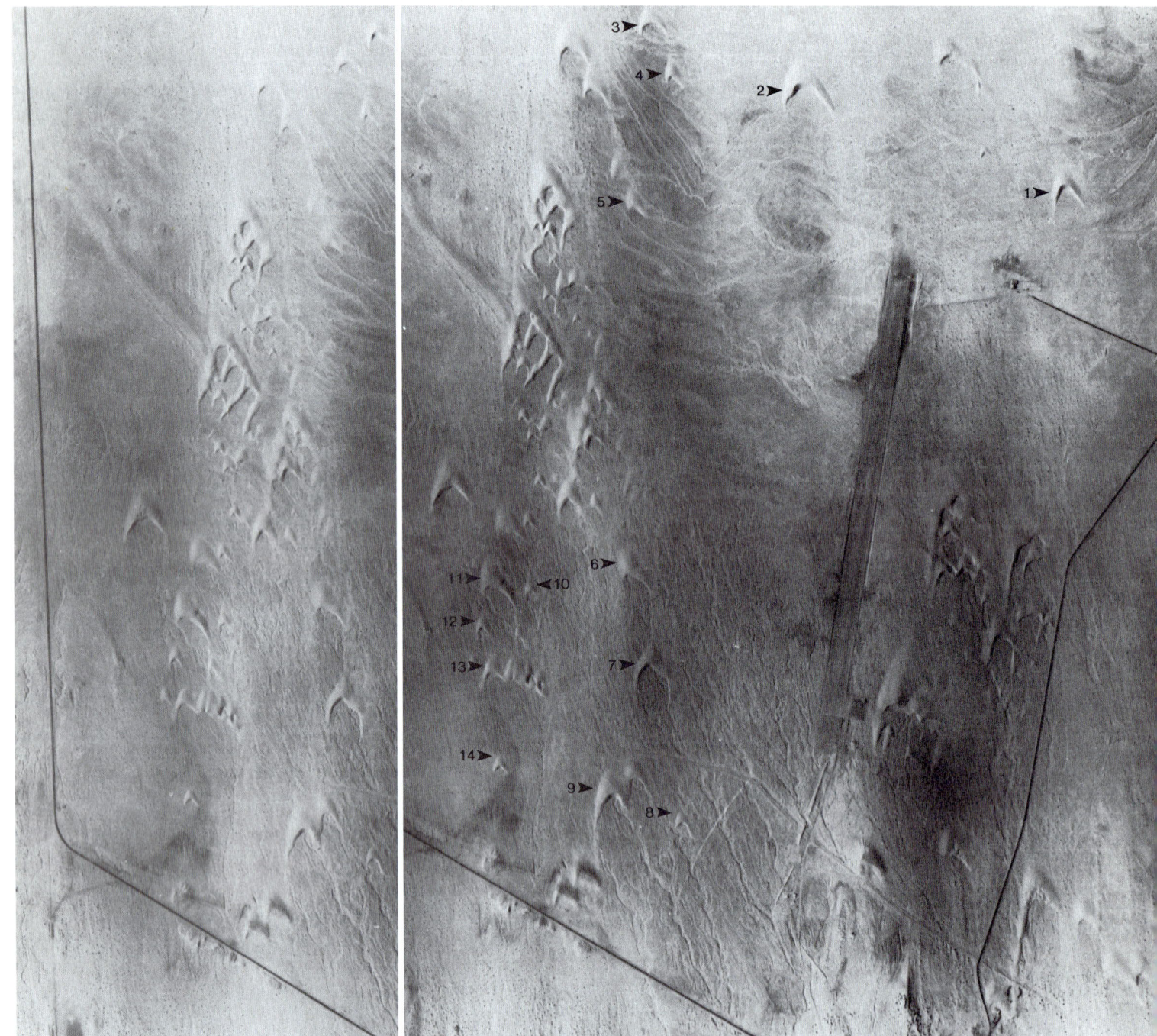

FIGURE 12.13
Stereopair of area southwest of Salton Sea, California, for Problem 5. Scale 1:20,000; photos taken November 10, 1959. North is to the right.

FIGURE 12.14
Graph paper for Problem 5c.

TABLE 12.1
Data for Problem 5
Source: Data from J. T. Long and R. P. Sharp, *GSA Bulletin*, 1964.

Dune Number On Map	Distance Moved in 7 Years (Meters)	Height of Slip Face (Meters)
1	137	4.3
2	107	12.2
3	183	6.4
4	183	6.1
5	236	4.0
6	168	6.4
7	160	7.3
8	229	3.4
9	152	8.2
10	244	4.6
11	137	8.2
12	274	4.3
13	160	7.6
14	274	3.0

6.  Search on "White Sands National Monument" or go to *www.nps.gov/whsa/* (or link to it through *www.mhhe.com/jones5e*—see Preface), and from information contained there or in links from there (such as under *In Depth*) answer the following questions about the dunes at White Sands National Monument.

   a. The dunes at White Sands National Monument are unusual because of the mineral of which they are composed; it is the largest dune field in the world made of that mineral. What is that mineral and what is its hardness?

   From what you know about that mineral, or can learn from its description in Chapter 2, would you expect that sand grains made of it could have been transported very far from their source?

   According to Chapter 4 in the lab manual (for example, Table 4.5), how do most sedimentary deposits made of this mineral form?

   b. What sequence of events led to the formation of the dunes?

   How long ago did the dunes begin to form?

   c. In what direction does the prevailing wind blow?

   How far do some dunes migrate during a year?

   What types of dunes are present, and how large do some get?

   Another site of interest is the National Park Service Death Valley site *(www.nps.gov/deva/)* and the links therein.

In Greater Depth

7. There are two dune fields near Stovepipe Wells, Death Valley, California (Figure 12.9). The upper dune field was used in the filming of Star Wars to represent the desert planet that R2D2 and C3PO landed on in the early scenes of the movie.

   a. What type of dunes are found near the bottom of the photographs?

   Given their shape, which way is the predominant wind direction (north is to the left)?

   Do these types of dunes form in places with abundant or sparse sand supplies?

   b. The dune field at the top is more complex, but you can see dunes that have a similar shape to those at the bottom. What is the predominant wind direction affecting these dunes?

c. Are the wind directions from the two fields parallel or do they intersect? Use this fact to speculate about the conditions needed to produce the complex dune shapes in the upper dune field.

d. The complex dune forms in the upper field are called **star dunes.** Consult a geology textbook, an encyclopedia, or do an on-line search (using, for example, *www.google.com*) to see if your speculations above were accurate. List your source below; be sure to use a reliable one!

PART V
Geologic Time and Sequences

Cross-cutting relations established by dikes intruding a granite, Coahuila, Mexico.

CHAPTER 13

Geologic Age

Materials Needed
- Pencil and eraser
- Metric ruler
- Calculator

Introduction

Rocks exposed at the Earth's surface often have long and complex histories. To understand the formation of these rocks, and thus to unravel the history of the Earth, it is necessary for geologists to use rigorous techniques to figure out not only what but also when things happened. For centuries, geologists have determined the *relative* ages of different events using simple observational techniques: this sandstone was deposited first, then came a limestone, and finally a granite intruded both. Only since World War II have geologists developed the sophisticated methods required to determine numerical ages. Thus, a sandstone was deposited 250 million years ago, a limestone 245 million years ago, and a granite intruded both 40 million years ago. The problems in this chapter illustrate the techniques used to determine both relative and numerical ages.

Geologists working in mountain ranges are regularly confronted with the complexity of the Earth's past. Instead of seeing merely horizontal layers of sedimentary rock, we often see sedimentary layers that are folded or steeply tilted (Fig. 13.1A, B). Other layers may be abruptly offset by fractures called faults (Fig. 13.1B). And sometimes igneous rocks have clearly intruded sedimentary rocks, or sedimentary rocks were deposited on top of older, cooled igneous rocks (Fig. 13.1C, D). These geologic relationships record the forces and events that help shape the Earth.

It was in the late 1700s that James Hutton, the father of modern geology, realized that the events recorded in the rock record must have taken a very long time to unfold. He and his contemporaries used careful field observations and scientific principles to recognize many different types of events, including deposition and burial of sediment, igneous activity, rock deformation, and uplift and erosion of preexisting rocks. They also were able to arrange these events in a *relative* sequence from oldest to youngest. However, they were unable to assign exact dates to anything that occurred before the beginning of recorded history. It was like knowing only that you are younger than your mother, and that she is younger than your grandfather, but not knowing anyone's actual age.

Geologists now are able to determine quite precisely the dates, or **numerical geologic ages** (also known as *absolute geologic ages*), of many types of geologic events. Determining a numerical geologic age is complex and expensive, so numerical ages are not always readily available. However, many thousands of numerical ages have been determined so that if the *relative* geologic age is known, an estimate of the numerical age can be made. Both types of ages, relative and numerical, are important pieces of information for interpreting and understanding the geology of an area, unraveling the complexities of past tectonic-plate movements and interactions, and reconstructing the history of climate change and the long pageant of prehistoric life.

Relative Geologic Time

General Concept

Although it is important to know the numerical age of a particular geologic event, it is equally important to be able to indicate

A. Folded rocks

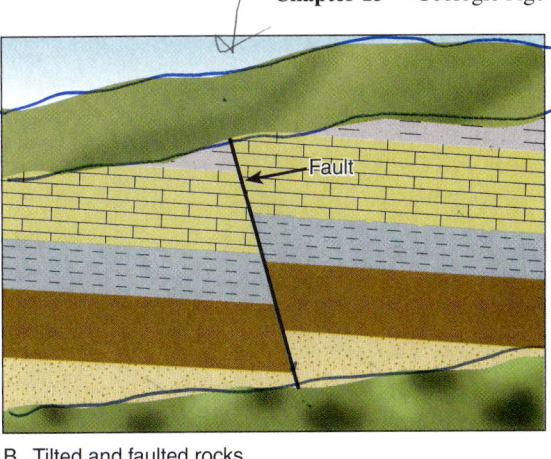
B. Tilted and faulted rocks

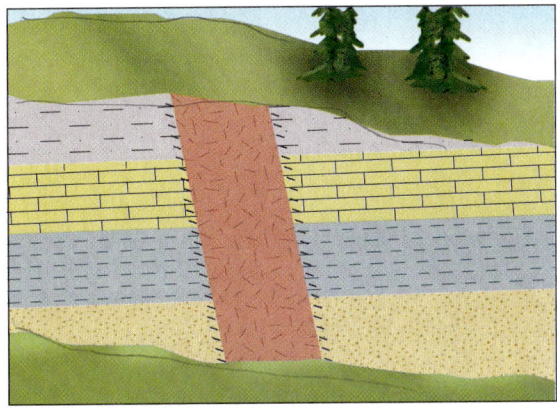
C. Igneous dike that intruded older sediments

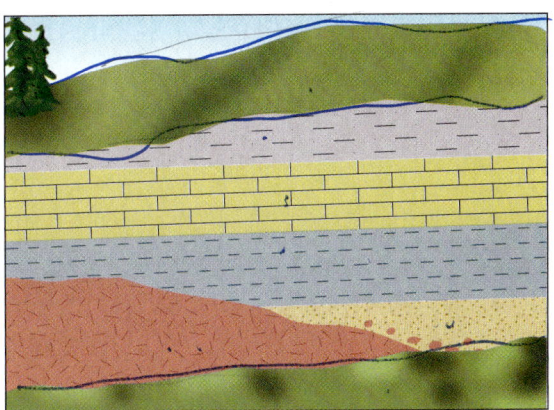
D. Igneous rock upon which sediments were deposited

FIGURE 13.1
Perspective sketches of some geologic structures that might be observed in roadcuts or cliff faces. A. Folded rocks. B. Tilted rocks with a fault offsetting the sedimentary layers. C. Intrusive igneous rock, in this case a dike (a tabular body filling an opened fracture). The crosshatches along the igneous/sedimentary rock contact indicate a contact metamorphic rock called hornfels. D. Intrusive igneous rock, exposed to erosion and later buried by sediment. No contact metamorphism is apparent.

whether that event occurred before or after another event. This is the basis for a **relative geologic age**—the age of one event relative to another.

A description of the geology of an area includes a list of the geologic events that took place, in the sequence in which they occurred, from oldest to youngest. Geologic events that are commonly described include deposition of sedimentary units, extrusion or intrusion of igneous rocks, metamorphism, folding, faulting, uplift, and erosion.

To understand how a sequence of events is determined, consider Figure 13.2, a sketch of a roadcut in a mountainous area. The sketch shows three inclined layers of sedimentary rock—sandstone, shale, and limestone—intruded by a granite dike. What geologic events must have occurred to produce what is seen in the roadcut, and in what order did they occur?

Start with the simplest, the dike. Because it cuts (intrudes) all three sedimentary layers, it must be younger than all of them.

Next, what are the relative ages of the sedimentary layers? The one on the bottom must have been deposited first (assuming the layers are not upside down, a possibility that is considered later), and the one on top last.

So far we have the following sequence:

intrusion of granite dike (**youngest**)

deposition of limestone

deposition of shale

deposition of sandstone (**oldest**)

But there is more. Was the original sediment of the sedimentary rocks deposited as inclined layers? Not likely. The layers must have been tilted or folded after deposition.

And the dike—was it intruded before or after the beds were tilted? Can you tell? Not without more information, of a kind that is too detailed to discuss here. So we must be satisfied with two possibilities for now; the dike could be older or younger than the tilting event.

What about the present-day surface, at the top of the roadcut? It is still forming, by erosion. Because the original sediments must have accumulated in a low place, and because they must have been buried under younger sediment in order to become lithified, the erosion must have been accompanied by uplift of the whole area.

FIGURE 13.2
View of rocks exposed in a vertical roadcut in a mountainous area. Sedimentary rocks are limestone (brick pattern), shale (parallel dashes), and sandstone (dots); granite dike shown with random dashes. What sequence of events led to what we see?

Now we can list the entire sequence in order of relative age:

uplift and erosion (**youngest**)

intrusion of granite dike (or tilting)

tilting of beds (or granite dike)

deposition of limestone

deposition of shale

deposition of sandstone (**oldest**)

As you can see, there is nothing magical or mystical about the way in which this sequence was determined. Logic and a few basic principles are all that's needed.

Basic Principles

From the preceding example, you can see that several fundamental principles exist to help interpret the relative time relations among rocks. Among the more useful are the following, the first three of which were formulated by Nicholas Steno in the late seventeenth century.

Steno's Principle of Original Horizontality (Fig. 13.3): Sediments are deposited in horizontal or near-horizontal layers. Therefore, non-horizontal layers have generally been folded or tilted from their original positions.

Steno's Principle of Superposition (Fig. 13.3): In any succession of sedimentary rock layers lying in their original horizontal position, the rocks at the bottom of the sequence are older than those lying above.

Steno's Principle of Original Lateral Continuity (Fig. 13.3): Sediments are deposited in layers that continue laterally in all directions until they thin out as a result of nondeposition, or until they reach the edge of the basin in which they are deposited. This means that if you find a layer that abruptly ends, something has cut this layer after it was deposited. Faults, dikes, and erosion can all truncate otherwise laterally continuous layers.

Principle of Cross-Cutting Relations (Fig. 13.4): Any geologic feature (intrusive igneous rock, fault, fracture, erosion surface, rock layer) is younger than any feature that it cuts.

Principle of Inclusions (Fig. 13.5): An inclusion in a rock is older than the rock containing it. Examples of inclusions are pebbles, cobbles, or boulders in a conglomerate, or xenoliths (pieces of other rocks) in igneous rocks.

Relative Ages Based on Fossils

These basic principles establish age relations among rocks that occur together in a local area. William Smith, working in the 1790s, examined the nicely layered sedimentary

FIGURE 13.3
Sedimentary rocks viewed from Dead Horse Point State Park, Utah, illustrate the principles of Original Horizontality (sediment is deposited in horizontal layers), Superposition (younger beds are deposited on older beds), and Original Lateral Continuity (sedimentary layers do not abruptly end unless they hit the edge of a basin or are cut by younger features).

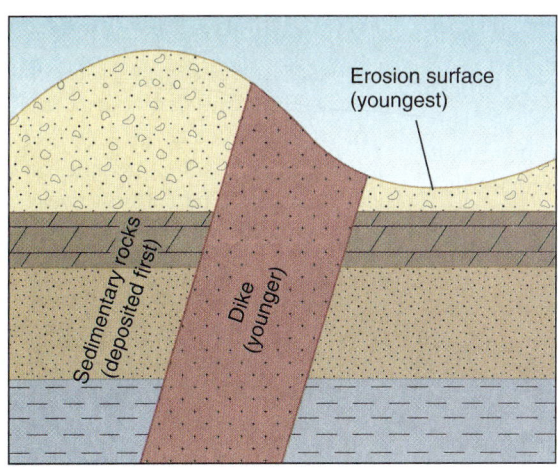

FIGURE 13.4
Principle of Cross-Cutting Relations. This cross section shows a dike cutting preexisting layers of sedimentary rock; the dike is younger than the rocks it cuts. The erosion surface cuts both the sedimentary rocks and the dike, so it is the youngest.

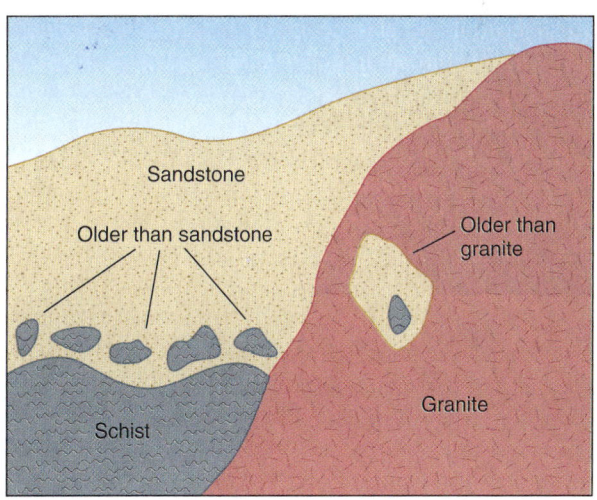

FIGURE 13.5
Principle of Inclusions. This cross section shows boulders of schist at the base of a sandstone; the boulders must be older than the sandstone containing them. Similarly, the xenolith of sandstone in the granite must be older than the granite.

rocks and their fossils across England. (**Fossils** are the preserved remains, traces, or imprints of ancient plants and animals). Employing the Principle of Superposition, he collected fossils from successive layers and discovered that he could use the fossils to determine the age-equivalence of widely separated sedimentary units, much as archaeologists recognize different historical periods in their excavations based on distinctive coins or pottery. Smith's work helped establish two important principles:

Principle of Fossil Succession: Fossil organisms succeed one another in time in a definite and recognizable order. Each distinct organism existed for a specific interval of time and not at older or younger times. The fossils in a sedimentary unit therefore can define a specific, unique interval of geological time.

Principle of Fossil Assemblages: Characteristic groups of fossil organisms also define unique geologic ages.

Fossils are an exceptionally useful means of determining relative time because they establish age relationships among widely separated rocks and because the sequence of fossil organisms is known over a very long interval of geologic time. The dinosaur *Tyrannosaurus rex,* for example, lived for a relatively short period of time. Any rocks found on any continent that contain its bones must date to this same short interval of time. Because the position of *T. rex* is known relative to the rest of the long fossil record, these rocks are securely located within all of geologic time.

Testing Hypotheses

In the example in which the sequence of events illustrated in Figure 13.2 was hypothesized, some assumptions were made that should not have been made without further observations.

The first assumption, a very reasonable one, was that the dike cut the layers of sedimentary rock. There is a *very* remote possibility that the dike was there first, standing as an inclined rock wall, while sediments were deposited on both sides of it. How would you tell? One way is to carefully examine the sedimentary rocks adjacent to the contact with the dike. For example, were they metamorphosed by the dike? If so, the dike must be younger. Do they contain any inclusions of the dike, such as pebbles, that might have been eroded from it? If so, the dike is older. Are there any small fingers of granite extending from the dike into the sedimentary rock that might have squeezed into weak places during intrusion of magma? If so, the dike is younger. Notice that the sedimentary layers cannot be projected straight across the dike but appear to be offset. Why? Because when the magma was intruded, it forced the rocks apart at 90° to the dike margins.

A second assumption was that the layers, though not horizontal, have not been completely overturned. If they have, then the limestone is the oldest, and the sandstone the youngest. How can you tell? One way is to look for sedimentary structures that allow you to tell top from bottom, up from down. Some useful sedimentary structures, shown in Chapter 4, are:

- **Cross-stratification** (see Fig. 4.4)—cross beds are commonly cut off on the top of the bed and become parallel to adjacent layers on the bottom.
- **Oscillation ripple marks** (see Fig. 4.5B) —symmetrical, wave-like features whose crests point to the top of the bed.
- **Graded beds** (see Fig. 4.6)—grain size commonly becomes progressively finer upward.
- **Mud cracks** (see Fig. 4.7)—in cross section, wider at the top than at the bottom.

Unconformities

The last event in the example shown in Figure 13.2 is erosion. What if sea level rose, or the land surface fell, and that erosion surface eventually was buried under younger sedimentary rock? The erosion surface would then be an **unconformity,** a surface that represents a substantial gap in the geologic record. It may be an ancient erosion surface, or it may be a surface on which neither erosion nor deposition occurred for a long period of time. If it is an erosion surface, it is recognizable because it cross-cuts older rocks, and its relative age can be determined by the Principle of Cross-Cutting Relations. If neither erosion nor deposition occurred, the unconformity may be difficult to recognize without studying fossils and applying the Principle of Fossil Succession. At any rate, there is no rock record of the time interval between the underlying rocks and those deposited on the unconformity.

Unconformities are of three principal types, each of which reflects distinct geologic events (Fig. 13.6). Figure 13.6A shows an **angular unconformity,** an erosion surface separating rocks whose layers are not parallel. Layers above and below meet at an angle. This implies that the underlying rocks were folded or tilted *and* eroded before the overlying layers were deposited. A **disconformity** is either an erosion surface or a surface of nondeposition separating rocks whose layers are parallel. An erosion surface is uneven and cuts layers of underlying rocks (Fig. 13.6B); a surface of nondeposition parallels underlying rock layers (Fig. 13.6C). A **nonconformity** (Fig. 13.6D) is an erosion surface separating sedimentary rocks from older plutonic or massive metamorphic rocks (that is, crystalline rocks that are not layered).

Numerical Geologic Time

Numerical ages or dates can be determined in several ways. For example, you can tell how old a tree is by counting the number of growth rings. Or you can determine how long some glacial meltwater lakes existed by counting the number of varves—annually deposited sets of layers—present in the lake sediment. For obvious reasons, these and similar methods have limited applicability. Radioactivity, however, provides a much more widely applicable method for determining numerical dates.

Radioactivity

Some isotopes of elements are **radioactive;** that is, their nuclei spontaneously break down or decay. In these reactions, a radioactive isotope, or **parent,** decays to form a different isotope, the **daughter.**

Isotopes are varieties of an element containing different numbers of neutrons in their nuclei. For example, the element uranium has two common radioactive isotopes, ^{235}U and ^{238}U. The superscripts 235 and 238 are the atomic masses or **mass numbers** (number of protons plus neutrons in the nucleus) of the isotopes. Each isotope has 92 protons in its nucleus—92 is the **atomic number** of uranium—but ^{235}U has 143 neutrons (235 − 92 = 143), whereas ^{238}U has 146 neutrons.

Ages from Isotopes

Laboratory measurements show that each radioactive isotope decays at a constant rate that is not affected by geologic age or surrounding temperature, pressure, or chemical conditions. The constant decay rate of each isotope enables us to calculate the numerical age of geologic samples. The decay rate is often described in terms of a **half-life,** which is the time required for half the atoms of any starting mass of a radioactive isotope to decay away. After one half-life, half of the parent atoms are gone, having produced an equal number of daughter atoms. After two half-lives, only half of one-half, or one quarter, of the original parent remains; three-quarters of the parent has converted to daughter, so there are three times more atoms of daughter than parent.

The age of a mineral is determined from the relative amounts of parent and daughter isotopes it contains. The relative amounts of parent and daughter atoms are measured with an instrument called a mass spectrometer. If one knows the relative number of atoms of daughter and parent isotopes that are present, the ratio of these two combined with the **decay constant** can be used to determine the age. The decay constant, which is directly related to the half-life, describes the proportion of a starting mass of a given isotope that will decay

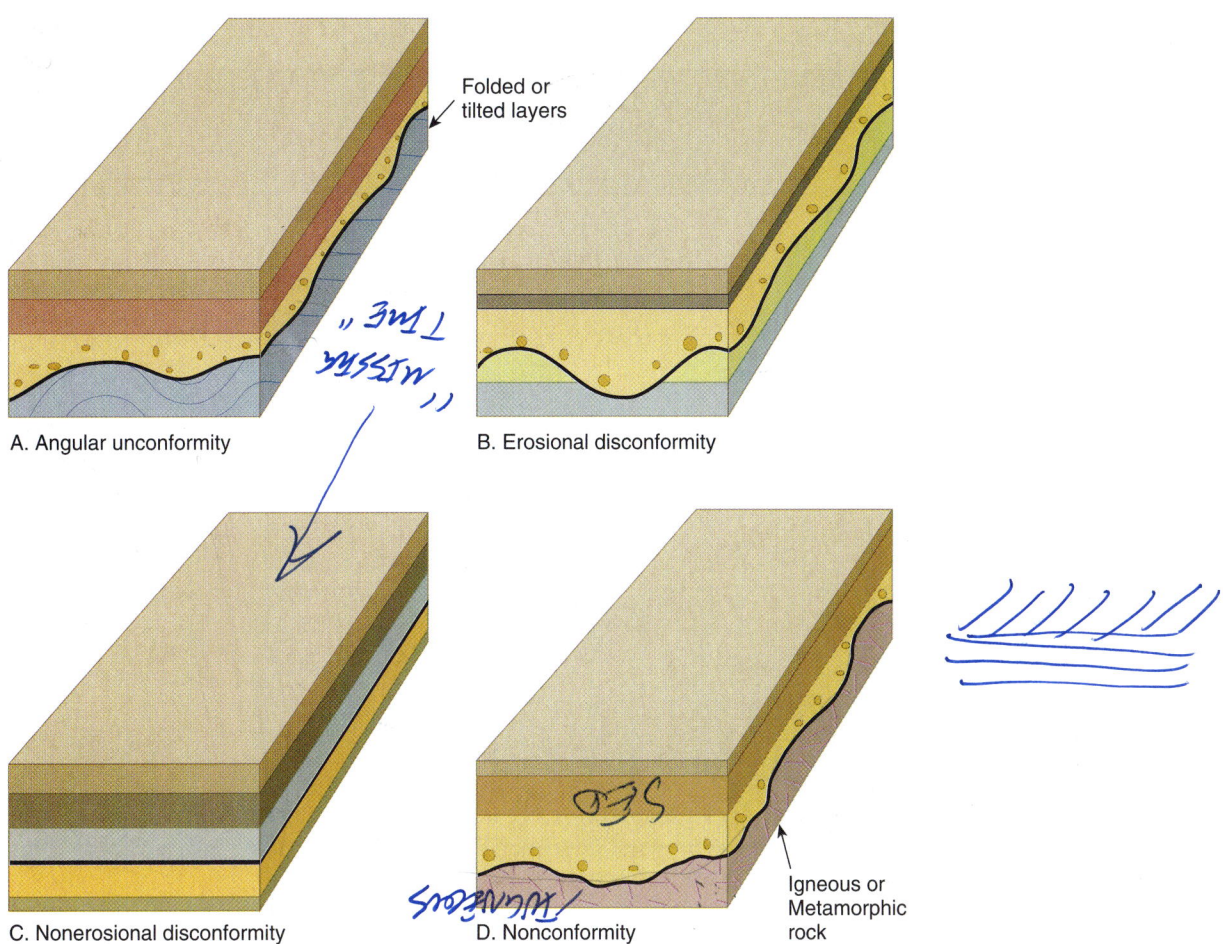

FIGURE 13.6
Three-dimensional diagrams of: A. angular unconformity; B. erosional disconformity; C. nonerosional disconformity; D. nonconformity. The unconformity is an ancient erosion surface in A, B, and D, but in C it is a surface on which no erosion or deposition occurred for a long period of time.

away in a year. The age can be determined graphically or mathematically, as illustrated by the problems at the end of the chapter.

The age is calculated mathematically using the following equation:

$$t = ln(N_D/N_P + 1)/\lambda$$

where t is time in years, N_D and N_P represent the number of atoms of daughter and parent, respectively, and λ is the decay constant in units of 1/year. The natural logarithm function, ln, is available on most scientific calculators.

As an example, if $N_D = 4000$ atoms, $N_P = 5000$ atoms, and the decay constant for the parent isotope is $\lambda = 1.40 \times 10^{-8}$/year, then:

$$t = ln(4000 atoms/5000 atoms + 1)/1.40 \times 10^{-8}/year$$

$$t = ln(0.8000 + 1)/1.40 \times 10^{-8}/year$$

$$t = ln(1.8000)/1.40 \times 10^{-8}/year$$

$$t = 0.5878/1.40 \times 10^{-8}/year$$

$$t = 4.20 \times 10^7 \text{ years or } 42,000,000 \text{ years}$$

Note that in practice two complications arise. First, daughter isotopes are commonly present in minerals when they first form. Second, surface weathering or metamorphism can cause parent or daughter elements to be gained or lost from minerals within a rock. Both types of complications are easily handled by modern methods, but we will not discuss them here.

Geologic Time Scale

The **geologic time scale** (Fig. 13.7) subdivides Earth's history into unequal intervals based on distinctive fossil assemblages and globally important geologic features, such as widespread unconformities. The scale was developed first as a *relative time scale,* using the principles discussed previously, but numerical ages are now known for all parts of the time scale. The largest subdivisions of geologic time are **eons;** eons are divided into **eras,** eras into **periods,** and periods into **epochs.** The Phanerozoic Eon has many subdivisions because well-preserved shelly fossils are present throughout and are found worldwide. That is not true for the time preceding the Phanerozoic, when organisms were simple life forms without easily preserved shells. Instead, subdivisions of that time, called the **Precambrian** (Fig. 13.7), are based on major geologic features or events and on numerical ages.

Geologic Time Scale

Eon	Era	Period		Epoch	Age (millions of years ago)	Event (Problem 4)
Phanerozoic	Cenozoic (Cz)	Quaternary (Q)		Recent or Holocene		
					0.01	
				Pleistocene		
					1.8	
		Tertiary (T)	Neogene (N)	Pliocene		
					5.3	
				Miocene		
					24	
			Paleogene (Pε)	Oligocene		
					34	
				Eocene		
					55	
				Paleocene		
					65	
	Mesozoic (Mz)	Cretaceous (K)				
					144	
		Jurassic (J)				
					206	
		Triassic (TR)				
					248	
	Paleozoic (Pz)	Permian (P)				
					290	
		Carboniferous (C)	Pennsylvanian (IP)			
					323	
			Mississippian (M)			
					354	
		Devonian (D)				
					417	
		Silurian (S)				
					443	
		Ordovician (O)				
					490	
		Cambrian (€)				
					543	
Precambrian	Proterozoic					
					2500	
	Archean					
					3800	
	Hadean					
					4500	

FIGURE 13.7

The geologic time scale, with symbol abbreviations in parentheses.
Adapted from web page of the Museum of Paleontology of the University of California at Berkeley.

Hands-On Applications

Rocks are the pages in the Earth's history book. Unfortunately, there is no single place on Earth where the whole book is intact. Pages, or even whole chapters, have been locally removed by erosion or bent or torn apart by deformation. The geologist's job is to read the book the way it is, and that requires sorting out the time relations using sound scientific methods. In this lab, you first will learn how basic principles are used to establish the order in which rocks formed or geologic events occurred. Additional problems illustrate how numerical geologic ages are determined.

Objectives

If you complete all the problems, you should be able to:

1. Determine relative ages and the sequence of geologic events as illustrated in a geologic cross section.
2. Use daughter-parent ratios, half-life, and the decay constant to determine the numerical age of a rock or mineral graphically and mathematically.

Problems

1. Figures 13.8 and 13.9 are geologic cross sections. Determine the sequence of events that led to the present situation, and list them in order, from oldest to youngest, in the blanks provided. Use the letters on the illustrations to specify rock units or events. Identify the rock types represented by each letter from the symbols given in Figure 13.10; identify other events by name (for example, "folding" or "uplift and erosion"). The number of blanks equals the number of events required.

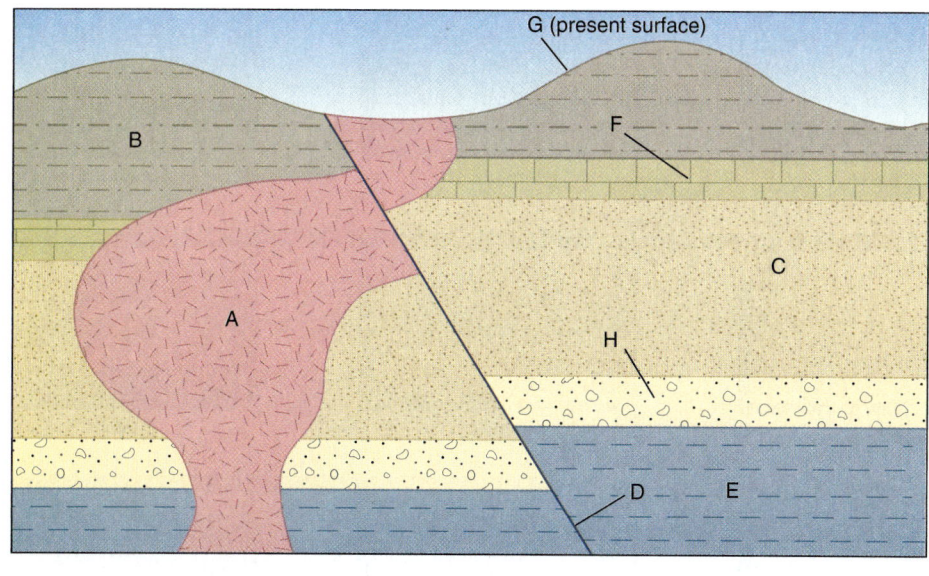

Youngest G - Siltstone
D - Fault
B - Siltstone
A - Granite
F - Limestone
C - Sandstone
H - Conglomerate
E - Shale
Oldest

FIGURE 13.8

For Problem 1, determine the relative ages of the lettered geologic features illustrated in this cross section, and list them chronologically in the spaces provided. D (dark line) is a fault, and G is the present surface.

242   Part V   Geologic Time and Sequences

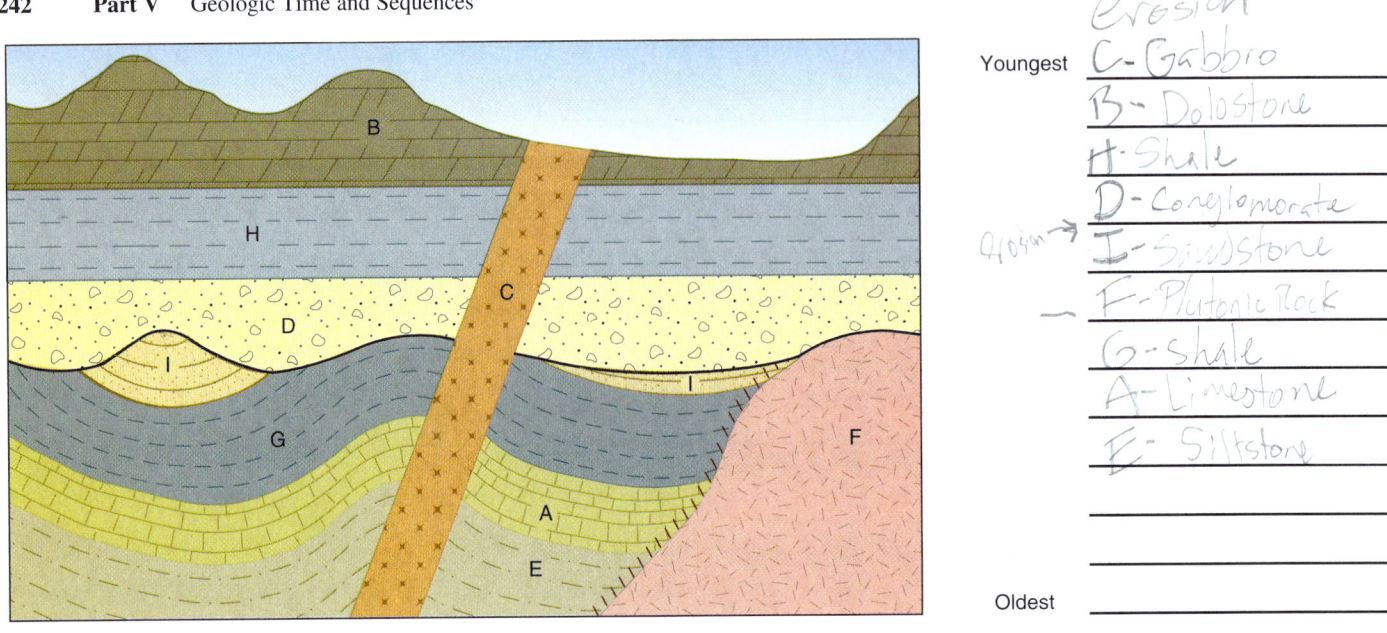

Youngest
Erosion
C - Gabbro
B - Dolostone
H - Shale
D - Conglomorate
Erosion → I - Sandstone
F - Plutonic Rock
G - Shale
A - Limestone
E - Siltstone

Oldest

FIGURE 13.9
For Problem 1, determine the relative ages of the geologic features illustrated in this cross section, and list them chronologically in the spaces provided. Letters indicate rock units; *other events are not labeled.* Also see Problem 6.

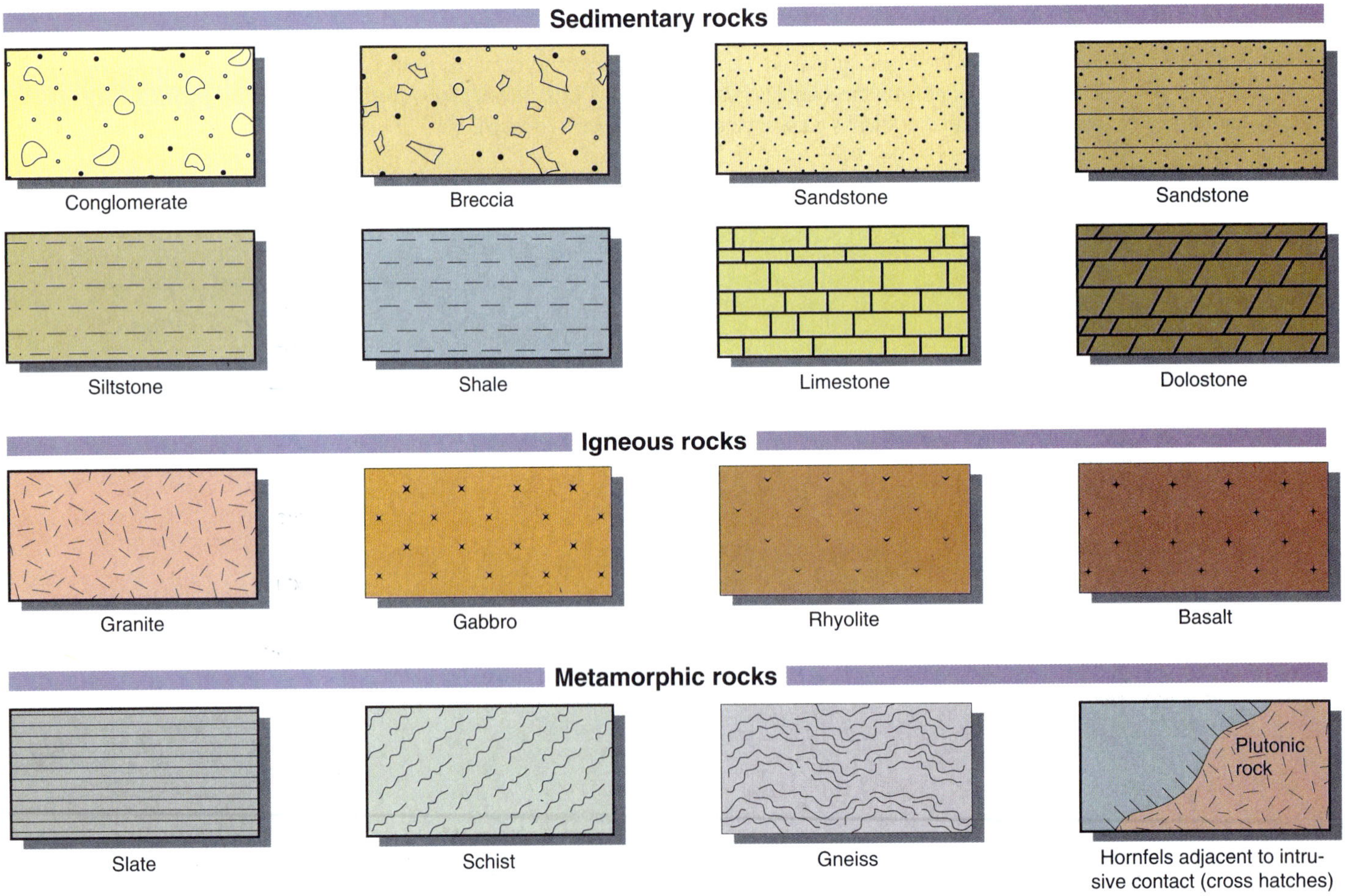

Sedimentary rocks

Conglomerate | Breccia | Sandstone | Sandstone

Siltstone | Shale | Limestone | Dolostone

Igneous rocks

Granite | Gabbro | Rhyolite | Basalt

Metamorphic rocks

Slate | Schist | Gneiss | Hornfels adjacent to intrusive contact (cross hatches)

FIGURE 13.10
Symbols commonly (but not universally) used to show different kinds of rocks.

2. The purpose of this problem is to illustrate how numerical ages based on radioactivity can be determined graphically.

   a. Complete columns A and B in the table that follows. For example, after one half-life, the parent fraction is 0.5 and the daughter fraction is 0.5; after two half-lives, the parent fraction is 0.25 and the daughter fraction is 0.75. Remember, the sum of the two fractions must equal 1.0.

   b. Next, complete column C in the table by dividing values in Column B by corresponding values in Column A. For example, for one half-life elapsed, (Col. B)/(Col. A) = 0.5/0.5 = 1.

Half-Lives Elapsed	A. Parent Fraction	B. Daughter Fraction	C. Daughter/Parent Ratio
0	1	0	0
1	0.5	0.5	1
2	0.25	0.75	3
3	.125	.875	7
4	.0625	.9375	15
5	.03615	.96875	31

   c. Using Figure 13.11 and the data in the preceding table, construct a graph in which the ordinate (vertical axis) is "Daughter-Parent Ratio" and the abscissa (horizontal axis) is "Half-Lives Elapsed." The point representing one half-life is already plotted. Plot the rest, and draw a *smooth* curve connecting the data points; that is, do not connect the points with straight-line segments, but estimate the curvature between points as best as you can so that the entire curve bends smoothly.

   d. For samples 1–3 in the following table, first calculate and record the daughter-parent ratio, N_D/N_P. Then, using your graph in Figure 13.11, determine the number of half-lives that have elapsed for each sample and write your answer in the "Half-Lives Elapsed" column.

Sample Number	Atoms of Parent N_P	Atoms of Daughter N_D	N_D/N_P	Half-Lives Elapsed	Age In Years
1	2135	3203	1.5	1.3	10,660
2	4326	10,815	2.5	1.77	13,940
3	731	14,620	20	4.44	36,080

   e. If the half-life is 8200 years, calculate the age in years of the samples in the preceding table and write your answer in the "Age in Years" column. Show your work.

3. The purpose of this problem is to illustrate how numerical ages based on radioactivity can be determined mathematically. Show all your calculations.

   a. If the decay constant for radioactive decay of ^{40}K is $\lambda = 5.543 \times 10^{-10}$/year, use the equation on p. 239 to calculate the ages of the samples in the following table, and write them in the "Age in Years" column.

Sample Number	Atoms of Parent N_P	Atoms of Daughter N_D	N_D/N_P	Half-Lives Elapsed	Age In Years
1	6439	2303			
2	4395	1303			
3	8763	1893			

244   Part V   Geologic Time and Sequences

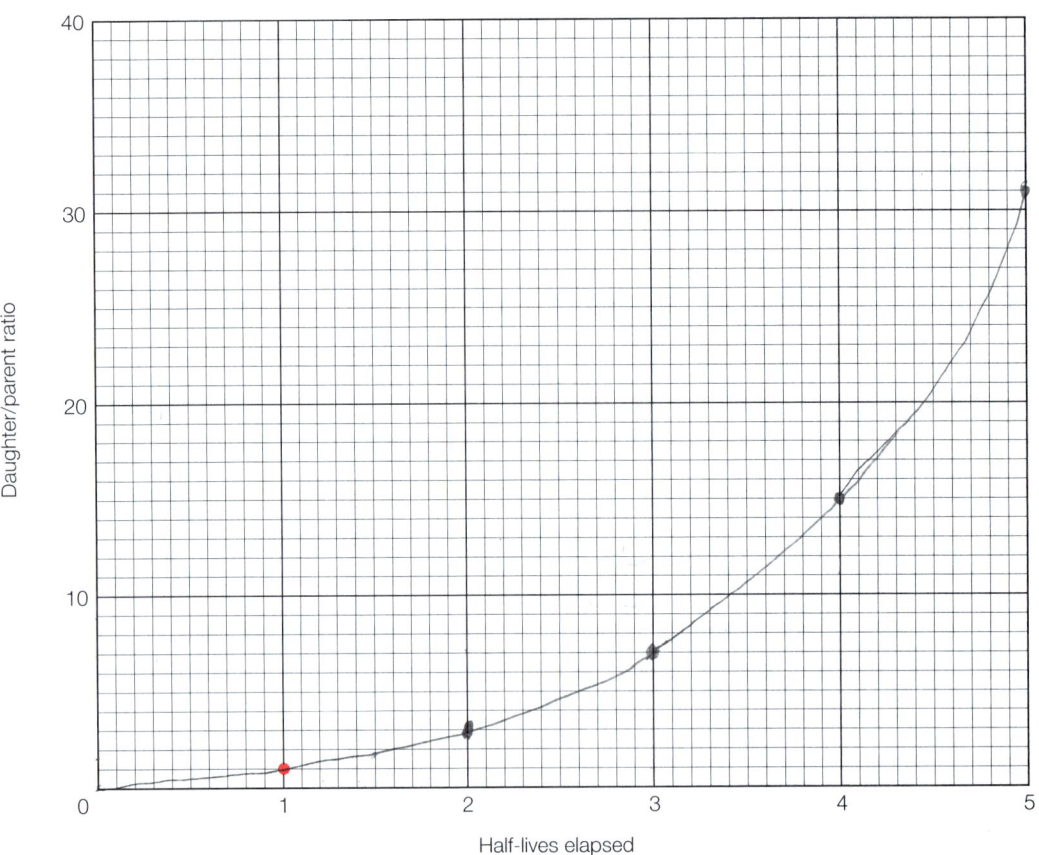

FIGURE 13.11
Graph for Problem 2c.

b. The half-life, $t_{1/2}$, of ^{40}K can be determined from the decay constant using the following relation:

$$t_{1/2} = 0.693/\lambda \text{ (see footnote 1)}$$

What is the half-life of ^{40}K? _____

c. Using the half-life calculated in Problem 3b, complete the column labeled "Half-Lives Elapsed" in the previous table.

4. Go to *www.kaibab.org/geology/gc_layer.htm* (or link to it through *www.mhhe.com/jones5e*—see Preface) and scroll down to the schematic geological section at the bottom of the page. Concentrate on the Precambrian part of the section, that is, everything below the Tapeats Sandstone, and answer the following:

   a. Is the Zoroaster Granite older or younger than the Vishnu Schist?

      The Bass Limestone?

   b. What is the oldest sedimentary rock?

---

[1] The equation $t_{1/2} = 0.693/\lambda$ is derived from the equation $t = \ln(N_D/N_P + 1)/\lambda$ as follows: If $t = t_{1/2}$, then $N_D/N_P = 1$, $\ln(N_D/N_P + 1) = \ln 2 = 0.693$, and $\ln(N_D/N_P + 1)/\lambda = 0.693/l = t_{1/2}$

c. What name is applied to the surface separating the Vishnu Schist and the Bass Limestone (be specific)?

What surface separates the Bass Limestone and the Tapeats Sandstone?

d. Knowing that the oldest Precambrian sedimentary rock in the Grand Canyon is about 1,250 m.y. (million years) old and the youngest about 825 m.y., use maximum thicknesses given for the Precambrian sedimentary rocks (that is, exclude the Cardenas Lavas) and calculate the "average" rate of accumulation in feet/m.y.

Evaluate the significance of this number; that is, why might this number be meaningful or meaningless?

Next go to The University of California, Berkeley Museum of Paleontology at *www.ucmp.berkeley.edu/index.html* and click on *Discover the History of Life* or *History of Life (www.ucmp.berkeley.edu/historyoflife/histoflife.html)*. Then, under *Geologic Time,* link to *Geologic Time Scale (www.ucmp.berkeley.edu/help/timeform.html)*. Use this site to determine when the things in the following list occurred. To do this, click on *Precambrian, Paleozoic Era, Mesozoic Era,* and *Cenozoic Era.* For the three eras, you will have to further click on the *Stratigraphy* buttons to find the information needed. Write the answers below <u>and</u>, to get an overall sense of history, in the right-hand column of the Geologic Time Scale (Fig. 13.7).

Oldest known rocks?

Transition to atmosphere with oxygen?

Start of breakup of Pangaea?

Oldest known fossils?

First animals?

Age of dinosaurs?

First:
   Birds?
   Reptiles?
   Mammals?
   Insects?
   Fish?
   Flowering plants?

Largest mass extinction?

An <u>excellent</u> interactive site that will help you understand how isotopes are used to determine numerical ages is at *www.sciencecourseware.com/VirtualDating/.* Give it a try.

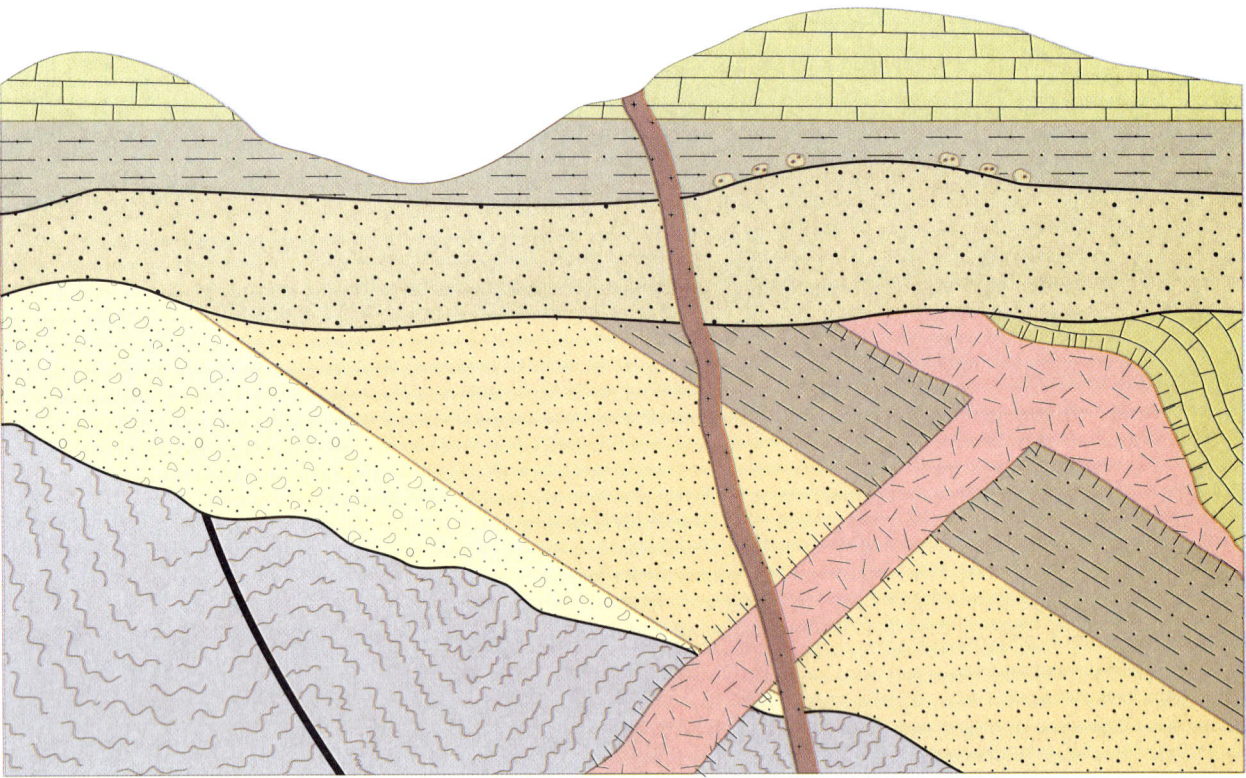

FIGURE 13.12
For Problem 5, determine the sequence of events illustrated in this cross section. Identify rock units from their symbols, and list all the events that led to the present situation. Wide line in lower left-hand unit is a fault.

In Greater Depth

5. Using Figure 13.12, identify rock units from their symbols, and identify and list in sequence *all* the events that led to the present situation. On the cross section, write number 1 on the oldest rock or event, number 17 on the youngest, and intermediate numbers on rocks or events in between. If you cannot determine which of two or more rocks/events is the older, explain why not.

6. Refer to Figure 13.9.
   a. The daughter-parent isotope ratios (N_D/N_P) for rocks F and C are 16.0 and 0.92, respectively. Assume that the half-life of the parent is 6.5×10^7 years, and use the curve you derived for Problem 2 to determine the numerical ages of each of these rocks.

   b. During what geologic periods were these rocks intruded (see Figure 13.7)?

   c. Assume, for illustrative purposes, the highly unlikely scenario in which each of the lettered rock units A through I represents a geologic period of the Phanerozoic. What period (or periods, if more than one answer is possible) does each represent?

   d. During what geologic period or periods did folding and erosion occur?

PART VI
Internal Processes

Folded layers in the banded iron formation, Jasper Knob, Ishpeming, Michigan ($\times 2$)

CHAPTER 14

Structural Geology

Materials Needed
- Pencil and eraser
- Protractor
- Metric ruler
- Silly Putty® (provided by instructor)
- Scissors
- Removable tape (optional)

Introduction

Stresses within the Earth bend and break rocks, producing folds, faults, and other geologic structures. These structures are three-dimensional, and they are commonly too large or too poorly exposed to be easily recognized from a single vantage point on the ground. For these reasons, basic geologic data must be collected in the field and plotted on a map to understand the structure fully. This chapter will help you visualize geologic structures in three dimensions so that you can understand them when seen in two dimensions on a geologic map, aerial photograph, or on the Earth's surface. You will learn how the orientations of planar features, such as sedimentary bedding, are described and used to portray the geologic structure on a geologic map. Vertical cross sections of geologic maps illustrate what you expect to see below the surface, and block diagrams combine maps and cross sections to provide a three-dimensional view. A clear understanding of subsurface structures is vital for an understanding of groundwater resources, pollution migration, oil reserves, ore deposits, and the stability of road cuts, tunnels, bridge footings, and building foundations.

Structural geology is the study of the form, arrangement, and internal structure of rocks. It is especially concerned with understanding how rocks are deformed. Evidence for deformation ranges from rocks shattered during earthquakes, to rocks long ago tilted, folded, and shoved over one another, to the foliation seen in slates, schists, and gneisses. The structural geologist puts together all such evidence to unravel the remarkable complex interactions of the Earth's tectonic plates.

Deformation is caused by **stress,** a force acting on an area of a body that tends to change its size or shape. Stress is a force per unit area, so the magnitude of stress depends not only on the magnitude of the force, but also on the area over which the force is applied. Stress will be greater if a given force is applied to a small, rather than large, area; a person wearing high-heeled shoes is more likely to punch a hole in a rotten floorboard than a person wearing tennis shoes. *Compressive stress,* or **compression,** causes shortening (Fig. 14.1A), and is an important type of stress at convergent tectonic plate boundaries. *Tensional stress,* or **tension,** the principal stress at divergent boundaries, causes stretching (Fig. 14.1B). **Shear stress** causes one side of a body to slide past the other (Fig. 14.1C), and is the principal type of stress along transform-fault plate boundaries.

Strain, the physical deformation that occurs in response to stress, is a change in size or shape. Strain may be temporary or permanent. **Elastic strain** is temporary; if the stress is removed, the object snaps back to its original shape. A stretched rubber band and a bouncing ball both undergo elastic strain. **Plastic strain** is permanent; if the stress is removed, the original size or shape is not restored. If you bend a piece of soft wire, the wire undergoes plastic strain; the bend is permanent. If it is bent back and forth at the same place, it eventually breaks or **ruptures.**

A rock undergoing elastic strain can suddenly break: this is **brittle deformation.** A dry stick of wood also behaves this way: it bends only so far before it suddenly breaks (ruptures), and each piece elastically recovers its original straight shape. A rock that undergoes much plastic strain without rupturing is **ductile.** The aforementioned bent wire is ductile.

Pressure, temperature, and rate of deformation determine brittle versus ductile

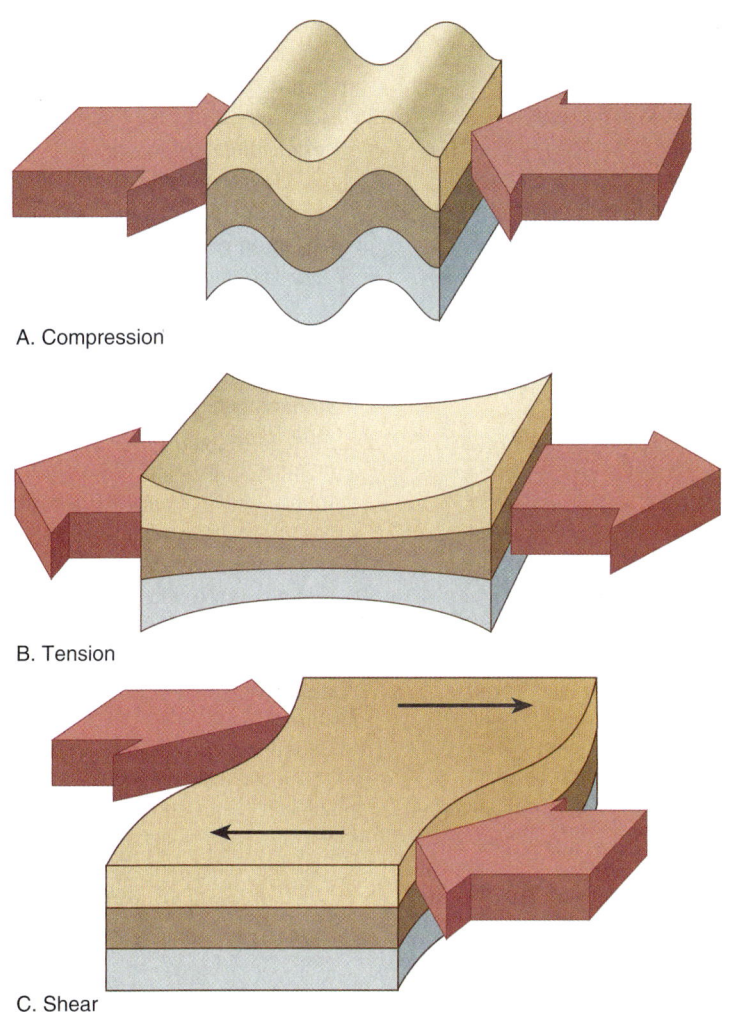

FIGURE 14.1
The three principal types of stress: A. compression; B. tension; C. shear.

TABLE 14.1
Conditions Required for Brittle versus Ductile Deformation of Rocks

Surrounding Conditions	Brittle	Ductile
Pressure	low	high
Temperature	low	high
Stress rate	rapid	slow
Stress magnitude	high	low to high

Chapter 14 Structural Geology

bend into folds. Eventually, erosion of overlying rock material and gradual uplift bring the deformed rocks to the surface where we can view them.

Recognizing and Describing Deformed Rocks

A geologist must be able to visualize rock structures in three dimensions. In some parts of the world, rocks are so well exposed at the surface that it is easy to see and understand their structure. In most areas, however, the rocks are buried beneath soil or young, unconsolidated sediment, and are exposed only in a few **outcrops** located far apart. In such places, it is difficult to understand the three-dimensional aspects without carefully examining the rocks, measuring their orientation, and plotting these data on a map.

What to Look for: Stratification, Contacts, and Fault and Joint Surfaces

How can you tell if sedimentary rocks have been deformed? It generally is easy—just look at the orientation of the strata. The original sediments were deposited as continuous horizontal or nearly horizontal **strata** (layers). If the strata are still horizontal and not broken, then they have not been deformed, although they may have undergone gentle uplift or subsidence. If strata are broken or not horizontal, then they have been deformed. Deformation takes many forms, from slight tilting of strata to complex folding, as described in a later section.

Contacts are surfaces separating adjacent bodies of rock of different types or ages. Understanding the nature of the contact is important, because the contact records a change, an event. There are several kinds of contacts.

Depositional contacts separate older bodies of rock from younger sedimentary rocks deposited upon them. In a sequence of sedimentary rocks deposited continuously, or **conformably,** in a basin of deposition, the contacts between adjacent layers are planar **bedding planes** (see Fig. 4.3); they are parallel to—and, in fact, define—the bedding or stratification. Some depositional contacts are ancient

behavior (Table 14.1). Cold rocks at the Earth's surface that experience a rapidly applied stress (like a hammer blow) tend to experience brittle fracture. Ductile deformation is more likely under the high temperature and pressures found deep underground, especially when stresses are applied slowly. The folded rocks visible in many mountain ranges were deformed well below the surface by forces acting over long periods of time (millions of years). The combination of depth and time enables deep rocks that would be brittle at the surface to behave plastically and gradually

FIGURE 14.2
The steeply inclined surface in the center of the photograph is a fault contact. It cuts the sandstone and mudstone layers of the Haymond Formation (Pennsylvanian) near Marathon, Texas.

erosion surfaces. These **unconformable** contacts can be uneven rather than planar, and can juxtapose very different rocks with very different orientations (see *Unconformities* in Chapter 13).

An *intrusive contact* separates intrusive igneous rock and the rock that it intruded; intrusive contacts are either *concordant* (parallel to layers in the intruded rock) or *discordant* (cut across layers in the intruded rock).

Rocks separated by a fault are in ***fault contact*** (Fig. 14.2). In the simplest case, the fault is exposed as a flat surface, and its orientation is apparent. However, faults commonly are expressed as zones along which rocks are brecciated (broken into angular fragments), or even pulverized, so that fault orientation may be difficult to determine. Even more commonly, faults, like other types of contacts, are buried beneath soil or sediment. In such places, their orientation,

A. Outcrop

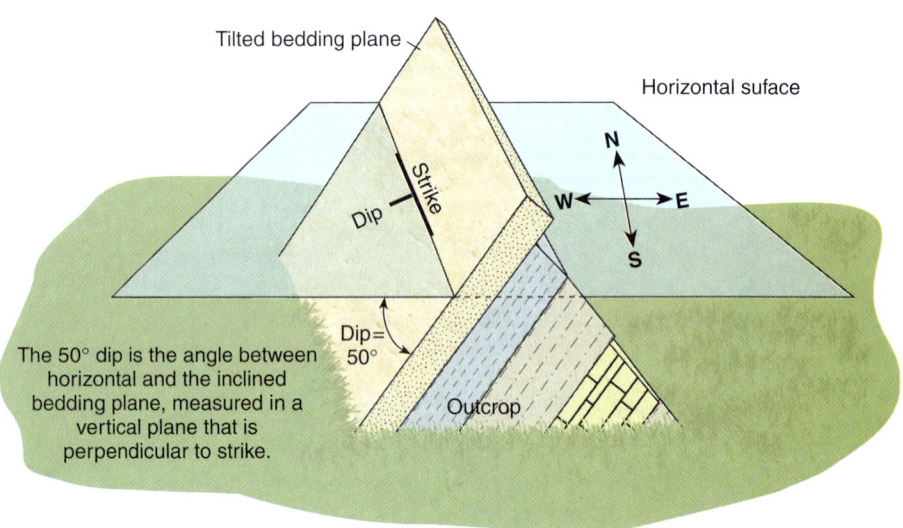

B. Perspective view

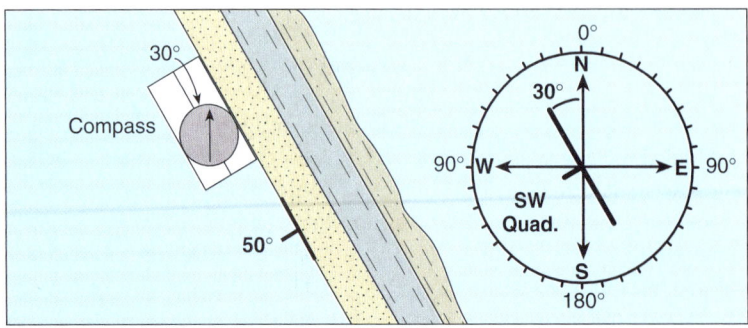

C. Map (overhead) view

FIGURE 14.3
A. Outcrop of Mississippian Bluefield Formation near Covington, Virginia. The sedimentary layers are inclined to the left, or southwest. B. Simplified diagram showing the inclined layers protruding through a horizontal plane. The intersection of the inclined layers with the horizontal plane creates a line that defines the strike of the bed. This line trends 30° west of the north arrow. The beds dip 50° below the horizontal plane. C. When viewed from above, as on a map, the strike and dip of the bed are shown with a T-like symbol. The top of the T parallels the strike, and the leg of the T points toward, and parallels, the dip direction. The dip angle is 50°. Comparison of the strike and dip symbol with the compass directions shows why the attitude of this bed is written as N30°W, 50°SW.

and even their presence, must be inferred from data gathered from many outcrops and plotted on a map.

Describing Surface Orientation

You will see on geologic maps curious little T's called **strike and dip** symbols (Fig. 14.3). These are used to describe the orientation of bedding planes and other planar features so that the overall geometry of tilted and folded layers can be described and visualized on maps. The long top part of the T parallels the **strike,** which represents the intersection of a tilted bedding plane with a *horizontal* plane (Fig. 14.3B). It gives the trend of that plane. For example, if you hold a credit card at a 45° angle to a table and draw a line along where it touches the table, you get the strike of the plane defined by your credit card. Hold the card against the strike line and notice how the strike indicates which way a plane trends, but not which way it tilts. The **dip direction** of a plane is shown by a short line drawn perpendicular to the strike and pointing downhill (Fig. 14.3B). Draw a tick perpendicular to your strike, away from the credit card, and you have the dip direction. Write, for example, 45° (the **dip angle**) at the end of your dip line and you uniquely define the orientation of your credit card. You can easily visualize the orientation of a plane described by a strike and dip symbol by holding your hand parallel to the strike and tilted down toward the dip. Try it on the perspective drawings later in the chapter!

A geologist measures the orientation, or **attitude,** of a tilted bedding plane by holding a compass in a horizontal position, with a straight side against the bedding plane (Fig. 14.3C). While the compass arrow points north, the compass body points in the direction of strike, a certain number of degrees east or west of north. Figure 14.3C shows a strike line pointing 30° west of north, which is written as N30°W. Strike measurements are commonly given with the number of degrees less than 90. Thus, N95°E becomes N85°W, and N90°E is written E-W.

The dip angle, or **angle of inclination,** is measured from the horizontal plane down to the bedding plane (Fig. 14.3B). The measuring device must be held in a vertical plane that is perpendicular to the strike in order to be accurate. Because a dip can point either direction from a strike line, it is necessary when recording dips to specify the dip direction. Since the dip is always perpendicular to strike, it is necessary only to indicate the compass quadrant toward which the surface dips (that is, NW, NE, SE or SW). For example, with a strike of N30°W, the dip direction could either be NE or SW (Fig. 14.3C). In the example used in Figure 14.3B, the complete strike and dip of the bedding plane is written N30°W, 50°SW.

Geologists also measure the attitudes of fault planes, fracture surfaces, and other geologic features. Figure 14.4 shows special symbols for vertical, horizontal, and overturned sedimentary layers, plus symbols for folds and faults. The arrows used in

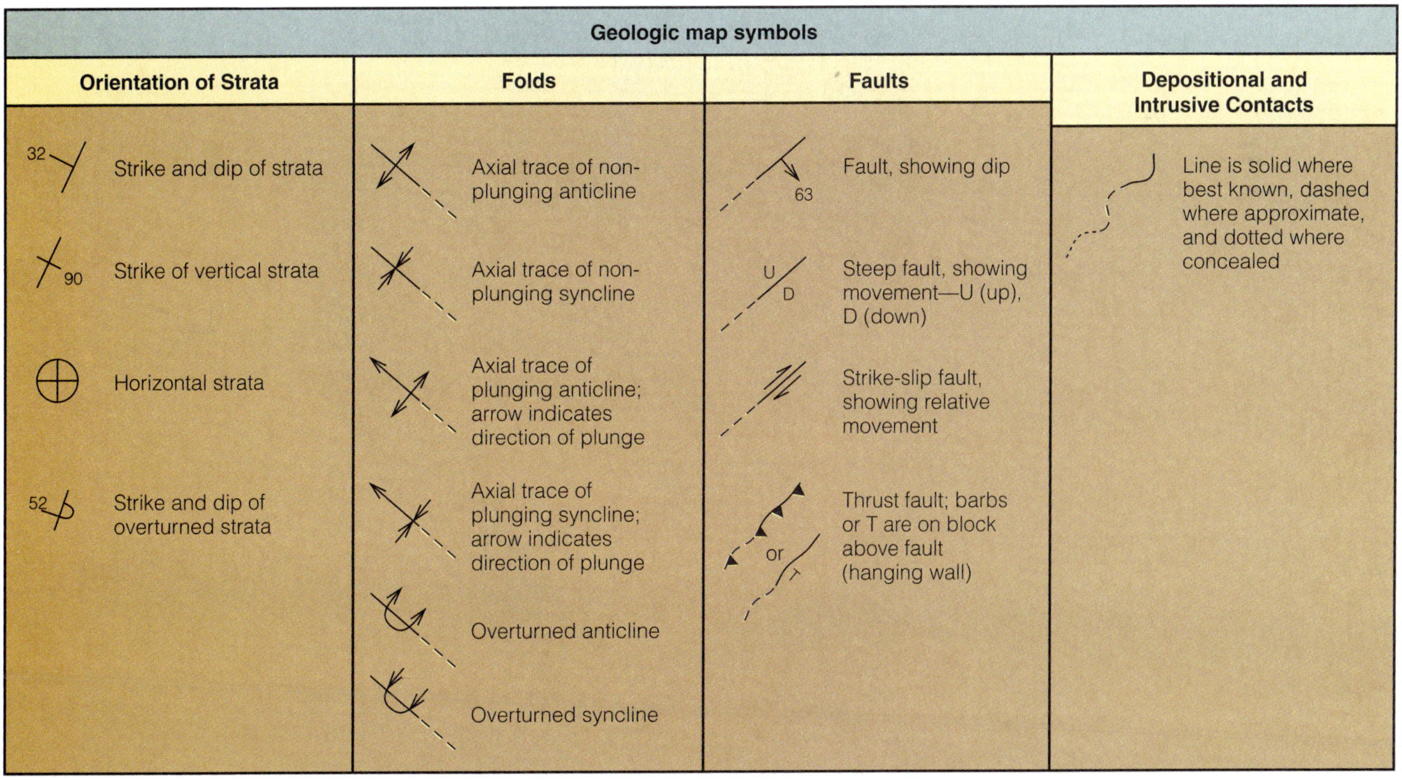

Geologic map symbols			
Orientation of Strata	**Folds**	**Faults**	**Depositional and Intrusive Contacts**
32 — Strike and dip of strata	Axial trace of non-plunging anticline	Fault, showing dip — 63	Line is solid where best known, dashed where approximate, and dotted where concealed
90 — Strike of vertical strata	Axial trace of non-plunging syncline	Steep fault, showing movement—U (up), D (down)	
— Horizontal strata	Axial trace of plunging anticline; arrow indicates direction of plunge	Strike-slip fault, showing relative movement	
52 — Strike and dip of overturned strata	Axial trace of plunging syncline; arrow indicates direction of plunge	Thrust fault; barbs or T are on block above fault (hanging wall)	
	Overturned anticline		
	Overturned syncline		

FIGURE 14.4
Geologic map symbols.
Source: Data from American Geological Institute and others.

the fold and fault symbols point in the direction of dip.

Geologic Maps, Cross Sections, and Block Diagrams

Geologic information is compiled and displayed in three common ways, shown in Figure 14.5: geologic maps, geologic cross sections, and block diagrams.

Geologic maps show the distribution of rock units at or very near the surface of the Earth (Fig. 14.5A). They also portray the structure of the rock units by means of strike and dip symbols. An accompanying legend, or **explanation,** lists and briefly describes the rock units in order of age; major periods of erosion are also listed. Topography may be shown with an overlay of contour lines. The next chapter focuses on geologic maps.

Geologic cross sections are vertical slices that show how the rocks would appear if they could be viewed on a cliff face (Fig. 14.5B). They are topographic profiles with the area below the surface filled in with representations of rocks. Because such cut-away views do not usually present themselves in nature, they are interpretations based on information obtained on the surface or from wells or geophysical data.

Block diagrams combine maps and cross sections to show the three-dimensional aspects (Fig. 14.5C). The three-dimensional blocks are drawn as two-dimensional perspective drawings, with the map view shown on the top and cross sections on the ends and sides. Most block diagrams have a flat upper surface and do not attempt to show topography, but some, drawn by clever people, manage to illustrate topography as well as geology.

Folds

A fold is a bend in a layer of rock due to ductile (plastic) strain. Upward bends, or upfolds, with older rocks in the center, are called **anticlines;** downward bends, downfolds, with younger rocks in the center, are called **synclines** (Fig. 14.6). The sides or flanks of a fold are the **limbs.** The limbs join at the **hinge line** of the fold, which is the line of maximum curvature (Fig. 14.6). The **axial surface** (or **axial plane** if it's not curved) of a stack of folded layers passes through the hinge lines and most nearly divides the fold into two equal parts (Fig. 14.6). An **upright fold** has a vertical

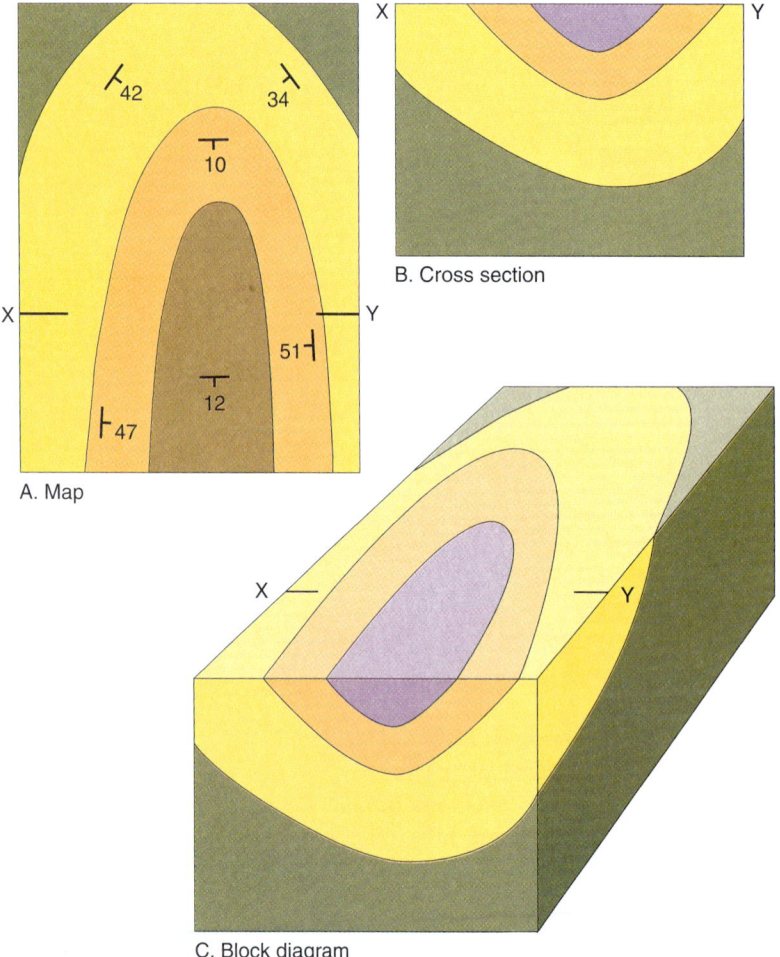

FIGURE 14.5

Three ways in which geologic information is illustrated: A. A geologic map shows rock units and geologic structure as they appear on the surface. You can use your hands to visualize the orientation of the bedding described by the strike and dip symbols. B. A geologic cross section of the map from X to Y is like a vertical slice into the Earth. C. Block diagrams combine map and cross-section views in a three-dimensional perspective drawing.

or nearly vertical axial surface; beds on opposite limbs have similar dips, though in opposite directions (Fig. 14.7A). An **inclined fold** has an axial surface that is neither vertical nor horizontal (Fig. 14.7B). An inclined fold is **overturned** if beds on opposite limbs dip in the same direction; beds on the overturned limb are upside-down, as they have been rotated more than 90° (Fig. 14.7C). A **recumbent fold** has an approximately horizontal axial surface (Fig. 14.7D).

A **non-plunging fold** has a horizontal or nearly horizontal hinge line (Figs. 14.8A, B). A **plunging fold** has an inclined hinge line (Figs. 14.8C, D).

Figure 14.9 illustrates folds using block diagrams in which the top surface is horizontal and flat. Note that when non-plunging folds intersect a horizontal

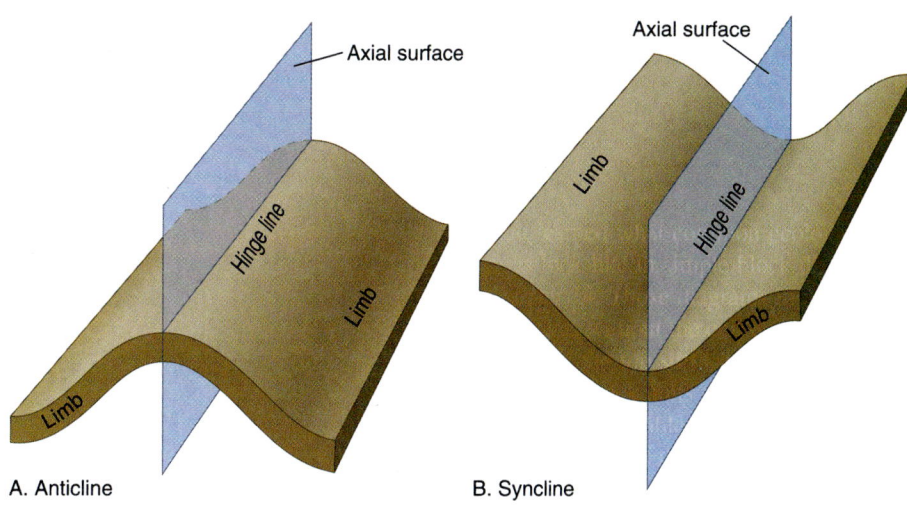

FIGURE 14.6
Folds are described in terms of limbs, hinge line, and axial surface, shown here for an anticline (A) and a syncline (B).

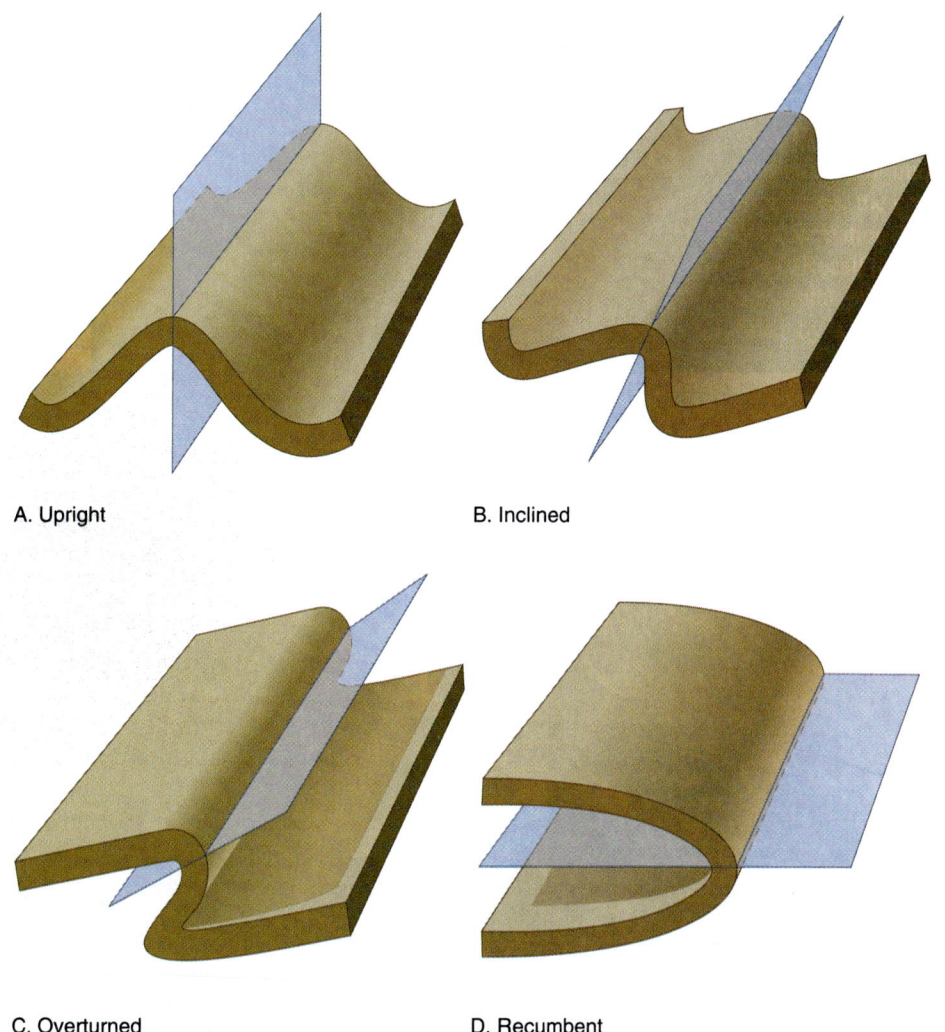

FIGURE 14.7
The axial surface of a fold can be: A. Vertical in upright folds; B. inclined in inclined folds; C. inclined so much that opposite limbs dip in the same direction in overturned folds; D. horizontal in recumbent folds.

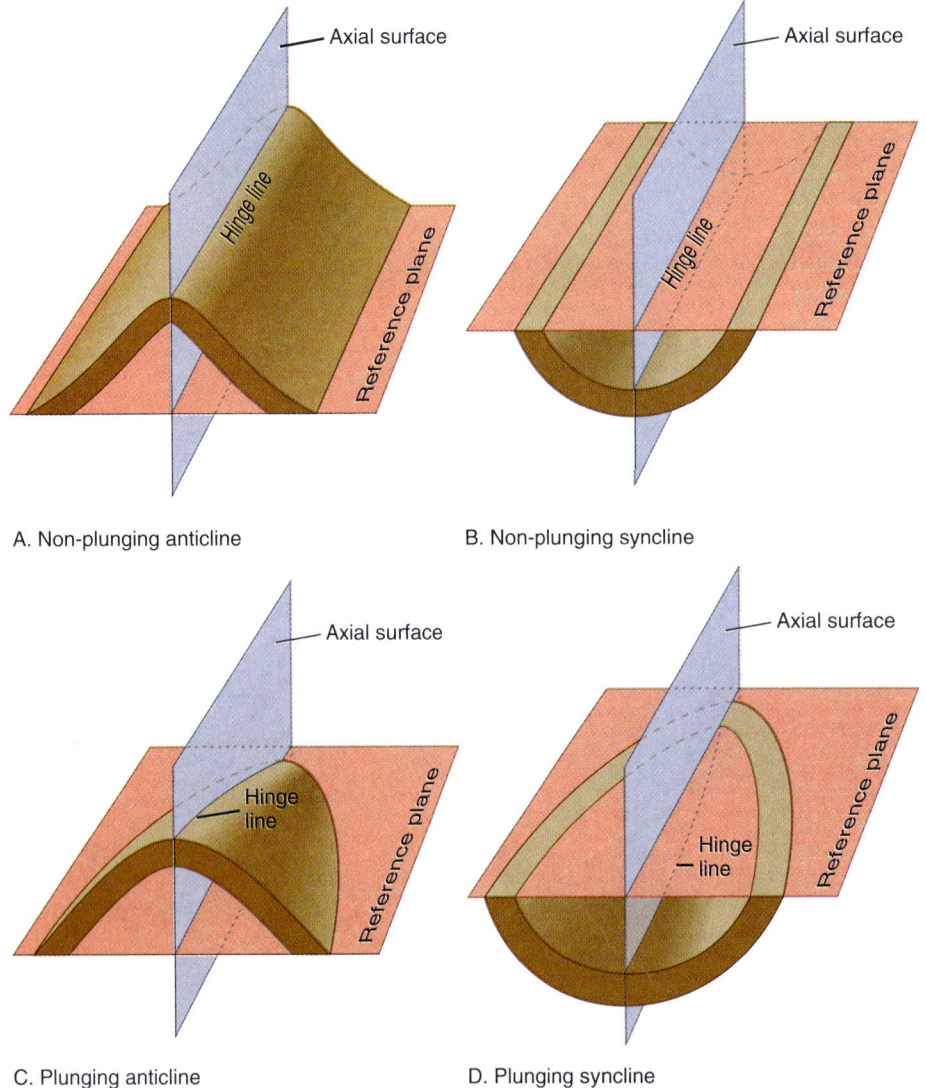

FIGURE 14.8
Non-plunging folds have horizontal hinge lines, as in a non-plunging anticline (A) and a non-plunging syncline (B). Plunging folds have inclined hinge lines, as in a plunging anticline (C) and a plunging syncline (D). Axial surfaces are vertical, and reference plane is horizontal.

surface, such as that approximated by the surface of the Earth, the contacts between adjacent beds are straight lines (Figs. 14.9A, B). When plunging folds intersect a horizontal surface, the contacts between adjacent beds are curved lines. In a plunging anticline, the contacts bend so that they "point" in the direction of plunge, whereas in a plunging syncline, they "point" in the direction opposite the plunge (Figs. 14.9C, D).

Using the Principle of Superposition (Chapter 13), you can see that in an eroded syncline, the *youngest* beds appearing on the surface are in the center of the syncline (Figs. 14.9B, D). In an eroded anticline, the *oldest* (originally deepest) beds are in the center (Figs. 14.9A, C).

Domes and **basins** have cross sections like anticlines and synclines, respectively, but are approximately circular or oval in map view (Fig. 14.10). Beds in domes dip outward in all directions, and like anticlines, the oldest rocks are found in the center. Similarly, the youngest rocks in basins are in the center, just as they are in synclines, and beds dip inward. Domes and basins may be quite large, with diameters of 100 km or more.

Faults

Faults are breaks or fractures in the Earth's crust along which movement has occurred. The rocks on one side of a fault have moved relative to those on the other side. Relative movement may be up and down, sideways, or a combination of the two. The amount of movement ranges from centimeters to kilometers.

Faults with inclined fault planes have **footwalls** and **hanging walls** (Fig. 14.11). The **footwall** is under the fault plane: if you dug a mine along the fault plane, you would stand on the footwall. The **hanging wall** is above the fault plane: in a mine, it would hang above your head.

Faults are classified according to their relative motion. When the hanging wall has dropped relative to the footwall, it is a **normal fault** (Fig. 14.11A). When the hanging wall has been pushed up relative to the

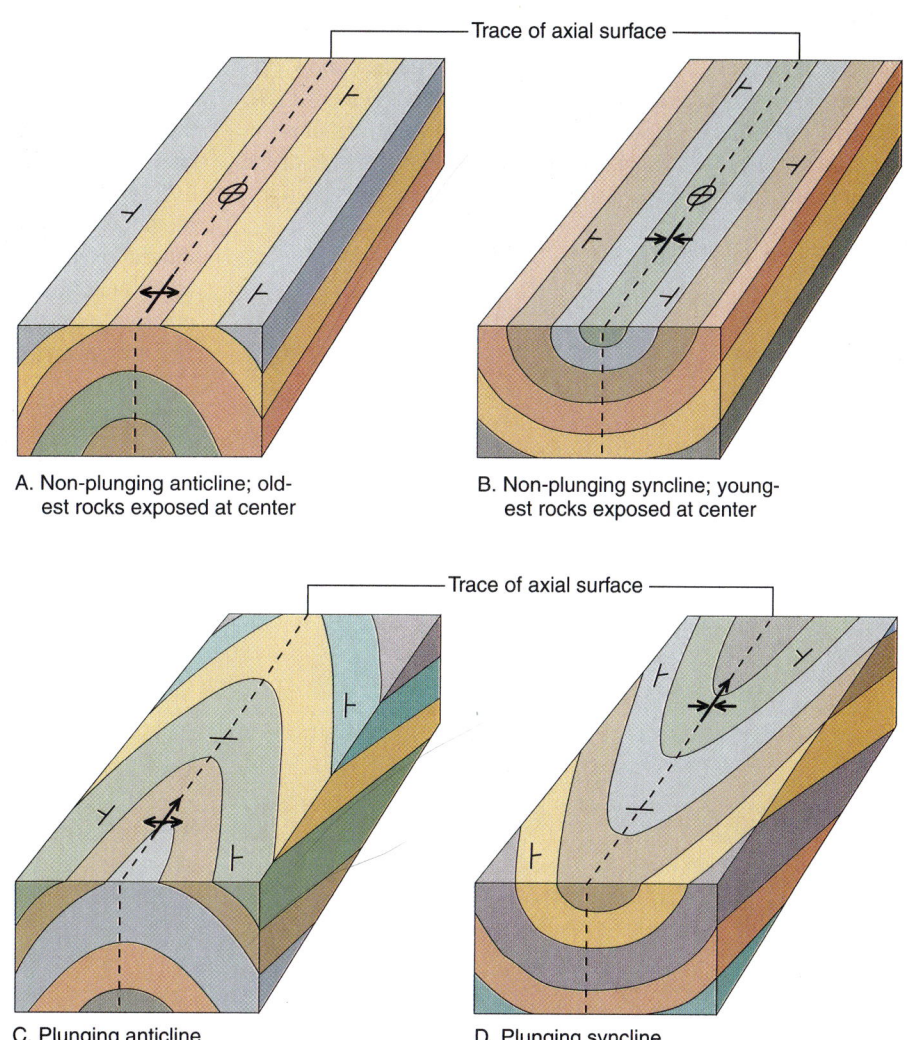

FIGURE 14.9
The top surface of these block diagrams, which is analogous to the Earth's surface, shows that older beds are exposed in the centers of eroded anticlines and younger beds are exposed in the centers of eroded synclines. Contact lines on the top surface are straight if folds do not plunge, but bend if they do. In plunging anticlines, the contact lines bend and point in the direction of plunge; in plunging synclines, they bend and point away from the direction of plunge. In C and D, both folds plunge into the page. The symbols on the top surface show the strike and direction of dip at various locations and the axial traces and type of folds. Note how these symbols help you visualize structure from map view alone.

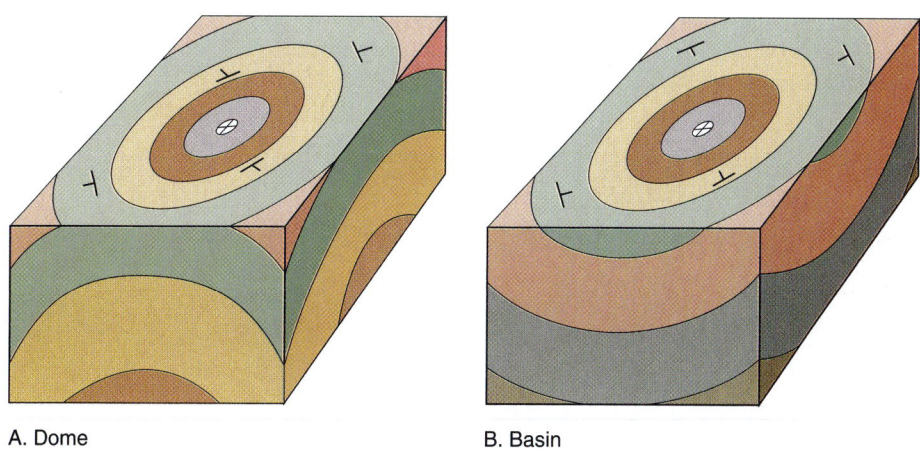

FIGURE 14.10
These block diagrams show that older beds appear in the center of an eroded dome, while younger beds appear in the center of an eroded basin. The symbols on the top surfaces show the strike and direction of dip at various locations.

footwall, it is a **reverse fault** if the fault plane is steeply dipping (more than 45°; Fig. 14.11B) or a **thrust fault** if the fault plane has shallow dip (less than 45°; Fig. 14.11C). **Lateral** or **strike-slip faults** have almost vertical fault planes and show offsets in a horizontal plane (Fig. 14.11D). *Right-lateral strike-slip faults* are those in which the block on the opposite side of the fault has moved relatively to the right.

Thus, if you stood on one side of a right-lateral strike-slip fault, and looked across the fault, the rocks on the other side would be displaced to the right, as in Figure 14.11D. In *left-lateral strike-slip faults,* the block on the opposite side of the fault has moved relatively to the left. Whereas the other faults are most obvious in vertical outcrops, offsets along strike-slip faults are best seen from the air.

Clues to the Past

Folds and faults record the history of geologic stresses in a given area. Such stresses may result, for example, from continental blocks pulling apart, sliding by one another, or colliding. Table 14.2 summarizes the types of stress and strain that are recorded by folds and the different types of faults. Assessing the type of stress and the directions from which it was applied helps us to unravel the long and complicated series of events that have shaped and reshaped large areas of especially eastern, southern, and western North America.

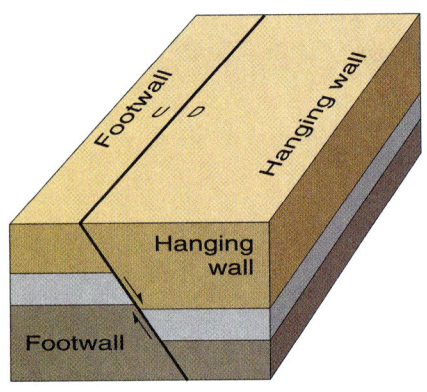

A. Normal fault

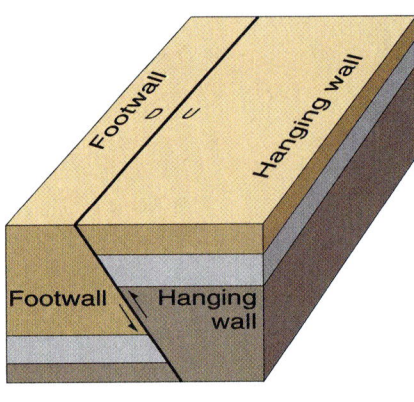

B. Reverse fault

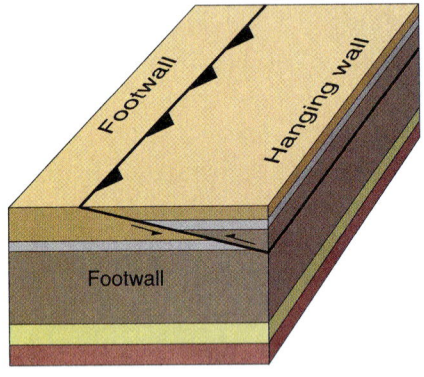

C. Thrust fault

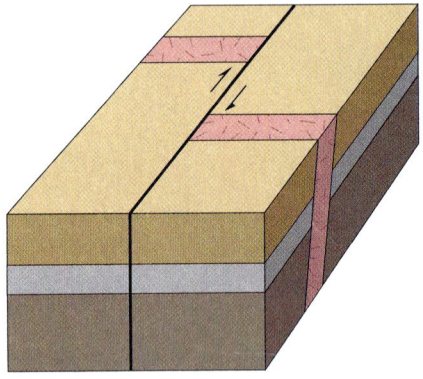

D. Right-lateral strike-slip fault

FIGURE 14.11
Four types of faults are illustrated in these block diagrams. If the fault plane is inclined, the side above the fault is the hanging wall, and the side below is the footwall. Displaced marker beds show the direction of movement. Note that in a map view, shown by the top surfaces of A, B, and C, the faults dip under the hanging wall, and in C the barbs or "teeth" are on the hanging wall.

TABLE 14.2
Stress/Strain Conditions of Common Geologic Structures

Strain	Stress		
	Compression	*Tension*	*Shear*
Brittle strain or rupture	Reverse fault or thrust fault	Normal fault	Strike-slip fault
Ductile or plastic strain	Folds (anticlines and synclines)	Not discussed in text	Not discussed in text

Hands-On Applications

If geologists are to understand a dynamic Earth, they must understand the nature of the stresses within the Earth. One way to learn about stress is to study the strain, or deformation, that it has produced. This type of study is based on descriptive and interpretive analysis of folds, faults, and other deformational features. Using drawings and three-dimensional models, this lab focuses mainly on the geometric aspects of deformed rocks. You will learn how observations and measurements made at the surface are used to predict the structure of rocks in the subsurface.

Objectives

If you complete all the problems, you should be able to:

1. Distinguish between elastic and plastic strain, and list the conditions that might favor one over the other.
2. Recognize structures formed by compression, tension, and shear.
3. Identify conformable and unconformable depositional contacts, intrusive contacts, and fault contacts on a geologic map, cross section, or block diagram.
4. Define dip and strike, plot them on a map, and determine their general orientation on a geologic map or top surface of a block diagram.
5. Define, sketch, and recognize—on a geologic map or a block diagram—a dome or basin and plunging and non-plunging anticlines and synclines.
6. Determine the direction of plunge of plunging anticlines and synclines on a geologic map or block diagram.
7. Define, sketch, and recognize—on a cross section or block diagram—normal, reverse, thrust, and strike-slip faults.
8. Distinguish the hanging wall and footwall of a normal, reverse, and thrust fault on a cross section, geologic map, or block diagram.

Problems

1. To demonstrate how the same material can respond to stress (deform) differently under different conditions, try the following with Silly Putty™. First, roll it into a ball and see if it bounces. Next, roll it into a thick cigar shape and slowly stretch it out. Finally, roll it into a thick cigar shape again and pull it apart as fast as you can. In each case, the Silly Putty behaves differently.

    a. Name the type of behavior exhibited in the three cases, using the terminology given in the introduction to this lab.

    b. How did the conditions under which deformation occurred differ in the three cases?

Your instructor may want you to do either Problem 2 (cut-out blocks) or Problems 3 and 4 (perspective drawings).

2. The following problems use the cut-out block diagrams A, B, and C, which are found at the end of the lab manual. The blocks should be cut out before beginning the problem in which they are used. When constructing the box, be very careful to make the folds exactly along the lines, or the problems may not work out as they should. It will be easier to draw on the boxes if you do not tape the sides together until after you have completed the problem. Even then, it is recommended that you use removable tape.

    a. Refer to Block A with Side 1 up. When Side 1 is up, Block A is similar to Figure 14.9A, and the top surface represents the Earth's surface.

       (1) Fill in the blank faces on the sides of the block.

       (2) Measure the dip with a protractor at several places on the front surface (note that the dip changes below the surface and that the angle you should measure is right at the top surface) and put strike and dip symbols at the places indicated with dots on the top (map) surface.

258   Part VI   Internal Processes

- (3) Illustrate the position of the axial surface of the fold by sketching its trace on the top and ends of the block (the **trace** is the imaginary line formed where the axial surface intersects other surfaces).

- (4) When Side 1 is up, what is the age of the beds in the center, or core, of the fold relative to the other beds?

- (5) What kind of fold is this (anticline or syncline; upright, inclined, overturned, or recumbent; plunging or non-plunging)?

b. Refer to Block A with Side 2 up. This is a new situation in which Side 2 now represents the Earth's surface.

- (1) Put strike and dip symbols at the places indicated with dots on the top (map) surface.

- (2) When Side 2 is up, what is the age of the beds in the center, or core, of the fold relative to the other beds?

- (3) What kind of fold is this?

- (4) Sketch the trace of the axial surface (axial trace) on the top of the block using one of the fold symbols in Figure 14.4.

c. Refer to Block B with Side 1 up. When Side 1 is up, Block B is similar to Figure 14.9C, and the top surface represents the Earth's surface.

- (1) Fill in the blank faces on the side of the block.

- (2) Put strike and dip symbols at the places indicated with dots on the top (map) surface.

- (3) Illustrate the position of the axial surface of the fold by sketching its trace on the top and ends of the block.

- (4) When Side 1 is up, number the beds from oldest (number 1) to youngest. Where are the oldest beds as seen on the top (map) surface?

- (5) What kind of fold is this?

d. Refer to Block B with Side 2 up. This is a new situation in which Side 2 now represents the Earth's surface.

- (1) Put strike and dip symbols at the places indicated with dots on the top (map) surface.

- (2) When Side 2 is up, what is the age of the beds in the center, or core, of the fold relative to the other beds?

- (3) What kind of fold is this?

- (4) Sketch the trace of the axial surface (axial trace) on the top of the block using one of the fold symbols in Figure 14.4.

e. Refer to Block C, with Side 1 up.

- (1) Complete the sides of the block.

- (2) Put strike and dip symbols of beds at the dots on the top (Side 1) surface.

- (3) Put a dip symbol (Fig. 14.4) on the fault (dark line).

- (4) Indicate the relative movement of the fault by placing arrows on opposite sides of the fault on the side faces.

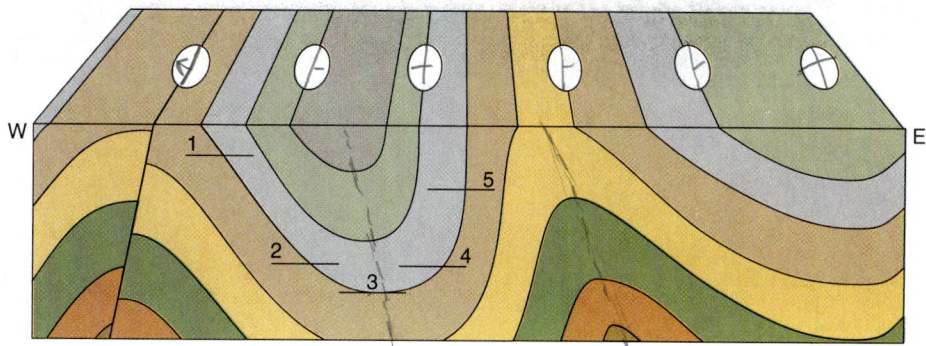

FIGURE 14.12
Block diagram to be completed for Problem 3.

(5) Label the hanging wall and footwall on the top and side faces.

(6) With Side 1 up, number the beds in order of age, with 1 the oldest. Are the beds at the surface of the upthrown block generally older or younger than those at the surface of the downthrown block? Why?

(7) What kind of fault is this?

3. Figure 14.12 is a block diagram showing a fault and a number of folds. The front face of this diagram is perpendicular to the strike of these features, so you can easily measure the dip of the fault and bedding planes where indicated. North is into the page, west and east are marked.

  a. Use a protractor to measure the dip angles along a single fold at points 1 through 5 on the front face of the diagram. Horizontal lines indicate the precise location for each measurement.

  1. 52   2. 42   3. 0   4. 55   5. 88

  b. A geologist would take strike and dip measurements at convenient outcrops on the surface. Take dip measurements along the front of the block diagram, as close to the surface as possible. Put the appropriate strike and dip symbols in the six white ovals along the top of the diagram, and write each dip measurement in the appropriate spot.

  c. Draw dashed axial traces for the three folds not cut by a fault. Put them along both the vertical and horizontal surface, and include the appropriate symbol from Figure 14.4.

  d. What type of fault is illustrated? Be sure the correct map symbol is used to indicate dip. Put arrows on the vertical surface and map symbols on the horizontal surface to indicate the relative sense of motion across the fault. Reverse

4. Partially completed block diagrams are shown in Figure 14.13. Using the information provided for each diagram, fill in the blank faces on the blocks and answer the questions.

  a. The vertical column on the front face of Figure 14.13A shows the different rock units encountered in an old oil well.

  (1) Use the available information to help you complete the diagram.

  (2) What kind of fold is this (anticline, syncline; upright, inclined, overturned, or recumbent; plunging or non-plunging)? Anticline Non plunging

  (3) Illustrate the position of the axial surface of the fold by sketching its trace on the top and end of the block (the **trace** is the imaginary line formed where the axial surface intersects another surface).

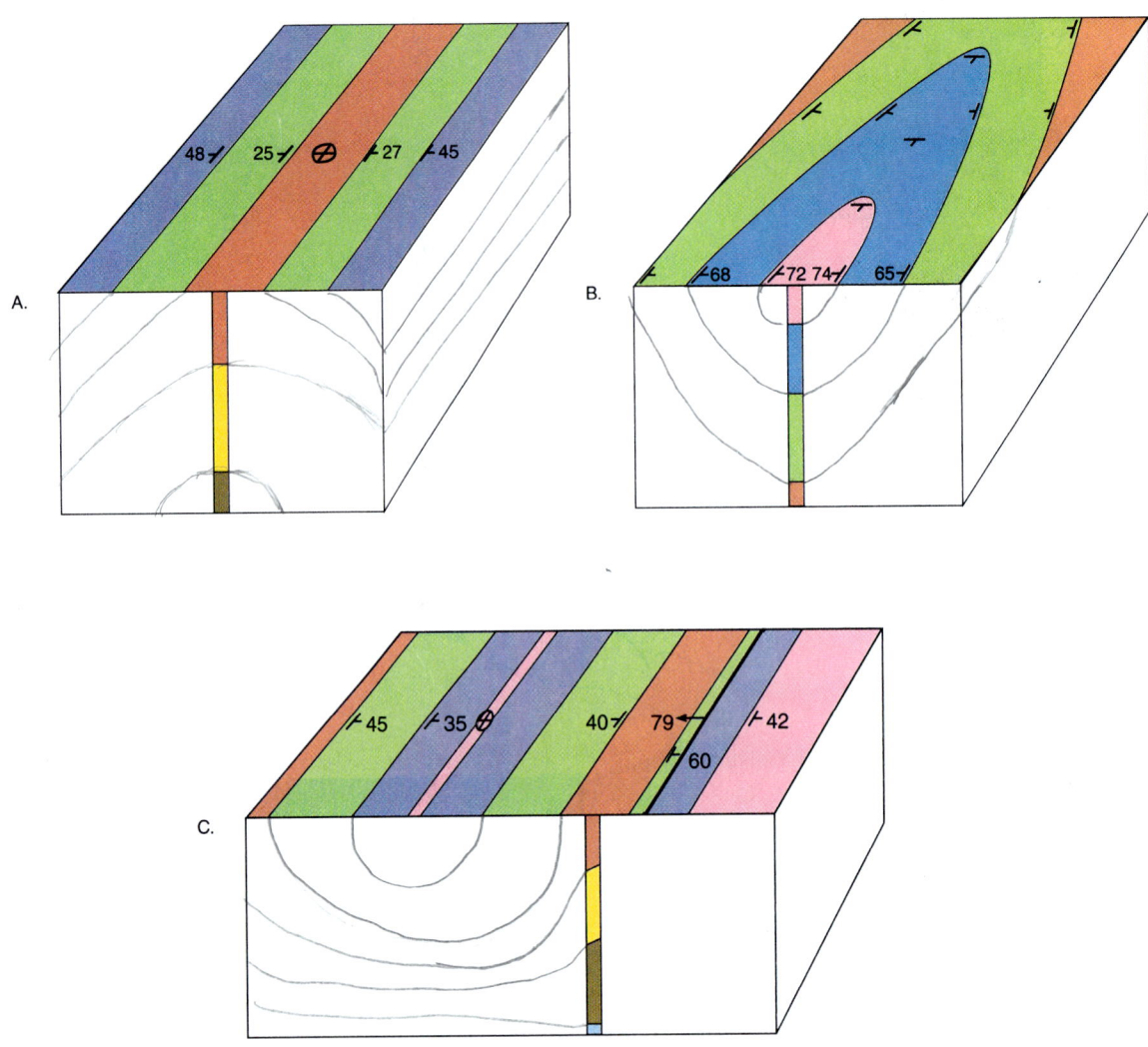

FIGURE 14.13
Block diagrams to be completed for Problem 4.

b. Again, the vertical column on the front face of Figure 14.13B shows the different rock units encountered in an old water well.

  (1) Use the available information to complete the front and as much of the side face *as possible*. (Don't go beyond what the observations allow!)

  (2) Much of the side face is blank even after using the information provided. Propose a site for another well that would provide the maximum information about that blank side, and fill out the side assuming that two new units were encountered in your well.

  (3) What kind of fold is this (anticline, syncline; upright, inclined, overturned, or recumbent; plunging or non-plunging)? *Plunging Syncline*

  (4) Illustrate the position of the axial surface of the fold by sketching its trace.

c. The vertical column on the front face of Figure 14.13C shows the different rock units found in a recently drilled oil well. A fault is shown on the top face.

  (1) Use the available information to help you complete the diagram.

  (2) Label any folds on the diagram using the usual terminology.

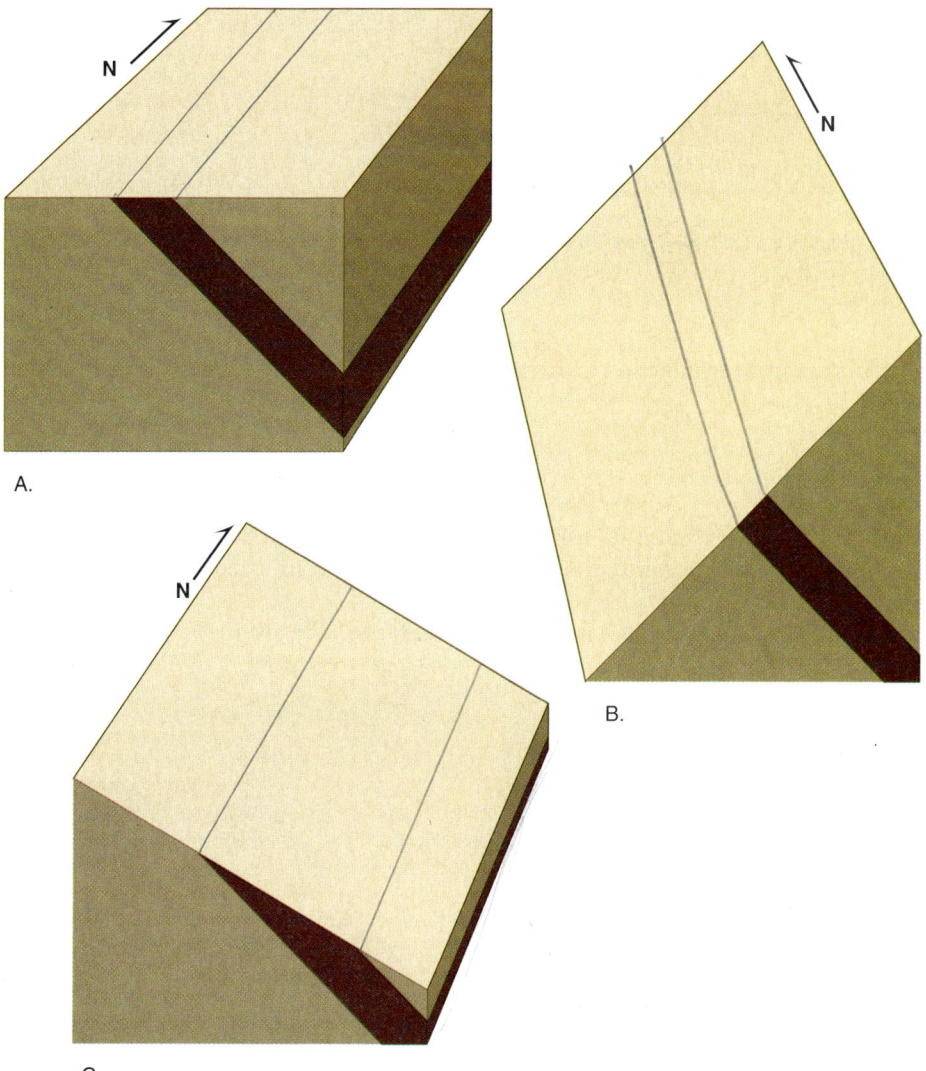

FIGURE 14.14
These block diagrams for Problem 5 show a bed dipping at 45°E, but from slightly different perspectives (note north arrows). A. The upper surface is horizontal, so it intersects the bed at 45°. B. The upper surface slopes at 45°W, so it intersects the bed at 90°. C. The upper surface slopes 30°E, so it intersects the bed at 15°.

(3) Indicate the sense of motion along the fault by putting arrows on the vertical surface. Which is the hanging wall and which the footwall?

(4) What type of fault is indicated (normal, reverse, thrust, strike-slip)?

5. You may have noticed on the various faces of the cut-out models or on the block diagrams of Figure 14.9 that the width of a given layer changes depending on the angle at which the layer intersects the face. This problem is intended to show you why. The three block diagrams in Figure 14.14 show a north-south-striking bed of the same thickness dipping at 45°E. In addition to being drawn from slightly different perspectives, the diagrams also differ because in A the upper surface is horizontal and intersects the bed at 45°, in B the top surface slopes 45°W so that it intersects the bed at 90°, and in C it slopes 30°E so that it intersects the bed at 15°.

   a. Complete the upper surface of the diagrams by extending the bed across that surface. The resulting band on the surface represents the outcrop of that bed.

b. Measure the following with a ruler (the outcrop width is best measured at the top of the front face of the box, where the front face intersects the upper surface):

   (1) width of the outcrop on A; 6 mm

   (2) width of the outcrop on B; 5 mm

   (3) width of the outcrop on C; 23 mm

   (4) the true thickness of the bed.

c. Why does the width of the outcrop differ in each of the diagrams when the true thickness of the bed is the same?

   Because of the angles

6. Partially completed block diagrams are shown in Figure 14.15. Complete each of these by filling in the blank faces. Place general dip and strike symbols on the top surface if not already present. Lowest numbers correspond to oldest rocks.

   a. In Figure 14.15A, what name is given to the uneven feature that runs diagonally across the map surface? Erosional Surface

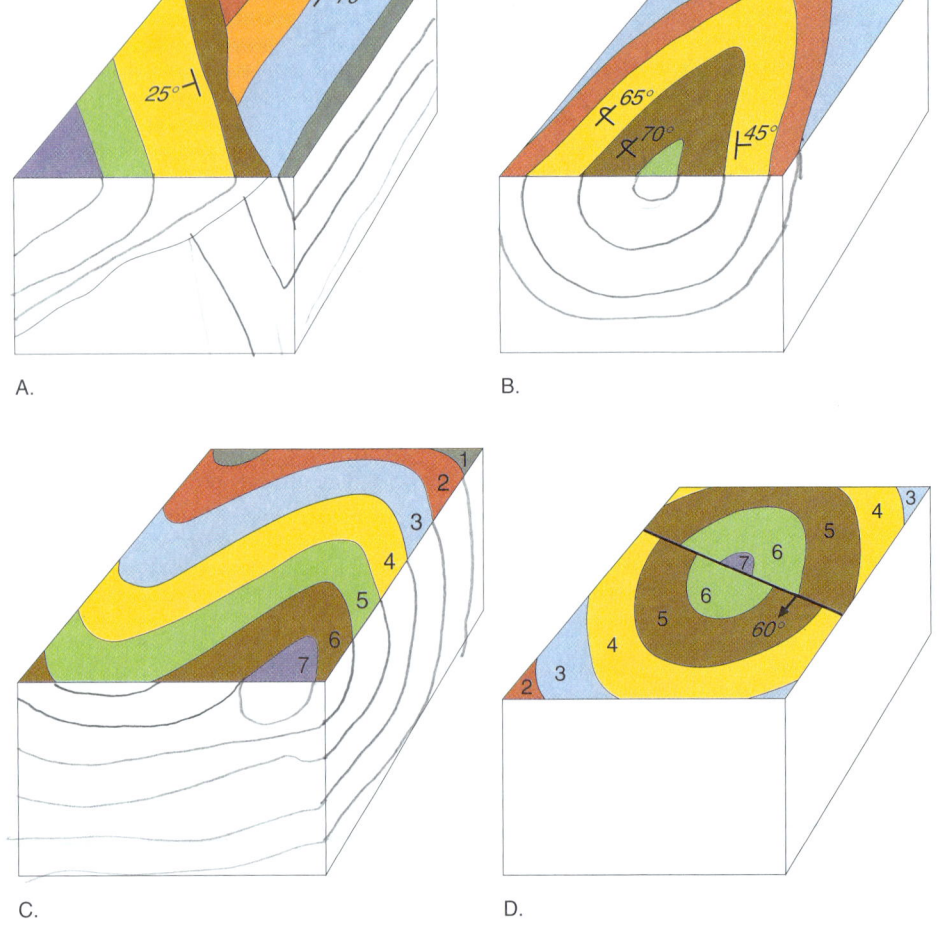

A.   B.   C.   D.

FIGURE 14.15
Block diagrams to be completed for Problem 6. Lowest numbers are oldest units.

b. What structure is shown in Figure 14.15B?

c. Describe the structure shown in Figure 14.15C.

d. What kind of fault is shown in Figure 14.15D, and what kind of structure does the fault cut?

7. Fill in the blanks to formulate a series of rules to help in the interpretation of geologic maps and cross sections.
   a. The following rules refer to relative ages of sedimentary beds. Choose *older* or *younger* for your answers.
      - When anticlines and domes are eroded, the beds exposed in the center are __older__ than the beds exposed away from the center.
      - When synclines and basins are eroded, the beds exposed in the center are __younger__ than the beds exposed away from the center.
      - On a geologic map, __younger__ beds dip under (or toward) __older__ beds, unless they are overturned. (Hint: Look at the upper surfaces of your cut-out block diagrams or Figure 14.9.)
   b. The following rules refer to beds and contacts between beds, as seen in map view.
      - When __non-plunging__ folds have been eroded, the contacts between beds, as seen on a horizontal surface, are straight and parallel. Choose *plunging* or *non-plunging*.
      - When plunging anticlines have been eroded, the contacts between beds, as seen in map view, bend so as to point __away__ the direction of plunge. Choose *toward* or *away from*.
      - When plunging synclines have been eroded, the contacts between beds, as seen on the surface, bend so as to point __towards__ the direction of plunge. Choose *toward* or *away from*.
      - The outcrop width of a bed __increases__ as the angle between the surface and the dip of the bed decreases. Choose *increases* or *decreases*.

8. Many state geological surveys post geologic and other maps on the Web. Go to www.ohiodnr.com/geosurvey/ (or link to it through www.mhhe.com/jones5e—see Preface) and link to *Geology of Ohio*, then *Maps*, and finally *The Old State Bedrock Map*. Because of the scale of the map and the low relief of Ohio, the map is comparable to the top surface of a block diagram. As indicated by the explanation, the rocks are grouped into the geologic time periods in which they were deposited. If you've forgotten the relative ages of geologic periods, consult Figure 13.7.
   a. Where are the youngest units in the state and what is their geologic age?

   Where are the oldest units in the state and what is their geologic age?

   b. Which way do the rock units in the eastern half of the state dip and how do you know?

   Which way do the rock units in the westernmost part of the state dip and how do you know?

   c. What is the large structural feature centered in the western half of the state and how do you know?

In Greater Depth

9. The following problems refer to cut-out block diagram D, found at the end of the lab manual.

   a. Complete the following:
      (1) Fill in the blank side panels.

      (2) Put strike and dip symbols at the places indicated with dots on the Side 1 surface.

      (3) Measure the dip with a protractor on:
         (a) the small, already filled-in, front panel;

         (b) the two corner panels that you filled in;

         (c) the side surface. The small front panel is perpendicular to strike, so the dip measured there is the *true dip;* the other surfaces are not perpendicular to strike, so the dips observed and measured there are not true dips but *apparent dips.*

      (4) Fill in Side 2.

      (5) Imagine that the top half of the box (either side up) was eroded away and you could see the new surface exposed. In what direction, in terms of strike or dip, would the outcrops have shifted from their pre-erosion positions?

   b. Fill in the blanks to formulate additional rules to help in the interpretation of geologic maps and cross sections.
      - The apparent dip of beds seen on vertical surfaces _____ as the angle between the strike of the beds and the strike of the vertical surface decreases (choose *increases* or *decreases*); the true dip can be seen only on vertical surfaces that are _____ to the strike of the beds (choose *parallel* or *perpendicular*).
      - As erosion continuously lowers the level of the land surface, a dipping bed or fault, as seen in map view, migrates _____ the direction of dip (choose *in* or *away from*).

10. The three cross sections in Figure 14.16 show three outcrops of the same graywacke (sandstone) unit, all of which have beds that dip about 60° to the west. The surrounding rocks, probably shale, are easily eroded and are covered with soil.

    a. Formulate three hypotheses to explain the relations among the outcrops by completing the cross sections in three different ways. Use dashed lines to indicate contacts above the cross section that have been eroded away.

    b. What kind of information might the three outcrops contain that would enable you to distinguish among your multiple hypotheses? (Hint: see *Sedimentary Structures* in Chapter 4.)

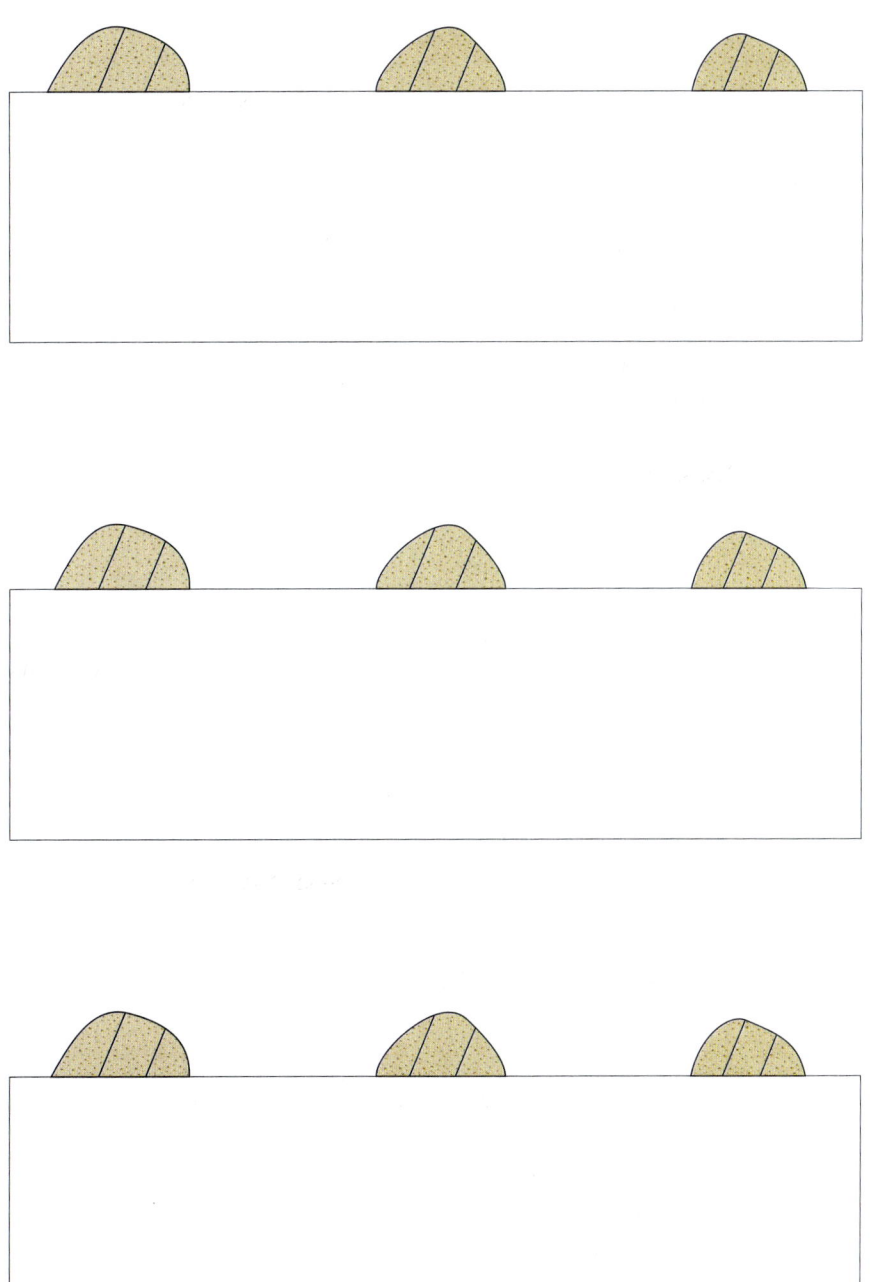

FIGURE 14.16
Cross sections for Problem 10.

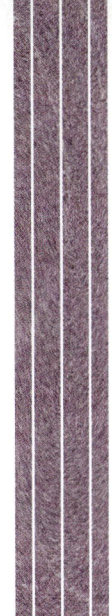

CHAPTER 15

Geologic Maps

Materials Needed
- Pencil and eraser
- Ruler
- Protractor
- Calculator
- Stereoscope (provided by instructor)

Introduction

Geologic maps are a visual expression of the basic field data of geology. A good geologic map shows not only what kinds of rocks are present, but also their ages and structure. Because geologic maps are two-dimensional views of three-dimensional rock sequences, they can appear quite complex. However, simple observations can help you understand what is shown on a geologic map. You will learn: how geologic maps are made; how to recognize whether rock formations are horizontal, inclined, folded, or faulted, even if dip and strike symbols are absent; how to determine the chronologic order of geologic events; and how to visualize the third dimension by drawing a geologic cross section.

A geologic map shows the distribution and structure of rock units on or very near the Earth's surface. An accompanying *explanation* lists the rock units present and explains special symbols. Geologic maps commonly are drawn on contour maps so that geology and topography are easily compared.

The rock units shown on geologic maps are usually *formations,* although in some cases, other types of rock units or time-rock units (rocks formed during a specific time interval) are shown. A **formation** is a mappable body of rock. That is, it has an appearance or composition that distinguishes it from other nearby rocks, plus distinct boundaries (**contacts**) that make it easy to separate from adjacent rock units. Formations have two-word names. The first word normally refers to the place where the formation originally was described. If the formation consists principally of one type of rock, the rock name is used as the second part; if not, the word "formation" is used. Examples: St. Peter Sandstone, Trenton Limestone, Morrison Formation.

Making a Geologic Map

Making a geologic map requires:

1. collecting the basic data in the field;
2. correlating among outcrops;
3. making a geologic cross section of the area to show the three-dimensional relation of one rock unit to another; and
4. preparing an *explanation.*

Collecting the basic data is by far the most important and time-consuming task —and the most fun. Exposures, or **outcrops,** of bedrock are located by studying aerial photographs and by walking the countryside to locate where the rock actually can be seen. In the western United States, rock is exposed over large areas. In more humid areas, like the U.S. Midwest, outcrops are commonly small and widely scattered; natural outcrops commonly are found in stream valleys, but rock is also exposed in roadcuts, mines, quarries, and construction sites. The outcrops are examined in the field and plotted on a topographic map or an aerial photograph. A description is recorded on the map or in a field notebook and includes such information as rock type, formation name, strike and dip of beds, faults, folds, or other structural features, fossils present, and anything else deemed important by the geologist. Figure 15.1A is an example of a map showing outcrops; the area between outcrops is covered with soil.

The next step is correlation among outcrops; this is usually done in the field. Notice in Figure 15.1A that strikes and dips are similar in all outcrops and that the sandstone labeled *Pc,* in particular, occurs in a series of outcrops strung out in the same orientation as the strike. The simplest and most reasonable assumption to make is that the four outcrops of this sandstone belong to the same formation and should be

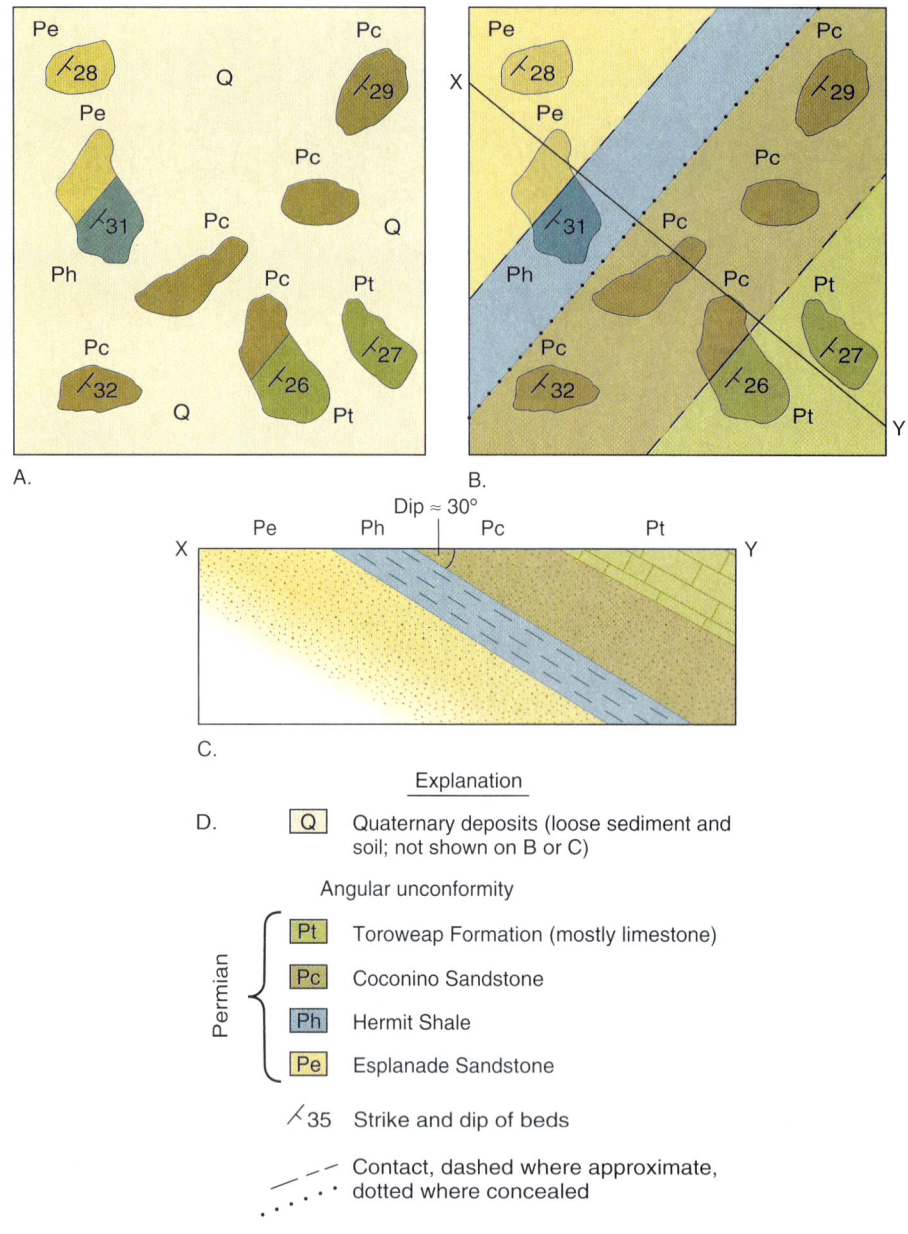

FIGURE 15.1
Making a geologic map. See text for discussion. A. Outcrop map. B. Contacts drawn in. C. Cross section along line *X-Y*. D. Explanation, with rock units and angular unconformity listed chronologically.

correlated, as illustrated in Figure 15.1B. Identical contacts are joined from one outcrop to another, using dashed lines to indicate they are not actually visible. In some places, no contacts are actually exposed, but their presence can be inferred. For example, say one outcrop is sandstone and an adjacent one is shale; the two must be in contact someplace, but the contact is buried. In such places, a "best-guess" dotted line is drawn to represent the contact (Fig. 15.1B).

One of the best ways to visualize what is shown on the map, and to double-check an interpretation, is to draw one or more geologic cross sections. The cross section in Figure 15.1C was drawn along the line *X-Y* on the map. That particular cross section was chosen because it best illustrates the overall structure. Note that the dip and strike symbols tell the angle and direction in which the layers are inclined at the surface. Correlation of the units below the surface completes the structure. On many maps, the Quaternary units form a thin layer of soil and loose sediment resting on the underlying bedrock. In Figure 15.2, they are sediments deposited in a river valley by a river. Because they are often thin and unrelated to the main geologic structures, they are often not drawn on cross sections (Fig. 15.1).

Finally, rock units and unconformities are listed, with the oldest at the bottom and the youngest at the top, in the *Explanation* (Fig. 15.1D).

Appearance of Geologic Features from the Air and on Maps

You learned what common geologic structures look like in simple block diagrams in Chapter 14. Those diagrams illustrate ideal structures without the complexity of topography on the top surface. Even so, outcrop patterns of rock units seen on maps or from the air do resemble what you saw on the upper surface of the block diagrams. Hills and valleys complicate the smooth and regular appearance of formation contacts somewhat, but the basic pattern is there. The purpose of this section is to help you interpret what you see from an airplane or on a geologic map.

Horizontal Strata

If horizontal strata are exposed on a horizontal surface, only one formation will be seen. However, if the surface is not flat and erosion has cut into two or more horizontal formations, the contact lines between adjacent formations will parallel the contour lines. (Recall that contour lines show where horizontal planes, such as potentially defined by a rising sea level, intersect the

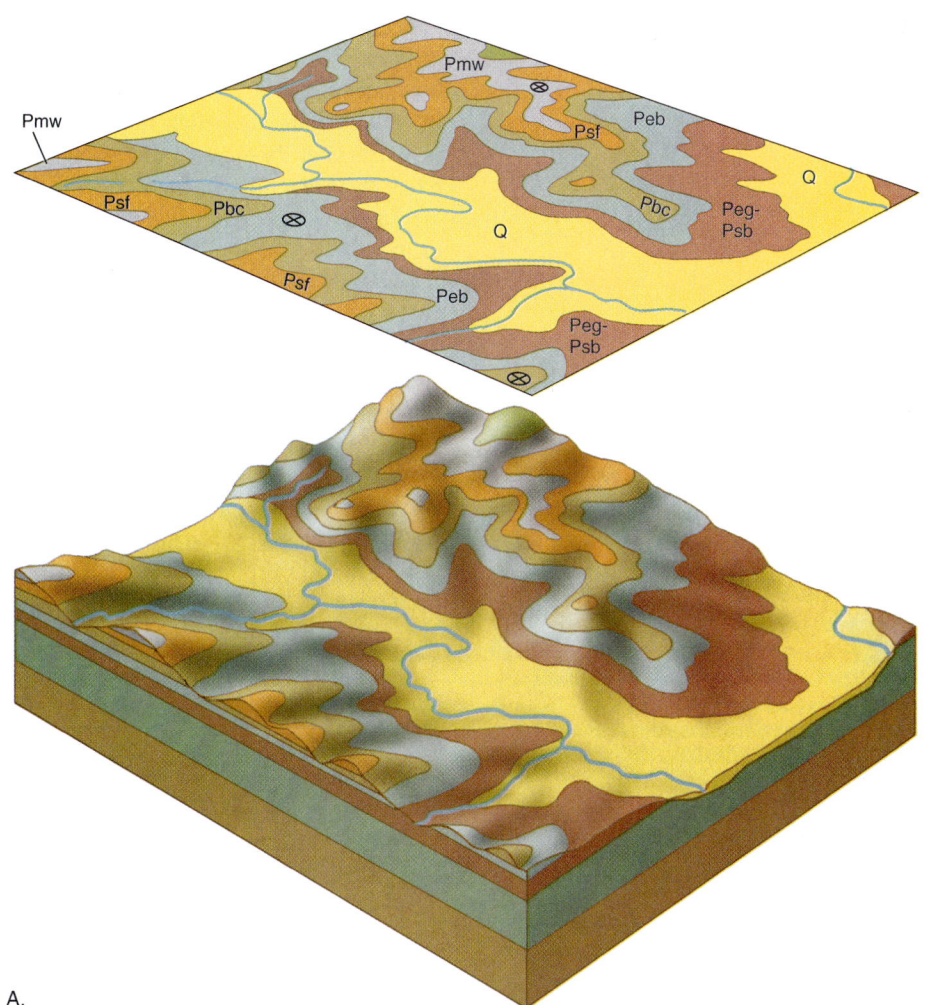

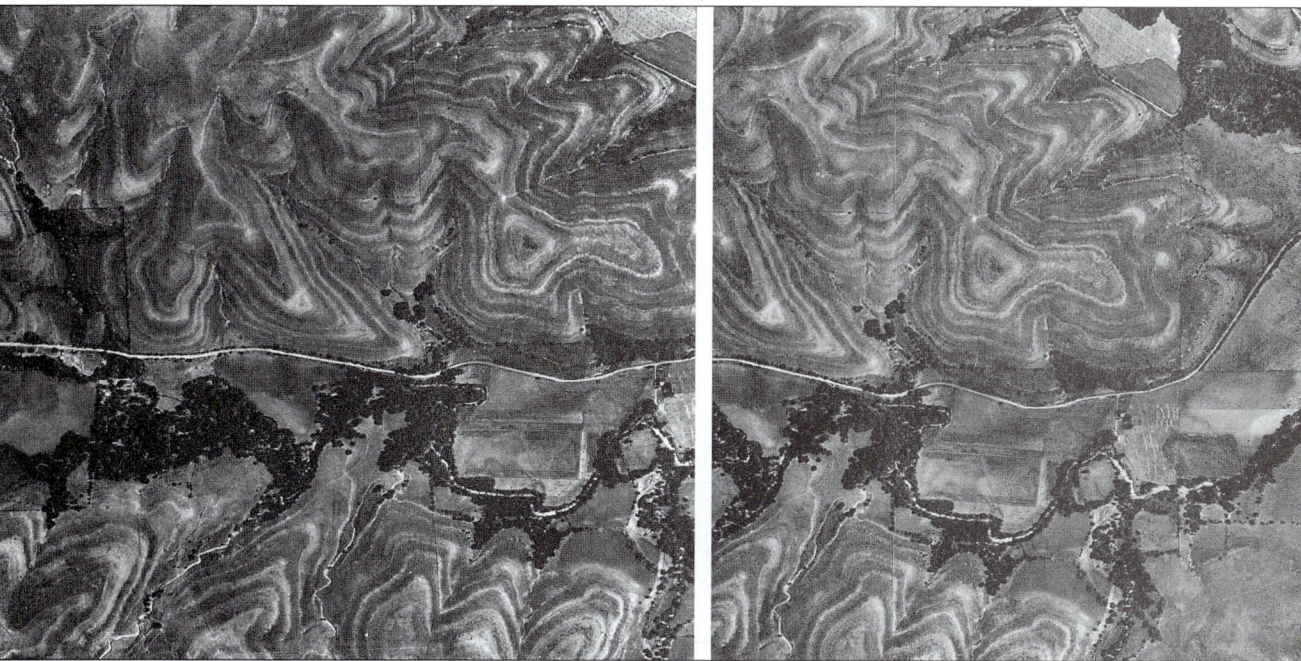

FIGURE 15.2
Horizontal Permian strata in the Flint Hills of northeastern Kansas. The geologic map and block diagram in A represent the area covered by the aerial photographs (scale 1:20,000) in B. *Q* on the map represents Quaternary deposits, and *Pmw, Psf, Pbc, Peb, Psb,* and *Peg* are symbols for Permian formations. The white bands in the photographs (B) are limestone; the gray bands are shale.

topography (Chapter 6)). Like contours, contacts bend so as to point upstream when crossing valleys, as shown in Figure 15.2. In an area with little relief, few strata will be exposed at the surface. The greater the relief, the deeper the land surface cuts into the underlying strata, exposing more of them. Because of differential erosion, resistant beds have steep slopes (narrow expression in map view), and easily eroded beds are marked by gentle slopes (wide expression in map view). If beds are not quite horizontal—that is, the dip is very low—contacts cross contour lines at small angles. In other words, *the flatter the dip, the more closely geologic contacts parallel contour lines.* Similarly, *the steeper the dip, the greater the divergence between contacts and contours.*

Inclined Strata

Strata inclined in one direction appear as roughly parallel bands on a map or aerial photograph, as shown in Figure 15.3. If the relief is moderate, the bands will be essentially parallel to the strike of the beds. In general, with increasing steepness of dip, the outcrop bands will be straighter and the contacts more regular, and the contacts will more closely parallel strike. Differential erosion may produce ridges and valleys that parallel the units. The direction of dip can be determined by the Rule of Vs, as described later, or in some places by differences in hill slope on opposite sides of ridges. Hills sloping in the direction of dip generally are gentler than those cutting across bedding (Fig. 15.3B).

Once the direction of dip is known, relative ages can be determined, because *older beds dip under (or toward) younger beds (assuming beds are not overturned* (Fig. 15.3A)).

Folded Strata

Complete folds are rarely exposed at the surface. Instead, erosion removes the crests of anticlines, leaving behind outcrop patterns that resemble the top surfaces of the block diagrams of Chapter 14 (Fig. 14.9, 14.10). If the folds are plunging, like those shown in Figure 15.4, the eroded, folded strata appear as winding or curved bands on geologic maps or aerial photographs. If non-plunging, the folded strata present roughly parallel bands, like those in Figure 15.3. Differential erosion commonly

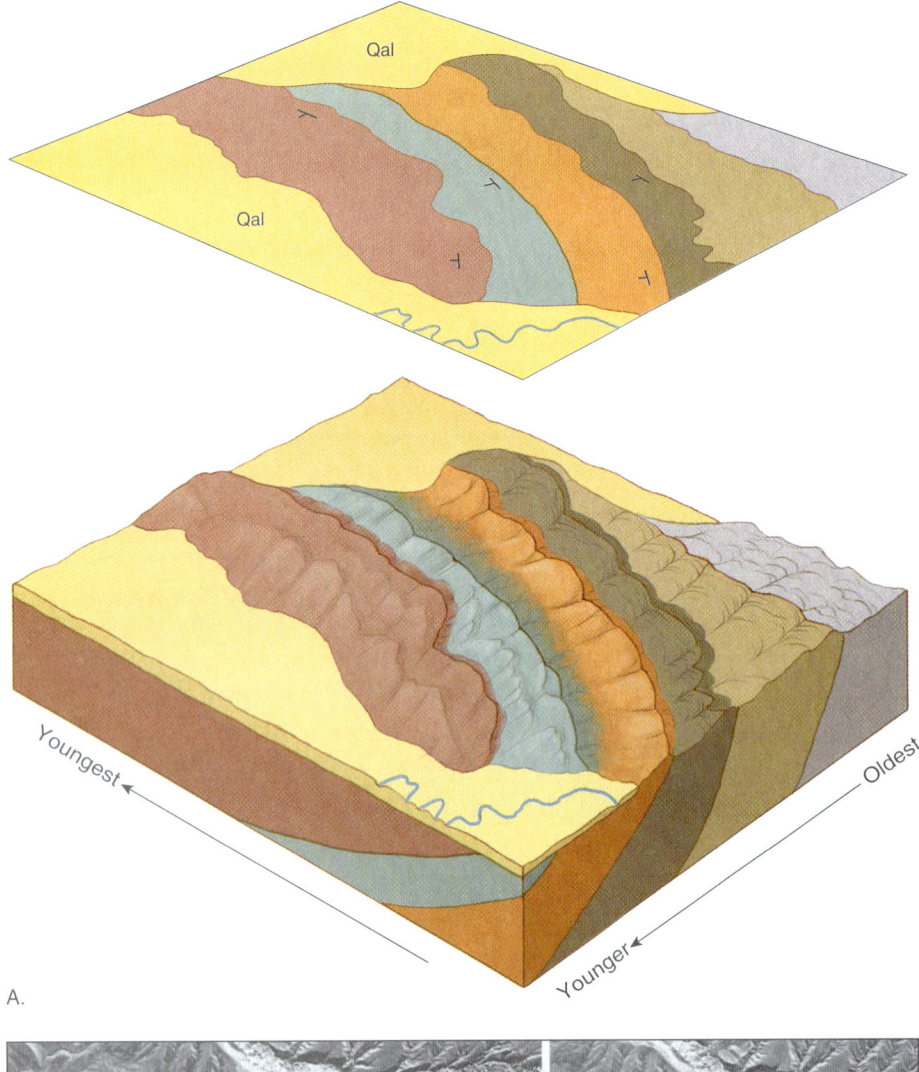

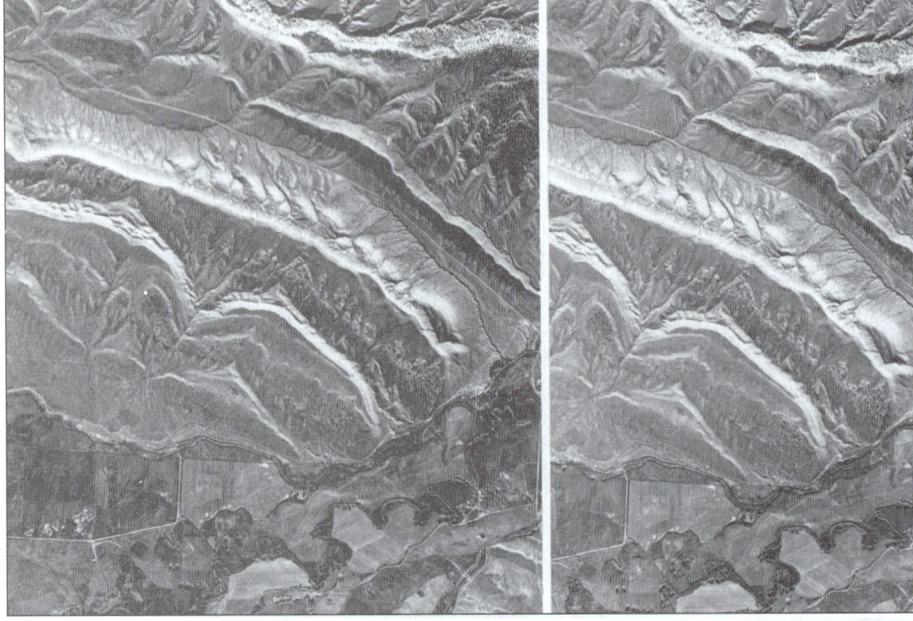

FIGURE 15.3
Inclined Cretaceous strata along Owl Creek, Hot Springs County, Wyoming. Ridges are sandstone, and valleys are shale. The geologic map and block diagram in A represent the area covered by the aerial photographs (scale 1:48,500) in B.

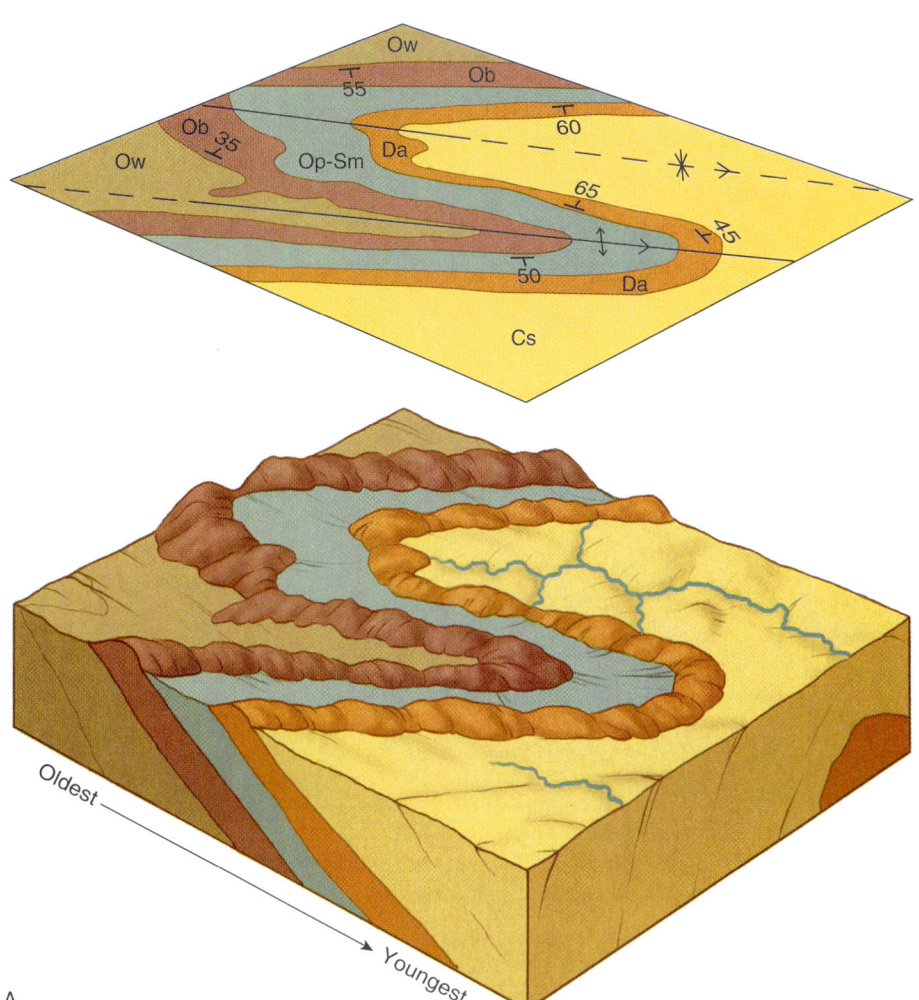

produces ridges and valleys that mimic the outcrop patterns.

As you learned from Problem 7 in Chapter 14, several things are worth remembering when observing eroded folds *in map view*.

1. When anticlines or domes are eroded, older rocks appear in the center and are surrounded by younger rocks.
2. When synclines or basins are eroded, younger rocks appear in the center and are surrounded by older rocks.
3. The contacts between beds in a plunging anticline, as seen in map view, bend so as to "point" in the direction of plunge.
4. The contacts between beds in a plunging syncline, as seen in map view, bend so as to "point" in the opposite direction from which it plunges.

Faulted Strata

Faults are shown on geologic maps with thick, black lines and symbols indicating the nature of the fault, if that is known (see symbols key, Fig. 14.4, in the previous chapter). Faults generally offset beds and contacts and thus commonly juxtapose quite different kinds of rock. Differential

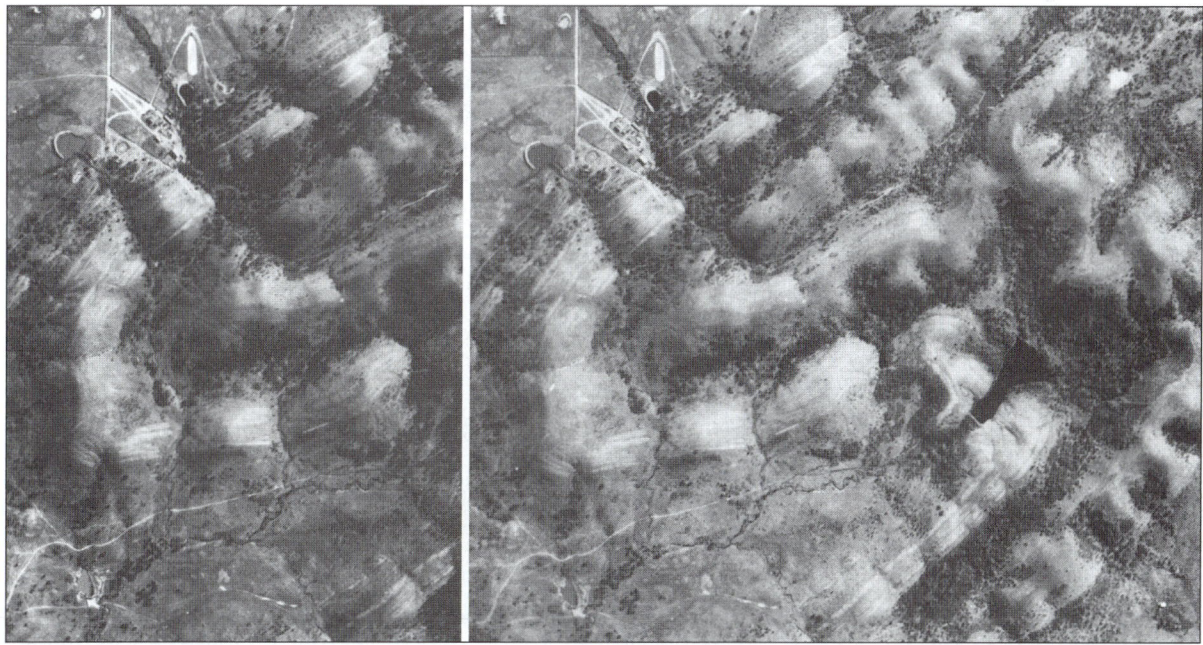

FIGURE 15.4
Folded strata in the Ouachita Mountains, Atoka County, Oklahoma. Ridges are Ordovician (*Ob*) and Devonian (*Da*) chert; low areas are Ordovician, Silurian, and Carboniferous shale (*Ow, Op, Sm, Cs*). The geologic map and block diagram in A represent the area covered by the aerial photographs (scale 1:17,000) in B. Note that without strike and dip or relative age information it is difficult to tell which fold is a plunging anticline and which is a plunging syncline.

erosion may produce a cliff on one side of the fault or a narrow valley marking the fault itself. Fault valleys typically appear as straight, dark lines on aerial photographs, because the vegetation along them is commonly more lush than that surrounding them. On a geologic map, contacts, formations, and structures may end or be offset at faults.

Steeply dipping faults have straight traces (*trace* is the intersection between the fault plane and the surface) on maps or aerial photographs, whereas gently dipping faults may have irregular traces, depending on the topography. Faults dip under their hanging walls; therefore, on a geologic map, the direction of dip is toward the hanging wall. In map view, *normal faults* (Fig.15.5) have older rocks on the footwall side. In contrast, *reverse faults* have younger rocks on the footwall side. This is because *erosion of the topographically higher side (the upthrown side) eventually exposes deeper and (usually) older rocks* (see Fig. 14.11). The traces of *strike-slip faults* are straight because of their common near-vertical dips. They are more likely than the other types of faults to be marked by straight, narrow valleys.

Thrust faults may have very irregular traces because of their low dips, often only

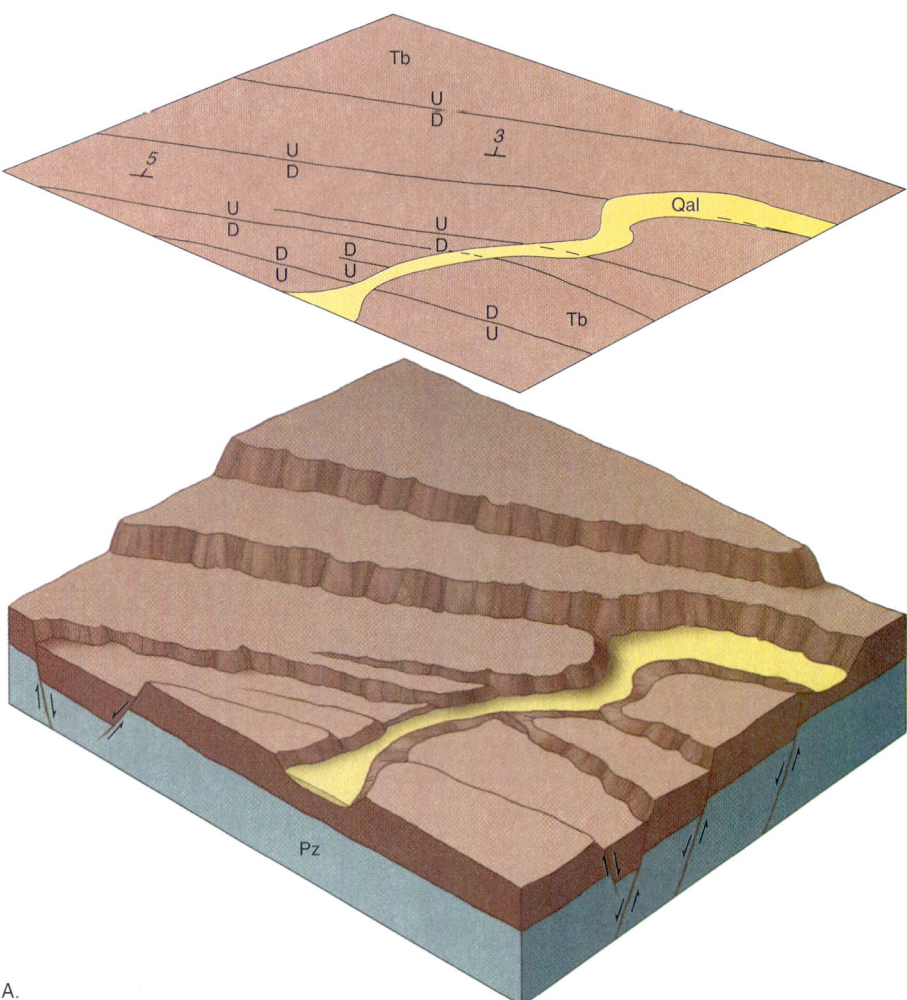

A.

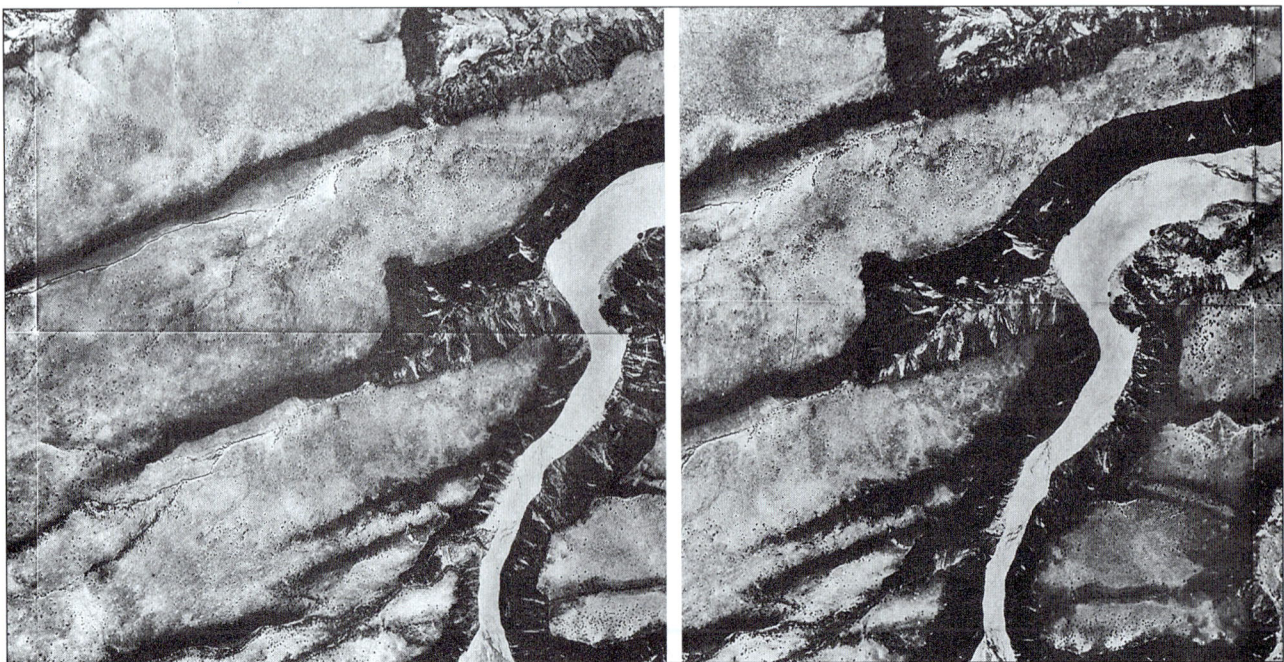

B.

FIGURE 15.5

Basalt flows of Tertiary age (*Tb*) cut by normal faults, Sandoval County, New Mexico. Displacement ranges from near zero to about 50 m. Fault blocks are tilted eastward. The geologic map and block diagram in A represent the area covered by the aerial photographs (scale 1:31,680) in B. *Qal* indicates Quaternary alluvium, Pz (in block diagram) indicates rocks of Paleozoic age.

a few degrees. They are recognizable on a geologic map when structures above and below the fault do not correspond (for example, formations are not the same, or strikes and dips differ), and when the rocks in the hanging wall are older than those in the footwall. See Figure 14.11.

Unconformities (see also Chapter 13)

Angular unconformities (Fig. 13.5A) are recognized most easily on aerial photographs and geologic maps by nonparallel layers in rocks above and below the unconformity, as shown in Figure 15.6. Angular unconformities with low dips appear similar to low-angle thrust faults, except that rocks above the unconformity are younger than those below it.

Nonconformities (Fig. 13.5D) separate nonbedded, crystalline rocks from younger sedimentary rocks. They can be confused with intrusive igneous contacts. In nonconformities, the igneous contact is usually with the same formation throughout the map area, whereas intrusive contacts may cut across more than one formation. Relative ages established in the field are most important; igneous rocks are younger than the rocks they intrude, but older than rocks

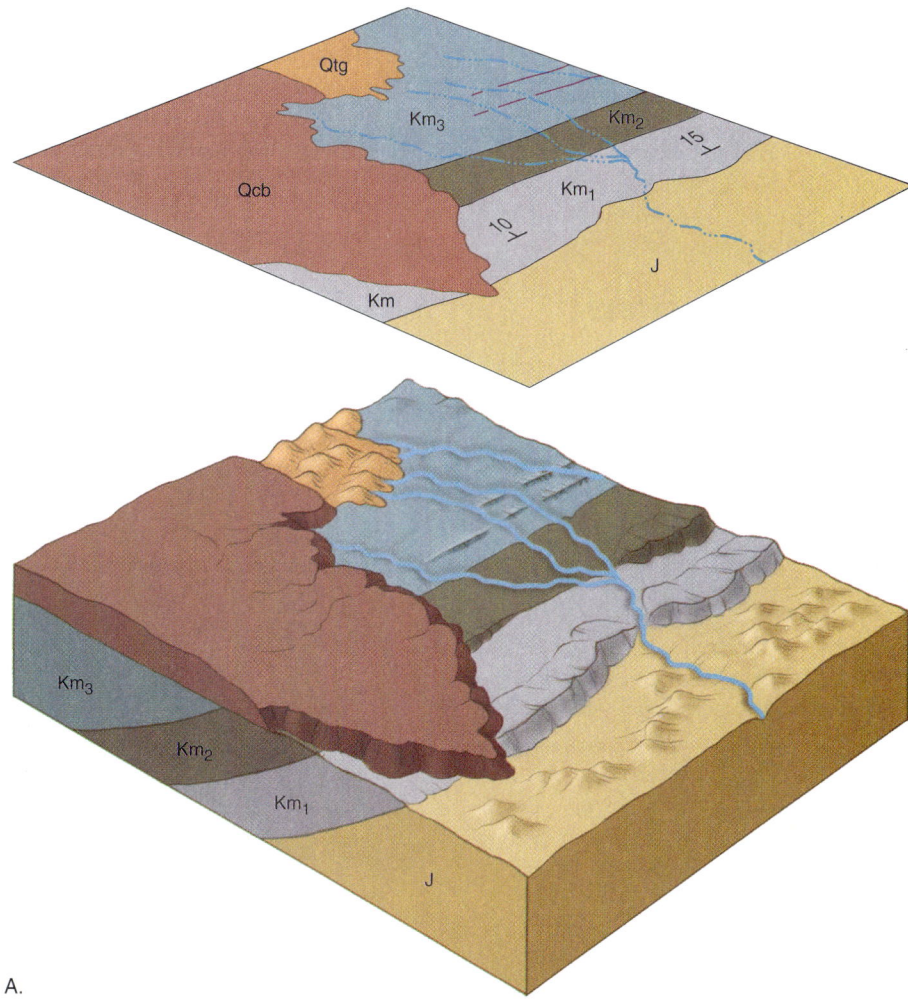

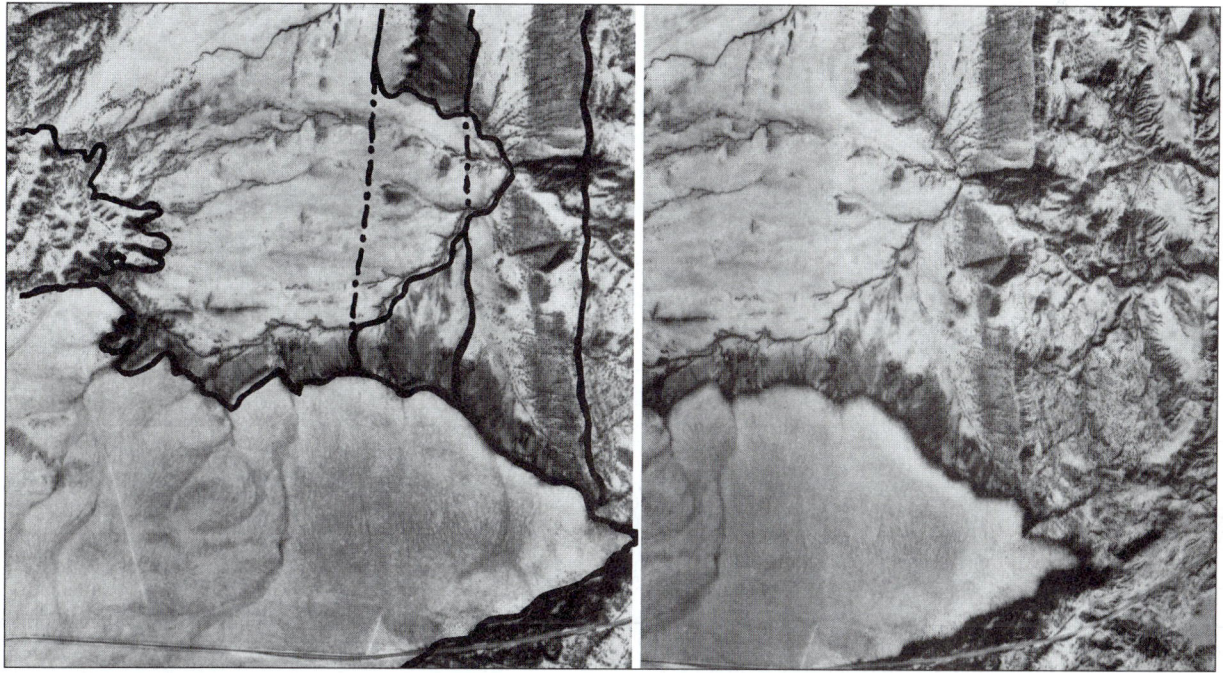

FIGURE 15.6
Angular unconformity between nearly flat-lying Quaternary basalt lava flows (*Qcb*) and gently dipping Jurassic (*J*) and Cretaceous (*Km1, Km2,* and *Km3*) sedimentary rocks, Santa Fe and Sandoval Counties, New Mexico. The red lines on the map are dikes; *Qtg* is Quaternary gravel. The geologic map and block diagram in A represent the area covered by the aerial photographs (scale 1:54,000) in B.

deposited on their eroded, *nonconformable* surfaces.

Disconformities (Fig. 13.5B, C) are much more difficult to recognize on a geologic map. Variable thickness of a particular formation, or inexplicable contacts between otherwise conformable (parallel) layers, are clues.

Igneous Intrusions

Irregular-shaped **plutons** (bodies of intrusive igneous rock) show up on the surface as **discordant** masses of igneous rock, meaning they cut across the layering of surrounding rocks. Plutons may be somewhat circular or elliptical in map view if the map scale is small enough to show the whole igneous body.

Dikes are thin, tabular intrusions that cut across bedding (Fig. 14.11D (pink); Fig. 15.6). They appear on maps as narrow, discordant lines, and commonly occur in sets (*swarms*) with parallel or near-parallel alignments. **Sills,** the other tabular intrusions, are **concordant** (parallel to bedding), so they cannot be recognized on maps unless it is known that the rock is intrusive.

Volcanoes and Volcanic Rocks

Young volcanoes with central **craters** (small depression at the top) or **calderas** (large depressions at the top), and with **lava flows** or **pyroclastic flows** (gas-charged flows that produce tuff) radiating outward, are easily recognized. Older, more deeply eroded volcanoes become increasingly difficult to recognize with age.

Extensive sheets of lava or tuff may blanket large areas and have a map appearance similar to sedimentary rocks, whether folded or horizontal. Like sedimentary rocks, they are given formation names (for example, San Juan Formation).

Geologic Cross Sections

Geologic cross sections are like topographic profiles, but with the subsurface geology shown schematically (Fig. 15.1). Geologic maps commonly are accompanied by one or more geologic cross sections, because they make it much easier to visualize the third dimension. Geologic cross sections are interpretive in nature, because geologic maps are based on a limited number of outcrops and most of what is portrayed is out of sight, below the surface. Cross sections help to clarify the author's interpretation.

Geologic cross sections are made so that they will provide the most information possible. A **line of section** must be chosen. This is the line, or set of connected line segments, along which the cross section will be made. For example, if the map shows an anticline, a logical line of section would be one perpendicular to the trace of the axial surface of the fold on the Earth's surface, because that section would best portray the anticline. Most cross sections are drawn perpendicular to the general strike. Not only do such sections convey the most information, they are a lot easier to draw because the actual dip can be shown.

To make a geologic cross section, a topographic profile is first drawn along the line of section. The thickness (or depth below the surface) represented on a cross section depends mostly on how much is known or can be inferred about what is below the surface. Some cross sections show thousands of meters, others only a hundred or so. It generally is less confusing if no vertical exaggeration is used on a cross section. Vertical exaggeration makes it more difficult to draw the geology below the surface, because all the angular relations also must be exaggerated (for example, with a vertical exaggeration of 10×, a 10° dip would appear as a 60.4° dip on the cross section). If all geologic contacts are horizontal, vertical exaggeration can be used with impunity.

The geology is next added along the line of section. Wherever the line of section crosses a geologic contact on the map, the contact is marked on the section, and a notation is made to indicate what rock units are present. The direction and approximate angle of dip of the contacts can be determined from the dip and strike symbols closest to the line of section, or from the Rule of Vs (explained later). Contacts are then extended below the surface to best represent the geology seen at the surface.

Using Geologic Maps

Map Symbols

A fairly standard set of symbols is used on geologic maps to depict various geologic features. The most common are shown in Figure 14.4 in Chapter 14.

Variations in Outcrop Patterns

Width of Outcrop

The outcrop width of a bed depends on the thickness of the bed, the angle at which that bed intersects the surface, and the slope of the surface. A bed intersecting the surface at a low angle has a wide outcrop; the outcrop width is equal to the thickness of the bed only if the bed intersects the surface at 90°. In *map view,* the apparent width of a bed equals its true thickness (as scaled on the map) only if the bed is vertical.

Because of this relation, horizontal beds have wider outcrop patterns where surface slopes are gentle, and more narrow patterns where slopes are steep, as shown in Figure 15.7A. Similarly, a folded bed of constant thickness but variable dip, such as illustrated in Figure 15.7B, has a variable outcrop width on a flat surface.

Rule of Vs

When viewed from above, the contact lines between formations are deflected as they cross a valley (Fig. 15.8). This leads to a general rule, called the **Rule of Vs,** that helps determine the general dip of a bed on a geologic map lacking strike and dip symbols: **Where a contact line crosses a valley, it forms a V whose point points in the direction of the dip of the contact between the two formations (Fig. 15.8A, B).**

The amount and direction of deflection depend mainly on the dip of the beds. A horizontal bed produces contacts that exactly parallel topographic contours (Fig. 15.8C; see Figure 15.2). Gently dipping beds produce relatively long and narrow Vs. As the dips become steeper, the Vs become wider and shorter. If beds are vertical there is no V. When viewed from above, contact lines are not deflected but cross a valley in a straight line (Fig. 15.8D). There is one exception to the rule of Vs that results from the fact that the amount and direction of deflection also depends on the slope of the hill: If beds dip down the valley, but *at an angle less than the slope of the valley,* the V points up-valley (Fig. 15.8E). To allow you to focus on the more common situations presented in Figures

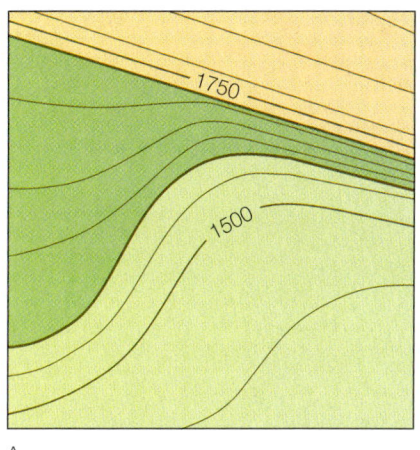
A.

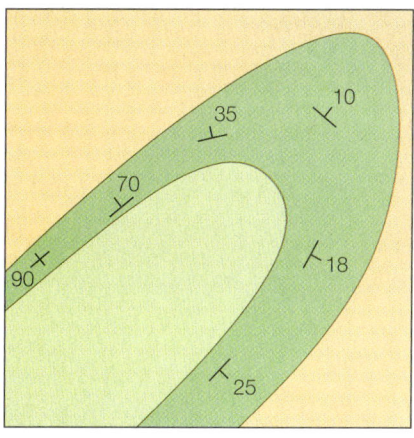
B.

FIGURE 15.7
A. Map of a horizontal layer, 150 feet thick, that forms a narrow outcrop where the slope is steep and a wide outcrop where the slope is more gentle. B. The outcrop width of a folded layer on a horizontal surface equals true thickness where the dip is 90° but is wider where the dip is less, as shown on this map.

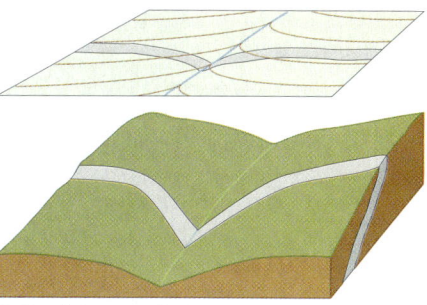
A. V points opposite to contours.

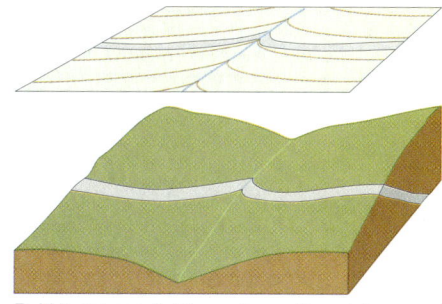
B. V is not parallel to contours; its point is at a **lower** elevation than its sides.

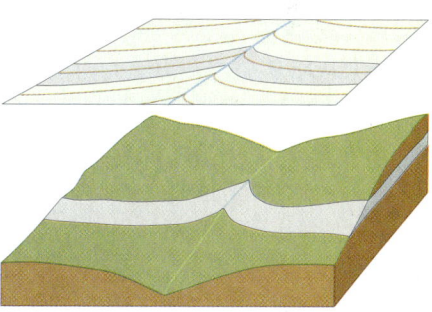
C. Contacts parallel contours.

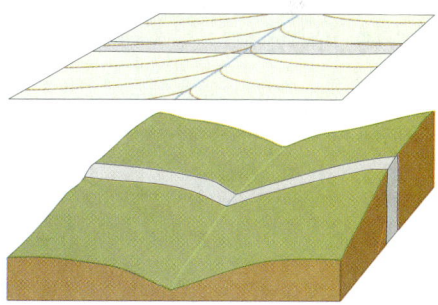
D. Contacts cut straight across contours.

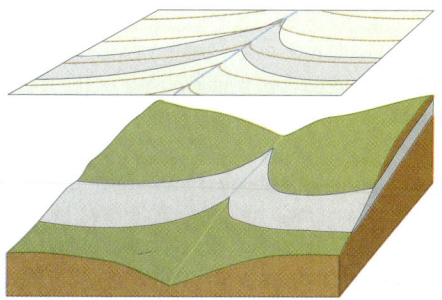
E. V is not parallel to contours; its point is at a **higher** elevation than its sides.

15.8A through D, we present no lab exercises dealing with this special situation.

In general, focus on the map relation between bedding contacts and contour lines to determine which situation you are viewing (Fig. 15.8). If there are no contours on the map, assume that the Vs point in the direction of dip, as illustrated in the following section.

Reconstructing Geologic History

A geologic map is used to reconstruct the geologic history of an area in the same way that geologic cross sections were used in Chapter 13. Figure 15.9 will serve as an example.

1. First, interpret the map. What is shown? If strike and dip symbols are not given, use the orientation of contacts and the Rule of Vs to determine the strike and dip directions, as has been done in Figure 15.9. *The strike generally approximately parallels the contacts, and the dip is in the direction of the Vs where the contacts cross valleys.* Put strike and dip symbols at enough places on the map so the structure is clear. Just show strike and direction of dip, not numbers, although you may be able to tell whether the dip is steep or shallow based on the size and shape of the Vs.

2. Remembering the *principle of superposition* (Chapter 13), and that older beds dip under (or toward) younger beds, determine the relative ages of any sedimentary rocks, and locate the oldest.

3. Using the *principle of cross-cutting relations* (Chapter 13), determine the relation between the sedimentary rocks and events such as faulting, folding, erosion, or igneous activity.

4. Finally, list all events, beginning with the oldest, in the sequence in which they occurred. Remember that unconformities usually require uplift and erosion, and that sedimentary rocks originally were deposited as horizontal layers.

FIGURE 15.8
Where dipping beds or contacts between them cross a valley, they generally form a V pattern on a map that points in the direction of dip. Illustrations show a block diagram with the corresponding contour and geologic map above it. A. Beds dip downstream. B. Beds dip upstream. Exceptions are (C) horizontal beds, which always form Vs that point upstream on a map and are parallel to contours; (D) vertical beds, which are not deflected on crossing a valley; and (E) beds that dip downstream but at an angle less than the slope of the valley.

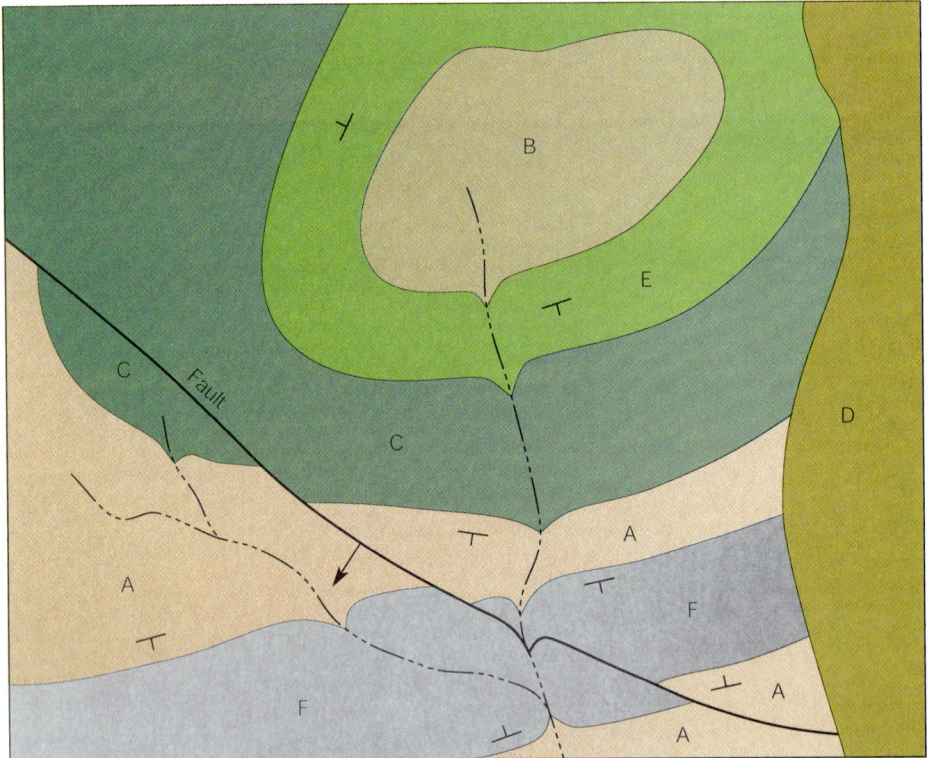

FIGURE 15.9
Reconstructing geologic history. Strike and dip symbols are based on contacts and their "Vs" where they cross valleys. Because older beds generally dip under younger, the oldest bed is B, and E, C, A, and F are successively younger. These beds are folded. The southwest-dipping fault is younger than F and the fold, but not D. An angular unconformity separates D from older units, and the present surface is being eroded. Events occurred as follows, from oldest to youngest: deposition of B, E, C, A, and F; folding; faulting; erosion; deposition of D; and erosion.

Prospecting

Geologic maps are the most important database for prospecting, whether it be for oil or gas, coal, or various mineral commodities. Geologists know from experience that fossil fuels and mineral commodities occur in some geologic settings and not in others. Geologic maps are used to target specific areas for closer examination.

For example, oil and gas form when organic materials that accumulate with marine mud are buried deeply. Once formed, the oil and gas are squeezed out of the mudstone and migrate upward (because they are less dense than water and therefore buoyant) through permeable rock, until they are trapped against an impermeable barrier. Knowing these characteristics of oil and gas deposits, geologists can quickly eliminate large areas from further exploration. For example, igneous and metamorphic rocks normally would not be very good places to look for oil, because they contain no suitable organic material. Nor would Precambrian rocks, most of which formed before organisms became sufficiently abundant to generate a large amount of oil or gas. Once possible source rocks are identified, the search can be concentrated where impermeable barriers exist that could trap petroleum.

Similar arguments apply to mineral commodities. Certain kinds of ore deposits are found only in certain kinds of rocks, or in association with certain geologic structures. Geologic maps are used to focus exploration projects.

Land Use

Geologic maps, together with topographic maps, are the bases for land-use maps. For example, in one case in Utah (the Sugar House 7 1/2 minute quadrangle, which covers part of Salt Lake City), *all* of the following kinds of maps were developed by the U.S. Geological Survey: surficial geology, age of faults, slopes, landslides, slope stability, construction materials, urban growth, thickness of saturated Quaternary deposits, depth to water in shallow aquifers, depth to top of principal aquifer, concentration of dissolved solids in water from principal aquifer, configuration of potentiometric surface, thickness of loosely packed sediments and depth to bedrock, flood and surface water, and earthquake stability. Other types of maps are developed for areas with different characteristics.

Such maps are used to determine how land can be used most effectively. For example, it would not be effective to build a shopping center in the middle of a landslide area, to route a buried pipeline through an area where slopes are unstable, or to put a landfill on a site where permeability is high.

Hands-On Applications

> Geologic maps summarize geologic field data and portray the mapping geologist's interpretation of that data. Geologic maps contain information on most of the aspects of geology you have already studied: minerals, rocks, surface processes, and rock sequences and structures. In this lab, you will learn how to apply the geologic principles you have learned thus far to extract that information from geologic maps.

Objectives

If you complete all the problems, you should be able to:

1. Construct a geologic map from an outcrop map on which rock type and structural information are given.
2. Recognize horizontal, inclined, folded, and faulted strata; unconformities; and igneous intrusions on geologic maps and aerial photographs.
3. Use the Rule of Vs to determine direction of dip for strata.
4. Draw a geologic cross section from a geologic map.
5. Determine the chronologic order of geologic events from a geologic map.

Problems

1. Figure 15.10 is an outcrop map. The symbols indicating rock types are those given in Figure 13.9. The outcrops are surrounded by loose sediment and soil of Quaternary age (indicated by the light tan color and labeled Q on the map).

   a. Using a protractor, plot strike and dip symbols on the map in Figure 15.10 for the outcrops listed below. Put the symbols on the black dots on each outcrop. To plot strike and dip, the boundaries of the map can be used as reference lines for the protractor. The sides of the map run north-south and the top and bottom run east-west.

Outcrop	Attitude	Outcrop	Attitude
A	N43°E, 25°SE	G	N30°W, 18°NE
E	N37°W, 26°SW	H	N53°E, 21°NW

   b. Make a geologic map by correlating the various rock types from one outcrop to another and by drawing contacts between adjoining units. See Figure 14.4 or 15.1 for correct symbols for contact lines.

   c. Using the principle of superposition, determine the relative age of each sedimentary unit and list them below, with the oldest units at the bottom and the youngest at the top.

   d. Make an east-west geologic cross section of the map along a line from X to Y in the space provided below the map. Assume the surface is flat. The vertical scale should be the same as the horizontal scale. Use a protractor to plot dips at the surface. Keep your lines short, for dip angles commonly change in the subsurface.

   e. What type of structure is present? If it is a fold, indicate whether it is an anticline, syncline, dome, or basin. If it is a fault, indicate whether it is a normal, reverse, thrust, or strike-slip fault; label it with the proper symbol from Figure 14.4; and label footwall and hanging wall, if appropriate, on your cross section.

   f. Using your cross section, determine the thickness of the shale.

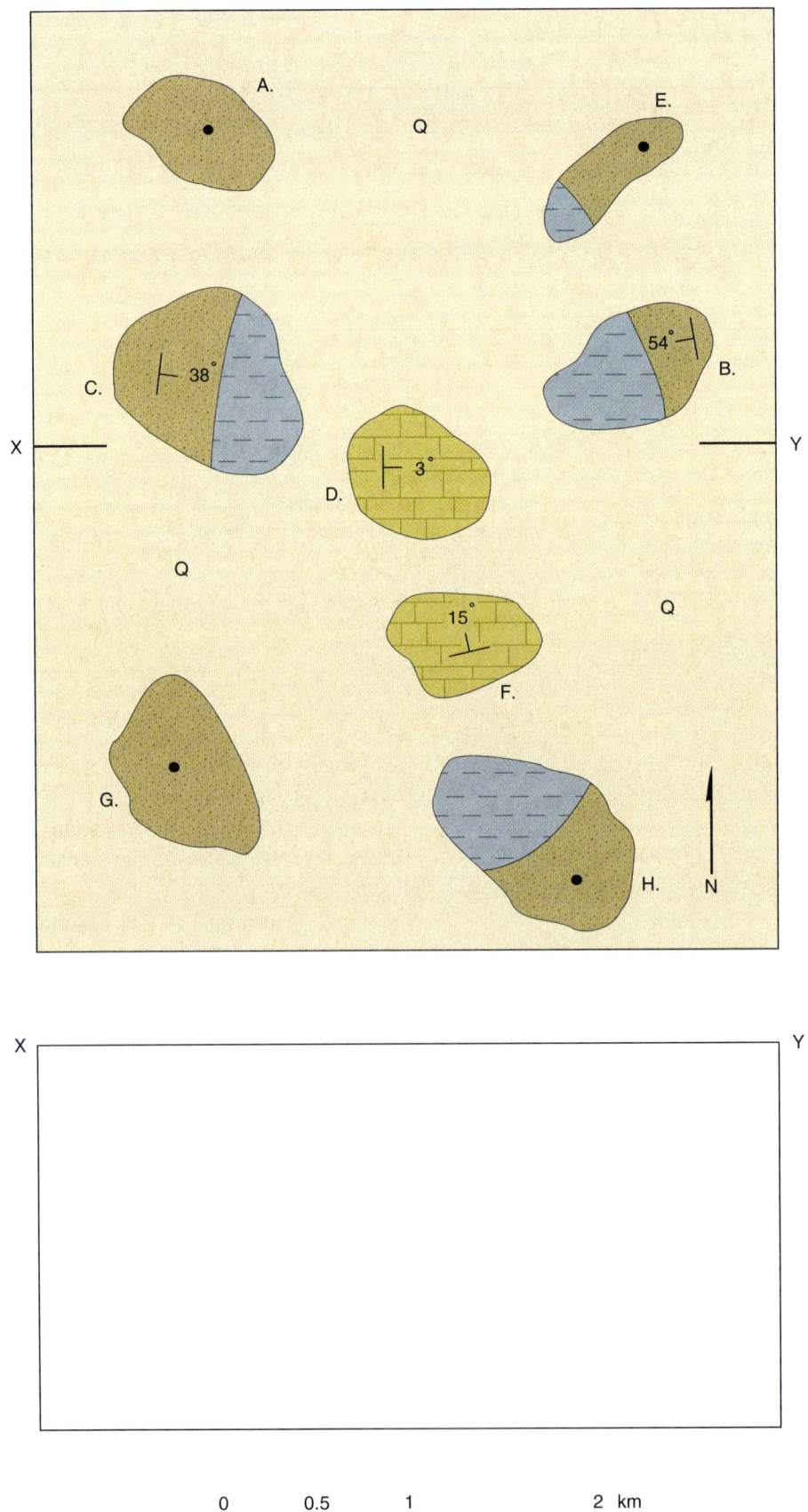

FIGURE 15.10

Outcrop map for Problem 1. Symbols of rock types are given in Figure 13.10.

Chapter 15   Geologic Maps   279

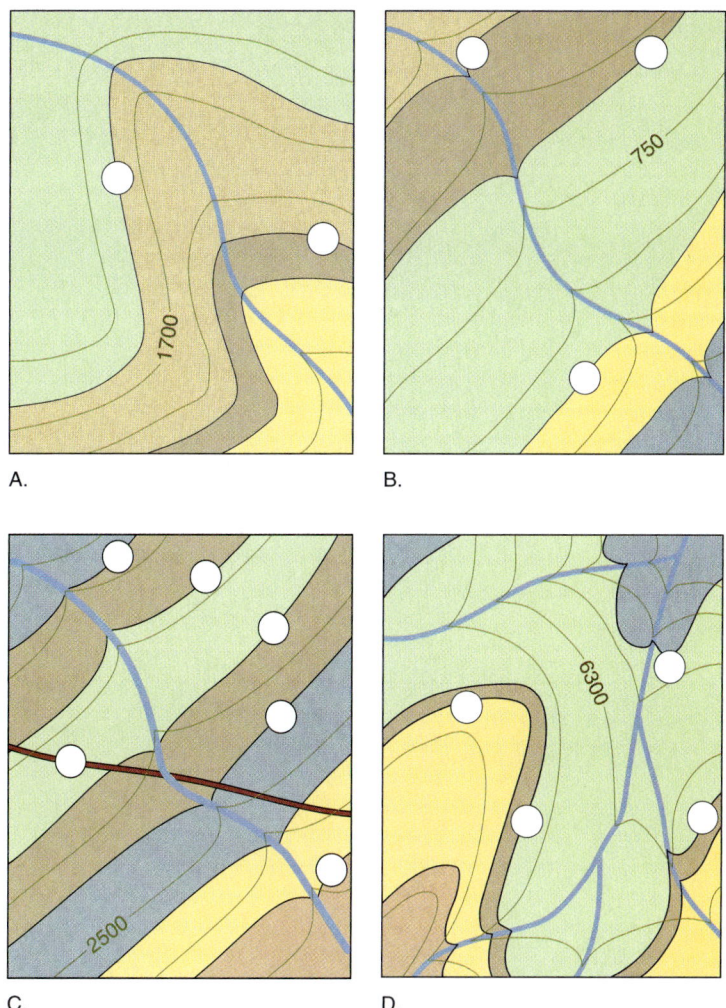

FIGURE 15.11
Geologic maps for Problem 2. Place strike and dip symbols in circles. Brown lines are contour lines, blue lines represent streams, and the red line is a dike.

2. Figures 15.11A, B, C, and D are simple geologic maps with contours superimposed. Place appropriate strike and dip symbols (see Figure 15.9) in the circles on each map; show the direction (not the angle) of dip. Use the Rule of Vs to determine the direction of dip. To determine which way is up- or downhill, note how the contours bend when they cross a stream valley (see Figure 6.7, Chapter 6). The red line in 15.11C is an igneous dike.

3. Figure 15.12 is part of a geologic map of the Gateway Quadrangle, Colorado, and provides the basic geologic data for that area. Answer the following standard geologic questions that might be asked about an area:

   a. What is the oldest rock represented, and where on the map does it occur?

   b. What is the general structure of the Mesozoic (Triassic, Jurassic, and Cretaceous) rock units? That is, are the strata horizontal (or nearly so), inclined, folded, or faulted?

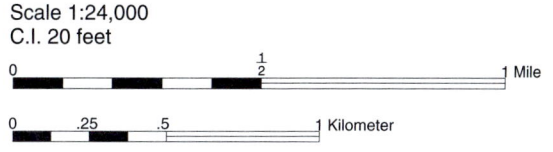

Scale 1:24,000
C.I. 20 feet

FIGURE 15.12
(Pages 280–281) Geologic map of portion of Gateway Quadrangle, Colorado, for use with Problem 3.
Source: Data from U.S. Geological Survey.

EXPLANATION

Alluvium
Light-red wind-deposited sand and silt on benches and mesa tops, reworked in part by water; Recent valley fill and stream deposits.

Pleistocene and Recent — QUATERNARY

UNCONFORMITY

Fanglomerate
Poorly sorted, in places rudely bedded, sand and angular fragments and boulders derived from older formations; somewhat indurated.

UNCONFORMITY

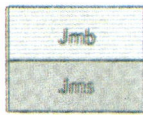

Burro Canyon formation
White, gray, and red sandstone and conglomerate with interbedded green and purplish shale.

Lower Cretaceous — CRETACEOUS

Morrison formation
Variegated shale and mudstone; white, gray, rusty-red, and buff sandstone; rusty-red conglomerate; local thin limestone beds. At the top the Brushy Basin shale member, Jmb, consisting largely of bentonitic shale but including some sandstone and conglomerate lenses, and at the base the Salt Wash sandstone member, Jms, with more numerous and thicker sandstone beds.

Upper Jurassic

Summerville formation
Thin-bedded red, gray, green, and brown sandy shale and mudstone.

San Rafael group

Entrada sandstone and Carmel formation undivided
Orange, buff, and white, fine-grained, massive and cross-bedded Entrada sandstone at the top. Red sandstone and mudstone of the Carmel formation at the base.

Middle Jurassic

UNCONFORMITY

Navajo sandstone
Buff and gray crossbedded, fine-grained sandstone.

Kayenta formation
Irregularly bedded, red, buff, gray, and lavender shale, siltstone, and fine- to coarse-grained sandstone.

Glen Canyon group — JURASSIC(?)

Wingate sandstone
Fine-grained reddish-brown, cliff-forming sandstone, thickbedded, massive and crossbedded.

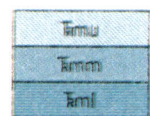

Chinle formation
Red to orange-red siltstone, with interbedded lenses of red sandstone, shale, and limestone-pebble and clay-pellet conglomerate. Lenses of quartz-pebble conglomerate and grit at base.

Upper Triassic

UNCONFORMITY

Moenkopi formation
Chocolate-brown ripple-bedded shale, brick-red sandy mudstone, reddish-brown and chocolate-brown sandstone, and purple and reddish-brown arkosic conglomerate. Local gypsum beds. The upper member, Tmu, consisting of thin- and ripple-bedded shale with interbedded sandstone; the middle member, Tmm, consisting of ledge-forming beds of shale, sandstone, and arkosic conglomerate; and the lower member, Tml, consisting of poorly sorted sandy mudstone and local gypsum beds near base.

Lower Triassic — TRIASSIC

Cutler formation
Maroon, red, mottled light-red, and purple conglomerate, arkose, and arkosic sandstone. Thin beds of sandy mudstone.

PERMIAN(?)

UNCONFORMITY

pϾ

Gneiss, schist, granite, and pegmatite

PRE-CAMBRIAN

Contact
Dashed where approximately located.

- - - - - -

Indefinite contact
Includes inferred contacts and indefinite boundaries of surficial deposits.

D / U

Fault
Dashed where approximately located. U, upthrown side; D, downthrown side.

20⟋

Strike and dip of beds

— 7500 - - -

Structure contours
Drawn on top of Entrada sandstone; dashes indicate projection above surface. Contour interval 100 feet. Datum is mean sea level.

Y

Adit

×

Prospect

✕

Mine

Source: Data from U.S. Geological Survey.

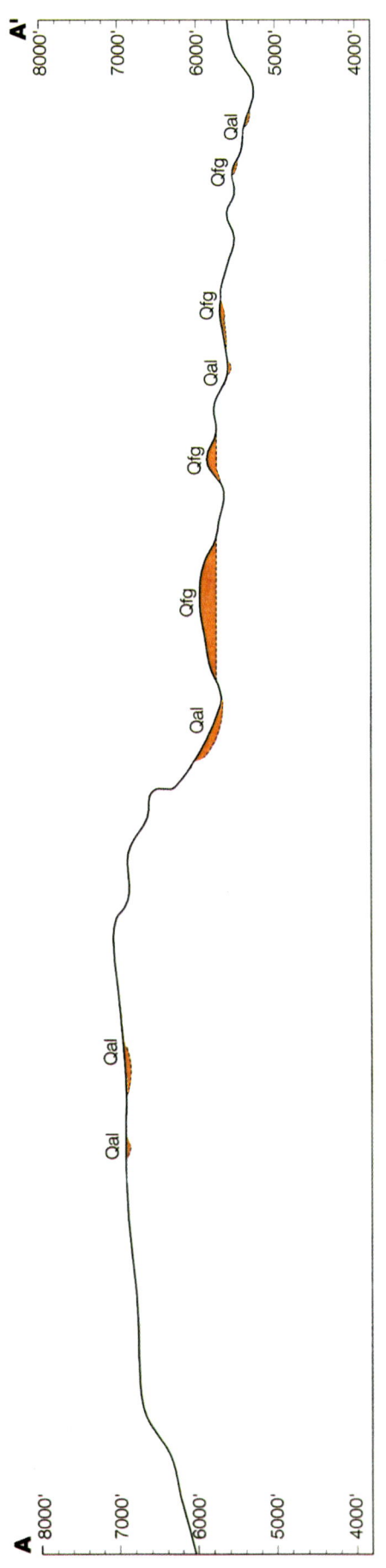

FIGURE 15.13
Cross section along A-A' of Gateway Quadrangle, Colorado, geologic map, for use with Problem 3c. Qal and Qfg are already located for you.

c. Construct a geologic cross section from A to A' using the topographic profile in Figure 15.13 as a basis. The vertical scale and horizontal scale are the same, one inch to 2000 feet, so there is no vertical exaggeration. The Qal and Qfg are surficial sediment deposits; they are already on the profile (Fig. 15.13).

d. What is the approximate thickness of the Mesozoic rock units?

e. Notice that one mine and a number of prospects (holes or pits dug to investigate for ore, which are shown with an "X" on the map) are located in the area. The elements of interest are uranium, vanadium, and radium. If you were to explore this area for minerals containing these elements, on what formation(s) would you concentrate your efforts? (Evidence is on the map.)

4. Refer to Figure 15.14, a geologic map of the Black Hills of South Dakota, and to Figure 15.15, a composite *Landsat* image.
   a. Study the outcrop pattern on the geologic map and the satellite image. What type of structure is present?

   b. Based on the overall geologic structure and the map explanation, what is the age of the oldest rock in the structure?

   c. Why are the areas underlain by the Minnelusa Sandstone and the Minnekahta Limestone—Opeche Formation wider on the west side of the structure than on the east?

   d. Are the Cenozoic igneous intrusions older or younger than the surrounding rocks? How could you tell this if you weren't given the age of the rocks?

   e. List the *sedimentary* units from the surface downward that would be penetrated by a well drilled at Edgemont in the southwestern part of the map.

   f. The big reddish area on the *Landsat* image more or less coincides with the heart of the Black Hills. Why is it reddish? (See Chapter 7.)

   g. Note the similarity in shape or pattern of the reddish areas of the satellite photo and some of the contact lines on the geologic map. This is due in part to elevation differences—that is, the elevation is higher within the red area, where rainfall is more plentiful—but that is not the only explanation. Examine the satellite image closely, compare with the geologic map, and suggest another possibility.

   h. What causes the narrow southeast-trending bright red streaks on the east side of the satellite image?

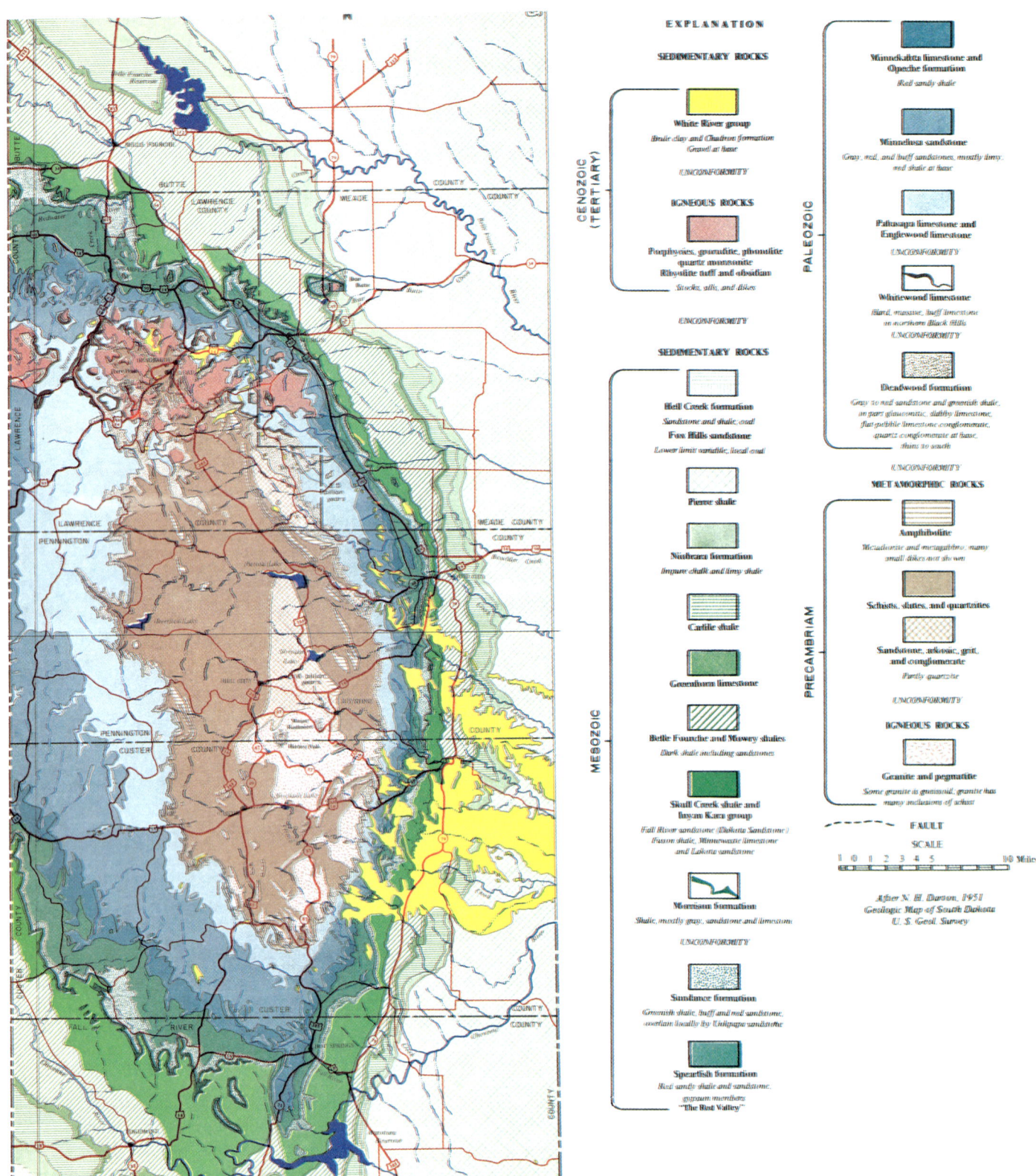

FIGURE 15.14
Geologic map of the Black Hills, South Dakota, for Problem 4.
Source: South Dakota Geological Survey.

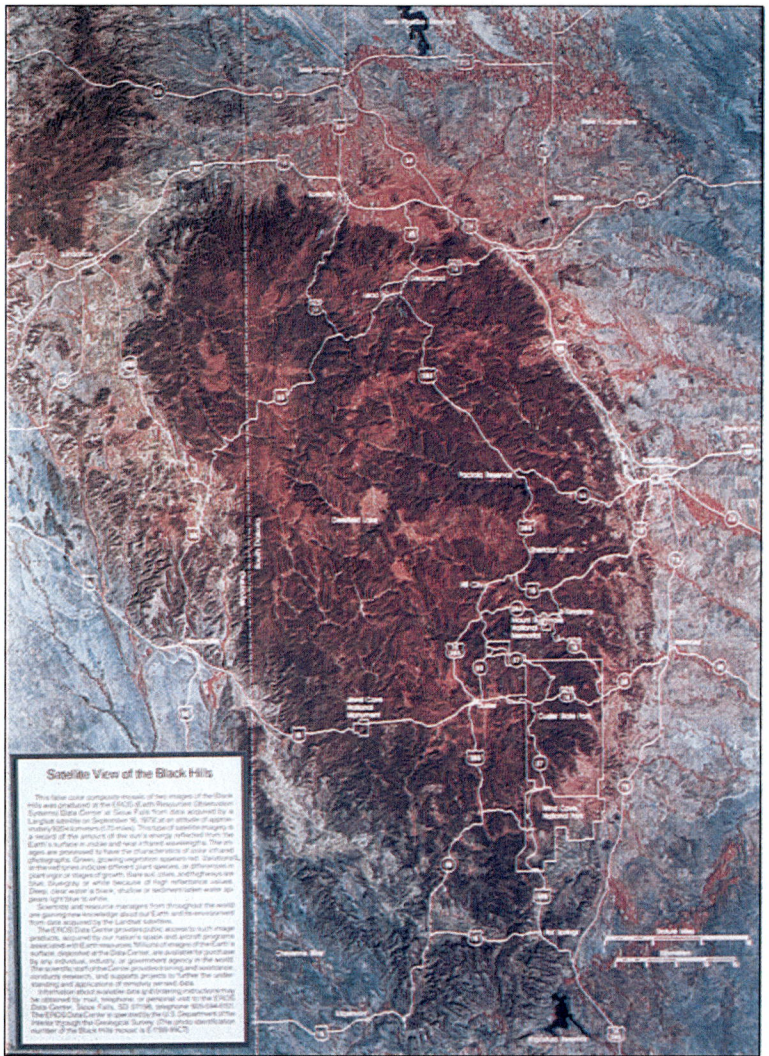

FIGURE 15.15
Composite false-color *Landsat* satellite image of the Black Hills, South Dakota/Wyoming, for Problem 4.

5. Figure 15.16 is a geologic map of an area that has undergone faulting, folding, igneous activity, and several periods of erosion. Using the Rule of Vs, place strike and dip symbols at several places on the map to help you visualize the structures.

   a. What is the general strike and dip (just directions, not numbers) of the gabbro dike in the northeast corner? How do you know?

   b. Is the gabbro older or younger than fault *A*? How do you know?

   c. Is the dip of fault *B* greater or less than fault *A*? How do you know?

   d. In what direction does fault *A* dip? How do you know?

   Which side of fault *A* is the hanging wall and which the footwall? How do you know?

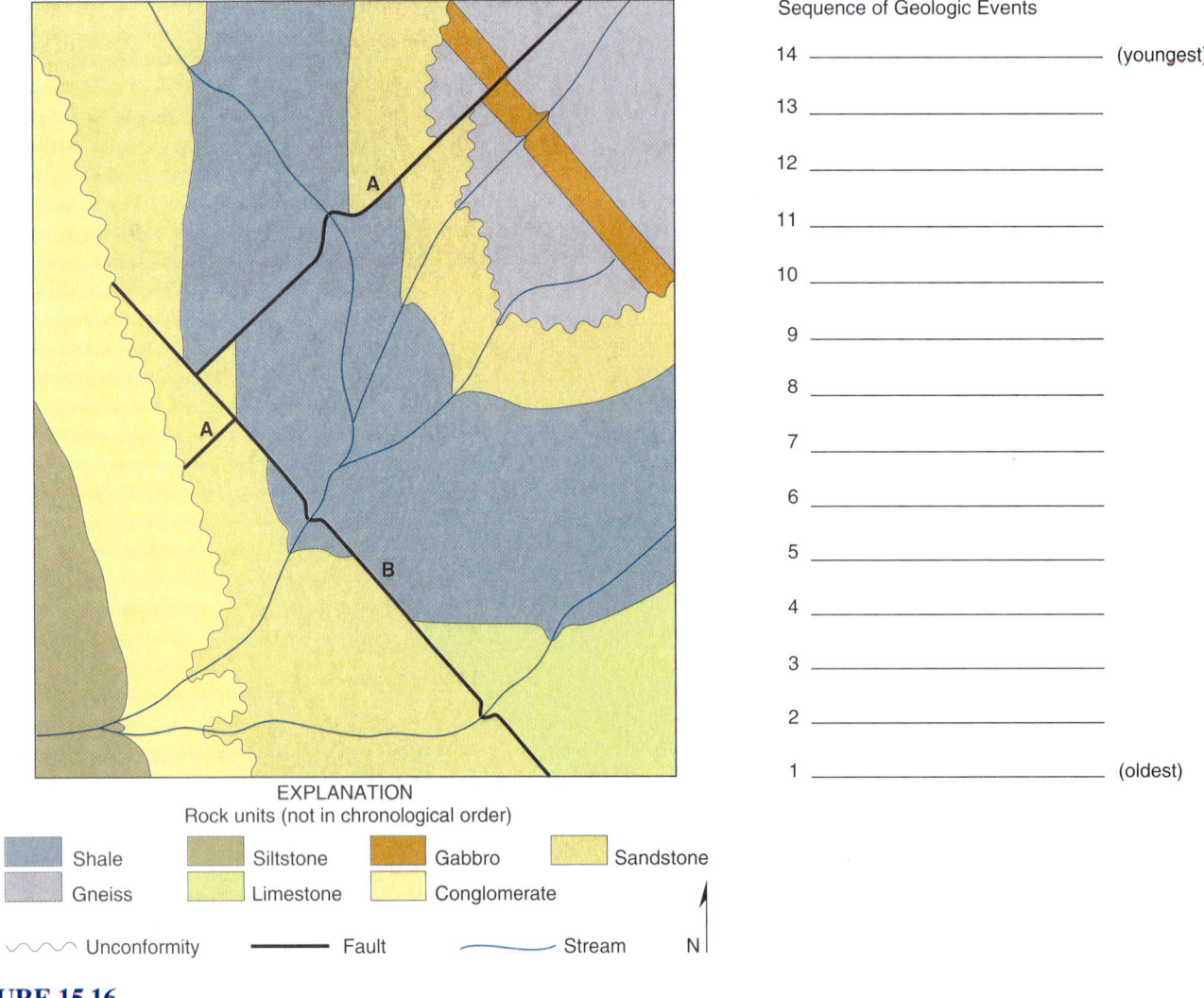

FIGURE 15.16
Geologic map for use with Problem 5.

e. Is fault *B* older or younger than fault *A*? How do you know?

Which side of fault *B* is the hanging wall and which the footwall?

f. Is the unconformity between the limestone and conglomerate an angular unconformity, a disconformity, or a nonconformity? How do you know? What type of unconformity is the one between the gneiss and the sandstone?

g. In the spaces to the right of the map, list the sequence of geologic events that produced the geologic relations shown on the map. Begin with the oldest.

6. Figure 15.17 is part of a geologic map of the Tazewell Quadrangle, Tennessee. The area is part of the Appalachian Mountain system, which here consists of Cambrian and Ordovician sedimentary rocks folded and cut by two prominent faults, the Wallen Valley thrust fault and the Hunter Valley thrust fault (extreme southeast corner).

Chapter 15   Geologic Maps   287

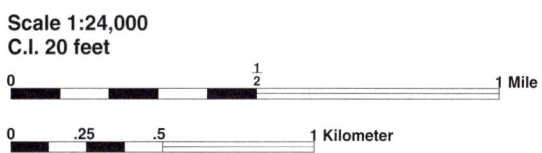

FIGURE 15.17
(Pages 288–289) Geologic map of portion of Tazewell Quadrangle, Tennessee, for use with Problem 6. Not all formations listed in the Explanation appear on this portion of the map. Source: U.S. Geological Survey.

a. What is the general structure of the rocks between the Wallen Valley (center of map) and Hunter Valley (southeast corner) thrust faults? That is, are the strata horizontal, inclined, folded?

b. Why are the contacts between *Ꞓcr, Oc, Ol,* and *On* wavy, whereas those farther southeast, such as between *Ow* and *Ohb,* are smooth?

c. In what direction do the two faults dip? Based on the outcrop pattern of the faults, is the dip steep or gentle? Compare the ages of the rocks immediately adjacent to each of the two faults. Do older rocks overlie or underlie younger rocks along the faults?

d. The outcrop pattern northwest of the Wallen Valley thrust fault is different from that to the southeast. Why? What features are present? Sketch an approximate cross section from the W in the word Wallen northwest to the edge of the map. You need not show the topography nor be too accurate in locating contact lines; the purpose is to see whether you understand the general structure.

e. Depressions, as indicated by depression contours, are abundant in some formations and absent in others. Explain. (Hint: In what kind of rock do the depressions occur?)

f. Draw a line on the map along which a cross section that best shows the geologic structures could be drawn. Label it A on one end and B on the other.

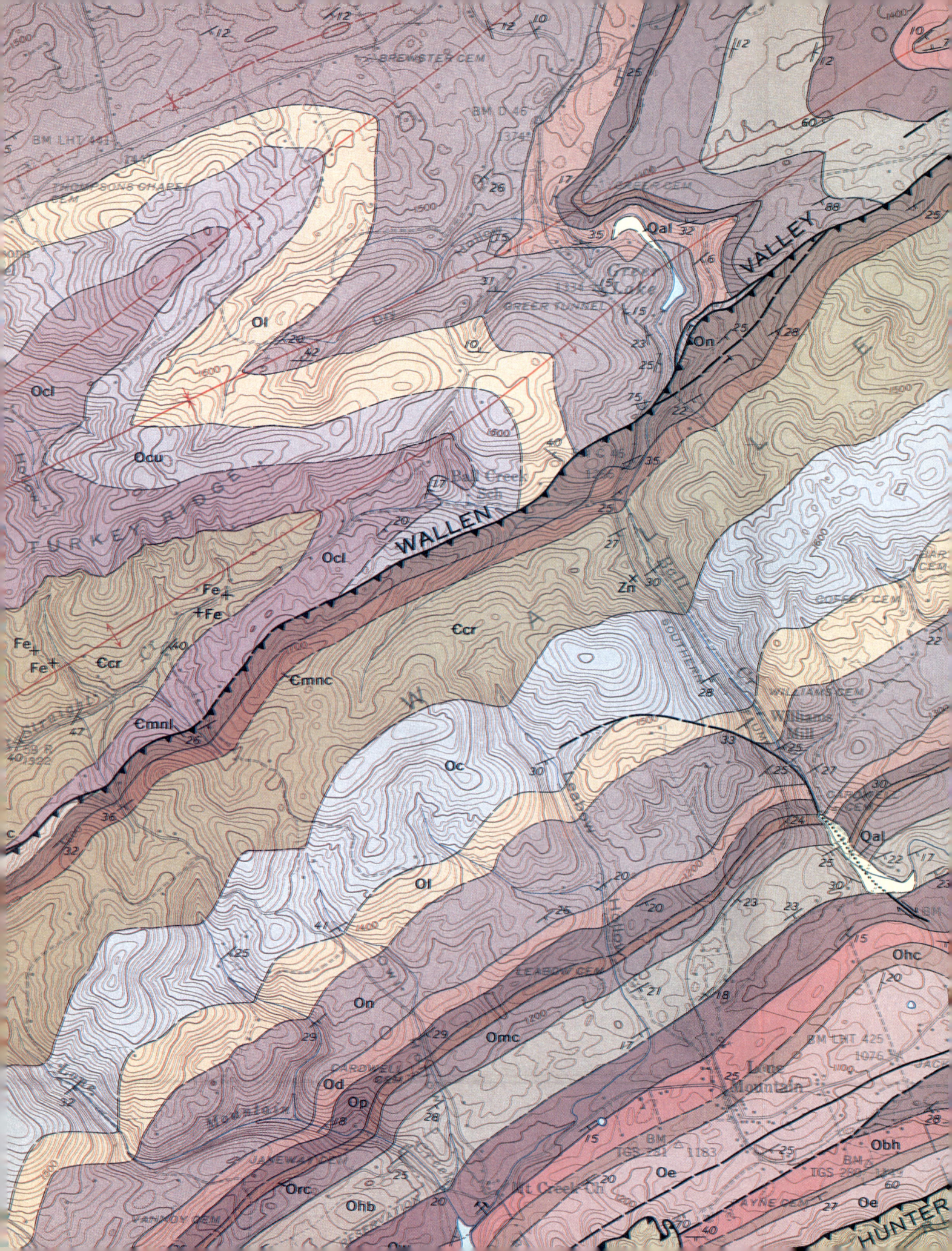

GEOLOGIC QUADRANGLE MAP
TAZEWELL QUADRANGLE, TENNESSEE
GQ-468

Explanation of symbols:

- Contact — Dashed where approximately located; dotted where concealed
- Fault — Dashed where approximately located; dotted where concealed. U, upthrown side; D, downthrown side; T, upper plate of minor thrust fault
- Tear Fault — Showing relative movement. Dashed where approximately located; dotted where concealed
- Major thrust fault — Dashed where approximately located; dotted where concealed
- Anticline — Showing trace of axial plane; dashed where approximately located; dotted where concealed
- Syncline — Showing trace of axial plane; dashed where approximately located; dotted where concealed
- Strike and dip of beds
- Strike and dip of overturned beds
- Horizontal beds
- Quarry
- Fe — Occurrence of limonite
- Zn — Zinc prospect pit

Stratigraphic units:

QUATERNARY
- Qal — Alluvium

SILURIAN
- Middle Silurian: Sct — Clinton Shale
- Lower Silurian: Sc — Clinch Sandstone (Scp, Poor Valley Ridge Member; Sch, Hagan Shale Member; Scs, Clinch Sandstone Member — Not exposed but inferred in subsurface)

ORDOVICIAN
- Upper Ordovician:
  - Os — Sequatchie Formation
  - Or — Reedsville Shale
- Middle Ordovician (Chickamauga Group of Swingle (1964)):
  - Ot — Trenton Limestone
  - Oe — Eggleston Limestone
  - Ohc — Hardy Creek Limestone
  - Obh — Ben Hur Limestone
  - Ow — Westway Limestone
  - Ohb — Hurricane Bridge Limestone
  - Omc — Martin Creek Limestone
  - Os — Rob Camp Limestone
  - Op — Poteet Limestone
  - Od — Dot Formation
- Lower Ordovician (Knox Group):
  - On — Newala Dolomite
  - Ol — Longview Dolomite
  - Oc — Chepultepec Dolomite (Ocu, upper member; Ocl, lower member)

CAMBRIAN
- Upper Cambrian (Knox Group):
  - Ccr — Copper Ridge Dolomite
  - Maynardville Formation (Cmne, Chances Branch Dolomite Member; Cmnl, Low Hollow Limestone Member)
- Middle Cambrian (Conasauga Group):
  - Cn — Nolichucky Shale
  - Cm — Maryville Limestone
  - Crv — Rogersville Shale
  - Crt — Rutledge Limestone
  - Cpv — Pumpkin Valley Shale
- Lower Cambrian:
  - Cr — Rome Formation (Crd, dolomite member)

289

7.  a. Go to *www.ogs.ou.edu/education/intgeol/index.htm* (or link to it through *www.mhhe.com/jones5e*—see Preface), and use the geologic map of Oklahoma and the cross sections to answer the following questions.

   (1) What is the age (give the name of the period) of the pre-Quaternary bedrock exposed near Oklahoma City? Near Tulsa?

   (2) Why do the Quaternary sediments (in yellow) form streaks across the state?

   (3) What is the age of the rocks exposed in the Wichita Mountains and what kinds of rocks are they?

   (4) The Wichita Mountains and the Anadarko Basin are separated by a steep fault. Which side has gone down relatively?

   Is the fault a normal or reverse fault according to the cross section?

   (5) What type of contact (see Chapter 14) separates the Precambrian and Cambrian igneous rocks from the Cambrian sedimentary rocks in the Anadarko Basin?

   (6) Many of the faults in the Ouachita Mountains are thrust faults that originally had low dips. The dips in cross section B-B' appear steeper than they really are, because the cross sections are vertically exaggerated about 10 ×. Based on cross section B-B', in which relative direction were the rocks above the thrust-fault surfaces moved, toward the north or south?

   Rocks of the Ouachitas have also been folded. Based on B-B', did folding occur before or after thrust faulting?

   b. You can find information about the geology of your state on the web at *http://geology.about.com/cs/geomapsusstates/*, or go to *www.consrv.ca.gov/cgs/index.htm* and link to your state geological survey to see what's available, or try some of the universities of colleges in your state.

In Greater Depth

8. Figure 15.18 is a geologic map of the Wetterhorn Peak Quadrangle, Colorado. This area is within the San Juan Mountains of Colorado, a range composed principally of Tertiary volcanic rocks. Tertiary units present are: (1) San Juan Formation (*Tsj* in gold), which consists of massive tuff, volcanic breccia, and lava flows of rhyolitic to andesitic composition; (2) porphyritic quartz latite, a fine-grained rhyolite-like rock with both plagioclase and K-feldspar (*Tql* in red); and (3) Potosi Volcanic Group, consisting mainly of welded pyroclastic-flow tuff (or ignimbrite) of rhyolitic composition (*Tp* in green).

   a. Is the porphyritic quartz latite (the red bodies) extrusive or intrusive igneous rock? Explain. What geologic term is given to the long, narrow bodies of *Tql*?

   b. List the three Tertiary units in order of age, with youngest first.

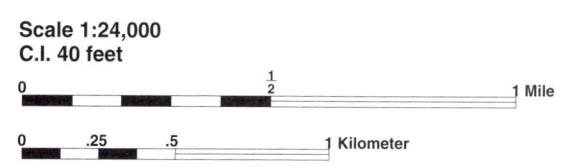

Scale 1:24,000
C.I. 40 feet

FIGURE 15.18
(Page 292) Geologic map of portion of Wetterhorn Peak Quadrangle, Colorado, for use with Problem 8.
Source: U.S. Geological Survey.

    c. Why do the long, narrow bodies of *Tql* follow such straight lines, even though the terrain is mountainous?

    d. Notice that the narrow bodies of *Tql* radiate outward from the irregular-shaped bodies of *Tql*, but the pattern is not random; certain orientations or strikes are preferred. Suggest one or more reasonable hypotheses to explain the preferred orientations.

    e. Five types of Quaternary surficial deposits are present on the map. All have *Q* as the first letter of their symbol. Two originated as stream deposits, the others as accumulations at the base of a cliff or steep slope, glacial deposits, or landslide deposits. Locate the five types on the map, and based on their locations, speculate on the origin for each, using its symbol to identify it.

    f. The red-dot pattern on Bighorn Ridge symbolizes a type of alteration, commonly associated with orebodies, that is caused by circulation of hydrothermal (hot water) solutions. Why might the alteration have occurred at that particular locality?

9. Refer to Figure 15.12 and answer the following questions.
    a. What kind of contact separates Precambrian rocks from the Cutler Formation? Be specific. Based on the nature of its lower contact, its stratigraphic position, and the description in the *Explanation,* give a hypothesis for the origin of the Cutler Formation.

    b. Note that the Navajo Sandstone is present in the southwest part of the map but does not continue to the north. Why? What kind of contact separates the Kayenta and Carmel Formations where the Navajo is absent? Be specific.

    c. The red lines running diagonally across the map from northwest to southeast are called *structure contours* because they are drawn on the top surface of a particular formation and show the shape of that buried surface. On this map, they are drawn on the top of the Entrada Sandstone (notice that they are dashed where the present-day surface is below the top of the Entrada). What is the strike of that surface? What is the gradient of that surface in feet per mile? Use trigonometry to calculate the angle of dip of the surface. (Hint: Convert the numerator and denominator of the gradient to the same units, then divide. The result is the *tangent* of the dip angle. The dip angle is the *arctangent* or *inverse tangent* on most calculators. If necessary, see your lab instructor for help.) Measure the dip on your cross section and compare it with the calculated dip.

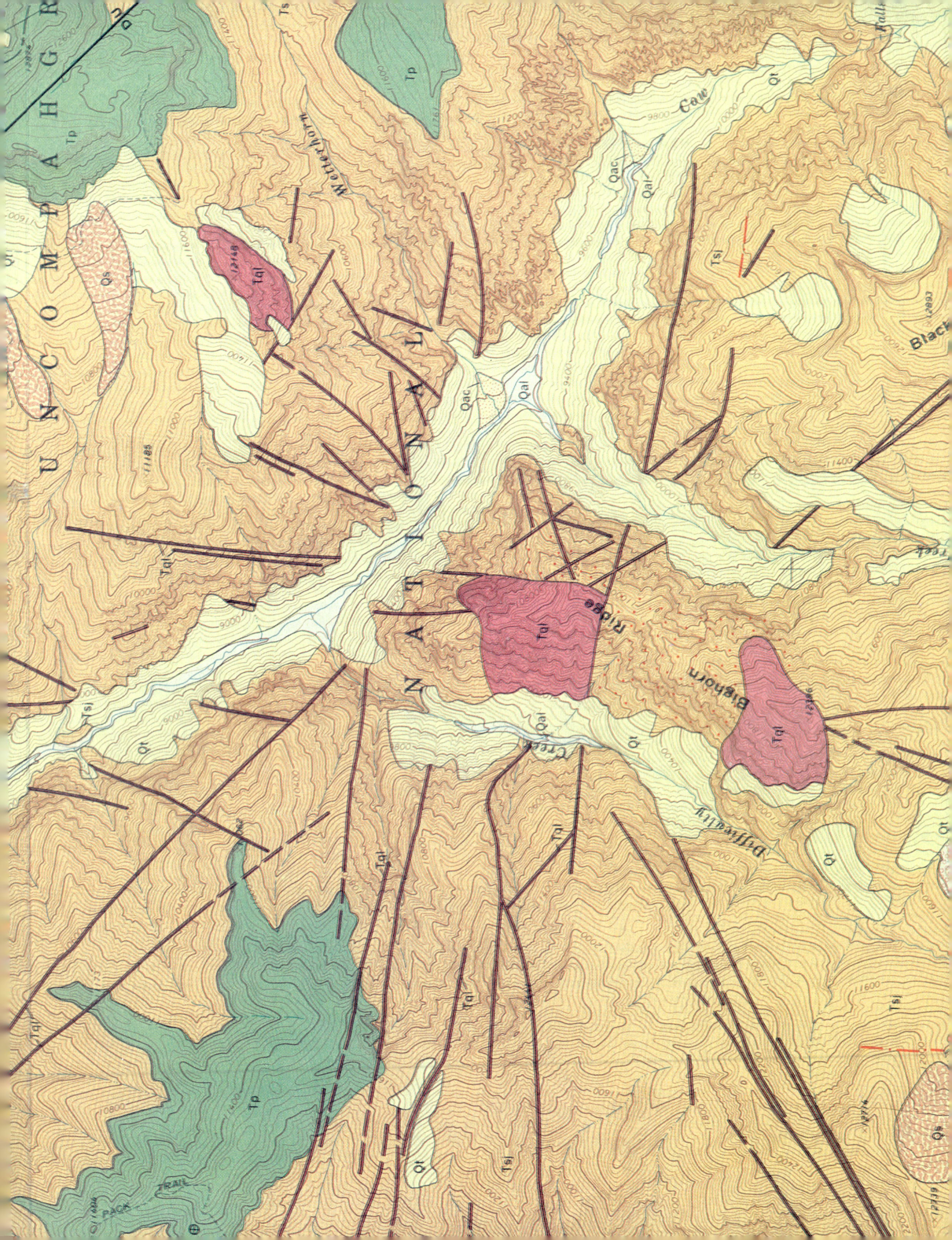

CHAPTER 16

Earthquakes

Materials Needed
- Pencil and eraser
- Ruler
- Drawing compass, or pencil on string, to draw circles
- Calculator

Introduction

Earthquakes are both destructive and informative. Energy from them travels through and around the Earth as three types of waves, called *P, S,* and surface waves. In this lab, you will learn to identify the three types of earthquake waves on a seismogram and to use the seismogram to locate an earthquake and determine its magnitude. You will also learn how the intensity or degree of shaking of an earthquake is measured and mapped and how intensity depends on the characteristics of the ground being shaken. Finally, a simple example will show how earthquakes are used to examine the Earth's interior.

Earthquakes are vibrations resulting from sudden movements within the Earth. They are most common at tectonic plate boundaries, as discussed more fully in Chapter 17. Most earthquakes result from slippage along faults, but some are caused by other kinds of movement. For example, the upward movement of magma into Mount St. Helens in 1980 caused earthquakes that first alerted geoscientists to the probability of an eruption. Huge landslides and collapse of caves can produce earthquakes, and vibrations caused by explosions, such as underground nuclear tests, are very similar to natural earthquakes. We can only presume that those who named the dinosaur *Seismosaurus* thought that, in its day, the huge dinosaur's footsteps must have generated some excitement.

Seismology is the study of earthquakes, and the branch of seismology known as earthquake seismology focuses specifically on the causes and effects of earthquakes, especially as earthquakes affect people. Because large earthquakes may result in death and destruction, a goal of earthquake seismologists is to be able to predict earthquakes. Other seismologists make use of the energy given off by earthquakes (or explosions) to see within the Earth, just as we process light energy with our eyes or sound energy with our ears to form images of our surroundings.

Seismic Waves

When a fault suddenly slips (**ruptures**), rocks on either side elastically snap back toward their original unstressed shapes. This elastic rebound releases a pulse of energy that radiates out in all directions (Fig. 16.1). The "point" below the surface at which rupture begins is called the **focus,** or **hypocenter,** of the earthquake. The focus is not just a point, but a small area along a fault that ruptures. Rupture may begin at one spot, but then propagates along the fault in an irregular, jerky fashion. In fact, a series of smaller **aftershock** earthquakes commonly rattle the region for a few days as stresses are relieved along the main fault or other nearby faults. The **epicenter** is the point on the Earth's surface directly above the focus. It is the epicenter that is first reported by the press, because it is of most concern to people and is easily determined.

Seismic energy is released in the form of **seismic waves,** which cause the Earth to vibrate or deform elastically as they pass by. The three principal kinds of seismic waves are as follows:

***P* wave (primary wave** or **compressive wave;** Fig. 16.2). The energy of a *P* wave travels through rock as a sequence of back-and-forth vibrations *parallel* to the direction of movement. Its passage causes alternate pushing (compression) and pulling (dilation) on particles in its path. *P* waves can travel in solids or liquids, and are the fastest of the three types of seismic waves.

***S* wave (secondary wave** or **shear wave;** Fig. 16.2). The energy of an *S* wave travels through rock as a sequence of vibrations *perpendicular* to the direction of movement. Its passage causes particles to vibrate up and

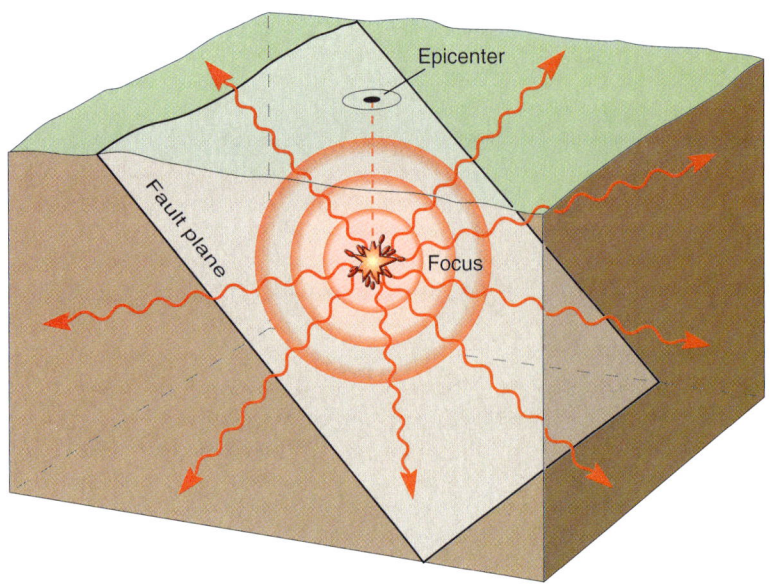

FIGURE 16.1
An earthquake originates at its *focus*, below the surface. The *epicenter* lies directly above the focus, on the surface. Seismic waves radiate outward in all directions from the focus.

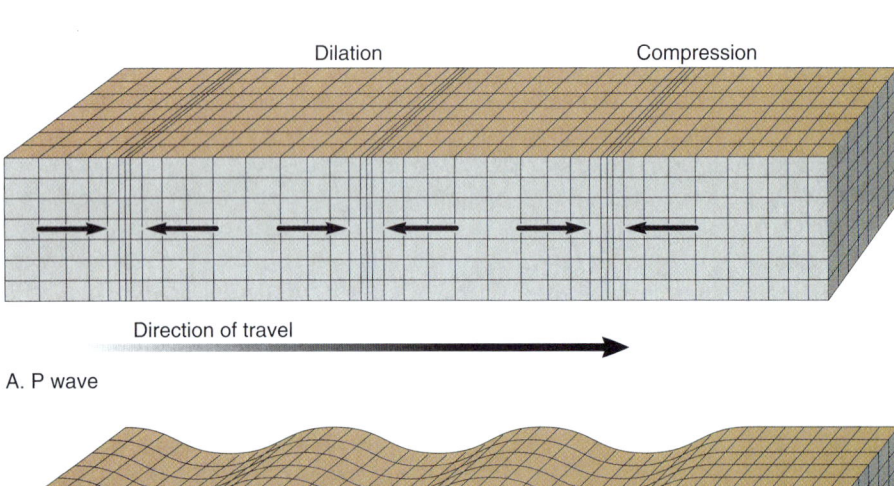

FIGURE 16.2
A. *P* waves vibrate back and forth, parallel to the direction of travel. B. *S* waves vibrate perpendicular to the direction of travel.

down or sideways. *S* waves can travel in solids but not in liquids, and their velocity is between that of the *P* and surface waves.

Surface waves. Surface waves travel along the surface of the Earth as two types of waves. One of these (called a Love wave) has a horizontal shearing motion similar to the motion of a snake across the ground, while the other (a Rayleigh wave) has a rolling up-and-down motion much like water waves. Surface waves are the slowest seismic waves, but typically these cause the most damage.

Seismograms

Earthquakes are very common; hundreds of thousands are detected each year. However, most are too small or too remote to be felt, and relatively few cause damage or loss of life. Earthquakes, even very small ones, are recorded with **seismographs.** A worldwide network of seismographs constantly monitors earthquake activity and provides information about an earthquake within minutes of its occurrence.

The record obtained with a seismograph is a **seismogram.** One type of seismogram is recorded on a rotating drum as a series of wiggly lines (Fig. 16.3). Tick marks or breaks in the lines are precisely spaced and represent time elapsed; actual time is Universal Coordinate (or Greenwich Mean) Time and, for most stations, is accurate to a few thousandths of a second. Figure 16.3 shows a simple seismogram recording the *P, S,* and surface waves.

A seismograph station may have three seismographs arranged to record the vertical component and two horizontal components of motion (for example, E-W and N-S). The three together allow one station to give an analysis of an earthquake.

Locating Earthquakes

The distance from an earthquake focus can be calculated if the seismic-wave velocity and the time it took to travel to the seismograph are known. Earthquake data from seismograph stations are combined with data on how fast seismic waves travel through specific rock units to produce *travel*

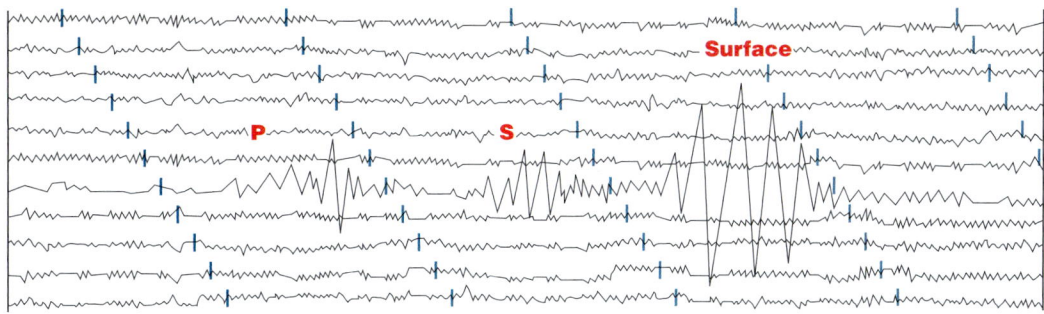

FIGURE 16.3
Simple seismogram with *P, S,* and surface waves. Blue tick marks indicate time. Originally written on a rotating drum, time moves forward from left to right in minutes, then skips down to the next line. One seismic event stands out above the background noise. Modern seismic data are recorded digitally.

time curves. **Travel time curves** show how the time required for various types of waves to travel a given distance varies with distance. In Figure 16.4, curves are shown for a *P* and an *S* wave. If the times at which an earthquake occurred and the *P* wave arrived are known, the distance from the earthquake can be read directly from the graph.

But what about the more common case in which the time at which the earthquake occurred is unknown? Because the farther a *P* wave travels, the more the *S* wave falls behind, the *difference* in *P*- and *S*-wave arrival times can be used to obtain the distance to an earthquake[1]. This is seen as an increasing vertical space between travel time curves in Figure 16.4. The distance to the earthquake epicenter can be determined from a seismogram by measuring the *difference* in *P*- and *S*-wave arrival times and finding the place on the travel time curves where they have the same time gap. The corresponding position on the horizontal axis gives the distance to the earthquake.

For example, let's say the seismogram in Figure 16.3 was recorded in Pierre, South Dakota, and the difference in arrival times of *P* and *S* waves is about 70 seconds, which on Figure 16.4 corresponds to a distance from the focus of about 500 km. Although the distance to the earthquake focus is known, the direction from the seismograph station to the focus is not known. Thus, on Figure 16.5, a circle of radius 500 km is drawn around Pierre. The earthquake, if a shallow one, must be located someplace on that circle. But where? One circle is not enough. Three intersecting circles for the same earthquake from three seismograph stations are needed, as illustrated in Figure 16.5. Thus a seismograph

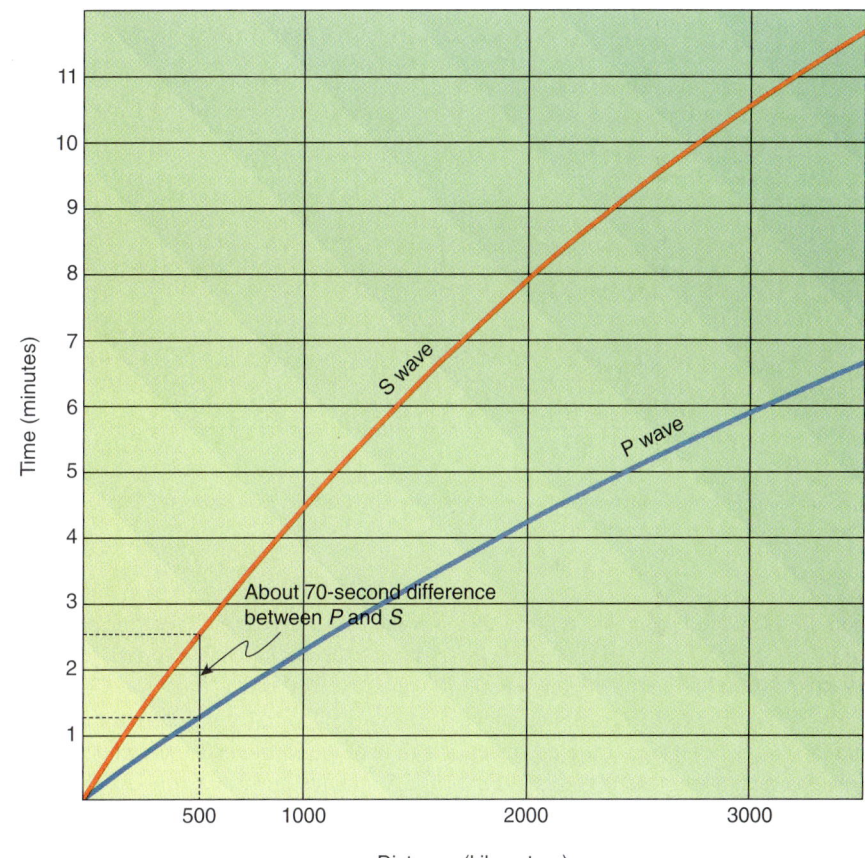

FIGURE 16.4
Travel-time curves for *P* and *S* waves. Note that the time interval between curves increases with distance. The seventy-second difference in arrival time between *P* and *S* waves indicates a distance to focus of about 500 km.

---

[1]The process is analogous to one you may have used to figure out how far away lightning was during a thunderstorm. You see a lightning flash instantaneously because the velocity of the light waves is very fast (3×10^5 km/sec), but the thunder takes awhile to get to you because the velocity of sound is much slower (about 0.34 km/sec). The farther away the lightning, the longer it takes for the thunder to get to you. If you count the seconds between lightning and thunder and multiply by the velocity of sound—0.34 km/sec—you know how far away the lightning was. A difference of 4 seconds means a distance of 4×0.34 km = 1.36 km, about 0.85 mile.

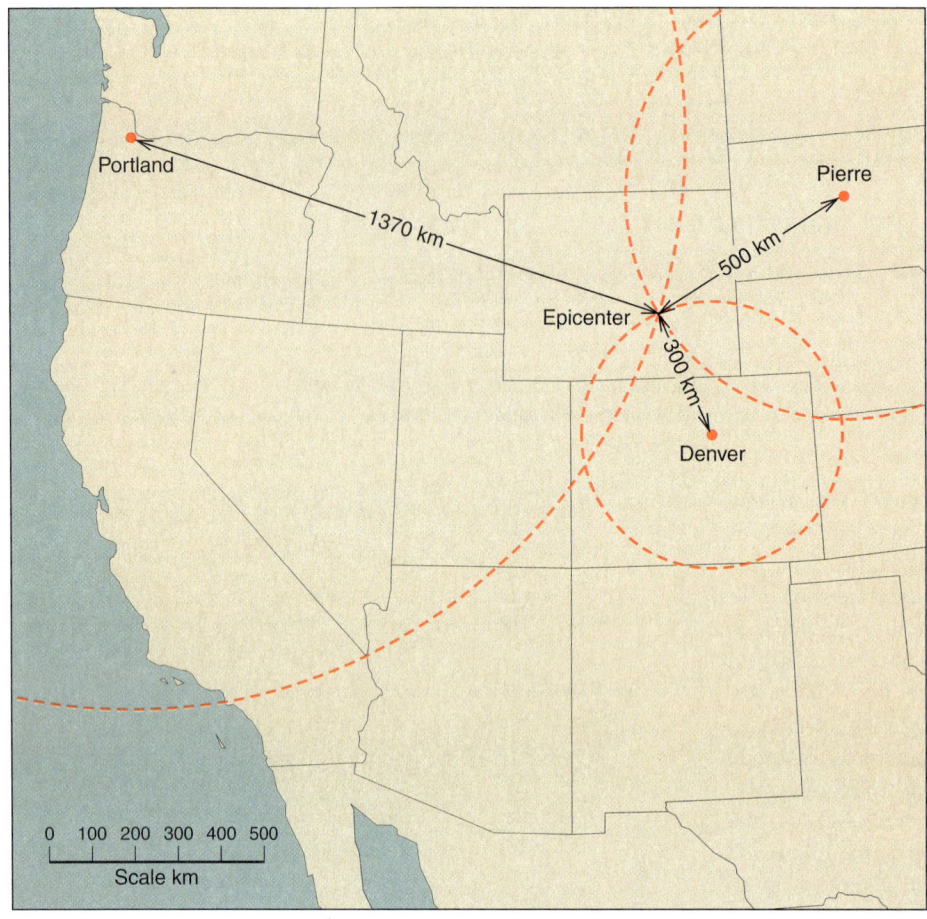

FIGURE 16.5
Seismograms from seismograph station at Portland, Oregon; Denver, Colorado; and Pierre, South Dakota are used to locate the epicenter near Douglas, Wyoming, with three intersecting circles.

station in Denver, Colorado, reports the earthquake as occurring 300 km away while a station in Portland, Oregon, reports it to be 1370 km away. Circles drawn about all three stations intersect at the epicenter, which lies near Douglas, Wyoming (Fig. 16.5).

When did the earthquake occur? On Figure 16.4, dotted horizontal lines are extended to the time axis from the point of intersection between the vertical line at a 500-km distance and the P- and S-wave curves. From this it can be determined that it took the P wave about 1.3 minutes, and the S wave about 2.5 minutes, to get from the earthquake focus to Pierre, South Dakota. Although not shown in Figure 16.3, the actual time represented by each blue tick mark is known, so the actual arrival time of P and S waves can be determined. For example, let's say that the P wave arrived at 13:45.3 Universal Coordinate Time and the S wave 1.2 minutes (70 seconds) later at 13:46.5. To find the time at which the earthquake occurred, subtract 1.3 minutes (the travel time for the P wave) from 13:45.3 to get 13:44 for the origin time of the earthquake. As a check, subtract 2.5 minutes (the travel time for the S wave) from 13:46.5 to get 13:44 again. Thus, the earthquake occurred at 13:44 Universal Coordinate Time. Since Douglas, Wyoming, is in the Mountain time zone, subtract 7 hours to get a local time of 6:44 AM.

Earthquake Intensity and Magnitude

The size of an earthquake is described in two ways: intensity and magnitude. **Earthquake intensity** is a measure of the destructive effects of an earthquake at a particular place and is a qualitative measurement based on people's experiences in an earthquake. Intensity is assessed with the *Modified Mercalli Scale* (Table 16.1), which has twelve categories (I to XII) that assess earthquake damage and human reactions to the quake. Questionnaires based on the scale are mailed out after an earthquake, and residents' responses are used to prepare a map of earthquake intensity. Lines of equal intensity, **isoseismals,** are like contours and outline the area where various levels of damage were recorded. Figure 16.6 illustrates the results for the Douglas, Wyoming, earthquake. The intensity felt throughout the area depends not only on the size and duration of the earthquake but also on the strength of the underlying ground. Areas underlain by loose sediment shook much more than those on solid bedrock. Notice that some people in Pierre felt the earthquake, but it was not felt in the surrounding area. These results illustrate why detailed earthquake intensity data are invaluable in preparing the earthquake

TABLE 16.1
Modified Mercalli Scale

I Not felt by people, except rarely under especially favorable circumstances.

II Felt indoors only by persons at rest, especially on upper floors. Some hanging objects may swing.

III Felt indoors by several. Hanging objects may swing slightly. Vibration like passing of light trucks. Duration estimated. May not be recognized as an earthquake.

IV Felt indoors by many, outdoors by few. Hanging objects swing. Vibration like passing of heavy trucks; or sensation of a jolt like a heavy ball striking the walls. Standing automobiles rock. Windows, dishes, doors rattle. Wooden walls and frame may creak.

V Felt indoors and outdoors by nearly everyone; direction estimated. Sleepers wakened. Liquids disturbed, some spilled. Small, unstable objects displaced, some dishes and glassware broken. Doors swing; shutters, pictures move. Pendulum clocks stop, start, change rate. Swaying of tall trees and poles sometimes noticed.

VI Felt by all. Damage slight. Many frightened and run outdoors. Persons walk unsteadily. Windows, dishes, glassware broken. Knickknacks and books fall off shelves; pictures off wall. Furniture moved or overturned. Weak plaster and masonry cracked.

VII Difficult to stand. Damage negligible in buildings of good design and construction; slight to moderate in well-built ordinary buildings; considerable in badly designed or poorly built buildings. Noticed by drivers of automobiles. Hanging objects quiver. Furniture broken. Weak chimneys broken. Damage to masonry; fall of plaster, loose bricks, stones, tiles, and unbraced parapets. Small slides and caving in along sand or gravel banks. Large bells ring.

VIII People frightened. Damage slight in specially designed structures; considerable in ordinary substantial buildings, partial collapse; great in poorly built structures. Steering of automobiles affected. Damage or partial collapse to some masonry and stucco. Failure of some chimneys, factory stacks, monuments, towers, elevated tanks. Frame houses moved on foundations if not bolted down; loose panel walls thrown out. Decayed pilings broken off. Branches broken from trees. Changes in flow or temperature of springs and wells. Cracks in wet ground and on steep slopes.

IX General panic. Damage considerable in specially designed structures; great in substantial buildings, with some collapse. General damage to foundations; frame structures, if not bolted, shifted off foundations and thrown out of plumb. Serious damage to reservoirs. Underground pipes broken. Conspicuous cracks in ground; liquefaction.

X Most masonry and frame structures destroyed with their foundations. Some well-built wooden structures and bridges destroyed. Serious damage to dams, dikes, embankments. Landslides on riverbanks and steep slopes considerable. Water splashed onto banks of canals, rivers, lakes. Sand and mud shifted horizontally on beaches and flat land. Rails bent slightly.

XI Few, if any masonry structures remain standing. Bridges destroyed. Broad fissures in ground; earth slumps and landslides widespread. Underground pipelines completely out of service. Rails bent greatly.

XII Damage nearly total. Waves seen on ground surfaces. Large rock masses displaced. Lines of sight and level distorted. Objects thrown upward into the air.

---

hazard maps that guide property development and regional engineering and construction standards.

Earthquake magnitude is a quantitative measure of the amount of energy released by the earthquake. The original magnitude scale was the *Richter Magnitude Scale*. It defined magnitude as the logarithm (base 10) of the amplitude[2], measured in micrometers (thousandths of a millimeter), of the largest waves recorded on a specified seismograph at a distance of 100 km from the epicenter. The distance must be specified because amplitude decreases with distance from the focus; the seismograph is specified to account for instrument variability. The assumption was that more energetic earthquakes would produce bigger wiggles, so that the height—or **amplitude** (top of Fig. 16.7)—of a selected wave would be proportional to the amount of energy released. There are no lower or upper limits to the Richter scale. However, the largest Richter-scale magnitude recorded thus far is 8.6. Because the Richter scale is a logarithmic scale, an increase of one unit in the scale corresponds to a 10-fold increase in amplitude. An increase in one unit in magnitude suggests an increase in the amount of energy released of about 32 times.

The original Richter scale was designed in 1935 for use with shallow focus earthquakes occurring in southern California, where the largest waves are common *S* waves. Since then, similar amplitude-based earthquake scales have been developed for more distant and/or deeper earthquakes. However, it was since found that these amplitude-based scales do not work well for the largest earthquakes, those with magnitudes greater than about 8.

Thus another kind of earthquake magnitude scale, called the **moment magnitude scale,** is now preferred by many seismologists because it is directly related to the amount of energy released by fault rupture and it is accurate for earthquakes of any size. It is based on the strength of the rock in which the fault occurs, the area ruptured by the fault, and the average displacement on the fault. The moment magnitude can be determined by detailed analysis of the seismogram, which is the usual practice.

---

[2] *The logarithm (base 10) of a number is the power to which 10 must be raised to equal that number. A logarithm can be a whole or fractional number. For example, $\log_{10} 100 = 2$; $10^2 = 100$. And $\log_{10} 5 = 0.699$; $10^{0.699} = 5$.*

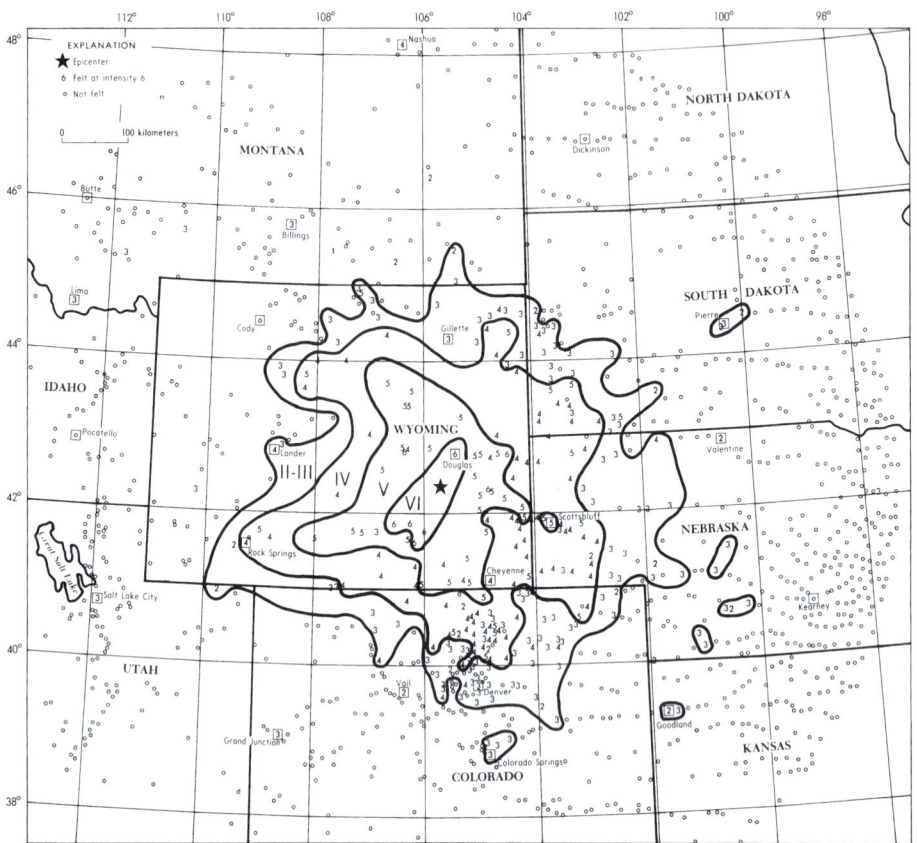

FIGURE 16.6
Intensity map for earthquake near Douglas, Wyoming, with isoseismal lines based on the Modified Mercalli scale. The contours are drawn as relatively simple, generally concentric shapes that are intended to summarize the most important patterns in the data. Intensely convoluted contours or contours encircling single data points are of little value.

Because the data necessary to calculate the moment magnitude require some time to gather, the first magnitude reported to the press might be an amplitude-based magnitude, which is later updated with the energy-based moment magnitude. Fortunately, the magnitudes of the amplitude-based scales are comparable to the moment-magnitude scale (at least up to about 8). Although the moment-magnitude scale is preferred, we focus on determining Richter magnitudes because the calculations are more straightforward.

Figure 16.7 illustrates the measurement of the Richter magnitude of an earthquake modified from the original Richter method (after Bolt, 1988). (1) Measure the amplitude of the largest surface wave in millimeters; in the example, it is 20 mm. (2) Use the graph to determine the Richter magnitude by running a straight-edge from the appropriate *Distance to epicenter* (assumed to be 60 km in the example) to the appropriate *Amplitude;* read the magnitude where this line crosses the *Magnitude* scale (about 5.6 in the example). This method applies only to local earthquakes less than 500 km away.

Using Seismic Waves to See Beneath the Surface

Seismic body waves (P and S) passing through the Earth do not follow straight-line paths but are bent (refracted) and reflected when they encounter rocks of contrasting densities (Fig. 16.8). Their behavior is just like light in this regard. When light travels from air (density A) to water (density B), some light is reflected while the rest is bent as it passes into the water. Figure 16.8 shows a P wave being reflected and traveling back to the surface. If the velocity and travel time of the P wave are known, the distance traveled can be calculated, and the depth to the reflecting horizon can be determined. Reflection seismology, using energy from human-produced explosions or vibrations, is used to map geologic structures below the surface and is one of the most important methods of exploration used in the oil industry.

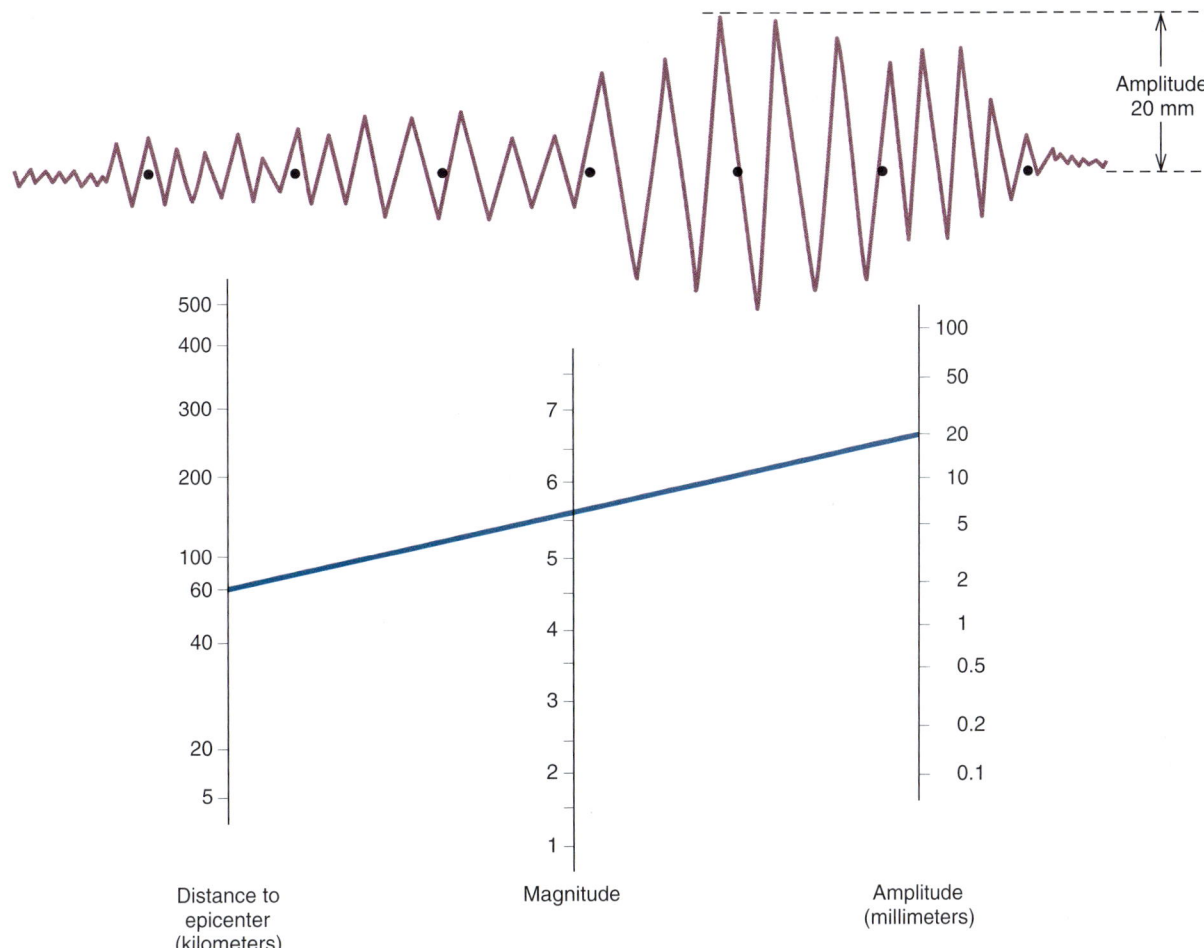

FIGURE 16.7
A line joining the amplitude of the largest wave on the seismogram and the distance to the epicenter crosses the magnitude scale at 5.6. This method is applicable to nearby earthquakes only.

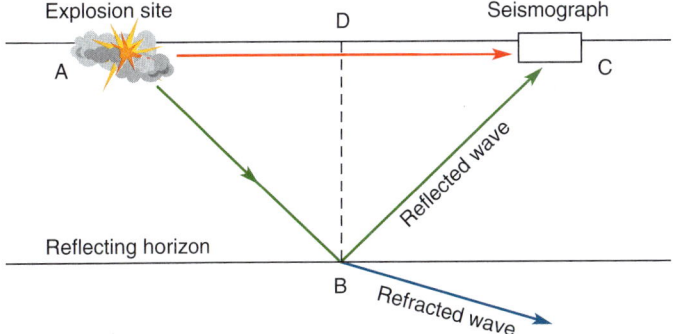

To determine depth to the reflecting horizon

1. Knowing the travel time and distance, ADC, from the explosion site to the seismograph, calculate the *P*-wave velocity (velocity = distance/time).

2. Knowing the travel time of the reflected wave and the velocity found in Step 1, calculate the distance, ABC, travelled by the reflected wave (distance = velocity × time).

3. Use the distances AD and AB to calculate the depth to the reflecting horizon, DB. (Remember the Pythagorean theorem? The square of the hypotenuse of a right triangle equals the sum of the squares of its legs—in this case, the hypotenuse is AB, and AD and DB are the legs, so $AB^2 = AD^2 + DB^2$.)

FIGURE 16.8
Reflection of a *P* wave and calculation of depth to reflector.

Hands-On Applications

The Loma Prieta Earthquake: Background information

As millions of Americans were tuned in to the beginning of the third game of the World Series (championship baseball) at Candlestick Park, San Francisco, California, on October 17, 1989, a large earthquake occurred. It was the largest to strike the San Francisco Bay region, home to almost 6 million people, since the great earthquake of 1906, when fewer than a million people lived there. The earthquake was felt over 400,000 square miles and resulted in 67 deaths, 12,000 people left homeless, and $6 billion in property damage. In terms of cost, it was one of the most expensive natural disasters ever in the United States, and it was the most costly earthquake since 1906.

The following problems illustrate how the epicenter was located, how the magnitude and intensities were determined, how the intensity was influenced by the nature of the surficial material, and how energy from the earthquake could be used to view the interior of the Earth. In each case, the scientific method is used to answer basic problems.

Objectives

If you complete all the problems, you should be able to:

1. Identify *P, S,* and surface waves on a simple seismogram.
2. Locate the epicenter of an earthquake using seismograms and travel-time curves.
3. Explain the basis for determining the magnitude and intensity of an earthquake, and draw isoseismals from intensity data.
4. Describe how different earth materials (mud, alluvium, bedrock) respond to seismic waves.
5. Use seismic reflection data to determine the depth to a reflector.

Problems

1. Figure 16.9 shows three simulated seismograms from the October 17, 1989, Loma Prieta Earthquake. Each was recorded at a different seismic station. Use them to answer the following:

    a. Where was the epicenter?

    (1) Find the differences in arrival time between *P* and *S* using Figure 16.9. Next determine the distance to the epicenter from each of the three stations using the travel-time curves in Figure 16.10.

    Distance from epicenter of: Station A

    Station B

    Station C

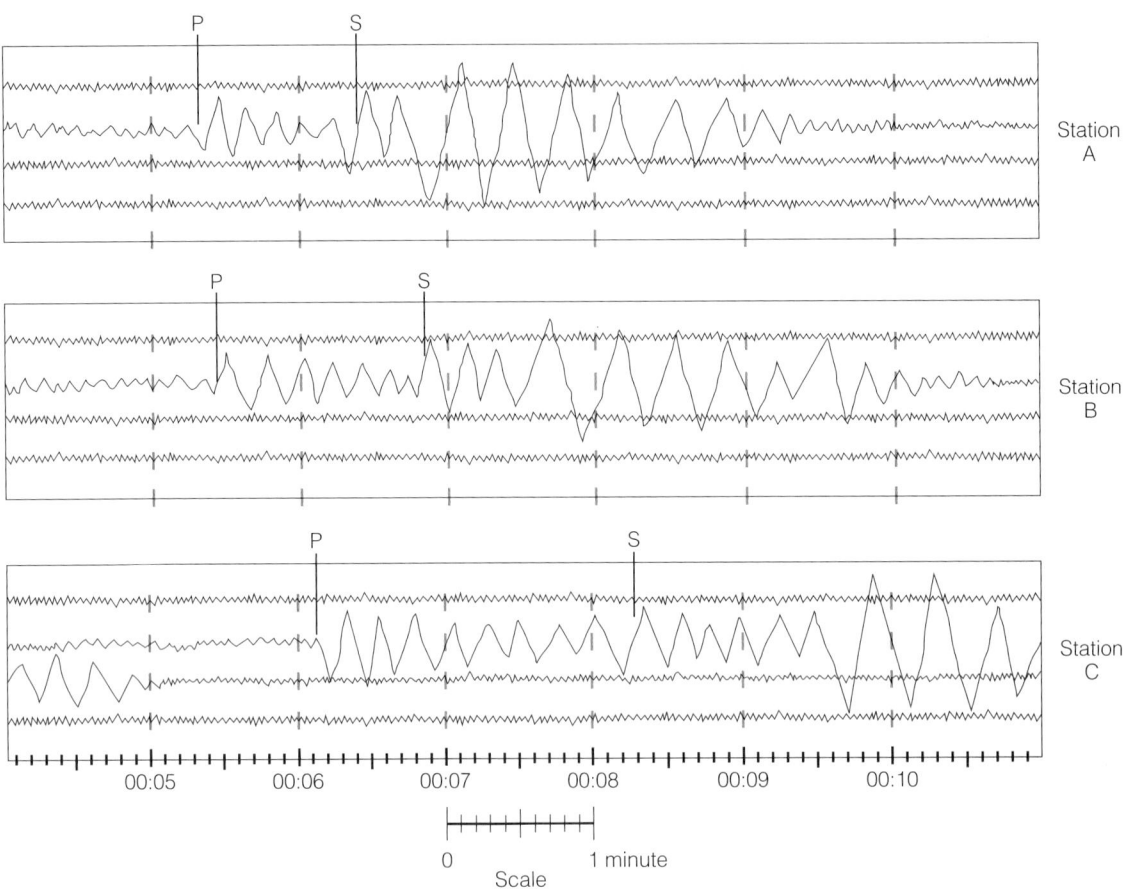

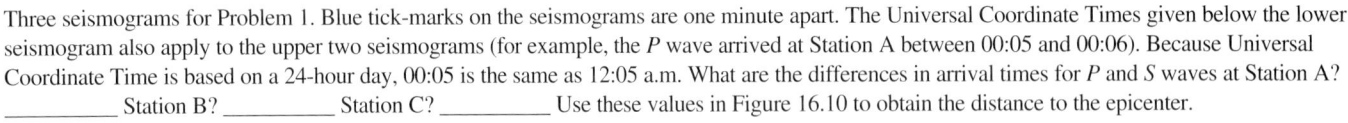

FIGURE 16.9
Three seismograms for Problem 1. Blue tick-marks on the seismograms are one minute apart. The Universal Coordinate Times given below the lower seismogram also apply to the upper two seismograms (for example, the *P* wave arrived at Station A between 00:05 and 00:06). Because Universal Coordinate Time is based on a 24-hour day, 00:05 is the same as 12:05 a.m. What are the differences in arrival times for *P* and *S* waves at Station A? _____ Station B? _____ Station C? _____ Use these values in Figure 16.10 to obtain the distance to the epicenter.

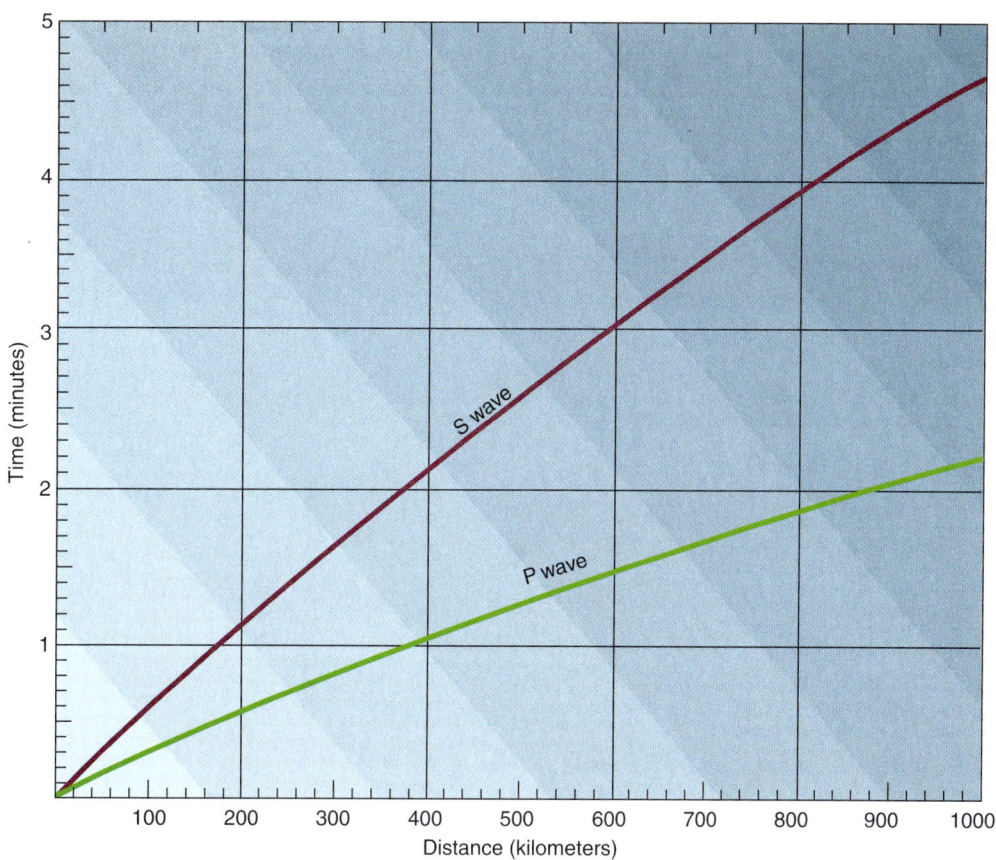

FIGURE 16.10
Travel time curves for *P* and *S* waves. This enlargement of Figure 16.4 has the same time scale as the seismograms in Figure 16.9 and is to be used for Problem 1. What is the distance from the epicenter of Station A? _____ Station B? _____ Station C? _____.

    (2) Plot the locations of the stations on the map in Figure 16.11. See the caption for Figure 16.11 for station locations.

    (3) Locate the epicenter of the earthquake on Figure 16.11 from intersecting circles centered at the three stations. Note that your epicenter and your neighbor's may not be exactly the same. Why not?

  b. When did the earthquake occur?
    (1) Use the travel-time curves and the time values indicated on the seismograms to determine the time of occurrence.

    (2) The time you determined is Universal Coordinate Time. What time would it have been in the Pacific Daylight time zone if there is a seven-hour difference?

  c. How big was the earthquake? Determine the Richter magnitude from seismogram A using the method shown in Figure 16.7.

FIGURE 16.11
Map for Problem 1. Plot stations A (lat. 40°08′N, long. 121°30′W), B (lat. 34°00′N, long. 118°00′W), and C (lat. 41°00′N, long. 114°30′W). Then locate the epicenter using the distances from the epicenter obtained from Figure 16.10.

2. Questionnaires designed for assigning Modified Mercalli scale intensities to the October 17 earthquake were returned from the locations shown on Figure 16.12. The responses were divided into five groups, as shown in Table 16.2. Using these responses, assign Modified Mercalli Scale intensities to the five groups in Table 16.2 using Table 16.1 as a guide. Next, draw isoseismals on the maps based on intensity distribution. How does the area of highest intensity compare with the epicenter you located in Problem 1?

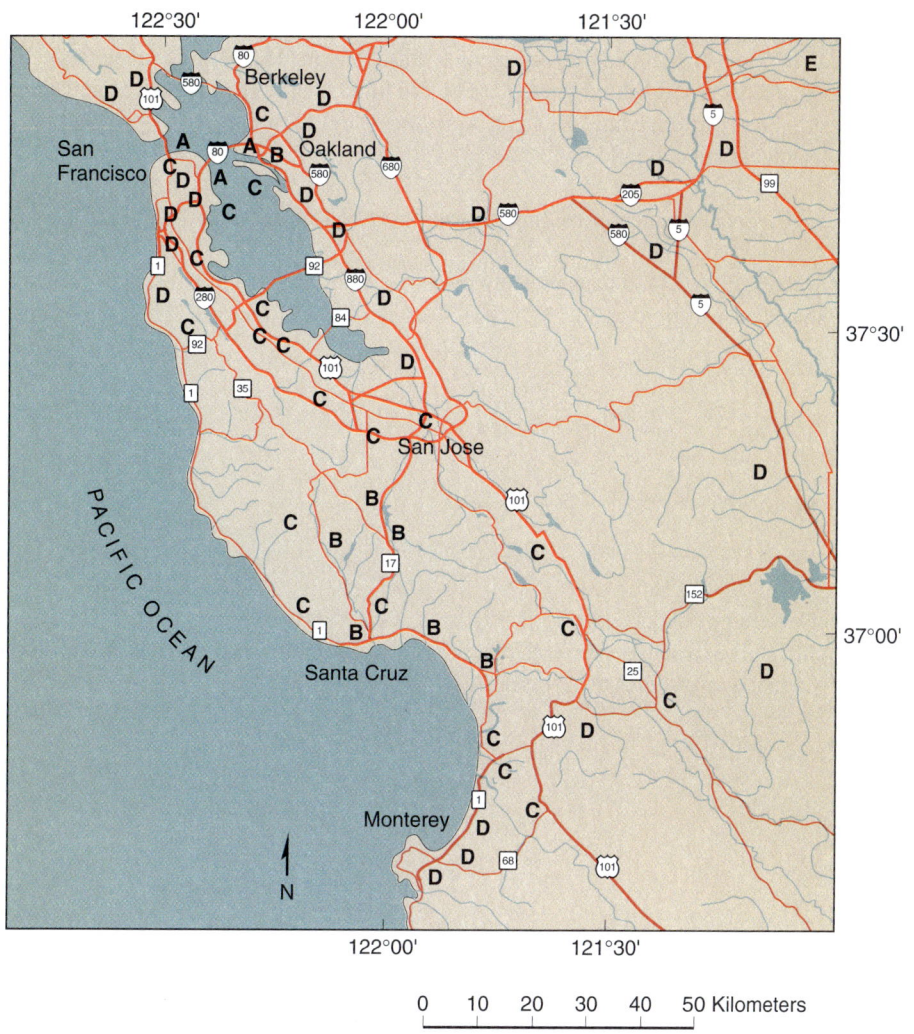

FIGURE 16.12
Map of San Francisco Bay area showing locations for which intensities have been assigned. Letters A through E correspond to earthquake intensities described in Table 16.2. For use with Problems 2 and 3.

TABLE 16.2
Earthquake Intensity Survey

Questionnaire Category Descriptions	Modified Mercalli Scale Equivalents
A. Felt by all, and people were extremely frightened. Drivers had great difficulty controlling vehicles. Some buildings collapsed, others very seriously damaged. Concrete bridge collapsed. Smell of gas in some areas. Ground cracked.	
B. Felt by all, most people were frightened. Drivers had some difficulty steering. Some reasonably well-built buildings partially collapsed, poorly built buildings badly damaged. Some water towers damaged. Large trees had tops broken. Some reports of furniture and people being thrown in air. Cracks in ground.	
C. Felt by all, some people knocked down. Noticed by some drivers. Damage to buildings moderate, except for those of poor construction. Windows broken, things knocked off shelves, plaster walls damaged. Trees and bushes shook. Caving noticed on some steep hillsides and along stream banks.	
D. Felt by all, some had trouble walking. Not noticed by most drivers. Damage to structures slight, but windows broken, things knocked off shelves, pictures knocked off walls. Plaster walls cracked.	
E. Felt by most people. Water in swimming pools disturbed. Some reported broken glassware. Pendulum clocks stopped.	

306   Part VI   Internal Processes

3. The area of highest intensity was near Oakland and San Francisco (Fig. 16.12). How can the intensity be so high, so far from the epicenter? Might it be related to the type of sediment or rock at the surface? Figure 16.13 is a *Landsat* false color image of the San Francisco Bay area. Bedrock appears as reddish brown to dark brown tones. Mud and fill occur around the margins of San Francisco Bay, indicated by either nearly black, green, white, or bluish white tones. Between these two is a light blue area speckled with red, which represents unconsolidated natural alluvium.

   a. Outline on Figure 16.13 the areas in which these three types of materials occur around San Francisco Bay.

   b. Compare your results with Figure 16.14, a map showing the expected effects of ground shaking in the San Francisco Bay area and with the earthquake intensity map you drew. Which of the three types of material shook the most and why?

   c. Sketch hypothetical seismograms obtained from seismographs, located at Oakland, that sit on (1) mud and fill, (2) unconsolidated soil (mainly alluvium), and (3) stable bedrock. (Because of proximity to the epicenter, *P, S,* and surface waves will not be separated, so just show relative amplitudes of vibrations.)

   d. On what famous fault was the October 17, 1989, earthquake centered (see Fig. 16.14 for named faults)?

FIGURE 16.13
Landsat image of San Francisco Bay area for use with Problem 3.

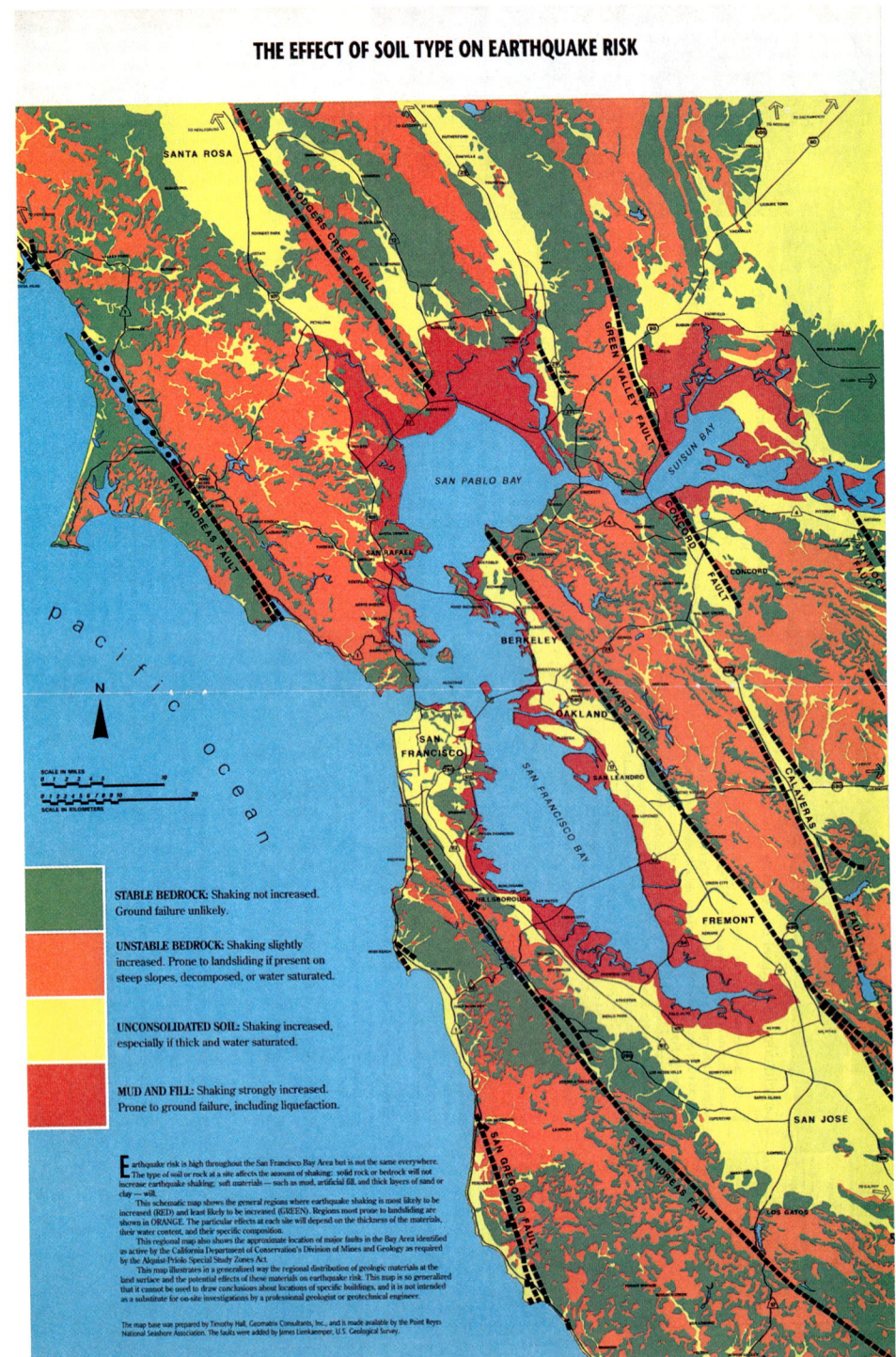

FIGURE 16.14
Map showing the expected effects of ground shaking in the San Francisco Bay area.

4. Go to www.eas.slu.edu/Earthquake_Center/earthquakecenter.html (or link to it through www.mhhe.com/jones5e—see Preface), select *Recent Earthquake Locations,* then select *Recent Earthquake Maps,* and choose the map of the area in which you live (e.g., *Continental US and Adjacent Canada*).

   a. When was the map last generated?

   b. Where was the nearest earthquake in the previous fourteen days?

   c. What was its magnitude?

   d. What was the depth of focus?

   e. If you are not looking at the map *Continental US and Adjacent Canada,* return to *Recent Earthquake Maps,* and select that map. Which state had the largest number of earthquakes?

   Where was the largest earthquake on the map, and what was its magnitude?

5. **Past Earthquake Records.** Visit http://quake.geo.berkeley.edu/cnss/ (or link to it through www.mhhe.com/jones5e—see Preface) and select *Earthquake Maps & Lists,* then *Maps of Earthquakes.*

   a. In the first row (World Map Mollweide Projection) and eighth column (30 years of records), click on $M > 6.5$ to see the location of all earthquakes with a magnitude greater than 6.5. Are shallow, intermediate, or deep focus earthquakes the most common? Which depth range is rarest?

   b. Around which ocean are most intermediate and deep earthquakes found?

   c. Now go to the Alaska Map row, 30 years of records column, and click on $M > 5.0$. Describe how the intermediate and shallow earthquakes are distributed relative to the string of Alaskan islands known as the Aleutian Islands and to each other.

   d. Go to the Western Hemisphere row, 30 years of records, and click on $M > 6.5$. The data are not as numerous here, and the view is from farther away, but can you see an earthquake focal-depth pattern analogous to that seen in Alaska? What is that pattern?

   Similar observations were important clues in the development of the theory of plate tectonics, as you will see in the next chapter!

You may want to check out www.sciencecourseware.com/VirtualEarthquake/. This is an excellent interactive site that leads you through the process of locating an earthquake and determining its Richter magnitude.

You also may find it interesting to look at the Global Earthquake Report by going to the www.geo.ed.ac.uk/quakexe/quakes site. There you find a list of the larger earthquakes that have occurred worldwide in the past week. You can access world and U.S. maps that show the epicenters, and you can zoom in on any area in those maps.

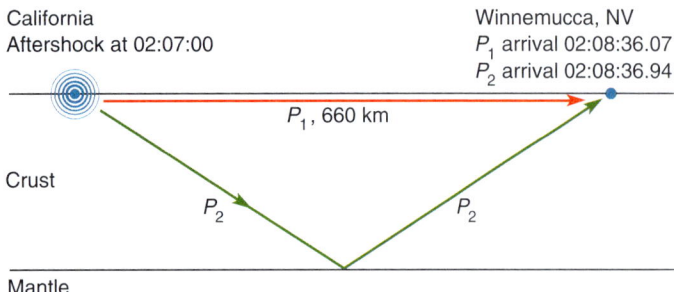

FIGURE 16.15
Cross section illustrating two paths for P waves from aftershock focus in California to Winnemucca, Nevada, 660 km away. Times given in hour:minute:second, Universal Coordinate Time. For use with Problem 6.

In Greater Depth

6. While the damage from earthquakes can be devastating, valuable information can be gained by studying them. The energy radiated from earthquakes can be used to look inside the Earth. An aftershock of the October 17 earthquake was recorded at Winnemucca, Nevada, a distance of 660 km away. Some of the P waves arriving in Winnemucca were reflected from the mantle, while others traveled there on a nearly direct path. Figure 16.15 illustrates, in cross section, two paths for the P waves. P_1 is the nearly direct path, and P_2 is reflected from the top of the mantle.

   a. If the earthquake occurred at 02:07:00 (Universal Coordinate Time), use the arrival time of P_1 (and the distance of 660 km) to calculate the velocity of P_1; show your work.

   b. Assume the velocity of P_2 is the same as P_1, and using its arrival time, calculate the distance traveled by P_2.

   c. Half of this distance will be the hypotenuse of a right triangle such as triangle ABD in Figure 16.8. Use the Pythagorean theorem ($c^2 = a^2 + b^2$) to calculate the depth to the mantle. Show your work.

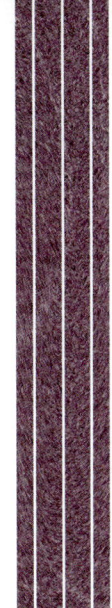

CHAPTER 17

Plate Tectonics

Materials Needed
- Pencil and eraser
- Metric ruler
- Protractor
- Calculator

Introduction

The plate tectonic theory is one of the most important and far-reaching theories in geology. It evolved from detailed observations made over many years and from individual hypotheses that required extensive and imaginative testing. For example, measurement of magnetism of the rocks on the seafloor was used as a test for seafloor spreading; the nature and distribution of earthquakes and volcanoes provided a test for plate convergence; and a variety of geological and geophysical studies proved the existence of transform faults. In this lab, a few predictions based on the plate tectonic hypothesis will be tested by analyzing appropriate data.

The theory of plate tectonics was more than 100 years in the making. Beginning in the nineteenth century, geologists began to compile basic observations about the Earth that could not be explained satisfactorily.

For example:

Fossils of the same kinds of land-dwelling plants and animals were found on opposite sides of oceans; how did these land dwellers cross the ocean?

Fossils of warm-weather plants and animals were found where climates are now cold, and evidence of glaciation was found in areas that now are warm; were Earth climates radically different in the past?

Earthquakes, volcanoes, and mountains are not randomly distributed about the Earth; why not?

Some interesting hypotheses were developed to explain these observations. To explain the distribution of fossils of land dwellers, it was supposed that continents were connected by land bridges that have since sunk into the oceans. To explain the rock-borne evidence of diverse climates, it was supposed that climates fluctuated widely in the geologic past. To explain the causes and distribution of earthquakes, volcanoes, and mountains, hypotheses involving contraction or expansion of the Earth were proposed; contraction would cause fold-and-thrust-faulted mountains, and expansion would cause fault-block mountains and provide conduits to the surface for magma. None of these hypotheses withstood testing, however, and eventually all were discarded.

The first detailed attempt at a unifying theory, one that could explain most of the major observations, was the continental-drift hypothesis of Alfred Wegener, proposed in a series of papers and books between 1910 and 1928. Wegener's hypothesis was severely criticized, especially by geologists in the northern hemisphere, on the grounds that it was mechanically impossible.

But more and more evidence seemed to suggest that continents indeed had moved relative to one another. Research that began in the late 1920s, and developed through the thirties and forties, indicated that a mechanism involving convection currents within the solid Earth might be possible. Additional geophysical evidence was especially persuasive, and by the early to mid-1960s, the framework of plate tectonics was in place.

At that time, plate tectonics was still a hypothesis, and there were many doubters, ready with observations that could not be readily explained by the existing model. The mark of a good hypothesis is that it can be tested, and the plate tectonics hypothesis offered many tests. Tests are conceived from predictions. That is, if the hypothesis is correct, then we can predict and test for certain consequent relations or results. In this lab, a series of predictions will be evaluated using real data.

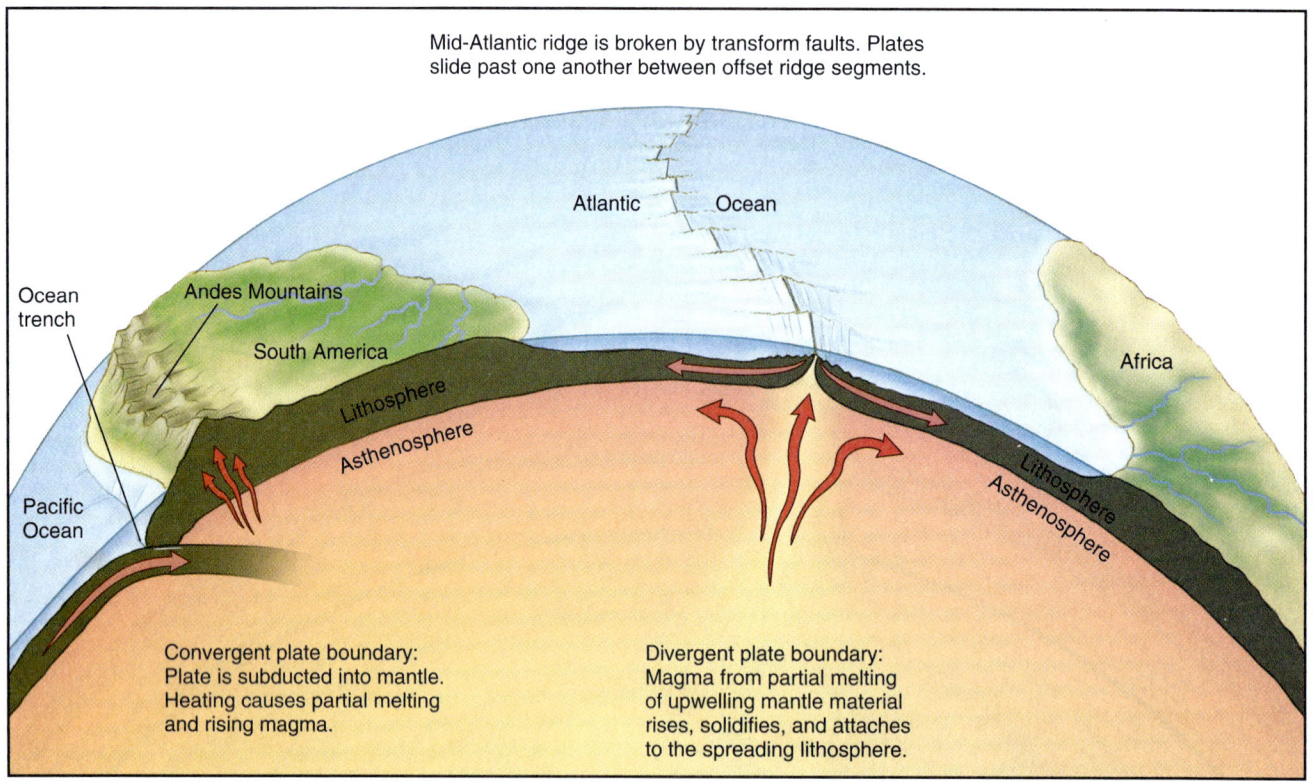

FIGURE 17.1
Cutaway view of the Earth, illustrating plates of lithosphere and the three types of plate boundaries. The thickness of lithosphere and asthenosphere is exaggerated.
Source: Data from USGS, *Professional Paper 1350.*

General Theory of Plate Tectonics

According to the theory of plate tectonics, the outermost part of the Earth is made of **lithospheric plates,** some 70 to 125 km thick, that move relative to one another (Fig. 17.1 and inside front cover). Plates are constructed by successive injections of magma at **divergent plate boundaries.** The plates move away from the divergent boundaries and, at **convergent plate boundaries,** they encounter other plates moving toward them in a relative sense. There, one of the plates moves beneath the other, and either descends into the mantle or becomes sutured (attached) to the overlying plate. A **transform-fault plate boundary** is a type of strike-slip fault; plates slide past one another, and lithosphere is neither created nor destroyed. Let's look at these three types of plate boundaries in a little more detail.

Divergent Boundaries

Plates move apart at divergent plate boundaries as magma from the mantle wells up to fill the gap between adjacent plates, as shown in Figure 17.2. Divergent boundaries in the oceans are expressed by ridges and rises, generally with a narrow rift valley marking the boundary between plates along a central ridge crest.

Perhaps the most critical piece of evidence for the plate tectonic hypothesis, or at least for the seafloor-spreading component, came in the early 1960s from the study of the magnetic properties of rocks on the floor of the North Atlantic Ocean. This story has two parts.

The first has to do with the way in which an igneous rock retains a record (its **remnant magnetism**) of the existing magnetic field as it cools from a molten state. Recall that some minerals, especially magnetite (Fe_3O_4), are strongly magnetic. However, if they are heated above a certain temperature, the **Curie point** (about 580°C for magnetite), the magnetism disappears. Conversely, magnetite, which forms at about 1100°C as magma crystallizes, becomes magnetic when the temperature drops below its Curie point. Its magnetism aligns with the existing magnetic field, as shown in Figure 17.3A, and is frozen into the rock as though the rock contained tiny bar magnets. Thus, the rock contains a record of the magnetic lines of force at that time and place; the north-seeking ends of the "magnets" point toward magnetic north, and the axes of their magnetic fields are parallel to the **inclination** of the magnetic lines of force. The angle of inclination is determined by the magnetic latitude at which the rock formed (Fig. 17.3A).

The second part of the story has to do with reversals of the magnetic field and **magnetic anomalies** (magnetic values

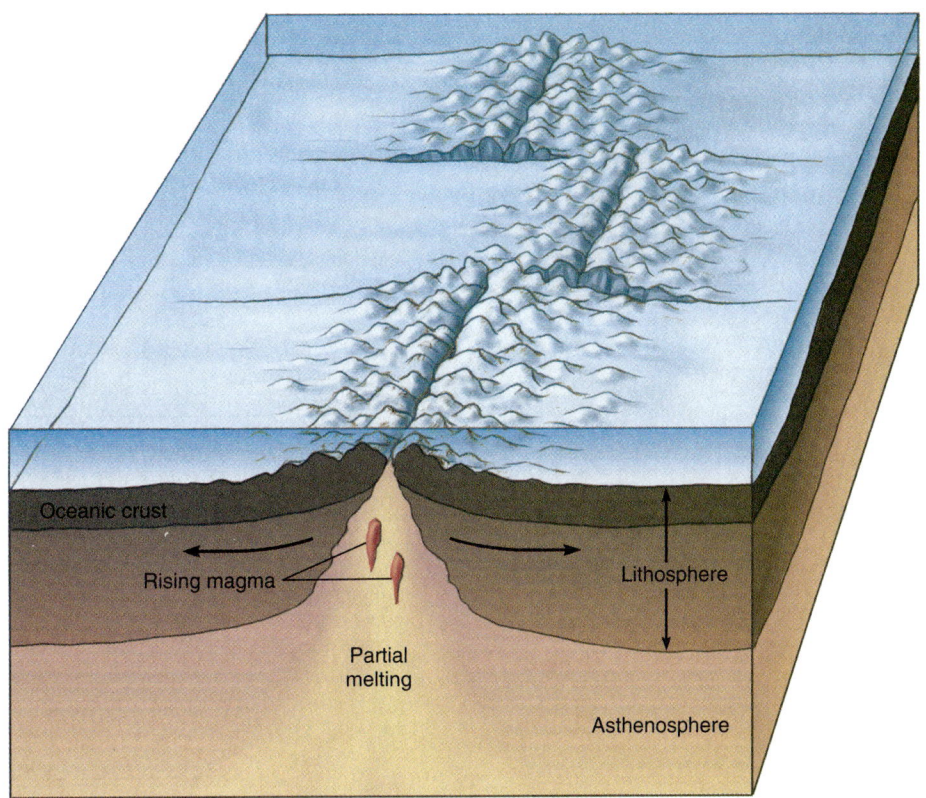

FIGURE 17.2
Block diagram of a mid-ocean ridge, a divergent plate boundary, at which seafloor spreading occurs.

greater or less than expected). In the present-day magnetic field, the north-seeking pole of a magnet points toward the north magnetic pole, which coincides approximately with the north geographic pole. But studies of remnant magnetism show that this has not always been the case. At many times in the geologic past, the orientation of the Earth's magnetic field has been reversed, as shown in Figure 17.3B. During these periods, the north-seeking poles of newly forming mineral "magnets" actually pointed toward the south geographic pole.

Imagine what could happen along a divergent boundary as magma rose, solidified, and took on the magnetic orientation existing at that time. During times of **normal magnetic polarity,** north-seeking remnant magnetic poles would point toward present-day north; during times of **reversed magnetic polarity,** north-seeking poles would point toward present-day south. As more magma is injected, the previously formed lithosphere (with one magnetic orientation) would be forced to move away from the plate boundary in both directions. Should the magnetic field reverse its polarity, newly injected magma would form rock that has a reversed polarity. Thus, as was hypothesized in the 1960s, alternating stripes or bands of rock with either normal or reversed remnant magnetism should form parallel to the divergent ocean ridges. Such magnetic stripes can be detected by a ship towing a magnetometer close to the seafloor. When passing close to rocks with normal remnant magnetism, their magnetic intensity is added to the modern magnetic field and a **positive magnetic anomaly** results. When passing over rocks with a reversed remnant magnetism, a weaker **negative magnetic anomaly** results.

It was an exciting time in geology when parallel bands of positive and negative magnetic anomalies were discovered along the Mid-Atlantic Ridge (Fig. 17.4). Here was a magnetic-tape-like record of past reversals of the magnetic field as recorded by basalt extruded at a divergent plate boundary. If the numerical age of a sample from one of the bands on Figure 17.4 were known (for example, from radioactive isotopes in the sample), then the age of the entire band could be inferred as could that of the symmetrically equivalent band on the opposite side of the divergent boundary.

If seafloor spreading occurs as described, and if magnetic anomalies result from magnetic polarity reversals, you should be able to predict the answers to the following questions and propose ways to test your predictions. (1) Will magnetic anomalies on opposite sides of a divergent boundary correlate; that is, will a negative anomaly indicating a specific reversal occur on both sides of the boundary? (2) Will magnetic data from other ocean ridges or rises show the same general pattern of parallel bands as shown in the North Atlantic? (3) Will specific magnetic anomalies correlate on a worldwide basis; that is, will the same anomalies be found worldwide?

314 Part VI Internal Processes

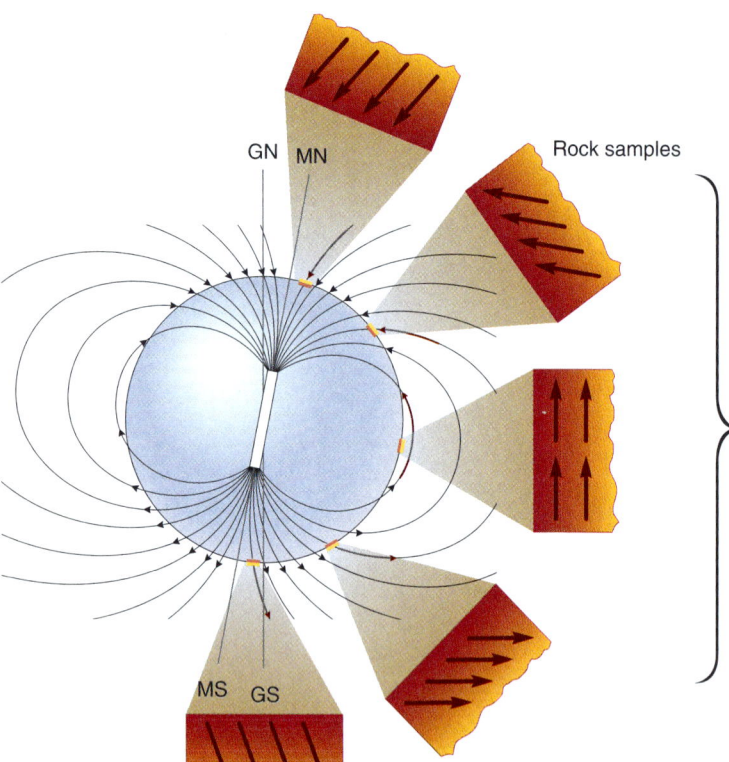

A. Normal geomagnetic polarity

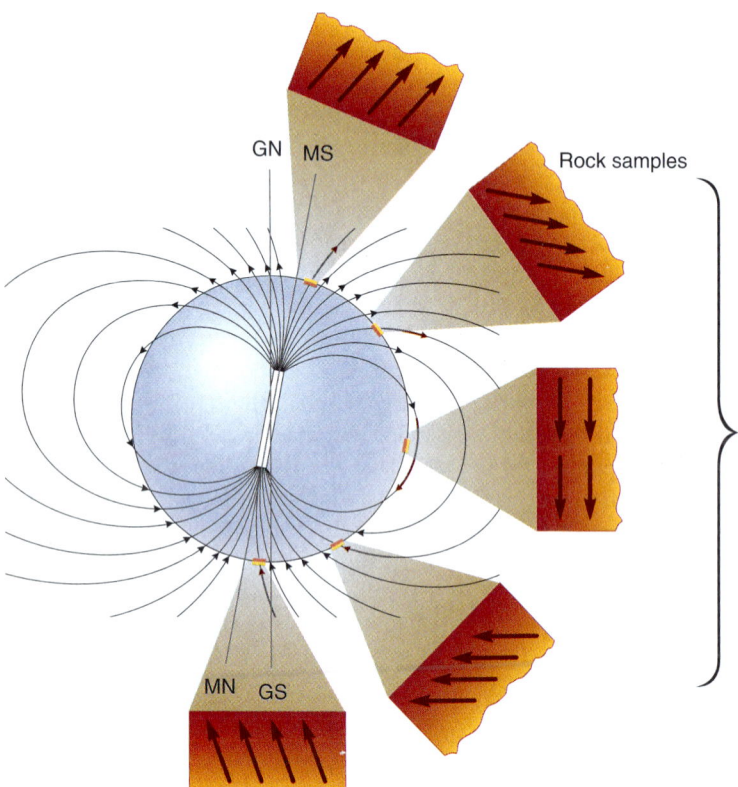

B. Reversed geomagnetic polarity

Basat samples from various latitudes have *normal* remnant magnetism. Arrows in samples are parallel to, and point in the same direction as, magnetic lines of force at each location and reflect the magnetic latitude at which rocks formed. Yellow marks upper surface of sample.

Basalt samples from various latitudes have *reversed* remnant magnetism. Arrows in samples are parallel to, and point in the same directions as, magnetic lines of force at each location and reflect the magnetic latitude at which rocks formed. Yellow marks upper surface of sample.

You also should be able to propose a way to use the magnetic anomalies to determine the rate at which seafloor spreading has occurred in a given area. Problems 1 and 6 deal with magnetic anomalies at divergent plate boundaries.

Convergent Boundaries

Plates come together at convergent boundaries, as illustrated in Figure 17.5. Converging plates contain either oceanic crust or continental crust at their leading edge. When two plates with oceanic crust converge, one dives, or is **subducted,** below the other (Fig. 17.5A). When an oceanic plate and a continental plate converge, the more dense oceanic plate is subducted beneath the continental plate (Fig. 17.5B). When two plates with continental crust collide, only partial subduction occurs. This is because continental crust contains rocks that are too buoyant (low density) to be forced down into the mantle. Oceanic rocks (basalt and gabbro) can subduct because they are only slightly less dense than the mantle. The partial subduction is accompanied by intense deformation and eventual suturing of segments of the subducted plate to the base of the overriding plate (Fig. 17.5C).

A subducted plate is comparatively cold and rigid when it begins its descent, but it warms and softens as it moves downward. Let's consider two of the many consequences implied by the converging process.

Earthquakes

Most of the world's major earthquakes take place on convergent plate boundaries. The

FIGURE 17.3

Earth's magnetic lines of force during (A) *normal* and (B) *reversed* geomagnetic polarity. During normal polarity, the magnetic north pole (MN) is near the geographic north pole (GN). When the magnetic pole is reversed, MN is near the geographic south (GS) and the magnetic south pole (MS) is near the geographic south. Lavas that cooled through the Curie point during normal polarity have a normal remnant magnetism with an orientation appropriate to the latitude at which they formed. This is illustrated in A by the red and yellow rock samples from various latitudes; samples with *reversed remnant magnetism* are shown in B.

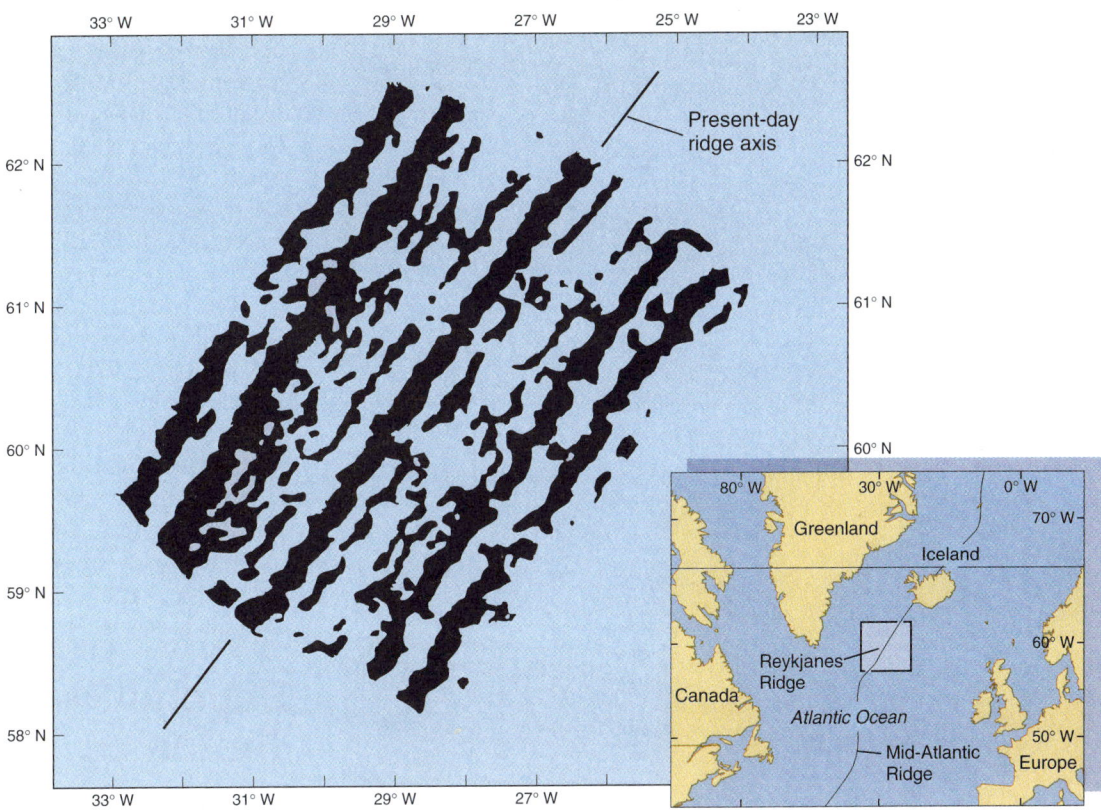

FIGURE 17.4
Symmetrical magnetic anomalies about the Reykjanes Ridge portion of the Mid-Atlantic Ridge, southwest of Iceland. Dark areas are positive magnetic anomalies, light areas between them are negative anomalies. Spreading is perpendicular to the ridge axis.
Source: "Symmetrical magnetic anomalies about the Reykjanes Ridge portion of the Mid-Atlantic Ridge" from "Spreading of the Ocean Floor: New Evidence" by F. J. Vine in *Science*, Figure 2, Vol. 154, 1966, p. 107. Copyright © 1966 American Association for the Advancement of Science. Reprinted with permission.

zone along which most occur is named the **Benioff zone,** after seismologist Hugo Benioff. The faulting that generates the earthquakes is both along the boundary between the two lithospheric plates and within the descending slab itself (Fig. 17.5).

If convergence does take place as outlined, you should be able to predict the answers to the following questions, and propose ways to test your predictions. (1) Are earthquakes likely to be shallow (0–70 km), intermediate (70–300 km), or deep focus (300–700 km), or will there be a range of focal depths? (2) Will earthquake foci show the angle at which the plate descends? Problem 2 deals with these predictions.

Volcanic Arcs

Volcanic, or magmatic, arcs are associated with convergent plate boundaries. The volcanoes usually occur 100 to 300 km away from the trench that marks the position of the subduction zone on the surface, and are located on the same side of the trench as the earthquake epicenters. The implication is that magma is generated as a result of the subduction process.

Given these general observations and the hypothesis of plate tectonics, (1) suggest a process by which magma forms as a result of subduction; (2) predict how the **arc-trench gap** (the horizontal distance between the arc and the trench) may be related to the dip of the Benioff zone; and (3) predict whether subducted lithosphere must reach a certain depth for magma to form. Problem 2 relates to volcanic arcs.

Transform Faults

Transform faults are so called because they transform motion from one plate boundary to another. For example, mid-ocean divergent boundaries are broken into numerous segments by transform faults (Fig. 17.6). Transform faults are strike-slip faults; the principal movement on them is horizontal, and fault planes are essentially vertical. Like strike-slip faults, the sense of movement is described as right-lateral or left-lateral, depending on whether rocks on one side of the fault appear to have moved to the right or left, when viewed from the other side of the fault. Notice that the *apparent* sense of movement on a transform fault may be different than the *actual* sense of movement. For example, in Figure 17.6, the apparent offset of the ridges is right-lateral, but because of the divergence process, the actual movement is left-lateral, as shown by the arrows. Furthermore, movement is taking place only between the two segments of the offset ridge.

As with any other kind of fault, movement along a transform fault results in offset of features on opposite sides. Suggest:

1. how the total amount of movement on a transform can be determined from offsets, and

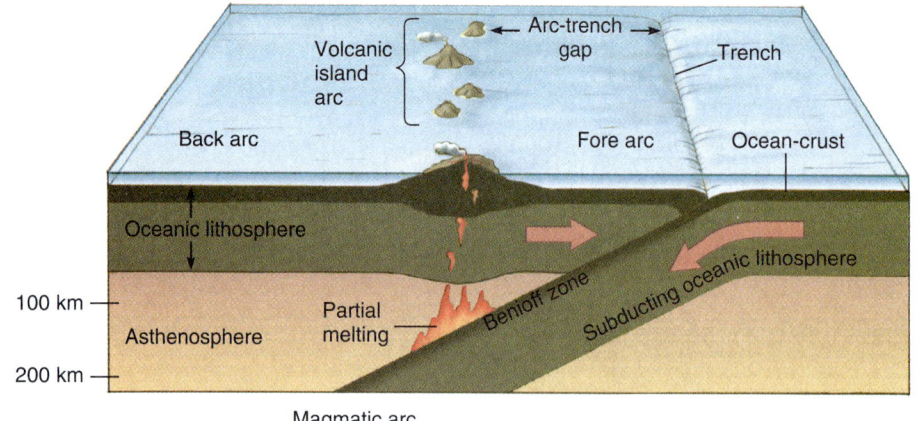

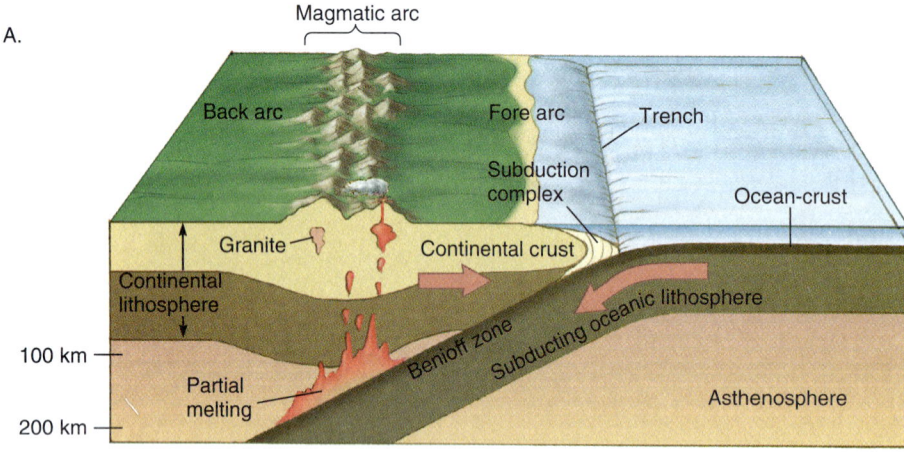

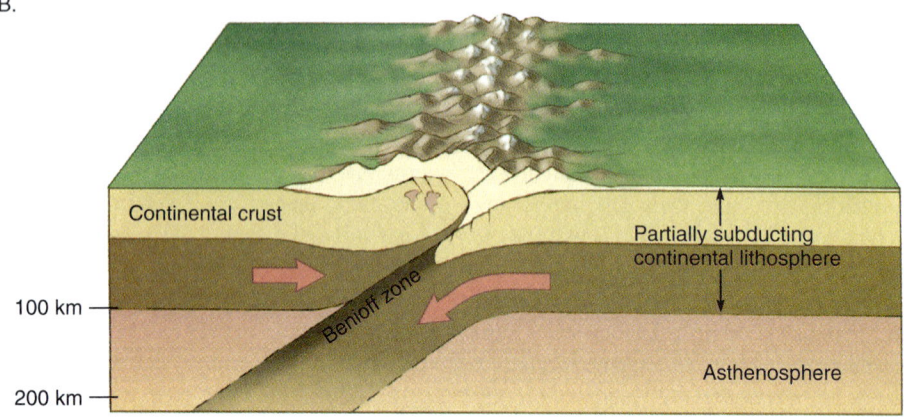

FIGURE 17.5
Block diagrams of convergent plate boundaries. A. Both plates contain oceanic crust. B. Subducting plate contains oceanic crust; overriding plate contains continental crust. C. Both plates contain continental crust. The subducted part of the slab in C is oceanic crust subducted before the continents collided.

2. how the amount of offset and the ages of the rocks offset can be used to determine the rate of movement and the time at which movement began.

Problem 5 relates to transform faults.

Plate Motion and Hot Spots

At more than 100 places on Earth, plumes of hot material rise through the mantle and penetrate the lithosphere. These **hot spots** are sites of persistent volcanic activity. Some, like Iceland, coincide with present-day divergent boundaries, but many do not. Those hot spots within plates may produce strings of volcanoes that form as lithosphere drifts over them (Fig. 17.7). The volcanoes are active when directly over the hot spot, but die out as the lithosphere drifts on.

The position of these plumes within the mantle is fixed, which makes them useful in testing several aspects of plate-tectonic theory. Predict how the ages of hot-spot-produced volcanoes should vary with distance from the hot spot, and suggest a way to test your prediction. Suggest how hot-spot tracks (strings of volcanoes) could be used to monitor the direction of plate movement through time. Problem 3 relates to hot spots.

Beyond Plate Tectonics

In its basic form, plate tectonic theory has rigid lithospheric plates drifting horizontally across the Earth's surface. Interactions between the plates cause intense deformation of their edges whereas the interiors remain largely undeformed. Mountains form when plates collide and pile up slivers of crust. Deep valleys and new oceans form when crust stretches, thins, and ultimately breaks. Plate tectonics is highly successful in explaining a huge number of geologic observations, but it does not explain everything. For example, why do some rare earthquakes occur in the middle of the great tectonic plates, far from any plate boundary? Another example: The Appalachians originally formed when Africa collided with North America some ~300 million years ago. However, erosion largely leveled the Appalachians by about 150 million years ago. Thus, the present-day mountains arose relatively recently, in the middle of a plate, independent of any collision. What caused this uplift? These and many other examples do not appear to invalidate the theory of plate tectonics, but they do show that there are still many things about the Earth that need better explanations.

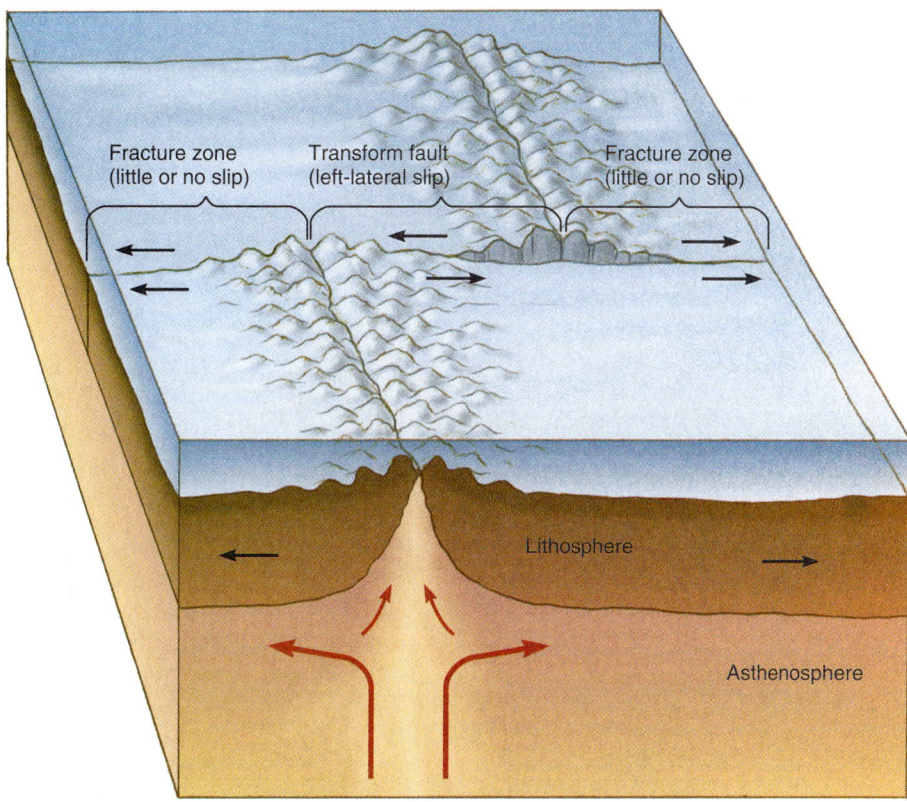

FIGURE 17.6
Transform fault joining segments of a mid-ocean-ridge, divergent-plate boundary. Movement on fault takes place between ridges but not on fracture-zone extensions.

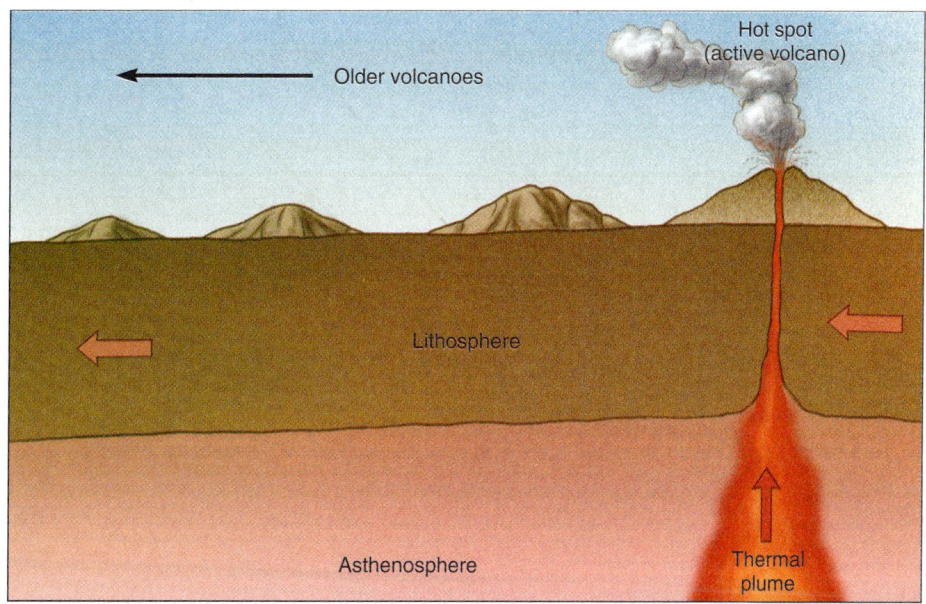

FIGURE 17.7
A fixed heat and magma source below the lithosphere causes a hot spot on the surface. As the lithosphere moves, a string of volcanoes is formed. Older volcanoes are lower and more rounded, because they have undergone more erosion.

Hands-On Applications

The development of the plate-tectonic theory is an excellent example of how science works. As data accumulated and relations were established, various models or hypotheses were developed and tested. The following problems illustrate some of the methods used to test important aspects of the theory.

Objectives

If you complete all the problems, you should be able to:

1. Explain how a hypothesis is tested.
2. Explain how magnetic profiles across divergent-plate boundaries are interpreted to show divergence and magnetic polarity reversals.
3. Correlate magnetic profiles (a) along a divergent boundary and (b) with the geomagnetic polarity time scale.
4. Calculate divergence rates from magnetic anomalies, given their ages.
5. Draw a cross section of a convergent plate boundary that shows the Benioff zone, volcanic arc, trench, arc-trench gap, forearc, and back-arc, and explain their distribution.
6. Use the ages and amount of offset of rocks or surface features to determine the average slip rate along a transform fault.
7. Explain how hot-spot volcanoes form, become extinct, and are used to determine speed and direction of plate movement.

Problems

1. Figure 17.8 shows magnetic profiles from data gathered in 1965 along four traverses in the South Pacific. The data were collected using a magnetometer (an instrument that measures magnetic intensity) towed behind a ship. Each profile crosses the Pacific-Antarctic Ridge, the crest of which is shown with a dashed line; water depth is 2 to 4 km. These data were among the first to test the hypothesis of seafloor spreading by comparing magnetic anomalies across and along ocean ridges. The peaks, or positive anomalies, on these profiles correspond to periods of normal magnetic polarity, and the valleys or negative anomalies correspond to times of reversed magnetic polarity.

    a. Four peaks that the authors of the original report found distinctive and useful are labeled on profile A. If the hypothesis of seafloor spreading is correct, similar peaks should be present at roughly equal distances from the ridge axis on the southeastern side of that profile. Identify peaks that are similar to 1, 2, 3, and 4 southeast of the ridge on profile A, and label them 1′, 2′, 3′, and 4′. In searching for matching peaks, use the surrounding peaks and valleys to help identify the numbered ones. To help you get started, 2′ is identified.

    b. A second requirement of the hypothesis of seafloor spreading is that the magnetic anomalies can be connected from profile to profile to form "stripes" parallel to the spreading axis (divergent-plate boundary). To test this, first find on profiles B and C the anomalies corresponding to the numbered ones on profile A. To help you, most are already located on profile D. Then connect those anomalies with the same numbers. You may have some difficulty locating the numbered anomalies. That's just the way it is—these are real data—so do your best.

    c. Explain how your results in *a* and *b* do or do not support the seafloor-spreading hypothesis.

    d. Peak 4 has been independently dated as being 7 million years old. What is the rate of seafloor-spreading at this ridge? Calculate the *half-spreading rate* (at which one plate moves away from the ridge crest) and the *full-spreading rate* (at which points on the two plates move away from each other) using profile D. Give your answers in cm/yr.

320 Part VI Internal Processes

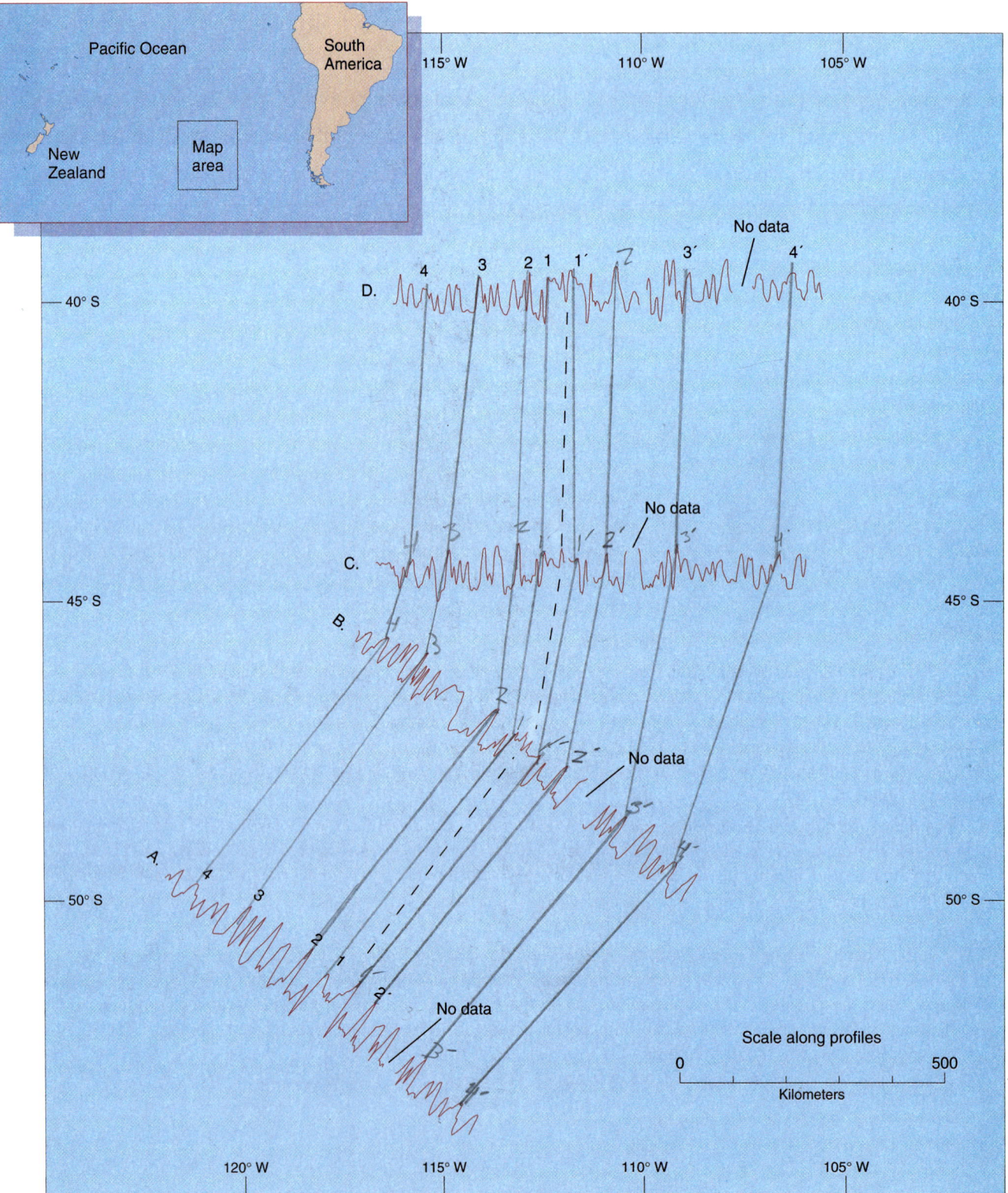

FIGURE 17.8
Four magnetic profiles across the Pacific-Antarctic Ridge. Dashed line marks the ridge crest. For use with Problem 1. Identify peaks 1–4 and 1′–4′ on each profile and connect them.
Source: "Four magnetic profiles across the Pacific-Antarctic Ridge" from W. C. Pitman III and J. R. Heirtzler in *Science*, 154:1164–71, 1966 Copyright © 1966 American Association for the Advancement of Science. Reprinted with permission.

EARTHQUAKE EPICENTERS

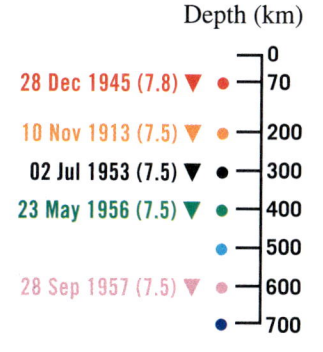

Focal depth Colored dots indicate 1964 through 1977 magnitude 5.0 through 7.4 events; triangles of matching colors labeled by date and magnitude indicate 1899 through 1985 events of magnitude 7.5 or greater (generally surface-wave magnitude)

VOLCANIC CENTERS

Active in historical time Generally within past 1000 yr. and documented during or soon after eruption; volcano named in color if active 1964 through 1985

Active in Holocene time Within past 10,000 yr; not documented historically

Holocene activity uncertain Includes volcano-related thermal springs and gas vents; no direct evidence for Holocene eruption

FIGURE 17.9

(Page 322) A plate boundary in the southwest Pacific Ocean near the Fiji, Samoa, and Tonga Islands. Shown on the map are bathymetry (water depth or depth to seafloor), active and recently active volcanoes, and epicenters of earthquakes of varying focal depths. Water depth is also shaded, with darker blue indicating deeper water. For use with Problem 2.

2. Figure 17.9 is a map showing a plate boundary in the southwest Pacific Ocean near the Fiji, Samoa, and Tonga Islands. Shown on the map are bathymetry (water depth or depth to seafloor, indicated by contour lines), active volcanoes, and epicenters of earthquakes of varying focal depths.

   a. What type of plate boundary is this? Given its location, what type of crust (oceanic or continental) do you think lies to the east? The west?

   b. The top part of Figure 17.10 gives a topographic profile of the seafloor along line A-A′ in Figure 17.9. The topographic profile has a vertical scale of 1 cm to 3,125 m. What is the vertical exaggeration if the horizontal scale is 1 cm to 97.5 km?

   c. The graph on the bottom part of Figure 17.10 is for you to show, with no vertical exaggeration, the focal depths of earthquakes along the line A-A′. Plot generalized earthquake foci on that cross section by placing circles at the appropriate depths below areas in which earthquakes of various focal groups (different color codes) occur. Look north and south of the profile line to ensure your circles are *representative of the trend as a whole*. What is the significance of the focal-depth pattern on your graph; that is, why are the earthquakes where they are?

   d. What is the dip of the earthquake zone? (Measure with a protractor.) Draw arrows on the graph to illustrate relative motion along the zone.

   e. Compare the two profiles. What topographic feature is present where the zone of earthquake foci intersects the surface?

   f. Label the trench, forearc, volcanic arc, and backarc on the topographic profile. Figure 17.5 may be helpful. What is the width of the arc-trench gap? Why are there volcanoes here—how does the magma form? (Hint: see Chapter 3.) At what depth range does it happen?

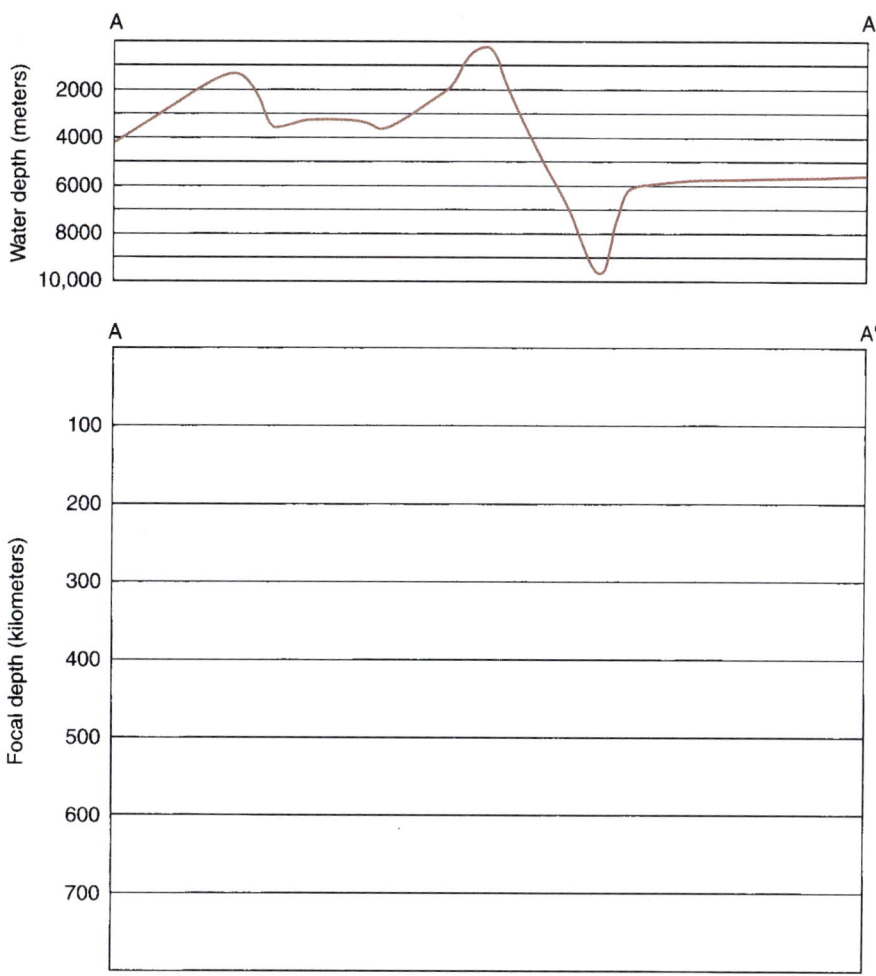

FIGURE 17.10
Section along line A-A′ of Figure 17.9, for use with Problem 2. A topographic profile appears on the top part of the diagram, and you can show earthquake foci on the bottom part. Horizontal scales are the same for top and bottom, but vertical scales are different.

3. Figure 17.11 is a map of the Hawaiian Islands and other islands and seamounts (submarine volcanoes) that form the Hawaiian-Emperor chain. All the features along the chain have a volcanic origin, and all are *younger* than the surrounding oceanic crust on which they sit. In 1963, J. Tuzo Wilson proposed that all the volcanoes in the Hawaiian chain had formed above the same hot spot or thermal plume. This idea was later extended to include the Emperor chain. If this hypothesis is correct, then (1) volcanoes should be older farther away from the hot spot, and (2) the distance-age relation can be used to measure the rate of plate movement.
Table 17.1 lists the ages of these volcanoes and their distance from Kilauea, a currently active volcano on the Island of Hawaii.

   a. Plot the data in Table 17.1 on Figure 17.12, and label each of the points with the name of the volcano. What does the graph indicate about the general relation between age and distance from Kilauea (located at 0 km)? Draw two best-fit straight-line segments, the first between Kilauea and Laysan, the second between Laysan and Suiko. How do they differ? *Moved quicker at first*

   b. Calculate the rate of plate movement in centimeters per year for each of the straight-line segments: divide the distance traveled by the time interval over which travel took place. Remember to convert to the proper units. Are the two values the same? What does this mean? Show your calculations.

   c. The chain bends where the Hawaiian and Emperor chains meet. What could have caused the bend, and when did it happen?

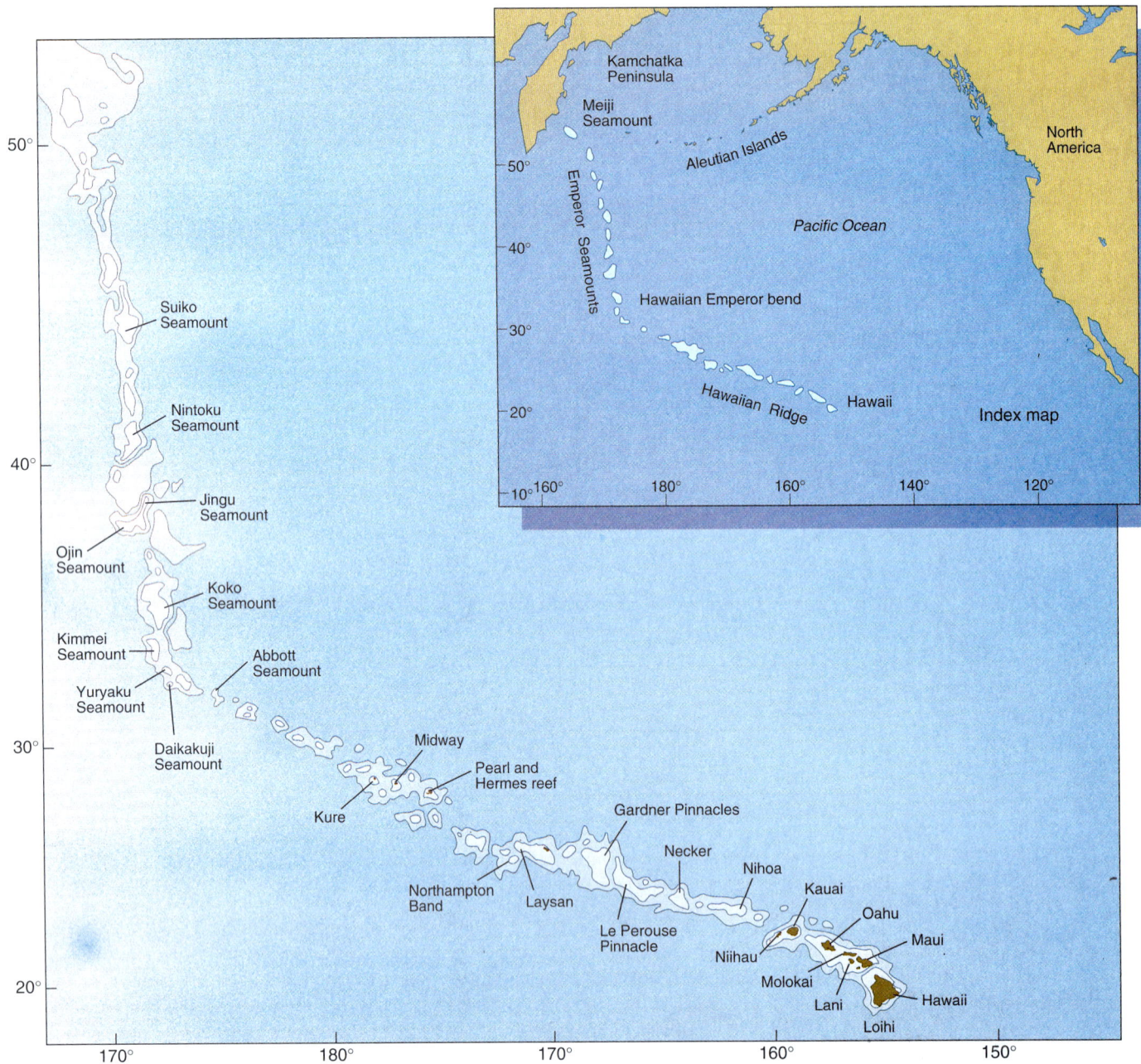

FIGURE 17.11
Hawaiian Islands and other islands and seamounts of the Hawaiian-Emperor chain, shown by 1- and 2-km bathymetric contours. For use with Problem 3.
Source: Data from USGS, Professional Paper 1350.

TABLE 17.1
Selected Volcanoes of the Hawaiian-Emperor Chain

Source: USGS Professional Paper 1350.

Volcano Name	Distance from Kilauea (Km)	Age in Millions of Years
Mauna Kea (on Hawaii)	54	0.375 ± 0.05
West Maui	221	1.32 ± 0.04
Kauai	519	5.1 ± 0.20
Nihoa	780	7.2 ± 0.3
Necker	1058	10.3 ± 0.4
La Perouse Pinnacle	1209	12.0 ± 0.4
Laysan	1818	19.9 ± 0.3
Midway	2432	27.7 ± 0.6
Abbott	3280	38.7 ± 0.9
Daikakuji	3493	42.4 ± 2.3
Koko (southern)	3758	48.1 ± 0.8
Jingu	4175	55.4 ± 0.9
Nintoku	4452	56.2 ± 0.6
Suiko (southern)	4794	59.6 ± 0.6
Suiko (central)	4860	64.7 ± 1.1

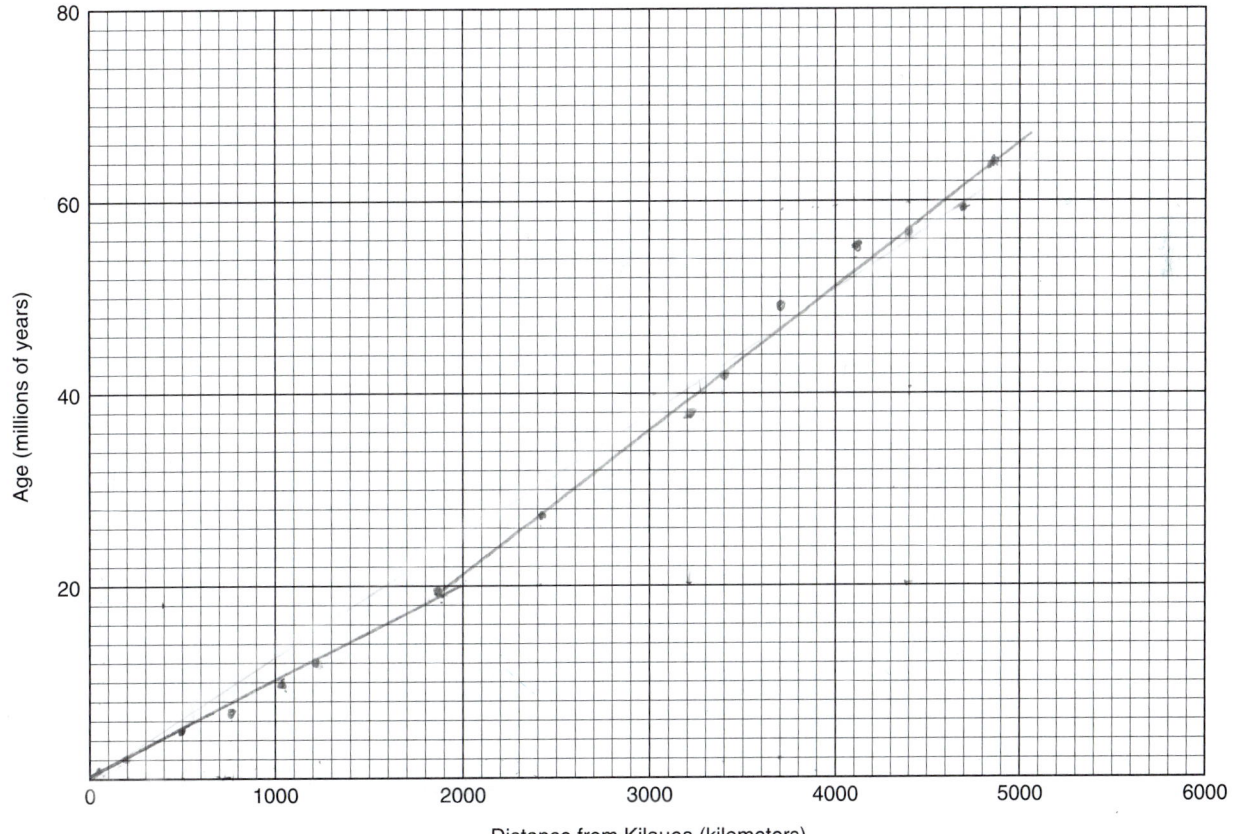

FIGURE 17.12
Graph of volcano age versus distance from Kilauea volcano (located at 0 km) along the Hawaiian-Emperor chain. Plot the data from Table 17.1, label points with the name of the volcano, draw best-fit straight lines between Kilauea and Laysan and between Laysan and Suiko, and answer the questions in Problem 3.

4. a. **Age of the Seafloor.** Go to *www.ngdc.noaa.gov/mgg/image/WorldCrustalAge.jpg* (or link to it through *www.mhhe.com/jones5e*—see Preface), which is a map showing the age of segments of the oceanic crust. Answer the following:

   (1) Which areas have the oldest ocean crust (indicate along which coasts or what parts of a given ocean)?

   (2) How old is the oldest crust? To what geologic period does this correspond?

   (3) Where is the youngest oceanic crust, and what is its age? Be more specific, if you can, than the key at the bottom of the image.

   (4) Explain the relation between the age of oceanic crust and its location in terms of plate tectonic theory.

   (5) Compare the age of the crust on the east and west sides of South America, and explain the age differences in terms of plate tectonic theory. Hint: look at a map that shows the tectonic plate boundaries (inside front cover).

   (6) Compare the ages of the Atlantic oceanic crust along the coasts of Greenland and northern Europe (north Atlantic), the United States and Africa (central Atlantic), and Argentina and southern Africa (south Atlantic). Which part of the Atlantic Ocean opened up first, and which part last?

   b. **Sediment on the Seafloor.** Go or link to *www.ngdc.noaa.gov/mgg/sedthick/sedthick.html,* which is a map showing the total thickness of sediment on the ocean floors. Answer the following:

   (1) Where is sediment the thickest, near the continents or in the deep ocean? Why?

   (2) The west side of South America is on a convergent plate boundary between a plate with oceanic crust and one with continental crust (e.g., Fig. 17.5B). There is an oceanic trench just offshore, and the Andes Mountains lie inland. The east side of South America is a trailing continental margin, which was once a divergent-plate boundary but now lies well within the South American plate.

   (a) On which side of South America is sediment the thickest: the west side, adjacent to the high Andes Mountains, or the east side, where the relief is comparatively low?

   (b) Formulate a hypothesis, based on plate-tectonic theory, to explain why this is so.

   (3) Visit *http://mahi.ucsd.edu/Gabi/sediment.html* to see a more complete, but lower resolution, sediment thickness map. Note the thickness of the sediment in the Bay of Bengal (east of India). This area is also close to a convergent plate boundary, but it is of a type like that in Figure 17.5C.

   (a) How does the sediment thickness in the Bay of Bengal compare with that on the west side of South America?

   (b) Formulate a hypothesis, based on plate-tectonic theory, to explain why this is so.

Also of interest is the U.S. Geological Survey site "Information on Plate Tectonics" *(http://geology.er.usgs.gov/eastern/tectonic.html),* which contains several useful links.

In Greater Depth

5. The San Andreas fault is a transform-fault boundary between the North America and Pacific plates in California. Figure 17.13 shows that it is really part of a network of faults collectively called the San Andreas fault *system*. The entire system accommodates movement between the two plates, although at present, most slippage occurs along the San Andreas fault itself. In this problem, you will learn how streams have been offset by the fault and how the study of offset streams and associated sediment can be used to determine the average rate at which movement has occurred in the recent past.

    Figure 17.14 is an aerial view of the San Andreas fault in the arid Carrizo Plain of central California (see Fig. 17.13 for location). A stream, informally named Wallace Creek, drains the Temblor Range to the northeast (look up *temblor* in your dictionary) and crosses the San Andreas fault. The photograph and the map in Figure 17.15 show two prominent stream valleys, both of which bend sharply as they cross the fault. The map also shows three kinds of surficial deposits. It is the two stream-channel alluvium deposits that are of interest here, although fan alluvium and slope-wash deposits cover most of the area. The older-channel alluvium ranges in age from 10,000 yr BP (years before present, with "present" defined as 1950) to 3680 yr BP. The ages are based on radiocarbon (^{14}C) dates from pieces of charcoal and organic matter found in the deposits.

    a. Why does the present-day stream channel bend sharply at the San Andreas fault? Can you think of two *possible* explanations? Which is much more likely, and why?

    b. Based on your answer to *a*, explain the map pattern of the older-channel alluvium.

    c. What evidence is there on the map or photograph for *vertical* movement on the San Andreas fault?

    d. The following statement was made in an article describing Wallace Creek: "Mr. Ray Cavanaugh, who farms at Wallace Creek, reported to us that water actually spilled over the edge [where Wallace Creek bends sharply northwest] in the winter of 1971–1972 or 1972–1973." Based on this statement and your answers to *a* and *b*, predict the future of Wallace Creek. (Note the orientation of the north arrow in the *Explanation* in Figure 17.15.)

    e. Movement along the San Andreas fault in this area is by "fits and starts." The last large earthquake here, in 1857, was the result of 9.5 m of instantaneous slip. The *average slip rate* of the fault can be determined for two periods in the Wallace Creek area, because two channels of Wallace Creek are displaced by the fault, and the time when the channels first formed is known.

    Calculate the two slip rates by measuring the offset across the fault for both channels and dividing by the greatest age of the channel deposits. Give your answers in centimeters/years. Show your calculations. Note: The map shows that the channel deposits in both channels just southwest of the fault line "bulge" toward the northwest, where the channels curve abruptly west. The deposits do this because the channels were eroded at the bends before all of the channel alluvium was deposited. Therefore, to get a better idea of the position of the channel when the oldest alluvium was deposited, use points farther downstream for your measurements.

    f. The best estimate of the total displacement along the San Andreas fault is about 320 km. This estimate was determined by matching distinctive groups of rocks that are offset by the fault. If the average slip rate you just calculated has been the same for the life of the fault, how old is the San Andreas fault? (Show your work, and remember to keep units consistent.)

328　Part VI　Internal Processes

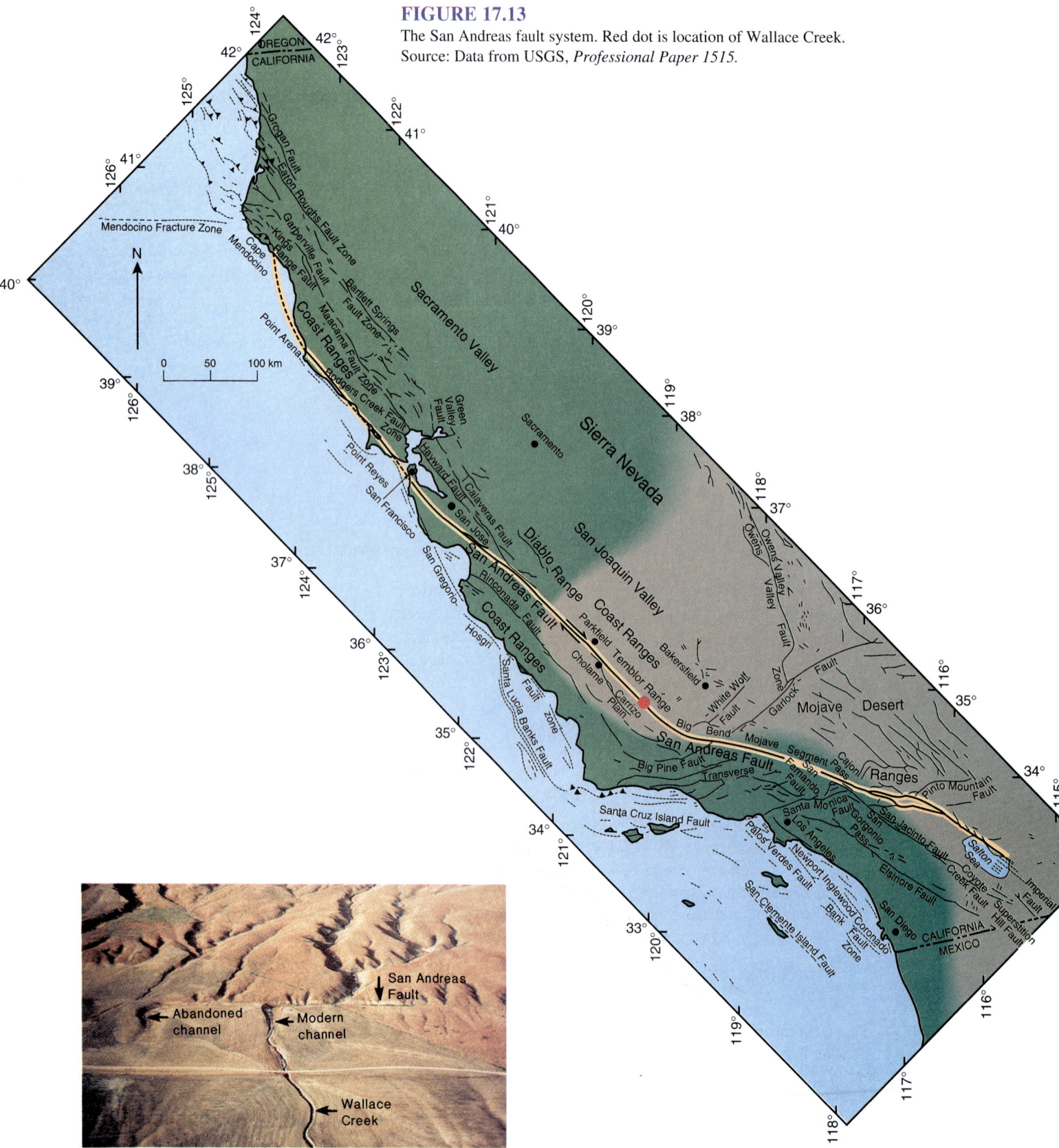

FIGURE 17.13
The San Andreas fault system. Red dot is location of Wallace Creek.
Source: Data from USGS, *Professional Paper 1515*.

FIGURE 17.14
Oblique aerial photograph of area shown on map in Figure 17.15. The modern and abandoned channels of Wallace Creek are visible on the foreground side of the San Andreas fault.

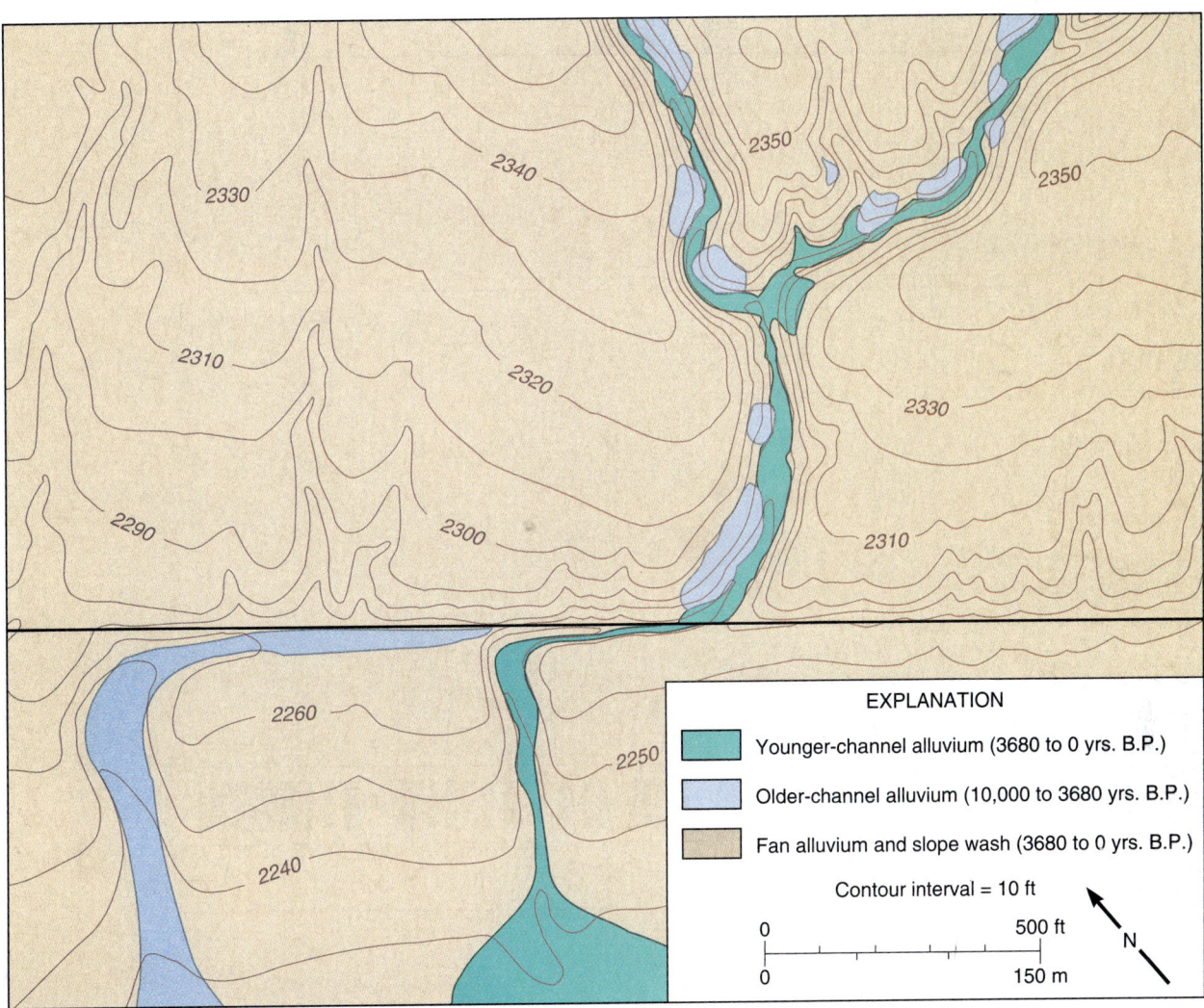

FIGURE 17.15
Map of Wallace Creek area, for use with Problem 5.
Source: Map of Wallace Creek from *GSA Bulletin,* Vol. 95, No. 8, 1984. Copyright © 1984 by Geological Society of America. Reproduced with permission of Geological Survey of America via Copright Clearance Center.

6. Figure 17.16 is an enlargement of profile A of Figure 17.8. If the hypothesis is correct—that peaks on the profile represent normal magnetic polarity, and valleys, reverse magnetic polarity—then peaks and valleys should match periods of normal and reversed magnetism known from well-dated rocks from other parts of the world.

   a. To test this, first fill in the bar above the profile by blackening the areas corresponding to peaks. The dark areas corresponding to large peaks should have widths equal to the width of the peak, as measured about halfway between the top and the bottom. Dark areas for the small peaks should be lines. Several of the trickier ones have been filled in to help you get started, and a few ambiguous parts of the profile have been simplified.

   b. Compare your bar diagram with the geomagnetic polarity time scale in Figure 17.17. Start at time zero and see if the patterns match, going back in time. If you are able to match the two, mark points on your diagram (Fig. 17.16) that correspond to time intervals of one million years.

   c. Compare the width (i.e., distance in km) of the spaces between the million-year points you marked above. Has the half spreading rate varied over time? Are half-spreading rates for the same time interval equal on both sides of the spreading center?

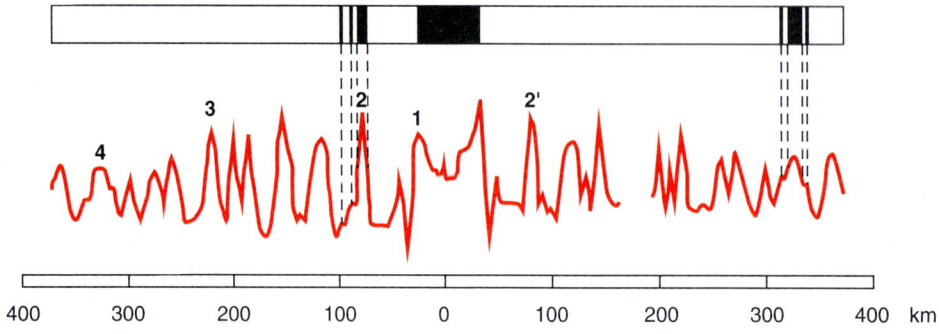

FIGURE 17.16
Enlargement of Profile A from Figure 17.8, for use with Problem 6.

FIGURE 17.17
Geomagnetic polarity time scale.

Credits

Part Openers
One: © McGraw-Hill Companies/Doug Sherman, photographer; **Two:** © Doug Sherman/Geofile; **Three:** U.S. Geological Survey, Reston, VA; **Four:** Courtesy of Earth Satellite Corporation (EarthSat); **Five:** © Norris W. Jones; **Six:** © Doug Sherman/Geofile.

Chapter One
Figures 1.1, 1.2A,B: © Doug Sherman/Geofile; **1.2C,D:** © Charles E. Jones; **1.2E:** © Doug Sherman/Geofile; **1.2F:** © The McGraw-Hill Companies/Dr. Parvinder Sethi, photographer; **1.2G,H, 1.3, 1.5, 1.6B, 1.7A, 1.9:** © Doug Sherman/Geofile; **1.10(left):** © Charles E. Jones; **1.10(right):** © Doug Sherman/Geofile; **1.11(both):** © Doug Sherman/Geofile; **1.12:** © Charles E. Jones; **1.13:** © Doug Sherman/Geofile; **1.14A,B:** © Charles E. Jones; **1.15, 1.16, 1.17A,B, 1.18A,B, 1.19, 1.20, 1.21:** © Doug Sherman/Geofile.

Chapter Two
Figures 2.1A–C, 2.2: © Doug Sherman/Geofile; **2.3:** © Charles E. Jones; **2.4A:** © Doug Sherman/Geofile; **2.4B:** © Charles E. Jones; **2.4C–F:** © Doug Sherman/Geofile; **2.5:** © Charles E. Jones; **2.6, 2.7, 2.8:** © Doug Sherman/Geofile; **2.9A:** © Charles E. Jones; **2.9B, 2.10, 2.11, 2.12:** © Doug Sherman/Geofile; **2.13:** © Charles E. Jones; **2.14, 2.15, 2.16, 2.17:** © Doug Sherman/Geofile.

Chapter Three
Figures 3.1A,B, 3.2A,B, 3.3, 3.4, 3.5, 3.7A,B, 3.8, 3.9A,B, 3.10A,B: © Doug Sherman/Geofile; **3.11A:** © Charles E. Jones; **3.11B:** © Doug Sherman/Geofile; **3.12A:** © The McGraw-Hill Companies/Dr. Parvinder Sethi, photographer; **3.12B:** © Doug Sherman/Geofile; **3.13A:** © The McGraw-Hill Companies/Dr. Parvinder Sethi, photographer; **3.13B:** © Doug Sherman/Geofile; **3.14:** © Charles E. Jones; **3.15A,B, 3.16A,B, 3.17:** © Doug Sherman/Geofile.

Chapter Four
Figure 4.3: © Doug Sherman/Geofile; **4.4A,B, 4.5B:** © Norris W. Jones; **4.6, 4.7:** © The McGraw-Hill Companies/Dr. Parvinder Sethi, photographer; **4.9, 4.11, 4.12A,B:** © Doug Sherman/Geofile; **4.13A:** © The McGraw-Hill Companies/Dr. Parvinder Sethi, photographer; **4.13B, 4.14A,B:** © Doug Sherman/Geofile; **4.15A–1:** © The McGraw-Hill Companies/Dr. Parvinder Sethi, photographer; **4.15A–2:** © Doug Sherman/Geofile; **4.15B(both):** © Doug Sherman/Geofile; **4.16A–E, 4.17, 4.18, 4.19, 4.20, 4.21:** © Doug Sherman/Geofile.

Chapter Five
Figure 5.3A: © Doug Sherman/Geofile; **5.3B:** © The McGraw-Hill Companies/Dr. Parvinder Sethi, photographer; **5.4, 5.5, 5.6, 5.7, 5.8, 5.9, 5.10A,B, 5.11A:** © Doug Sherman/Geofile; **5.11B:** Charles E. Jones.

Chapter Six
Figures 6.2, 6.4: U.S. Geological Survey; **6.16:** Massachusetts Office of Travel and Tourism.

Chapter Seven
Figures 7.3A,B: NASA/GSFC/MITI/ERSDAC/JAROS, and U.S./Japan ASTER Science Team; **7.4A–C:** U.S. Geological Survey, ESIC; **7.4D,E:** NASA/GSFC/MITI/ERSDAC/JAROS, and U.S./Japan ASTER Science Team; **7.4F:** U.S. Geological Survey, EROS Data Center; **7.6:** U.S. Geological Survey, ESIC; **7.7A:** U.S. Geological Survey, Denver, CO; **7.9:** U.S. Geological Survey, EROS Data Center; **7.10:** U.S. Geological Survey, ESIC; **7.11(all):** NOAA.

Chapter Eight
Figures 8.3, 8.4, 8.10: U.S. Geological Survey, Rolla, MO; **8.11:** U.S. Geological Survey, Denver, CO; **8.12, 8.13, 8.14:** U.S. Geological Survey, Rolla, MO; **8.16:** U.S. Geological Survey, Denver, CO.

Chapter Nine
Figures 9.8, 9.9: U.S. Geological Survey, Rolla, MO; **9.10:** Courtesy of James F. Quinlan and Joseph A. Ray, 1981, revised 1989.

Chapter Ten
Figure 10.1: Austin Post, U.S. Geological Survey, Tacoma, WA; **10.2:** Robert Felder, U.S. Geological Survey, Menlo Park, CA; **10.4:** © Norris W. Jones; **10.7:** U.S. Geological Survey, Denver, CO; **10.8:** Prepared from USC & GS Photography by the University of Illinois Committee on Aerial Photography; **10.9:** U.S. Geological Survey; **10.10:** The Geological Society of America; **10.11:** U.S. Geological Survey, ESIC; **10.12:** U.S. Geological Survey, Rolla, MO; **10.14:** USGS prof. Paper 1180, 1980, Courtesy U.S. Geological Survey, Denver, CO.

Chapter Eleven
Figure 11.8: © Doug Sherman/Geofile; **11.9A–C:** Prepared from USC & GS Photography by the University of Illinois Committee on Aerial Photography; **11.10, 11.11, 11.12:** U.S. Geological Survey, Denver, CO; **11.13A,B:** U.S. Geological Survey, NAPP.

Chapter Twelve
Figures 12.1, 12.2, 12.5, 12.6: © Doug Sherman/Geofile; **12.9:** Prepared from USC & GS Photography by the University of Illinois Committee on Aerial Photography; **12.10, 12.11, 12.12:** U.S. Geological Survey, Denver, CO; **12.13:** U.S. Geological Survey.

Chapter Thirteen
Figure 13.3: © Norris W. Jones.

Chapter Fourteen
Figure 14.2: © Norris W. Jones; **14.3:** © The McGraw-Hill Companies/Dr. Parvinder Sethi, photographer.

Chapter Fifteen
Figure 15.2B: U.S. Geological Survey, ESIC; **15.3B:** Prepared from USC & GS Photography by the University of Illinois Committee on Aerial Photography; **15.4B, 15.5B, 15.6B:** U.S. Geological Survey, ESIC; **15.12B,C:** U.S. Geological Survey; **15.14:** N.H. Darton, 1951. Geologic Map of South Dakota, U.S. Geological Survey; **15.15:** U.S. Geological Survey, EROS Data Center; **15.17B, 15.18B:** U.S. Geological Survey.

Chapter Sixteen
Figure 16.6: U.S. Geological Survey, Denver, CO; **16.13:** U.S. Geological Survey, EROS Data Center; **16.14:** Timothy Hall, U.S. Geological Survey.

Chapter Seventeen
Figure 17.9: American Association of Petroleum Geologists; **17.14:** U.S. Geological Survey, Menlo Park, CA.

Index

Note: An *f* or *t* following a page number indicates a figure or table that appears outside the related text discussion.

Absolute (numerical) geologic age, 234, 238–239
   exercises, 243–244, 246
Accessory and minor minerals, 17, 18*t*, 21–22
Accumulation, zone of, 173–174, 175*f*
Acid, 10
   acid mine drainage exercise, 34
   mineral exercise, 13
Acres, 101
Advance, 174
Aeration, zone of, 157, 158*f*
Aerial photographs
   distortion, 119
   exercises, 121*f*, 122*f*, 125–126
   geologic features, 269–270
   scale, 119
   stereoscopic viewing, 119, 121
   vs. topographic maps, 121–123
   types, 118–119, 120*f*
Aftershocks, 293
Agate, 19*t*
Age of seafloor exercise, 326. *See also* Geologic age
Alluvial fans, 217
Alpine glaciers, 173
   erosional effects, 174–175
   exercises, 178*f*, 179–183, 191–194
Amphiboles, 19, 84*t*
Amphibolite, 86*f*
Amplitude, 117, 118*f*, 297, 299*f*
Andesite, 42*t*, 43*f*, 45*f*
Angle of inclination, 251
Angular unconformities, 238, 239*f*, 273–274
Angular units of measure and conversions. *See* inside back cover
Antelope Peak exercise, 221, 225*f*
Anthracite, 84*t*, 86*f*
Anticlines, 252, 253*f*, 254, 255*f*
Apatite, 27*t*, 31*t*
Aphanitic texture, 39, 42*t*
Aquifers, 158, 159*f*
Archimedes, 16
Arc-trench gap, 315
Area units of measure and conversions. *See* inside back cover
Arêtes, 174
Arid-climate landscapes, 215–231
   depositional, 217–218
   erosional, 215–217
   exercises, 219–231
Arkansas River exercise, 144, 146*f*
Arkose, 67*f*
Artesian spring, 159
Artesian wells, 159
Assemblages, Principle of Fossil, 237–238
Assimilation (magmatic), 48
Atomic numbers, 238
Attitude, 251

Augite, 19, 20*f*
   chemical group and formula, 31*t*
   identification tables, 28*t*, 30*t*
   in igneous rock, 41*t*
Axial surface/plane, 252, 253*f*
Azurite, 27*t*, 31*t*

Backswamps, 136, 139*f*
Bajadas, 217
Barchan dunes, 217, 218*f*
Barite, 28*t*, 31*t*
Barrier islands, 198, 200*f*
   exercises, 203–206
Bars
   Baymouth, 198, 200*f*
   point, 136, 139*f*
Basalt
   classifying/identifying, 42*t*, 43*f*, 46*f*, 84*t*, 87
   crystalline texture, 39*f*, 40*f*
Base level of stream, 136
Baseline, 99
Basin of deposition, 57
Basins, geologic, 254, 255*f*. *See also* Drainage basins
Bathymetric features, map symbols, 103*f*
Bauxite, 24, 27*t*, 29*t*
Baymouth bars, 198, 200*f*
Beaches, 197. *See also* Sea coasts
Bed(s), 61*f*
Bedding planes, 61*f*, 249–250
Beheading, 140
Bench marks, 105
Benioff zone, 315
Bennetts Well exercise, 220, 222*f*–223*f*
Beryl, 30*t*, 31*t*
Biochemical sediment/sedimentary rocks, 60, 62, 65*t*
Bioclastic texture, 62
Biotite, 20
   chemical group and formula, 31*t*
   identification table, 28*t*
   in igneous rock, 41*t*
   in metamorphic rock, 83*t*, 85*f*
   in sedimentary rock, 64*t*
Black Hills exercise, 283, 284*f*–285*f*
Block diagrams, 252
   exercises, 257–265
   templates, 339–345
Blowouts, 217
Bogs, map symbols, 103*f*
Boothbay exercise, 208–210
Boundaries
   plate. *See* Plate tectonics
   topographic map symbols, 103*f*
Bowen, N. L., 41
Bowen's Reaction Series
   exercises, 50, 52
   magma crystallization and, 41, 46–47
Braided streams, 215, 216*f*
Breccia, 67*f*, 87
Brittle deformation, 248–249

Brittle strain, 256*t*
Buildings, map symbols, 103*f*
Buttes, 216
   imagery exercise, 121*f*, 122*f*, 125

Calcite, 21
   chemical group and formula, 31*t*
   identification table, 28*t*
   in metamorphic rock, 83*t*
   in sedimentary rock, 64*t*
Calderas, 274
Canals, map symbols, 103*f*
Carbonate minerals, 31*t*
Carbonate rocks, 160–161, 162*f*
Cascade Range exercise, 52–53
Caves
   groundwater and, 160–161, 162*f*
   map symbols, 103*f*
Cement, 62
Chalcedony, 19*t*
Chalcopyrite, 24, 26*t*, 30*t*
Chemical analysis exercises, 52–54
Chemical groups, mineral, 30*t*–31*t*
Chemical sediment/sedimentary rocks, 60, 62, 65*t*
Chemical symbols. *See* inside back cover
Chemical weathering, 57, 58*t*
Chert, 19*t*, 69*f*
   classifying/identifying, 84*t*
   properties, 64*t*
   texture, 66*f*, 86*f*
Chlorite, 20
   chemical group and formula, 31*t*
   identification table, 28*t*
   in metamorphic rock, 83*t*, 85*f*
Chromite, 26*t*, 31*t*
Cirques, 174, 175*f*
Classification
   igneous rock, 41, 42*t*, 43*f*–46*f*, 49–51
   metamorphic rock, 83–84
   minerals, 17–24
   sedimentary rock, 62–63, 65*t*
Clastic sediments, 59–60
Clastic texture, 62, 66*f*
Clay particles, 20
Clayborn Creek, 136, 137*f*–138*f*
Clays, 20, 64*t*
Claystone, texture, 66*f*
Cleavage and fracture, 5–8
   exercise, 13
   identification, 24, 26*t*–30*t*
Cleavage planes, 5
Coal, 69*f*
   classifying/identifying, 84*t*, 86*f*
   texture, 66*f*
Coarse-grained texture, 39, 42*t*, 43*f*
Coasts. *See* Sea coasts
Color
   false-color images, 118, 119*f*, 123
   of minerals, 2, 24, 26*t*–30*t*
Cols, 174

Index

Composition
  of magma, changes in, 48
  of rock, 38
    igneous, 40–41, 43f
    metamorphic, 83–84
    sedimentary, 65t, 66f
Compression, 248, 249f, 256t
Compressive waves. *See* P waves
Concordant, 274
Conductivity, hydraulic, 157
Cone of depression, 159
Confined aquifers, 158, 159f
Confining layers, 158, 159f
Conformable contacts, 249
Conglomerates, 63f, 84t, 87
Contact goniometer, 15f
Contact metamorphism, 80
Contacts, 249–251
Contamination of groundwater, 160, 161f
  exercises, 168, 169f, 170, 171f
Continental depositional environments, 61t
Continental glaciers, 173
  depositional effects, 175–177
  exercises, 184–185, 186f–189f
Continuous reaction series, 46–47
Continuous reactions, 41, 46
Contour intervals, 104–105
Contour lines, 102, 104
Contours
  exercises, 110, 111f, 112f
  intervals, 104–105
  lines, 102, 104
  map symbols, 103f
Convergent plate boundaries, 312, 314–315, 316f
  magma processes, 47–48
Conversions
  map scales, 102, 109
  units of measure. *See* inside back cover
Copper, 26t, 30t
Corundum, 29t, 31t
Craters, 274
Crillon Glacier exercises, 180–182
Cross sections, 252, 274
  exercises, 263, 264, 265f
Cross-bedding, 62f
Cross-beds, 62f
Cross-cutting relations, principle of, 236, 237f
Cross-stratification, 238
Crystal(s)
  igneous texture based on, 39–40, 43f
  mineral form, 8–9, 10f
    angle analysis exercise, 14–16
Crystalline texture, 39, 62, 66f
Crystallization
  Bowen's Reaction Series and, 41, 46–47, 52
  fractional, 48
Cultural features, map, 96
Curie point, 312
Current ripple marks, 62f
Cutbanks, 136, 139f
Cutoffs, 136, 139f
Cyanide spill exercise, 166, 169f

Dating techniques, 238–239. *See also* Relative geologic age
  exercises, 243–244, 246

Datum, 102
Daughter isotope, 238
Daughter-parent exercise, 243
Death Valley exercises, 219–220, 222f–223f, 230–231
Decay constant, 238–239
Declination, magnetic, 102
Deer Creek exercise, 154–156
Deformation. *See* Structural geology
Deltas, 198, 200f
Dendritic drainage pattern, 139, 141f
Density
  of fluids, 159–160
  of minerals, 11
Deposition
  sedimentary rock, 59–62, 63f
  stream action causing, 136, 141
Depositional contacts, 249–250, 251f
Depositional environments
  exercises, 74–75
  types and characteristics, 60–61
Depositional landforms
  in arid/semiarid regions, 217–218
  exercises, 224, 226, 228f, 229t, 230–231
  glaciated, 175–177
  in humid regions, 136, 141
  sea coasts, 197–198, 200f, 201f
Depression, cone of, 159
Depression contours, 104
Desert. *See* Arid-climate landscapes
Detrital sediment/sedimentary rocks, 59–60, 62, 65t, 66f
Devils Tower exercise, 113–114
Dikes, 235, 274
Diorite, 42t, 43f, 44f
Dip angle, 251
Dip direction, 251
Directed pressure, 81
Directions of cleavage, 5–8
Disappearing streams, 161, 162f
Discharge, 136
  watershed data exercise, 154
Disconformities, 238, 239f, 274
Discontinuous reaction series, 46–47
Discontinuous reactions, 41, 46
Discordant, 274
Distortion, in aerial photographs, 119
Divergent plate boundaries, 47, 312–314, 315f
Divides, 136
Dolomite, 21
  chemical group and formula, 31t
  classifying/identifying, 84t
  identification tables, 27t, 28t
  in metamorphic rock, 83t
  in sedimentary rock, 64t
Dolostone, 66f, 69f, 87
Domes, 254, 255f
Drainage basins, 136
  drainage patterns, 139, 141f
  exercises, 148, 149f, 155f, 156
  runoff and, 134–136
Drift
  glacial, 175
  longshore, 197–198, 200f, 201f
Drumlins, 176f, 177
Ductile deformation, 248–249

Ductile strain, 256t
Dunes, 217–218

Early-stage streams, 139
Earthquake intensity, 296–297
Earthquake magnitude, 297
Earthquakes, 293–310
  exercises, 301–310, 321, 322f, 323f
  intensity and magnitude, 296–298, 299f
  locating, 294–296
  plate tectonics, 314–315. *See also* Plate tectonics
  reflection seismology, 298, 299f
  seismic waves, 293–294
  seismograms, 294, 295f
Effective porosity, 157
Elastic strain, 248
Electromagnetic radiation, 117, 118f
Electromagnetic spectrum, 117, 118f
Elements, 30t
Elevation(s), 102
  bench marks, 105
  contour lines denoting, 102, 104
  vs. height and relief, 105, 106f
  reading, 105
Emergent sea coasts, 200, 201f
Emperor islands exercise, 323, 324f, 325f, 325t
End moraines, 175f, 176
Environments of deposition
  exercises, 74–75
  types and characteristics, 60–61
Eons, 239, 240f
Epicenter, 293
Epidote, 30t, 31t
Epochs, 239, 240f
Equilibrium line, 174
Eras, 239, 240f
Erosion
  glaciers causing, 174–175
  sedimentary rock, 57–59
  stream action causing, 136, 139–141, 142f
Erosional landforms
  in arid/semiarid regions, 215–217
  exercises, 219–224, 225f, 227f
  glaciated, 174–175
  in humid regions, 136, 139–141, 142f
  sea coasts, 197, 199f, 201f
Eskers, 176f, 177
Estuaries, 200
Explanation (map legend), 252
Extrusive rocks, 38, 52–53

False-color images, 118, 119f, 123
Fans
  alluvial, 217
  overwash, 198
Far infrared light, 118
Fault contacts, 250–251
Faulted rock
  aerial/map views, 271–273
  deformational processes, 254, 256
  exercises, 258–265
  geologic age, 234, 235f
  map symbols, 251f
  San Andreas fault exercise, 327, 328f–330f
  transform faults, 315–316, 317f

Feel of minerals, 11
   exercise, 13
Feldspars, 17, 18f
   in igneous rock, 41t
   in metamorphic rock, 83t, 85f
   in sedimentary rock, 64t
Felsic rock, 42t
Fine-grained texture
   igneous rock, 39, 42t, 43f
   sedimentary rock, 66f
Fiords, 175, 200
Fissile, 63
Flint, 19t, 69f
   in sedimentary rock, 64t
   texture, 66f
Flood-frequency curve, 139, 140f
   exercise, 150, 152t
Flooding, 136, 139, 140f
   exercises, 150–154
Floodplain, 136, 139f
Flow
   of groundwater, 157–158
   of lava, aerial/map views, 274
Fluid density, 159–160
Fluorescence exercise, 14
Fluorite, 21, 28t, 31t
Focus (earthquake), 293
Folded rock
   aerial/map views, 270–271
   deformational processes, 252–254, 255f
   exercises, 258–265
   geologic age, 234, 235f
   map symbols, 251f
Foliated metamorphic rocks, 81–83
   classifying/identifying, 83, 84t
   exercises, 89
Foliated texture, 81
Foliation, 81
Footwalls, 254, 256f
Formations, 267
Fossil record, 236–238
Fossiliferous rock, 66f
Fossils, 237
Fractional crystallization, 48
Fractional scale, 98f, 101, 102
   conversion exercise, 109
Fracture, cleavage and, 5–8
Fracture surfaces, 6, 8
Frequency, 117, 118f

Gabbro, 42t, 43f, 45f, 84t
Galena, 23, 26t, 30t
Garnet, 22
   chemical group and formula, 31t
   identification table, 29t
   in metamorphic rock, 83t, 85f
Gateway Quadrangle exercise, 279–283
Geographic Information Systems (GIS), 96
Geographic poles, 97
Geologic age, 234–246
   exercises, 241–246
   geologic time scale, 239, 240f
   numerical (absolute), 234, 238–239
   relative, 234–238
Geologic cross sections, 252, 274
   exercise, 263

Geologic maps, 267–292
   cross sections, 252, 263, 274
   defined, 96, 252
   exercises, 263, 277–292
   features
      faulted strata, 271–273
      folded strata, 270–271
      horizontal strata, 269–270
      igneous intrusions, 274
      inclined strata, 270
      unconformities, 273–274
      volcanoes and volcanic rocks, 274
   making
      requirements for, 267–268
      seismic waves as basis for, 298, 299f
   symbols
      geologic structures, 251f
      rock types, 242f
   using
      outcrop variation patterns and, 274–275
      as prospecting and land use databases, 276
      to reconstruct geologic history, 275
Geologic structures. See Structural geology
Geologic time scale, 239, 240f
   exercise, 245
Geomagnetic polarity time scale exercise, 329, 330f
Geostationary Operational Environmental Satellite (GOES) exercise, 129, 130f
Geostationary orbits, satellite, 123
GIS (Geographic Information Systems), 96
Glacial drift, 175
Glacial Lake Agassiz, 177
   exercises, 191
Glacier National Park exercises, 178f, 179–183, 191–194
Glaciers, 173–195
   depositional effects, 175–177
   erosional effects, 174–175
   exercises, 178f, 179–195
   formation, movement, and mass balance, 173–174, 175f
   map symbols, 103f
   transport of sediment, 59
   types, 173
Glassy texture, 39, 40f, 42t, 43f
Global warming, 200
G.N. (grid north), 102
Gneiss, 84t, 85f, 86f
Gneissic foliation, 83
GOES (Geostationary Operational Environmental Satellite) exercise, 129, 130f
Goethite, 26t, 31t
Goniometer, 15f
Grade of metamorphism, 80f, 81, 85f
Graded beds, 63f, 238
Gradient, 136
   hydraulic, 157
   topographic map, 108
Grain size and shape, 58, 59f
Grains, 58, 59f
Grand Canyon exercise, 73–74
Granite
   classifying/identifying, 42t, 43f, 44f, 84t
   crystalline texture, 39f
Granite Falls imagery exercise, 125, 126f
Graphic scale, 98f, 101

Graphite, 24, 26t, 27t, 30t
Gravity. See Specific gravity
Graywacke, 67f, 84t, 87
Great Salt Lake imagery exercise, 124f, 126–127
Greenstone, 84t, 86f
Grid north (G.N.), 102
Grinnell Glacier exercises, 191–195
Groins, 198, 201f
Ground moraines, 176
Ground-mass, 39–40
Groundwater, 157–171
   in carbonate rocks, 160–161, 162f
   defined, 157
   exercises, 163–171
   human use, 158–160, 161f
   occurrence and movement, 157–158
Gypsum, 69f
   chemical group and formula, 31t
   identification tables, 27t, 28t
   properties, 21, 64t, 66f

Half-life, 238
   exercises, 243–244
Halfway elevation, 105
Halides, 31t
Halite, 21
   atomic structure and cleavage, 5
   chemical group and formula, 31t
   identification table, 28t
   in sedimentary rock, 64t
Hanging valleys, 175
Hanging walls, 254, 256f
Hardness, 3–4
   exercise, 13
   identification, 24, 26t–30t
   Mohs Hardness Scale, 4t
Hawaiian Islands
   Maui imagery exercise, 128
   plate tectonics exercise, 323, 324f, 325f, 325t
Hawaiian-Emperor island chain exercises, 323, 324f, 325f, 325t
Head, 136, 157
Headlands, 197, 199f
Headward erosion, 139, 142f
Height, 105, 106f
Hematite, 23, 26t, 27t, 31t
High-grade metamorphic rocks, 81
Hinge lines, 252, 253f
Horizontal strata, 269–270
Horizontality, Steno's Principle of Original, 236, 237f
Hornblende, 19
   chemical group and formula, 31t
   identification tables, 28t, 30t
   in igneous rock, 41t
   in metamorphic rock, 83t, 85f
Hornfels, 84, 86f
Horns, 174
Hot spots, 316, 317f
Humans
   groundwater use, 158–160, 161f
   sea coasts and, 198, 200, 201f
Humid-climate landscapes, 134–156
   drainage patterns, 139, 141f
   exercises, 143–156
      drainage and erosion patterns, 148, 149f, 156

Humid-climate landscapes—*Cont.*
  stream features, 143–147, 154–156
  runoff and drainage basins, 134–139
  stream system evolution, 139–141, 142*f*
Hunter Valley exercise, 286–287, 288*f*–289*f*
Hydraulic conductivity, 157
Hydraulic gradient, 157
Hydrologic cycle, 135*f*
Hydroxides, 31*t*
Hypocenter, 293
Hypotheses, 238

Identification of rock types, 85–87
  exercises
    igneous rock, 49–54
    metamorphic rock, 89–90
    minerals, 33–34
    sedimentary rock, 71–75
  igneous rock, 41, 42*t*, 43*f*–46*f*
  metamorphic rock, 83–84
  mineral, 24–25, 26*t*–30*t*
  sedimentary rock, 63, 65*t*, 66*f*–69*f*
Igneous rocks, 38–56
  aerial/map views, 274
  classifying/identifying, 41, 42*t*, 43*f*–46*f*
  defined, 38
  exercises, 49–54
  formation, 38
  magma processes, 41, 46–48
  mineral content, 40–41
  rock symbols, 242*f*
  vs. sedimentary/metamorphic rock, 85–87
  textures, 39–40
Imaging techniques, 117–131
  aerial photographs, 118–123
  exercises, 125–131
  light properties and, 117–118, 119*f*
  radar images, 124, 128
  satellite images, 120*f*, 123–124
Inclination
  angle of, 251
  of magnetic lines of force, 312
Inclined folds, 253
Inclined strata, 270
Inclusions, principle of, 236, 237*f*
Index contour, 105
Infiltration loss, 134
Infrared light, 117–118, 119*f*
Inselbergs, 216*f*, 217
Intermediate rock, 42*t*
Intermittent streams, 215
Intrusions, 274
Intrusive contacts, 250, 251*f*
Intrusive rocks, 38
Invisible light exercise, 14
Iron oxide, 64*t*
Isograds, 85
Isoseismals, 296, 298*f*
  exercise, 304
Isotopes, 238–239

Jasper, 19*t*, 64*t*, 66*f*, 69*f*
Jordan quadrangle exercise, 150, 151*f*, 152*t*

Kaaterskill quadrangle exercise, 148, 149*f*
Kames, 176*f*, 177

Kaolinite, 20*f*, 27*t*, 31*t*, 64*t*
Karst topography, 161, 162*f*
Kettle lakes, 176*f*, 177
Kettles, 176*f*, 177
K-feldspar, 83*t*
Knoop hardness, 4*f*
Kyanite, 22
  chemical group and formula, 31*t*
  identification, 30*t*
  in metamorphic rock, 83*t*, 85*f*

Lakes
  exercises
    ancient floodplain, 191
    groundwater, 163–164, 165*f*
    imaging, 124*f*, 126–127
  glacial, 174, 176*f*, 177
  map symbols, 103*f*
  oxbow, 136, 139*f*
Lakeside quadrangle exercise, 163–164, 165*f*
Laminations, 61*f*
Land subdivisions
  latitude-longitude system, 96–97, 98*f*
  map symbols, 103*f*
  U.S. Public Land Survey System, 99–101
  UTM System, 97, 99, 100*f*
Land surface features, map symbols, 103*f*
Land use, geologic maps and, 276
Landsat satellites, 123–124
Land-use maps, 96
Lateral Continuity, Steno's Principle of Original, 236, 237*f*
Lateral faults, 256
Lateral moraines, 175*f*, 176
Late-stage streams, 140
Latitude, 96–97, 98*f*
Lava, 38
Lava flows, 274
Length units of measure and conversions. *See* inside back cover
Levees, natural, 136, 139*f*
Light
  imaging techniques and, 117–118, 119*f*
  minerals and, 5
  exercise, 14
Limbs, 252, 253*f*
Limestone, 66*f*, 68*f*
  classifying/identifying, 84*t*, 87
  Mammoth Cave exercise, 164, 167*f*
Limonite, 23, 27*t*
Line of section, 274
Lithification, 62
Lithospheric plates, 312
Lithostatic pressure, 81
Loma Prieta earthquake exercises, 301–308, 310
Longitude, 97, 98*f*
Longitudinal dunes, 217, 218*f*
Longitudinal profiles, 136, 138*f*
  Deer Creek exercise, 154–156
Longshore currents, 197, 199*f*
Longshore drift, 197–198, 200*f*, 201*f*
Low-grade metamorphic rocks, 81
Luster, mineral
  exercise, 13
  identification, 24, 26*t*–30*t*
  metallic *vs.* nonmetallic, 2–3

Mafic rock, 42*t*
Magma, 38. *See also* Plate tectonics
  compositional changes, 48
  crystallization, Bowen's Reaction Series and, 41, 46–47, 52
  formation, 47–48
  hot spots, 316, 317*f*
Magma mixing, 48
Magnetic anomalies, 312–313, 315*f*
  exercises, 319, 320*f*
Magnetic declination, 102
Magnetic north pole, 102
Magnetism
  exercises
    geomagnetic polarity, 329, 330*f*
    minerals, 13
    plate tectonics, 319, 320*f*
  of minerals, 10
  plate tectonics and, 312–314, 315*f*
Magnetite, 23, 26*t*, 31*t*
Malachite, 27*t*, 31*t*
Mammoth Cave exercises, 164, 166, 167*f*, 169*f*
Maps. *See also* Geologic maps; Topographic maps
  elements, 101–102
  symbols
    geologic structures, 251*f*
    rock types, 242*f*
    surface features, 102, 103*f*
  types, 96
Marble, 82, 84*t*, 86*f*, 87
Marine depositional environments, 61*t*
Marine terraces, 200, 201*f*
  exercise, 210
Mass balance, 174
Mass numbers, 238
Mass units of measure and conversions. *See* inside back cover
Massachusetts road map exercise, 114, 115*f*
Massive texture, 63
Matagorda Island exercise, 203–206
Matrix, 58, 59*f*
Maui imagery exercise, 128
Meander belts, 136, 139*f*
Meanders, 136, 139*f*
Measurement units and conversions. *See* Inside Back Cover
Mechanical weathering, 57, 58*t*
Medial moraines, 175*f*, 176
Menan Buttes imagery exercise, 121*f*, 122*f*, 125
Mercalli Scale. *See* Modified Mercalli Scale
Meridians, 97, 99
Mesas, 216
Metachert, 84*t*, 86*f*
Metaconglomerate, 84*t*, 86*f*
Metagraywacke, 84*t*, 86*f*
Metallic luster
  identification table, 26*t*
  of minerals, 2–3
  exercise, 13
Metamorphic reactions, 87
Metamorphic rocks, 79–93
  classifying/identifying, 83–84
  defined, 38, 79
  exercises, 89–91
  vs. igneous/sedimentary rock, 85–87

mineral content, 80–81, 83, 85f
parent rock, 83, 84t
rock symbols, 242f
textures, 81–83
Metamorphism, 79
changes during, 80–81
conditions, 79, 80f
grade, 80f, 81, 85f
metamorphic reactions, 87
types, 79–80
zones, 84–85
Micas, 20
Microcrystalline rocks, 63
Middle-stage streams, 140
Midwest glaciation exercise, 184–185, 186f–189f
Migmatite, 83, 84t, 86f
Mineral(s), 1–36
chemical groups and formulae, 30t–31t
crystallization, 41, 46–47
defined, 2
exercises, 13–16, 33–34
identification, 24–25, 26t–30t
metamorphism and, 80–81, 85f
properties, 1–16
cleavage and fracture, 5–8
color, 2
crystal form, 8–9, 10f
hardness, 3–4
in igneous rock, 40–41, 42t, 43f
luster, 2–3
in metamorphic rock, 83, 83t
physical, 2–11
in sedimentary rock, 64t
special, 10–11
streak, 4, 5f
types, 17, 18t
accessory and minor, 21–22
ore, 23–24
rock-forming, 17–21
Mineraloids, 2
Mines, map symbols, 103f
Minnesota River
flood frequency exercise, 150, 151f, 152t
Glacial Lake Agassiz and, 177, 191
Minor minerals, 17, 18t, 21–22
Mississippi River exercise, 144, 147f
Missouri River exercise, 144, 145f
Modified Mercalli Scale, 296, 297t
exercises, 304, 305f, 305t
Mohs Hardness Scale, 3–4
exercise, 13
identification tables, 26t–30t
vs. Knoop hardness, 4f
Moment magnitude scale, 297
Monuments, 216
Moraines, 175f, 176
Morro Bay exercise, 206–208
Mouth, 136
Mud cracks, 63f, 238
Mudstone, 63, 68f
classifying/identifying, 84t, 87
texture, 66f
Muscovite, 20
chemical group and formula, 31t
identification table, 28t
in igneous rock, 41t

in metamorphic rock, 83t, 85f
in sedimentary rock, 64t

Natural levees, 136, 139f
Near infrared light, 118
Negative magnetic anomalies, 313, 315f
Negative mass balance, 174
Nonconformities, 238, 239f, 273–274
Nonfoliated metamorphic rocks, 81, 83–84
Nonmetallic luster, 2–3
exercise, 13
identification tables, 27t–30t
Non-plunging folds, 253, 254f, 255f
Normal faults, 254, 256
Normal magnetic polarity, 313
exercise, 329, 330f
North, map conventions, 102
Numbers, atomic/mass, 238
Numerical geologic ages, 234, 238–239
exercises, 243–244, 246

Oakland area exercise, 306, 307f–308f
Oblique photographs, 118
Obsidian, 40f, 42t, 43f
Odor, 11
exercise, 13
Ohio bedrock exercise, 263
Oklahoma Web exercise, 290
Olivine, 20–21, 29t, 31t, 41t
Opal, 29t, 31t
Orbits, satellite, 123
Ore minerals, 17, 18t, 23–24
Oscillation ripple marks, 62f, 238
Outcrops
exercises
mapping, 90, 91f, 277, 278f
structural geology, 261–264, 265f
geologic mapping and, 267–268, 269f
Rule of Vs, 274–275, 279
width, 274
Outwash, 176
Outwash plains, 176
Overturned folds, 253
Overwash fans, 198
Oxbow lakes, 136, 139f
Oxides, 31t

P waves. See also Earthquakes
characteristics, 293, 294f, 295f
reflective seismology, 298, 299f
Parabolic dunes, 217, 218f
Parallels, 96–97, 98f
Parent isotope, 238
Parent rock, 83, 84t
Partial melting, 47–48
Paternoster lakes, 174
Pediments, 216, 217f
Pegmatites, 39, 44f
Pegmatitic texture, 39, 42t, 43f
Perennial streams, 215
Peridotite, 42t, 43f, 45f, 84t
Periods, 239, 240f
Permeability, 157–158
Petrified wood, 19t
Phaneritic texture, 39, 42t

Phenocrysts, 39
Phosphates, 31t
Phosphorescence exercise, 14
Photographs. See Aerial photographs
Phyllite, 82, 84t, 86f
Phyllitic foliation, 82
Piedmonts, 216
Pipelines, map symbols, 103f
Pixel, 123
Plagioclase feldspar, 17, 18f
chemical group and formula, 31t
identification table, 30t
in igneous rock, 41t
in metamorphic rock, 83t
Plastic strain, 248, 256t
Plate tectonics, 311–330
convergent plate boundaries, 314–315, 316f
divergent plate boundaries, 312–314, 315f
exercises, 319–330
general theory of, 312
magma processes, 47–48
plate motion and hot spots, 316, 317f
transform faults, 315–316, 317f
Plateaus, 215–216
Playas, 217
Plunging folds, 253, 254f, 255f
Plutons, 274
Point bars, 136, 139f
Polar orbits, satellite, 123
Poles
geographic, 97
magnetic, 102
Pore spaces, 157, 158f
Porosity, 157
Porphyritic texture, 39–40, 42t, 43f
Positive magnetic anomaly, 313, 315f
Positive mass balance, 174
Potassium feldspar, 17, 18f
chemical group and formula, 31t
identification table, 30t
in igneous rock, 41t
Potentiometric surface, 159
Powers of ten. See inside back cover
Precambrian, 239, 240f
exercise, 244–245
Precipitation, runoff and, 134–136
Preglacial lakes, 176f, 177
Primary waves. See P waves
Prime Meridian, 97
Principal meridian, 99
Principle
of fossil assemblages, 237–238
of fossil succession, 237
of inclusions, 236, 237f
of relative time relations, 237f
Profiles
longitudinal, 136, 138f
exercise, 154–156
topographic, 106–107
Prospecting, 276
Pumice, 42t, 43f, 46f
Pyrite, 24, 26t, 30t
Pyroclastic flows, 274
Pyroclastic texture, 40, 42t, 43f
Pyroclasts, 40
Pyroxenes, 19, 20f, 85f

Quadrangles, 97
Quartz, 18–19
   chemical group and formula, 31t
   crystal shape exercise, 14–16
   identification table, 29t
   in igneous rock, 41t
   in metamorphic rock, 83t, 85f
   in sedimentary rock, 64t
Quartz sandstone, 67f, 84t, 87
Quartzite, 82, 84t, 86f, 87

Radar images, 124
   exercise, 128
Radial drainage pattern, 139, 141f
Radiation, 117, 118f
Radioactive, 238
Radioactivity, 238
   dating exercises, 243–244, 246
   dating techniques, 238–239
Railroads, map symbols, 103f
Rainier, Mount, exercise, 51
Ranges, U.S. Public Land System, 99–101
Ratio (fractional) scale, 98f, 101, 102
   conversion exercise, 109
Reactions
   acid-mineral, 10
   continuous and discontinuous, 41, 46–47, 52
   metamorphic, 87
Rectangular drainage pattern, 139, 141f
Recumbent folds, 253
Recurrence intervals, 139, 140f
   exercises, 150, 151f, 152t, 154
Redondo Beach exercise, 211, 212f–214f
Reflected infrared light, 118
Reflection of light, minerals and, 5
Reflective seismology, 298, 299f
Refraction of waves, 197, 199f, 200f
Regional metamorphism, 79–80
Relative geologic age
   concept and principles, 234–236, 237f
   exercises, 241, 242f, 246
   fossils and, 236–238
   hypothesis testing and, 238
   unconformities and, 238, 239f
Relief, 105, 106f
Remnant magnetism, 312
Retreat, 174
Reverse faults, 256
Reversed magnetic polarity, 313
   exercise, 329, 330f
Rhyolite, 42t, 43f, 45f
Rhyolite tuff, 40f
Richter Magnitude Scale, 297
   exercise, 303
Ripple marks, 62f, 238
Rivers. *See* Streams and rivers
Road maps, 96
Roads, map symbols, 103f
Rock Creek flood-frequency curve, 140f
Rock gypsum, 66f, 69f
Rock salt, 66f, 69f
Rock-forming minerals, 17–21
Rocks, composition, 38. *See also specific type of rock*
Rounding/roundness, 58, 59f

Rule of Vs, 274–275, 279
Runoff, 134, 136. *See also* Streams and rivers
Ruptures, 248, 256t

S waves, 293–294, 295f. *See also* Earthquakes
St. Helens, Mount, exercise, 129, 130f, 131f
Salton Sea exercise, 224, 226, 228f, 229t
San Andreas fault exercise, 327, 328f, 329f, 330f
San Francisco Bay area exercise, 306, 307f–308f
Sand dunes, 217–218
Sandstone, groundwater exercise, 164, 167f
Satellite images, 120f, 123–124
   exercises, 124f, 126–127, 129, 130f, 131f
Saturation, zone of, 157, 158f
Scales
   aerial photographs, 119
   map
     converting, 102, 109
     exercises, 109
     types, 98f, 101–102
Schist, 84f, 84t, 86f
Schistose, 82
Scoria, 42t, 43f, 46f
Sea coasts, 197–214
   deposition, 197–198, 200f
   erosion, 197, 199f
   exercises
     barrier islands, 203–206
     features, 206–214
   human interaction with, 198, 200, 201f
   map symbols, 103f
   sea level rise and, 200
   submergent *vs.* emergent, 200, 201f
Sea level, 200
Sea stacks, 197, 199f
Seafloor exercises, 319, 326
Seawalls, 198
Secondary (S) waves, 293–294, 295f
Sections
   line of section, 274
   township, 101
Sedimentary breccia, 87
Sedimentary structures, 61, 62f–63f
Sediment/sedimentary rocks, 57–77. *See also* Erosional landforms; Streams and rivers
   classifying/identifying, 62–63, 65t, 66f–69f
   defined, 38, 57
   deposition and lithification, 59–62, 63f
   erosion and transportation, 57–59
   exercises, 71–75, 326
   mineral composition, 64t
   rock symbols, 242f
   sedimentary *vs.* igneous/metamorphic rock, 85–87
   types of sediment, 59–60
   weathering, 57
Seismic waves. *See also* Earthquakes
   characteristics, 293–294
   reflection seismology, 298, 299f
Seismograms, 294, 295f
   exercise, 301–303
Seismographs, 294
Seismology, 293
Serpentine, 22, 27t, 31t, 83t
Serpentinite, 84t, 86f

Shale, 63, 68f
   classifying/identifying, 84t, 86
   texture, 66f
Shape of sedimentary particle, 58
Shasta, Mt., topographic map, 98f
Shear stress, 248, 249f, 256t
Shear (S) waves, 293–294, 295f
Silicates, 17–21, 31t
Sillimanite, 30t, 31t, 85f
Sills, 274
Siltstone, 66f
Sinkholes, 161, 162f
Size of sedimentary particles, 57–58
   exercise, 75
Slate, 82, 84t, 86
Slaty foliation, 82
Snowfields, map symbols, 103f
Soapstone, 84t, 86f
Sorting, 58, 59f
Specific gravity
   calculation formula, 16
   exercises, 13, 16
   of minerals, 11
Sphalerite, 23–24, 27t, 30t
Spits, 198, 200f
Spot elevations, 105
Springs, 158, 159
Stage of flood, 136
Staurolite, 22
   chemical group and formula, 31t
   identification table, 29t
   in metamorphic rock, 83t, 85f
Steno, Nicolaus, 14
Steno's Law, 15
Steno's principles, 236, 237f
Stereoscope, 119, 121f
Stereoscopic views of aerial photographs, 119, 121
Stovepipe Wells exercises, 219, 220f, 230–231
Strain, 248, 256t
Strata, 57, 249
   aerial/map views, 269–273
   map symbols, 251f
Stratification, 61f
   cross-stratification, 238
   deformation and, 249–250
Stratified drift, 175
Streak, 4, 5f
   exercise, 13
   identification, 24, 26t–30t
Stream capture/piracy, 140, 142f
Stream terraces, 136, 139f
Stream valleys, 136, 139f
Streams and rivers, 134–156. *See also specific stream or river*
   braided streams, 215, 216f
   channels and features, 136, 138f, 139f
   disappearing, 161, 162f
   drainage patterns, 139, 141f
   exercises, 143–156
     drainage and erosion patterns, 148, 149f, 156
     stream features, 143–147, 154–156
   flooding, 136, 139, 140f
   map symbols, 103f
   runoff and drainage basins, 134–136
   systems, evolution, 139–141, 142f

Stress, 248, 249f, 256t. *See also* Structural geology
Striation, 10–11
Strike, 250f, 251
Strike-slip faults, 256
Structural geology, 248–265
    exercises, 257–265
    faults, 254, 256
    folds, 252–254, 255f
    illustrating. *See* Geologic maps
    strata and contacts, 249–251
    stress/strain conditions, 248–249, 256t
    surface orientation, 250f, 251–252
Subducted, 314
Submerged areas, map symbols, 103f
Submergent sea coasts, 200, 201f
Succession, Principle of Fossil, 237
Sulfates, 31t
Sulfides, 30t
Sulfur, 27t, 30t
Superposition, Steno's Principle of, 236, 237f
Surf zone, 197, 198f, 199f
Surface runoff, 134
Surface waves (earthquake), 294
Symbols
    chemical elements. *See* inside back cover
    geologic map, 251f
    rock types, 242f
    topographic map, 102, 103f
Synclines, 252, 253f, 254, 255f

Talc, 20
    chemical group and formula, 31t
    identification tables, 27t, 28t
    in metamorphic rock, 83t
Tarns, 174
Taste of minerals, 11
Tazewell Quadrangle exercise, 286–287, 288f–289f
Tectonics. *See* Plate tectonics
Temperature
    magma processes and, 41, 46–48
    units of measure and conversions. *See* inside back cover
Ten, powers of. *See* inside back cover
Tenacity, 11
Tension, 248, 249f, 256t
Terraces
    exercise, 210
    marine, 200, 201f
    stream, 136, 139f
Texture, 39
    igneous rock, 39–40, 42t, 43f
    metamorphic rock, 81–83
    sedimentary rock, 62–63, 66f
Thermal infrared light, 118
Thin sections, 39
Thrust faults, 256
Tidal deltas, 198, 200f
Tidal inlets, 198
Tiers, 99
Till, 175
Tilted rocks, 234, 235f
Time
    geologic time scale, 239, 240f, 245

    geomagnetic polarity time scale exercise, 329, 330f
    relative time relations, 236, 237f
    travel time curves, 295, 300, 303f
    Universal Coordinate Time exercise, 302f, 303
Tombolos, 198, 200f
Topaz, 30t, 31t
Topographic maps, 96–115
    vs. aerial photographs, 121–123
    contour intervals, 104–105
    contour lines, 102, 104
    coordinates and land subdivision
        latitude-longitude system, 96–97, 98f
        symbols, 103f
        U.S. Public Land Survey, 99–101
        UTM system, 97, 99, 100f
    defined, 96, 102
    elements, 101–102
    elevations, height, and relief, 105, 106f
    exercises, 109–115
    gradient, 108
    making, instructions for, 105–106
    profiles, 106–107
    Shasta, Mt., 98f
    symbols, 103f
Topographic profiles, 106–107
Topography, 96
Tourmaline, 29t, 31t
Township-Range System, 99–101
Townships, 99
Transform-fault plate boundaries, 312, 315–316, 317f
    exercise, 327, 328f, 329f, 330f
Transitional depositional environments, 61t
Transmission lines, map symbols, 103f
Transport of sediment, 57–59
Transverse dunes, 217, 218f
Travel time curves, 295
    exercises, 301, 303f
Trellis drainage pattern, 139, 141f
Triboluminescence exercise, 14
True geographic north, 102
Tuff, 40
    classifying/identifying, 42t, 43f, 46f
    vs. mudstone, 87

Ultramafic rock, 42t
Ultraviolet light exercise, 14
Unconfined aquifers, 158
Unconformable contacts, 250
Unconformities
    aerial/map views, 273–274
    relative geologic age and, 238, 239f
Universal Coordinate Time exercise, 302f, 303
Universal Transverse Mercator System (UTM), 97, 99, 100f
Unsaturated zone, 157, 158f
Upheaval Dome exercise, 224, 227f
Upright folds, 252–253
Urne quadrangle exercise, 154–156
U.S. Public Land Survey, 99–101
U-shaped valleys, 174, 175f

Valley trains, 176
Vegetation, map symbols, 103f

Venus imagery exercise, 128
Verbal scale, 101
    conversion exercise, 109
Vertical exaggeration, 107–108
Vertical photographs, 118
Vesicles, 40
Vesicular texture, 40, 42t, 43f
Visible light, imaging and, 117–118, 119f
Volcanic arcs, 315
Volcanic breccia, 87
Volcanoes. *See also* Plate tectonics
    aerial photos/views
        vs. topographic map, 121–123
        volcanic rock, 274
    arcs, 315
    exercises, 51, 321, 322f, 323f
    hot spots, 316, 317f
Vs, Rule of, 274–275, 279

Wallace Creek exercise, 327, 328f, 329f, 330f
Warren Valley exercise, 286–287, 288f–289f
Wastage, zone of, 174, 175f
Water
    depositional landforms and, 217
    resources, hydrologic cycle and, 135f
    transport of sedimentary particles, 57–59
Water table, 157, 158f
Water-pressure surface, 159
Watershed, 136. *See also* Streams and rivers
    exercise, 152, 154
Wave base, 197, 198f
Wave-cut cliffs, 197, 199f
Wave-cut platforms, 197, 199f, 201f
Wavelength, 117, 118f
Waves
    seismic, 293–294, 295f, 298, 299f
    water, refraction of, 197, 199f, 200f
Weathering of sedimentary rock, 57
    exercise, 74
Wells, 158–159
    exercise, 170, 171f
Wetterhorn Peak Quadrangle exercise, 290–291, 292f
White Sands National Monument exercise, 230
Wind
    depositional landforms and, 217–218
    exercises, 224, 226, 228f, 229t, 230–231
    transport of sedimentary particles, 59
Winnemucca earthquake exercise, 310
Wisconsin groundwater exercise, 168
Wood, petrified, 19t
Wyoming imagery exercise, 128

Yazoo streams, 136, 139f

Zones
    of accumulation, 173–174, 175f
    of aeration, 157, 158f
    Benioff, 315
    of metamorphism, 84–85
    of saturation, 157, 158f
    surf, 197
    unsaturated, 157, 158f
    of wastage, 174, 175f

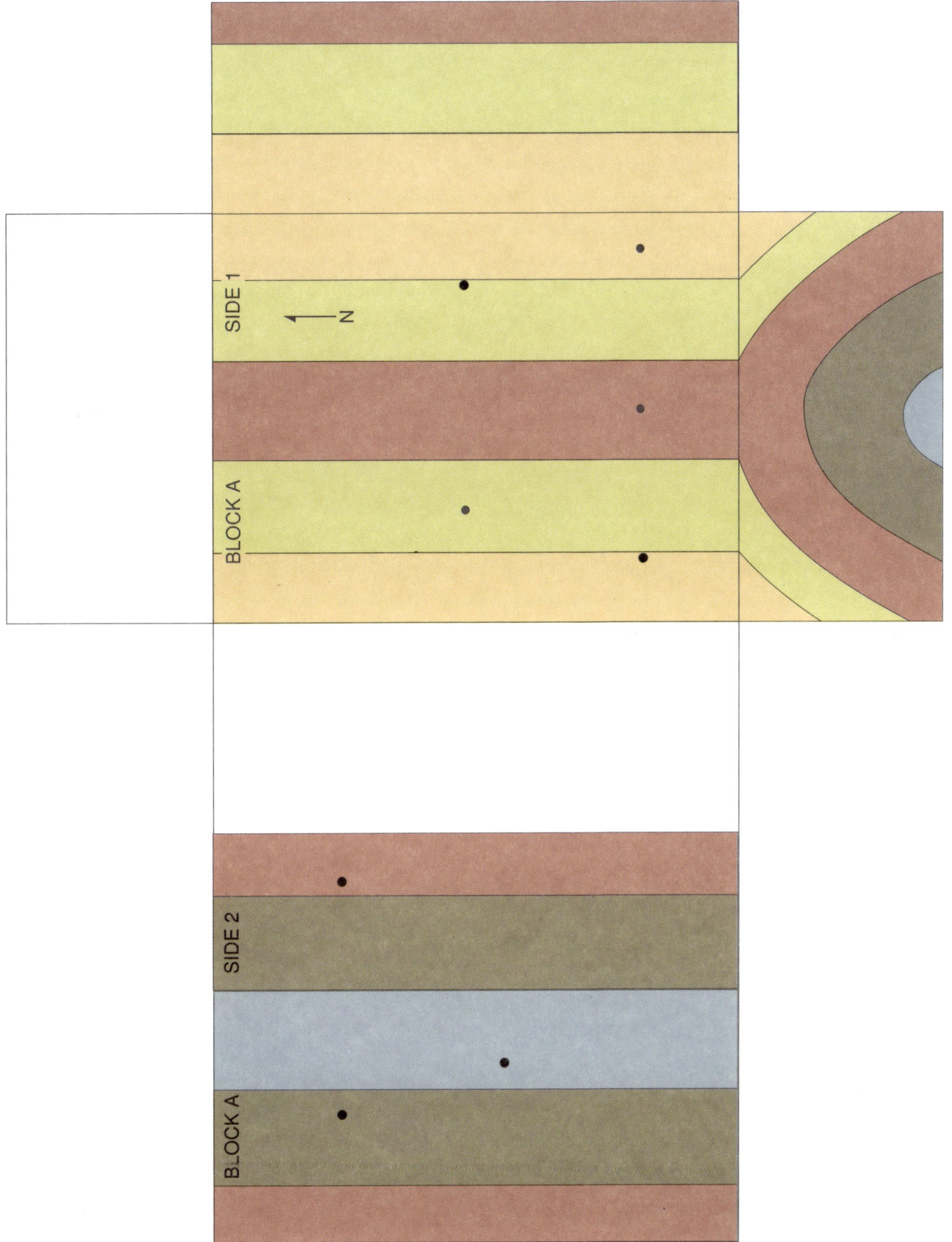

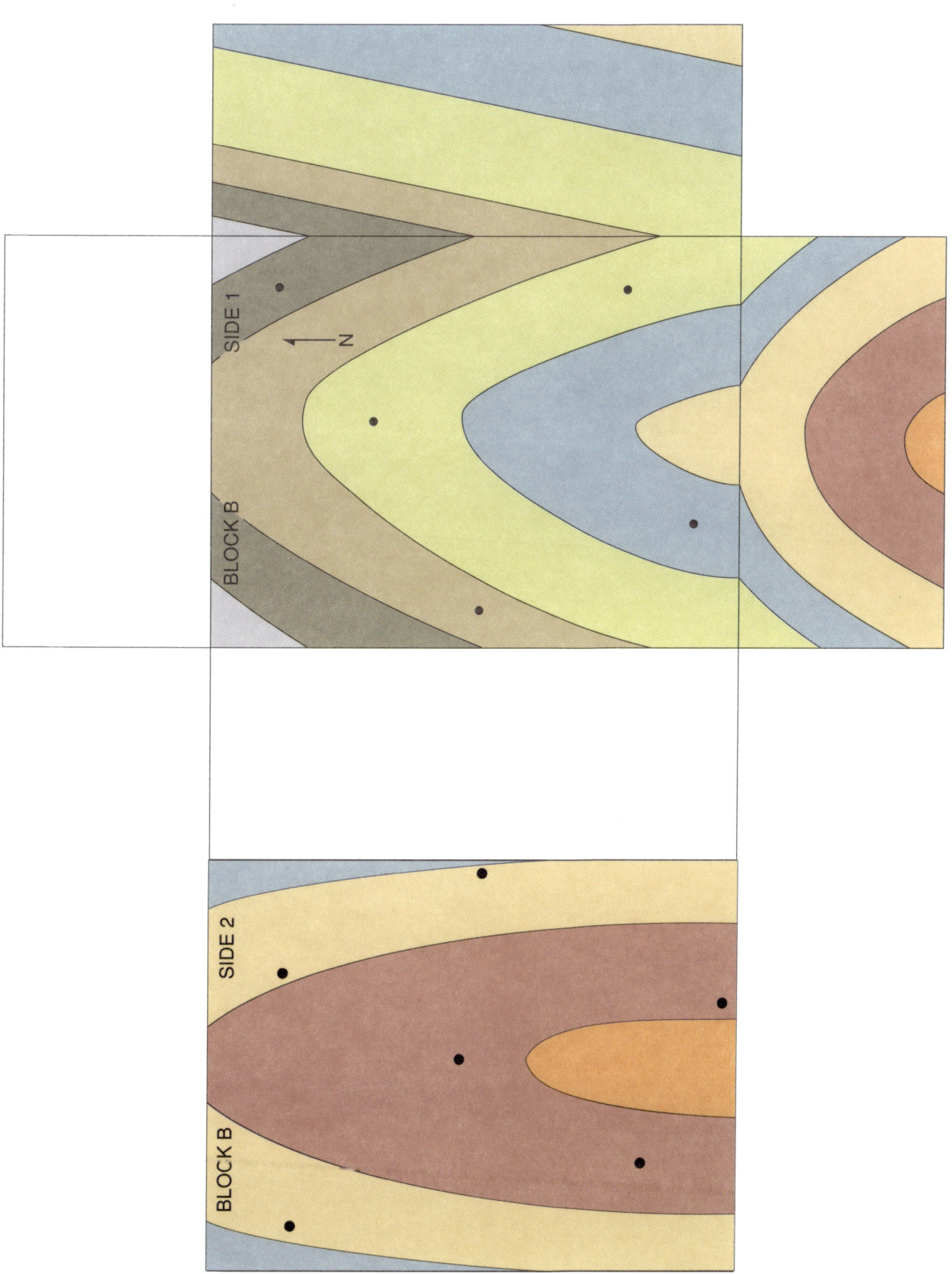

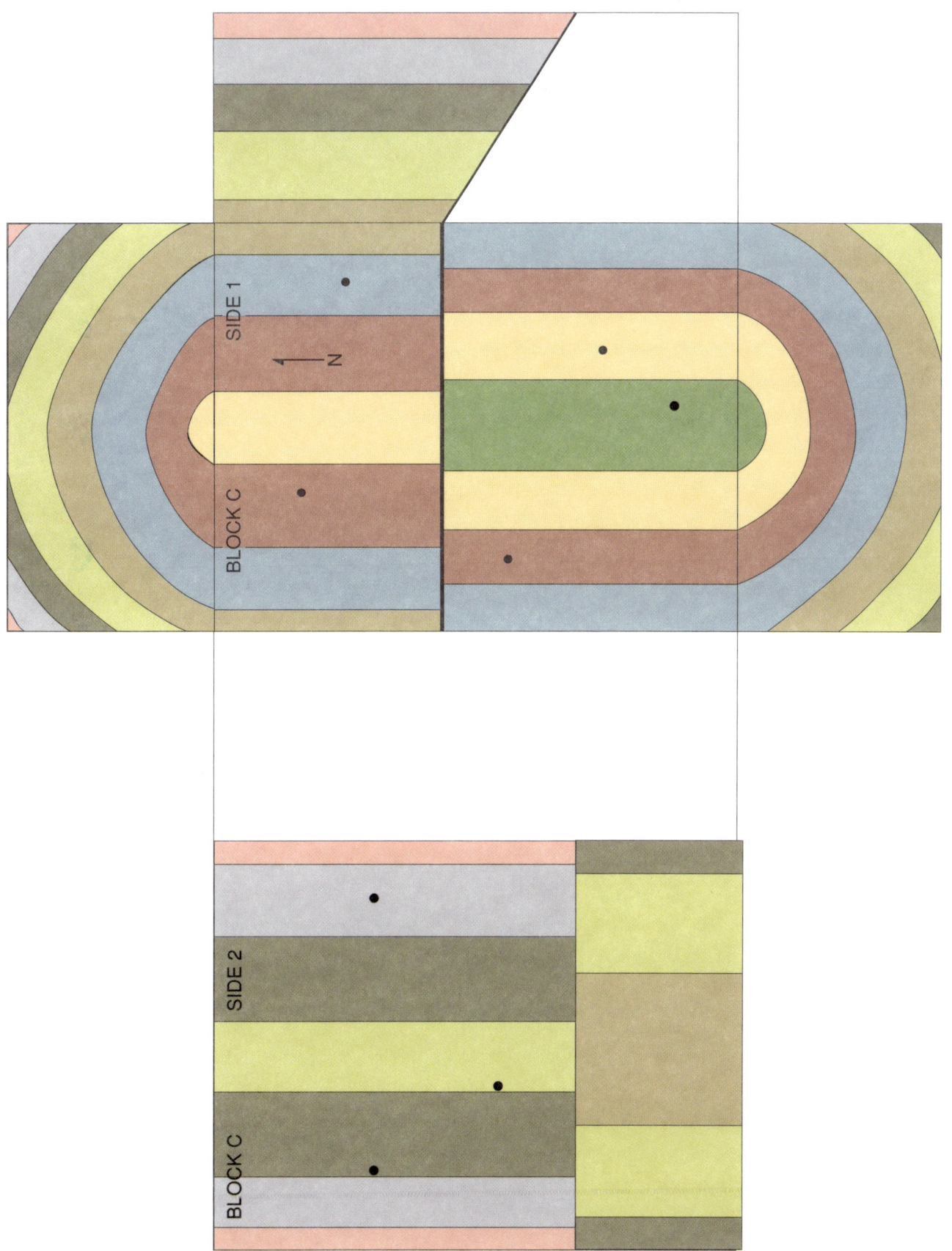

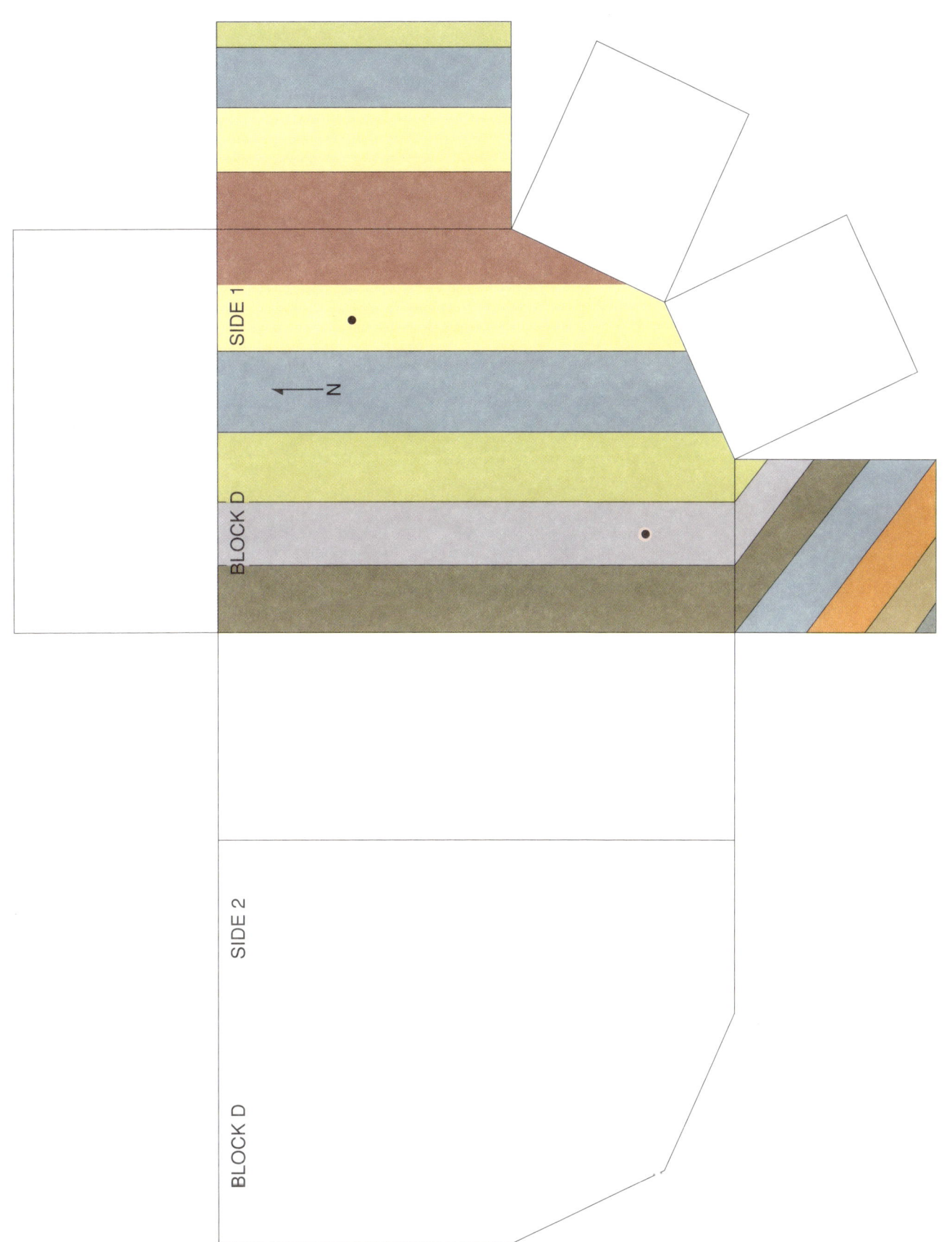